**CONCEPTS IN
EUKARYOTIC
DNA REPLICATION**

CONCEPTS IN EUKARYOTIC DNA REPLICATION

Edited by

Melvin L. DePamphilis
National Institute of Child Health
and Human Development
National Institutes of Health
Bethesda, Maryland

COLD SPRING HARBOR LABORATORY PRESS
Cold Spring Harbor, New York

CONCEPTS IN EUKARYOTIC DNA REPLICATION

© 1999 by Cold Spring Harbor Laboratory Press,
Cold Spring Harbor, New York
All rights reserved
Printed in the United States of America

Project Coordinator: Liz Powers
Production Editor: Dorothy Brown
Desktop Editor: Susan Schaefer
Interior Book Designer: Emily Harste
Cover Designer: Tony Urgo

ISBN 0-87969-557-9
ISSN 0270-1847
LC 98-73326

Cover: Replication fork proteins. Model for a dimeric DNA polymerase at the replication fork. (Adapted, with permission, from Stillman (*Cell 78:* 725 [1994] [copyright Cell Press].)

Authorization to photocopy items for internal or personal use, or the internal or personal use of specific clients, is granted by Cold Spring Harbor Laboratory Press, provided that the appropriate fee is paid directly to the Copyright Clearance Center (CCC). Write or call CCC at 222 Rosewood Drive, Danvers, MA 01923 (508-750-8400) for information about fees and regulations. Prior to photocopying items for educational classroom use, contact CCC at the above address. Additional information on CCC can be obtained at CCC Online at http://www.copyright.com/

All Cold Spring Harbor Laboratory Press publications may be ordered directly from Cold Spring Harbor Laboratory Press, 10 Skyline Drive, Plainview, New York 11803-2500. Phone: 1-800-843-4388 in Continental U.S. and Canada. All other locations: (516) 349-1930. FAX: (516) 349-1946. E-mail: cshpress@cshl.org. For a complete catalog of Cold Spring Harbor Laboratory Press publications, visit our World Wide Web Site http://www.cshl.org

Contents

Preface to *Concepts in Eukaryotic DNA Replication*, vii
Preface to *DNA Replication in Eukaryotic Cells*, ix
Acknowledgments, xiii

1 **Mechanisms for Replicating DNA**, 1
   G.S. Brush and T.J. Kelly

2 **Origins of DNA Replication**, 45
   M.L. DePamphilis

3 **Roles of Transcription Factors in DNA Replication**, 87
   P.C. van der Vliet

4 **Roles of Nuclear Structure in DNA Replication**, 119
   R. Laskey and M. Madine

5 **Mechanisms for Priming DNA Synthesis**, 131
   M. Salas, J.T. Miller, J. Leis, and M.L. DePamphilis

6 **Mechanisms for Completing DNA Replication**, 177
   D. Bastia and B.K. Mohanty

7 **Fidelity of DNA Replication**, 217
   J.D. Roberts and T.A. Kunkel

8 **DNA Excision Repair Pathways**, 249
   E.C. Friedberg and R.D. Wood

9 **Chromatin Structure and DNA Replication: Implications for Transcriptional Activity**, 271
   A.P. Wolffe

10 **Roles of Phosphorylation in DNA Replication**, 295
   K. Weisshart and E. Fanning

11 **Control of S Phase,** 331
K. Nasmyth

12 **Temporal Order of DNA Replication,** 387
I. Simon and H. Cedar

13 **Changes in DNA Replication during Animal Development,** 409
J.L. Carminati and T.L. Orr-Weaver

14 **Comparison of DNA Replication in Cells from Prokarya and Eukarya,** 435
B. Stillman

15 **DNA Replication in Eukaryotic Cells: 1996 to 1998,** 461
M.L. DePamphilis

Index, 503

Preface to *Concepts in Eukaryotic DNA Replication*

In 1996, the Cold Spring Harbor Laboratory Press published a monograph entitled *DNA Replication in Eukaryotic Cells* (M.L. DePamphilis, ed.). This book was arranged in three parts and covered all aspects of the subject. The first section ("Concepts") consisted of 14 chapters written by established investigators in this field. Concepts was intended for scientists, teachers, and students who want a synopsis of current ideas and information in a rapidly developing field. These chapters integrated results from a wide variety of biological systems that were published in literally thousands of papers during the past two decades. The second and third sections were designed for those who want additional details on specific subjects. They consisted of 26 chapters that reviewed specific replication proteins and experimental systems (listed on p. viii). The chapters provided a detailed, up-to-date description of what was known about the genetics, biochemistry, and molecular biology of DNA replication in the genomes of viruses, cells, and mitochondria up to that time. In addition, they provided extensive bibliographies.

Although this monograph was hugely successful, it was also expensive. Therefore, in an effort to provide a more affordable version, the first 14 chapters have been reproduced here in an abridged paperback version. They continue to provide a solid foundation as well as a current summary of the subject. Nevertheless, a significant amount of new information has emerged since they were first published, particularly in the area of initiation of DNA replication. Therefore, some of these advances have been summarized in a new chapter entitled "DNA Replication in Eukaryotic Cells: 1996 to 1998."

Preface to *Concepts in Eukaryotic DNA Replication*

The following chapters were not reproduced in this abridged version.

15	Cellular DNA Polymerases	T.S.-F. Wang
16	Viral DNA Polymerases	D.M. Coen
17	DNA Replication Accessory Proteins	U. Hübscher, G. Maga, and V.N. Podust
18	DNA Helicases	J.A. Borowiec
19	DNA Ligases	R. Nash and T. Lindahl
20	DNA Topoisomerases	A. Hangaard Andersen, C. Bendixen, and O. Westergaard
21	DNA Telomerases	C.W. Greider, K. Collins, and C. Autexier
22	SV40 and Polyomavirus DNA Replication	J.A. Hassell and B.T. Brinton
23	Papillomavirus DNA Replication	A. Stenlund
24	Adenovirus DNA Replication	R.T. Hay
25	Herpesvirus DNA Replication	M. Challberg
26	Epstein-Barr Virus DNA Replication	J.L. Yates
27	Poxvirus DNA Replication	P. Traktman
28	Parvovirus DNA Replication	S.F. Cotmore and P. Tattersall
29	Replication of the Hepatitis Virus Genome	C. Seeger and W.S. Mason
30	Geminivirus DNA Replication	D.M. Bisaro
31	Baculovirus DNA Replication	C.H. Ahrens, D.J. Leisy, and G.F. Rohrmann
32	DNA Replication in Yeast	C.S. Newlon
33	DNA Replication in *Tetrahymena*	G.M. Kapler, D.L. Dobbs, and E.H. Blackburn
34	DNA Replication in *Physarum*	G. Pierron and M. Bénard
35	Differential DNA Replication in Insects	S.A. Gerbi and F.D. Urnov
36	DNA Replication in *Xenopus*	J.J. Blow and J.P.J. Chong
37	DNA Replication in Mammals	N.H. Heintz
38	DNA Replication in Plants	J. Van't Hof
39	Mitochondrial DNA Replication	D.A. Clayton
40	Kinetoplast DNA Replication	A.F. Torri, L.J. Rocco Carpenter, and P.T. Englund

July, 1998

Melvin L. DePamphilis

Preface to *DNA Replication in Eukaryotic Cells*

Science should be made as simple as possible, but not simpler.
Albert Einstein (1879-1955)

In 1959, Arthur Kornberg received the Nobel Prize in medicine and physiology for his pioneering work on the biological synthesis of DNA, an event that marked the beginning of a tremendous effort to understand how cells and viruses replicate their genetic information. This effort produced a new field of scientific research known as DNA replication, accompanied by an extensive literature that has accumulated over the past 35 years. In 1991, Arthur Kornberg and Tania Baker dealt with this vast and complex literature by publishing a wonderful synopsis of DNA replication that covers all aspects of the subject, but whose particular strength lies in its clearly articulated description of the many manifestations of DNA synthesis observed in prokaryotic cells, their bacteriophages, and their plasmids. The goal of this book is to provide a more detailed treatment of the subject of DNA replication in eukaryotic cells.

During the past two decades, advances in technology have permitted analyses of viral, mitochondrial, and cellular genomes that previously had not been possible. Interest in this field accelerated as it became clear that regulation of DNA replication was central to understanding regulation of cell and viral proliferation, events that have a direct impact on our understanding of human diseases. Given the complexity of eukaryotic cells and their need to coordinate proliferation with differentiation during animal development, it is not surprising that the subject of DNA replication in eukaryotic cells is more complex than may have been anticipated. On the order of 30 to 40 proteins are involved directly in the process of DNA replication, as are components of the transcription machinery, chromosome structure, and nuclear structure. DNA replication in eukaryotes involves nuclear and mitochondrial genomes as well as viral genomes containing single-stranded or double-stranded DNA or RNA. It involves specific DNA sequences that determine where replication begins (replication origins) and that mark the ends of linear chromosomes (telomeres). It involves decisions of when, where, and how to initiate DNA replication, repair DNA damage that would otherwise interfere with replication, and ensure that one, and only one, complete and accurate copy of its genome is made before a cell attempts to divide. It utilizes many of the same principles found in prokaryotic cells but offers the possibility that new principles, uniquely suited to problems encountered in eukaryotes, still await discovery. Nevertheless, progress in this field has been impressive.

Our knowledge of events at DNA replication forks in eukaryotic DNA now rivals that in prokaryotic systems: All of the DNA and RNA replication intermediates have been identified. The proteins that drive replication and synthesize and repair DNA, as well as the genes that produce these proteins, are well on the way to full disclosure. Even the more subtle processes that initiate and regulate replication are yielding to investigation. Several viral origins have been characterized and shown to contain transcription elements and to function in soluble systems with purified components, behaving in many ways like prokaryotic replication origins. The genomes of mitochondria and single-cell organisms also contain well-characterized replication origins, and many of the proteins that interact with them have been identified. However, initiation of replication at these origins has not yet been made to occur outside of the cell. Intriguingly, the genomes of metazoan organisms contain more complex replication origins whose function depends on nuclear structure as well as soluble proteins. Many of the proteins that regulate DNA replication have been identified, and their chemical modifications and associations with other proteins have been monitored throughout the cell proliferation cycle. Thus, Bruce Stillman, Director of the Cold Spring Harbor Laboratory, and John Inglis, Executive Director of the Cold Spring Harbor Laboratory Press, believed that the time had come to collate this information and summarize our current understanding of DNA replication in eukaryotic cells. I was invited to orchestrate the effort.

The purpose of the book is threefold. First, it is for teachers who would appreciate a current survey of the concepts involved, but who do not have the time to sort through the vast amount of information on which these concepts are based. Second, it is for scientists already working in this and other fields who would appreciate a synopsis of the facts and concepts that have emerged so far. Finally, it is for those who would like to relate DNA replication to other biological problems in eukaryotes, but who may feel intimidated by the sheer size and complexity of the subject. Accordingly, we have organized the book into three parts: concepts, proteins, and systems.

The first part presents the major concepts involved in DNA replication as well as correlative subjects such as DNA repair, chromatin structure, protein phosphorylation, and cell cycle control. Each chapter collates results from many different experimental systems, distills the essential ideas from these results, searches for common underlying themes, and presents to the reader as coherent a view of the subject as possible. The second part describes each class of proteins that are directly involved in DNA replication. These chapters summarize the current infor-mation concerning the structure of these proteins, their catalytic activities, their mechanism of action, and useful inhibitors and genetic mutations. The third group of chapters describes what is known about how a particular genome replicates. Each chapter identifies the DNA replication intermediates, the specific DNA or RNA sequences that are required, and the replication proteins that are involved. The advantage of this organization is that the overall subject is presented from several points of view. Although this approach results in a certain amount of redundancy, it also allows the same material to be presented by different authors with different points of view.

To ensure that each aspect of the subject was accurately, completely, and clearly presented, 92 scientists in the field of DNA replication participated in assembling this book. Each of the 40 chapters was written independently by its respective author(s), then reviewed by the editor as well as one or more experts in that aspect of the subject. These included authors of other chapters as well as the reviewers acknowledged below. In addition, this work would not have been possible without the patient, cheerful, and professional editorial staff at the Cold Spring Harbor Laboratory Press. In particular, we are indebted to Patricia Barker and Joan Ebert for bringing this book to fruition. I also thank Dr. Ernst Winnacker, Director of the Gene Center at the University of Munich in Germany, and the Humboldt Society for enabling me to take a brief sabbatical in an excellent scientific environment that contributed to the success of this project.

On behalf of the authors, the external reviewers, and the members of the Cold Spring Harbor Laboratory Press, I express the hope that those who read this book will enjoy it as much as did those who wrote it.

March, 1996   **Melvin L. DePamphilis**

Acknowledgments

The following scientists generously donated their time and effort by reviewing individual chapters. We are indebted to them for sharing the burden of writing this book.

Avril Authur, *Institute for Biochemistry, Munich, Germany*
Kenneth Berns, *Cornell University Medical College, New York, New York*
Craig Chinault, *Baylor College of Medicine, Houston, Texas*
Louise Chow, *University of Alabama, Birmingham, Alabama*
Donald Ganem, *University of California Medical Center, San Francisco, California*
David Gilbert, *SUNY Health Science Center, Syracuse, New York*
David Glover, *University of Dundee, Dundee, Scotland*
Linda Hanley-Bowdoin, *North Carolina State University, Raleigh, North Carolina*
Chris Hutchison, *University of Dundee, Dundee, Scotland*
Laurie Kaguni, *Michigan State University, East Lansing, Michigan*
Takehiko Kobayashi, *National Institute for Basic Biology, Okazaki, Japan*
Hans Krokan, *University of Trondheim, Trondheim, Norway*
Robert Lehman, *Stanford University Medical School, Stanford, California*
Lawrence Loeb, *University of Washington, Seattle, Washington*
Grant McFadden, *University of Alberta, Edmonton, Alberta, Canada*
Neil Osheroff, *Vanderbilt University Medical School, Nashville, Tennessee*
Dan Ray, *University of California, Los Angeles, California*
J.B. Schvartzman, *Centro de Investigaciones Biológicas, Madrid, Spain*
José Sogo, *ETH, Zürich, Switzerland*
Allan Spradling, *Carnegie Institution of Washington, Baltimore, Maryland*
William Sugden, *University of Wisconsin, Madison, Wisconsin*
David Weaver, *Dana Farber Cancer Institute, Boston, Massachusetts*
Sandra Weller, *University of Connecticut Health Center, Farmington, Connecticut*

CONCEPTS IN EUKARYOTIC DNA REPLICATION

1
Mechanisms for Replicating DNA

George S. Brush and Thomas J. Kelly
Department of Molecular Biology and Genetics
The Johns Hopkins University School of Medicine
Baltimore, Maryland 21205

COMMON STEPS IN DNA REPLICATION

In the most fundamental sense, the general mechanism of DNA replication was first suggested by Watson and Crick as an immediate and obvious consequence of the complementarity of the two strands of the DNA structure. Thus, all replication processes simply involve the melting apart of the two strands followed by the polymerization of each complementary strand on the resulting single-stranded templates. However, when one looks a bit closer at the details of the process of genome duplication, one finds that cells, plasmids, and viruses have evolved a bewildering variety of particular solutions to the problem (Kornberg and Baker 1992). In many cases, the level of complexity of the enzymatic machinery for DNA replication is considerably greater than might have been expected, given that the information required to generate two daughter genomes is encoded in the structure of the parental genome in such a simple way. Such complexity presumably evolved to increase the efficiency and fidelity of DNA replication and to ensure that the duplication of the genome is coordinated with other events in the life of a cell. Below we attempt to distill the observed complexity down to the few basic processes that are common to most DNA replication pathways.

Initial Opening of the Duplex at Origins of Replication

DNA replication usually begins at one or more specific sites within the genome, referred to as origins of DNA replication. The first essential event in the initiation of DNA synthesis is the local opening of the duplex to provide access to the template strands. Origins of replication serve to increase the efficiency of initiation of DNA replication by providing loci for the assembly of multiprotein complexes that mediate DNA synthesis. If the individual components required for DNA synthesis were capable of interacting with random sites along the DNA, a sufficient local concentration of all of the essential factors might be achieved infrequently. Thus, the required components are usually brought together in one place by specific protein-DNA and protein-protein interactions. Origins of replication also provide specific points for the control of cellular DNA replication, ensuring that replication occurs at the right point in the cell cycle and that each segment of DNA is replicated precisely once. The initial opening of the DNA duplex at origins is generally mediated by specific initiator proteins and can be facilitated by certain structural features of the DNA (e.g., negative supercoiling, easily unwound sequences) and by certain accessory proteins (e.g., single-stranded DNA binding proteins [SSBs]).

Duplex Unwinding at Replication Forks

The initial opening of the duplex allows the establishment of a replication fork(s). The essence of this process is the loading of a DNA helicase on one or both of the exposed single strands. DNA helicases are enzymes that utilize the energy of ATP hydrolysis to translocate unidirectionally along a DNA strand, melting the duplex. At some origins, helicases are loaded onto both DNA strands, resulting in the establishment of two active replication forks (bidirectional DNA replication). In other cases, only a single fork is established (unidirectional DNA replication). An important characteristic of a given helicase activity is its polarity of translocation. Some helicases track along in the 3′ to 5′ direction of the so-called "leading strand" template of the replication fork, and others move in the 5′ to 3′ direction of the "lagging strand" template.

Priming of DNA Synthesis

In most DNA replication systems, the process of starting new DNA chains is distinct from the process of elongating established chains. Thus, all of the known DNA polymerases are incapable of starting chains de novo and require a primer to begin DNA synthesis. In eukaryotic cells, DNA synthesis is generally primed by short RNA chains. The separation of initiation from elongation and the use of RNA, rather than DNA, to initiate DNA synthesis are probably consequences of the requirement for extremely high fidelity in DNA replication. One major mechanism for achieving high fidelity involves proofreading the products of a given polymerization step before proceeding to the next polymerization step. A proofreading exonuclease built into most DNA polymerases recognizes and efficiently excises mismatched nucleotides at the primer terminus. It is likely that this type of proofreading mechanism would be rather inefficient in detecting errors at or near the beginning of a new DNA chain. This problem is apparently solved by "marking" sites of initiation with a chemically distinct RNA chain. The replication machinery can later remove the RNA and fill the resulting gap by extension of an upstream DNA chain by DNA polymerase with proofreading function. This is a relatively costly solution, but one that maintains the accuracy of the genome.

Although both nuclear and mitochondrial DNA replication make use of RNA priming, the enzymatic mechanism of primer synthesis is different in the two cases. The synthesis of RNA primers for nuclear DNA synthesis is accomplished by a primase enzyme that is a component of DNA polymerase-α (pol-α:primase). This primase appears to be the only enzyme that is capable of priming chromosomal DNA synthesis in eukaryotes and is distinct from the RNA polymerases involved in the transcription of nuclear genes. A few eukaryotic viruses, such as SV40, utilize the cellular priming apparatus for the replication of their genomes. There is no evidence for the presence of pol-α:primase in mitochondria, and it is believed that the synthesis of RNA primers for mitochondrial DNA replication is carried out by uniquely mitochondrial enzymes. Interestingly, initiation of DNA synthesis at the primary origin of DNA replication in the mitochondrial genome is mediated by mitochondrial RNA polymerase, the same enzyme that is responsible for mitochondrial transcription.

Some eukaryotic viruses have evolved mechanisms for priming DNA synthesis that do not involve oligoribonucleotide synthesis. Two examples are discussed below. The parvovirus genomes have self-complementary hairpin termini that allow the 3' termini of the genomic single strands to prime DNA synthesis. This example represents a very simple case, since the primers for DNA replication are incorporated into the viral genome itself. Adenoviruses make use of a protein to prime DNA replication. In this interesting case, the first phosphodiester bond is formed between the terminal nucleotide and a serine residue of a virus-encoded protein.

The distribution and frequency of priming events on the two parental strands determine the general pattern of DNA replication in a given system (Fig. 1). In continuous DNA replication there is only a single priming event per template strand, so each progeny DNA strand is synthesized continuously from one end to the other. Examples include mitochondrial DNA replication and the replication of the parvoviruses and adenoviruses. In semidiscontinuous DNA replication one progeny

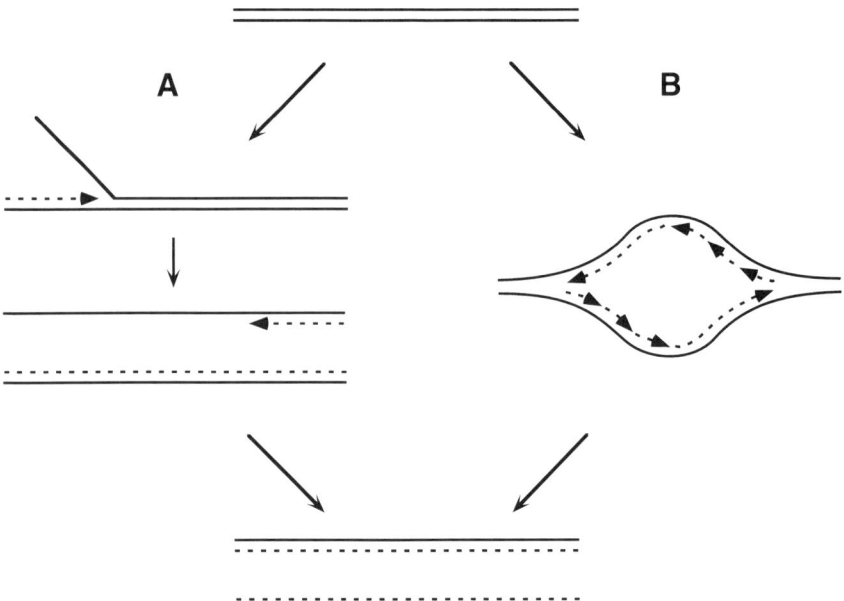

Figure 1 Continuous (*A*) versus semidiscontinuous (*B*) DNA replication. The two basic mechanisms for replicating DNA are shown, with nascent DNA (*broken lines*) synthesized from primers in the direction of the arrows.

strand (leading strand) is elongated continuously from a single primer, whereas the other progeny strand (lagging strand) is constructed from many short DNA chains elongated from multiple primers. Polymerization of the leading strand occurs in the same direction as replication fork movement, and polymerization of the lagging strand occurs in the opposite direction. Nuclear DNA replication and the replication of some viruses (e.g., SV40, herpesviruses) proceed by a semidiscontinuous mechanism. In semidiscontinuous DNA synthesis, completion of the lagging strand requires a repair system to remove the primers, fill in the resulting gaps, and join together the short nascent DNA strands.

Elongation of Nascent DNA Strands

Eukaryotic cells contain several DNA polymerase activities. The nuclear pol-α:primase complex mentioned above appears to be largely concerned with the synthesis of primers during semidiscontinuous DNA synthesis. The DNA polymerase activity of the enzyme is capable of extending the oligoribonucleotides synthesized by the intrinsic primase activity. However, pol-α:primase is relatively nonprocessive, so that only a short DNA chain is synthesized before the enzyme dissociates from the template. The relatively rapid turnover of the enzyme is consistent with its primary role in the synthesis of multiple primers on the lagging strand as the replication fork advances. The RNA-DNA primers synthesized by pol-α:primase can be extended by the highly processive DNA polymerases δ (pol-δ) and ε (pol-ε). Both of these enzymes can be assembled into complexes with the eukaryotic processivity factor proliferating cell nuclear antigen (PCNA) at primer termini. PCNA serves as a topological clamp, tethering pol-δ or pol-ε to the template, so that many thousands of nucleotides can be polymerized before the enzyme dissociates. It is likely that the bulk of chromosomal DNA synthesis is mediated by either pol-δ or pol-ε or both. In addition to their high processivity, which contributes to the efficiency of DNA replication, both enzymes have active proofreading exonuclease activities that enhance the fidelity of DNA replication.

As described in greater detail below, mitochondria possess a DNA polymerase activity distinct from the enzymes involved in nuclear DNA replication. Although some eukaryotic viruses, like SV40, utilize the resident cellular DNA polymerases, others, like the adenoviruses, encode DNA polymerases of their own.

Maturation of Nascent DNA Strands

A consequence of the widespread use of RNA priming mechanisms to initiate DNA synthesis is the requirement for an enzymatic machinery to remove the primers and replace them with DNA. In the case of nuclear DNA replication, this process is rapid and efficient, occurring soon after Okazaki fragment synthesis. Recent work suggests that primers are removed through the combined action of ribonuclease H (RNase H) and a 5′ to 3′ exonuclease. The resulting gaps are filled by DNA polymerase, probably pol-δ or pol-ε, and the final nick is sealed by DNA ligase I. Since this maturation process may result in the nearly complete removal of both RNA and DNA portions of the primers originally generated by the pol-α:primase complex, nearly all of the final DNA product is the result of polymerization by the more processive (and more accurate) DNA polymerases δ and ε. The process of primer replacement in mitochondria is probably mediated by a different set of enzymes, but little work has been done on the problem. A significant fraction of mitochondrial genomes contain residual ribonucleotides at the initiation sites of DNA synthesis, suggesting that removal of RNA primers may be relatively slow and/or inefficient within the organelle.

The DNA-priming mechanism utilized by the parvoviruses requires a maturation mechanism that is quite different from that needed for RNA priming. Since DNA synthesis is primed by a duplex hairpin at the terminus of the genome (see below), there remains the problem of replicating the hairpin primer sequence itself. This is accomplished by a novel and interesting mechanism in which the hairpin sequence first serves as a primer and then is transferred to the newly synthesized DNA strand where it can serve as a template for DNA synthesis (see below). This maturation scheme regenerates the full-length parvovirus genome. Although a similar scheme could, in principle, be used for the maturation of the ends (telomeres) of linear cellular chromosomes, it is now clear that telomeres are synthesized by a special enzyme, telomerase, which polymerizes short DNA repeats at chromosome termini using an RNA template that is intrinsic to the enzyme. Further discussion of telomerase is beyond the scope of this chapter.

CONTINUOUS DNA REPLICATION

As indicated above, mitochondrial DNA and a number of viruses that replicate in eukaryotic cells have evolved replication mechanisms in which both progeny strands are synthesized continuously from one end to the other. To illustrate this general mode of DNA replication, we summarize the replication pathways of two viral systems, the parvoviruses and the adenoviruses, and then discuss mammalian mitochondrial DNA replication. The examples chosen also illustrate the major modes of priming DNA synthesis: DNA self-priming, protein priming, and RNA priming.

Mechanism of Parvovirus (AAV) DNA Replication

The parvoviruses are the simplest of the viruses that infect animal cells (for reviews, see Berns 1990a; Muzyczka 1992; Berns and Linden 1995). They contain linear single-stranded DNA genomes that are converted to duplex replicating intermediates within infected cells. Progeny single strands are generated from the duplex intermediates by a self-priming, strand-displacement mechanism. Viral DNA replication is largely dependent on host enzymes, probably the factors normally involved in leading-strand synthesis at chromosomal replication forks. Only one viral gene, *rep*, is required for DNA replication, and its product(s) is mainly concerned with processing of the termini of the genome.

Two general classes of parvoviruses have been identified (Berns 1990a). Adeno-associated viruses (AAV) generally require coinfection with a helper virus for efficient replication, although the basis for this requirement is not understood. Other parvoviruses are capable of autonomous replication in the absence of any helper virus (Cotmore and Tattersall 1987). All members of the parvovirus group have a similar genetic organization and follow a roughly similar replication pathway. The replication mechanism of AAV is discussed here as prototypical of the group.

AAV Genome

The AAV genome contains two large open reading frames (ORFs) (Srivastava et al. 1983). The 5' ORF, referred to as the *rep* gene, encodes four distinct, but overlapping, polypeptide chains generated by alternate promoter utilization and alternate splicing. The two largest gene products, Rep78 and Rep68, have been implicated in AAV DNA replication by genetic studies (Hermonat et al. 1984; Tratschin et al. 1984; Yang et al. 1992). It is not known whether they play different roles in DNA replication in vivo, but either protein appears to be capable of supporting AAV DNA replication in vitro. The Rep proteins have site-specific endonuclease and helicase activities that mediate the novel processing reaction responsible for regenerating AAV termini (Ashktorab and Srivastava 1989; Im and Muzyczka 1990; Snyder et al. 1990a,b; McCarty et al. 1994; Ni et al. 1994).

Replication Pathway

Current evidence indicates that AAV DNA replication is completely continuous (Berns and Hauswirth 1979; Challberg and Kelly 1989; Muzyczka 1992; Berns and Linden 1995). The termini of the viral genome contain 145-bp palindromic sequences, referred to as inverted terminal repeats (ITRs) (Lusby et al. 1980). When folded in such a way as to maximize base-pairing, the ITRs are capable of forming T-shaped hairpin structures containing only seven unpaired bases (Fig. 2). The 3' end of the genome represents a primer terminus that is used initially to convert the infecting single-stranded genome to a linear duplex form whose strands are covalently linked at one end via the terminal hairpin (Fig. 2) (Hauswirth and Berns 1979). The hairpin terminus is converted to an open duplex through the action of the Rep protein(s) and host enzymes (Tattersall and Ward 1976; Hauswirth and Berns 1979; Snyder et al. 1990b). This processing event, called terminal resolution or hairpin

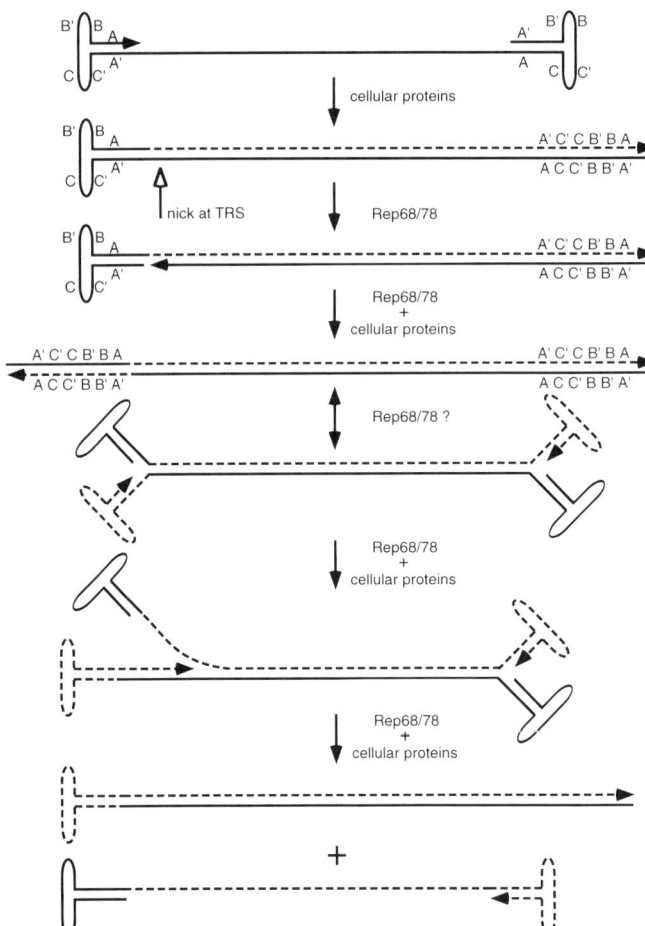

Figure 2 Model for AAV DNA replication. The single-stranded AAV genome contains self-complementary sequences at the termini (indicated by the letters A,A′,B,B′,C,C′) and folds to form the T-shaped hairpin structures shown. The 3′-OH end of one of the two hairpins serves to prime AAV DNA synthesis by host enzymes. Newly synthesized DNA is indicated by broken lines. The terminal resolution site (*trs*) is nicked by the AAV Rep protein to begin the terminal resolution process. DNA synthesis proceeds from the newly created 3′-OH terminus at the *trs* and continues to the end of the genome, resulting in the formation of an open duplex replication intermediate. The two strands of this intermediate are identical to the infecting AAV genomes except for inversion of the terminal palindromic sequence. After rearrangement of the termini to reform the hairpin structures, DNA synthesis proceeds by a displacement mechanism. Each round of DNA replication produces a progeny single strand and a duplex that can undergo terminal resolution to produce the open duplex replication intermediate.

transfer, is initiated by a single endonucleolytic cleavage at a site opposite the original 3' terminus of the genome. Cleavage at the terminal resolution site (TRS) is carried out by the Rep protein, which becomes covalently linked to the 5'-phosphoryl terminus at the TRS cleavage site (Im and Muzyczka 1990; Snyder et al. 1990a). The 3'-OH terminus is then extended by a cellular polymerase(s) to replicate the terminal hairpin. It is likely that the helicase activity of the covalently bound Rep protein facilitates this process by unwinding the hairpin. The result of the terminal resolution reaction is the generation of a linear duplex replication intermediate whose strands are identical to those of the infecting AAV genome except for inversion of the terminal palindromic sequence (Lusby et al. 1981). The intermediate is then replicated via a self-priming, strand-displacement mechanism (Fig. 2). Priming of this reaction probably involves a rearrangement of the termini reforming the hairpin structures. The resulting structure resembles a replication fork and is presumably a competent substrate for the cellular elongation machinery. It is not clear whether rearrangement of the termini is catalyzed by Rep or a cellular protein or whether it occurs spontaneously. The products of the second stage of the AAV replication reaction are a displaced single strand and a linear duplex with covalently joined termini. The latter can be resolved to an open duplex by Rep and the process can be repeated. Since the two termini of AAV are essentially equivalent, equal numbers of the two complementary strands are synthesized. Both strands are packaged into progeny virus particles (Berns 1990b).

Enzymology of AAV DNA Replication

Several in vitro systems that carry out various aspects of the AAV DNA replication reaction have been developed (Hong et al. 1992; Ni et al. 1994; Ward et al. 1994). When duplex DNA molecules containing functional AAV origins are used as templates, either Rep68 or Rep78 is required for extensive DNA synthesis (Ni et al. 1994). It has also been reported that DNA synthesis in vitro can be stimulated significantly when the extracts are prepared from adenovirus-infected cells, suggesting that a protein(s) encoded or induced by the helper may play some role in AAV DNA replication (Ni et al. 1994). However, it seems likely that most of the proteins involved in viral DNA replication are derived from the host cell. Recent work indicates involvement of the cellular replication proteins PCNA, replication factor C (RF-C), and replication protein A (RP-A) (N. Muzyczka, pers. comm.). The DNA polymerase responsible for AAV DNA chain elongation has not yet been identified with certainty, but, given the involvement of the processivity factor PCNA, pol-δ and pol-ε are likely candidates. Thus, AAV is probably replicated mainly by the apparatus responsible for leading-strand synthesis at cellular replication forks (see below). It is not known whether movement of the AAV replication fork is catalyzed by the helicase activity of Rep68/78 or whether a cellular helicase mediates this function. Since the AAV genome is linear and rather short, there is no apparent need for a DNA topoisomerase, consistent with the observation that antibodies against mammalian topoisomerase I and II do not inhibit DNA replication in vitro (N. Muzyczka, pers. comm.).

Mechanism of Adenovirus DNA Replication

Adenovirus Genome

The mechanism of adenovirus DNA replication has been well characterized because of the early development of a cell-free replication system (Challberg and Kelly 1979). The most extensively studied viral serotypes are the closely related Ad2 and Ad5 (for reviews, see Challberg and Kelly 1989; Hay and Russell 1989; Stillman 1989; Salas 1991; Van der Vliet 1991; Kornberg and Baker 1992). The Ad2/5 genome is a linear double-stranded DNA molecule of 36 kb with two novel structural features: (1) The nucleotide sequences at the ends of the genome are identical for the first 103 nucleotides and (2) the 5' end of each strand is covalently linked to the virus-encoded, 55-kD terminal protein (TP). Viral DNA replication requires three viral proteins: an 80-kD precursor to the terminal protein (pTP), a 140-kD DNA polymerase (Ad pol), and a single-stranded DNA-binding protein (Ad DBP). The efficiency of DNA replication is increased significantly by several cellular proteins, including the transcription factors nuclear factor I (NFI) and octamer-binding protein 1 (Oct-1) and a DNA topoisomerase (NFII) (Nagata et al. 1982; Pruijn et al. 1986; Rosenfeld and Kelly 1986; Rosenfeld et al. 1987). Initiation of adenovirus DNA replication takes place by a protein-priming mechanism at the termini of the genome, and each daughter strand is synthesized continuously from one end to the other (Lechner and Kelly 1977; Rekosh et al. 1977; Challberg et al. 1980; Ikeda et al. 1982).

Initiation of Adenovirus DNA Replication

The adenovirus origin of replication, encompassing roughly the first 50 bp of the viral genome, is the locus of binding of several proteins. NFI and Oct-1 bind to specific recognition sites within the region between nucleotides 19 and 51 (Nagata et al. 1983b; Rawlins et al. 1984; Pruijn et al. 1986; Rosenfeld and Kelly 1986; Rosenfeld et al. 1987; Wides et al. 1987). A complex of the two viral proteins, pTP and Ad pol (pTP-pol), appears to recognize sequence elements within the first 18 nucleotides that constitute the minimal essential origin (Ikeda et al. 1982; Mul and Van der Vliet 1992; Temperley and Hay 1992). Binding of pTP-pol to the viral origin of replication is the critical first step in the initiation reaction (Fig. 3). This step is greatly facilitated by protein-protein interactions between pTP-pol and the bound cellular proteins, NFI and Oct-1 (Bosher et al. 1990; Chen et al. 1990; Mul et al. 1990). The binding reaction may also be stimulated by the presence of the 55-kD terminal protein at the 5' end of one of the parental strands (see Fig. 3). Once the pTP-pol is bound at the terminus of the genome, DNA synthesis is initiated by the formation of a phosphodiester bond between the first nucleotide in the new DNA chain, dCMP, and the β-hydroxyl group of a serine residue in the pTP (Challberg et al. 1980). This novel protein-priming reaction is presumably catalyzed by the adenovirus DNA polymerase, which then functions to extend the nascent chain (Challberg et al. 1980; Ikeda et al. 1982). An early intermediate in the elongation reaction of Ad5 is a trinucleotide covalently attached to the pTP (pTP-CAT) (King and Van der Vliet 1994). Interestingly, it has recently been demonstrated that the template for synthesis of pTP-CAT is not the first three nucleotides in the parental strand, but nucleotides 4–6. The sequence at the terminus of the template strand (3' GTAGTAGTTA...5') is repetitive, so it is thought that the pTP-CAT, synthesized opposite residues 4–6, translocates to positions 1–3 to start elongation. The function of this rather baroque mechanism is not yet clear, but it has been suggested that it may serve to protect the integrity of the terminal sequences of the viral genome during DNA replication (King and Van der Vliet 1994).

Elongation of Adenovirus DNA Strands

Following initiation at one of the termini, DNA synthesis proceeds by a displacement mechanism, producing a daughter duplex and a free single strand (Fig. 3) (Lechner and Kelly 1977). The latter can cyclize via self-complementary termini to form a panhandle structure. Since the duplex panhandle is identical to the termini of the original genome, initiation of DNA synthesis can presumably occur by the same mechanism, leading to the completion of a second daughter duplex. In vitro studies have demonstrated that the elongation of nascent adenovirus DNA strands re-

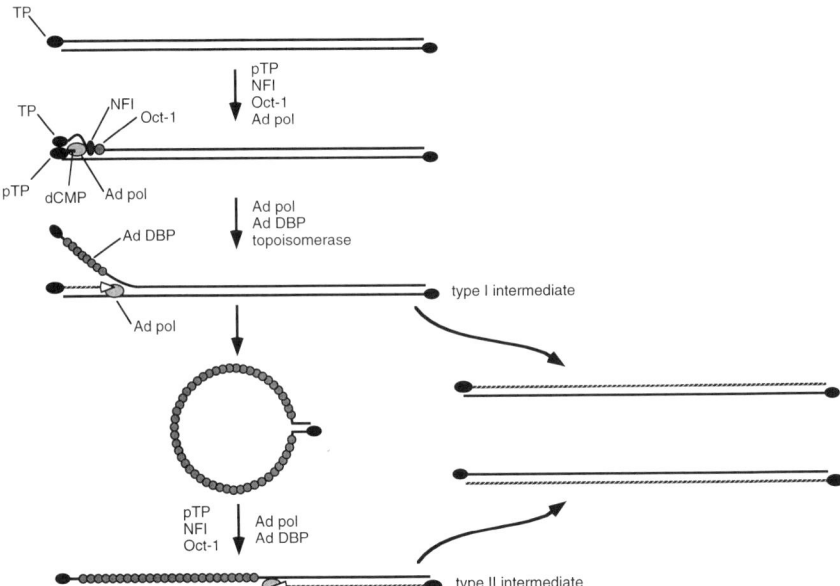

Figure 3 Model for adenovirus DNA replication. The 5' ends of the adenovirus genome are covalently linked to the 55-kD terminal protein (TP). Initiation of DNA synthesis requires the 80-kD preterminal protein (pTP), the 140-kD adenovirus DNA polymerase (Ad pol), and the cellular proteins nuclear factor I (NFI) and octamer-binding protein 1 (Oct-1). Initiation is facilitated by the adenovirus DNA-binding protein (Ad DBP), which enhances NFI binding. Initiation involves the formation of a phosphodiester bond between the β-OH of a serine residue in pTP and the 5' phosphoryl group of dCMP (see text for details). Elongation of nascent adenovirus DNA chains (*dotted lines*) requires the Ad pol, Ad DBP, and a cellular topoisomerase (NFII). Displacement synthesis on type I intermediates leads to the generation of a progeny duplex and a displaced single strand. The displaced single strand circularizes via self-complementary termini and serves as a substrate for a second initiation event, identical to the first. Completion of DNA synthesis on type II intermediates results in the formation of a second progeny duplex.

quires only two proteins, Ad pol and Ad DBP (Lindenbaum et al. 1986). The adenovirus DNA polymerase is a highly processive enzyme that extends DNA primers much more efficiently than RNA primers (Field et al. 1984; Lindenbaum et al. 1986). Its activity is specifically stimulated by the Ad DBP; prokaryotic or other eukaryotic DBPs cannot replace Ad DBP in the replication reaction (Lindenbaum et al. 1986). Movement of the displacement fork during adenovirus DNA replication does not appear to require a separate helicase activity. The Ad DNA pol, acting in concert with the Ad DBP, is sufficient to unwind the parental duplex. The energy required for unidirectional fork movement is derived exclusively from hydrolysis of the deoxyribonucleoside triphosphate precursors for DNA synthesis (Pronk et al. 1994). Thus, adenovirus appears to have evolved a highly efficient two-protein engine for DNA synthesis that combines the functions of helicase and polymerase. Interestingly, in vitro studies have suggested that the replication of the adenovirus genome is facilitated by a cellular DNA topoisomerase (NFII) even though the genome is linear (Nagata et al. 1983a). Replication proceeds efficiently in the presence of Ad DNA pol and Ad DBP until replication is about 25% complete, after which fork movement is slow unless topoisomerase activity is present. The basis for this requirement is not completely clear, but presumably it reflects some hindrance to the free rotation of the unreplicated parental duplex in adenovirus replication intermediates.

Mechanism of Mitochondrial DNA Replication

General Features of Mitochondrial DNA Replication

Much of the early work on the mechanism of mitochondrial DNA replication was done in mammalian systems, but studies in other organisms, particularly yeast, have been increasingly important for identifying and characterizing required replication proteins (for reviews, see Clayton 1991, 1992; Kornberg and Baker 1992; Schmitt and Clayton 1993). The available evidence suggests that the basic features of mitochondrial DNA replication have been conserved from yeast to man, so the following description focuses primarily on mammalian cells (Clayton 1982, 1991). The mammalian mitochondrial genome is a closed duplex circle of about 16 kb. Cells contain on the order of 10^3–10^4 mitochondrial genomes with an average of 5–10 per organelle. Unlike the nuclear DNA, mitochondrial genomes are replicated throughout the cell cycle, and templates appear to be drawn at random from the pool of genomes. Thus, in a given cell cycle, some mitochondrial genomes are replicated more than once and some are not replicated at all. The regulatory mechanisms that determine the total number of mitochondrial genomes per cell are not understood.

Studies of the structures of replication intermediates have provided strong evidence that the replication of the mitochondrial genome occurs by a completely continuous mechanism (Fig. 4). In mammalian cells there are two origins of DNA replication, one for each strand, which are located about 11 kb apart (Clayton 1982). At each of these origins, DNA synthesis is primed by specific RNA molecules (see below). A single priming event apparently suffices for the complete synthesis of each strand. The origin for heavy (H) strand synthesis (O_H) is activated first, resulting in the establishment of a replication fork that moves unidirectionally (Gillum and Clayton 1979; Chang and Clayton 1985). At this fork a new H strand is continuously elongated, and the parental H strand is displaced. The origin for light (L) strand synthesis (O_L) is activated only when it has been rendered single-stranded by the passage of the H-strand replication fork (Wong and Clayton 1985a,b). It is likely that a specific secondary structure that forms in the displaced strand is recognized by the protein(s) responsible for RNA primer synthesis. The L strand is continuously elongated from the primer in what is essentially a large gap-filling reaction. Since the two origins of DNA replication are separated by approximately two-thirds of the genome, the replication of the two strands of mitochondrial DNA is quite asynchronous (see Fig. 4). When the synthesis of the H strand is completed, a process that takes about an hour, two products are generated. One product is a duplex circle with a full-length progeny H strand, and the other is a gapped circle with an incomplete progeny L strand, which is subsequently extended to full length.

The primary event in mitochondrial DNA replication is the activation of the H-strand origin, since the activation of the L-strand origin is secondary to the establishment of the H-strand replication fork. Thus, one likely point of regulation of mitochondrial DNA synthesis is the generation of the initial RNA primer. There may also be a second control point that operates at the level of elongation of the H strand. A majority of mitochondrial genomes in mammalian cells contain a so-called D loop in the vicinity of O_H (Gillum and Clayton 1979). This structure consists of a short nascent H DNA whose 5' end is located at the origin for H-strand synthesis. The nascent H strands in D loops appear to be metabolically labile and are turned over at a rate that significantly exceeds the rate of replication of mitochondrial genomes. The existence of the short nascent H strands suggests that there may be a barrier to chain elongation just downstream from O_H (Clayton 1982; Madsen et al. 1993). It is possible that mechanisms exist that regulate whether or not the H-strand replication fork can pass this barrier, but at present there is no evidence on this point.

Less work has been done on the replication of the mitochondrial DNA of yeast, but it is likely that DNA synthesis is largely continuous, although there may be more than one priming event per strand (Blanc and Dujon 1980; Baldacci et al. 1984; de Zamaroczy et al. 1984; Schmitt and

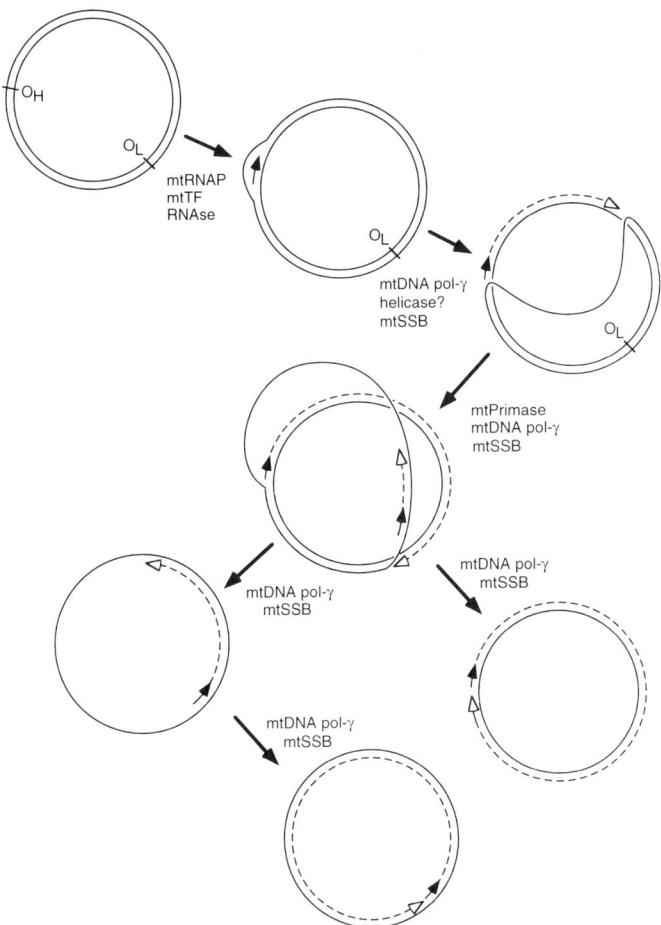

Figure 4 Model for mammalian mitochondrial DNA replication. The two mitochondrial origins O_H and O_L are located about two-thirds of the genome apart. Initiation of DNA replication begins with the synthesis of a transcript by the mitochondrial RNA polymerase (mtRNAP) and an essential transcription factor (mtTF). The transcripts are either processed by an endoribonuclease (RNase) to form a primer for mitochondrial DNA replication or continue to be extended to form mRNA. If cleavage occurs, a nascent H strand (*broken line*) is synthesized by the mitochondrial DNA polymerase (mtDNA pol-γ). Other proteins, including a putative mitochondrial helicase and SSB (mtSSB) probably participate in the chain elongation process. Synthesis of the L strand is initiated by a mitochondrial primase (mtPrimase) after the displacement fork passes O_L. Completion of H-strand synthesis and separation of the two progeny molecules occurs before completion of L-strand synthesis.

Clayton 1993). At least four putative origins of replication have been located in the yeast mitochondrial genome. It is likely that more origins are required because of the large size of the yeast mitochondrial genome. The structural features of yeast origins are similar to those of higher eukaryotes (Schmitt and Clayton 1993). However, unlike mammalian origins, the sites of initiation of synthesis of the two strands are in close proximity rather than being separated by many kilobases.

Priming of Mitochondrial DNA Synthesis

Transcription of the mammalian mitochondrial genome is directed by two divergent promoters located within the D-loop region (Chang and Clayton 1984). Transcripts initiated at one of these promoters (LSP) also serve to prime H-strand synthesis during mitochondrial DNA replication (Chang and Clayton 1985; Chang et al. 1985). A fraction of the LSP transcripts are cleaved by an endoribonuclease at one of several discrete sites at O_H, approximately 100 nucleotides from the promoter. The 3′-OH termini of the transcripts are then elongated by the mitochondrial DNA replication apparatus. Transcription from the LSP requires the mitochondrial RNA polymerase and a specific transcription factor, mtTFA (Fisher and Clayton 1985; Parisi et al. 1993). An RNase capable of processing nascent transcripts at sites near the transition from RNA synthesis to DNA synthesis has been identified in mammalian cells and yeast (Chang and Clayton 1987; Schmitt and Clayton 1992; Stohl and Clayton 1992). The enzyme, called RNase MRP, contains a required RNA moiety that may be involved in recognition of the specific cleavage sites. Recently, it has been suggested that a second mitochondrial nuclease, endonuclease G, may also play a role in processing nascent transcripts to generate primers for DNA replication (Cote and Ruiz-Carrillo 1993).

Priming of L-strand synthesis at O_L appears to be mediated by a mitochondrial primase that recognizes a specific stem-loop structure which forms in the H-strand template (Wong and Clayton 1985a,b; Hixson et al. 1986). The enzyme initiates the synthesis of a short oligoribonucleotide at a run of T residues in this loop. The transition from RNA synthesis to DNA synthesis occurs immediately adjacent to the base of the stem. The mitochondrial primase appears to require an RNA component for activity, and the most likely candidate is cellular 5.8S RNA (Wong and Clayton 1986). Thus, the priming mechanisms for the synthesis of both strands of the mitochondrial genome require ribonucleoprotein enzymes.

Mechanism of Chain Elongation

Although some of the required enzymes have been identified, the detailed mechanism of continuous DNA synthesis in mitochondria is not well understood. This is mainly due to the absence of an efficient cell-free replication system. All of the components involved in mitochondrial DNA replication, including both the priming and elongation stages, are encoded by nuclear genes and imported into the organelle. A highly processive mitochondrial DNA polymerase, referred to as DNA polymerase-γ, has been purified from several sources (Wernette and Kaguni 1986; Wernette et al. 1988; Insdorf and Bogenhagen 1989a,b; Gray and Wong 1992). The human enzyme consists of subunits of 140 kD and 54 kD. The larger subunit contains the polymerase active site as well as a potent 3' to 5' exonuclease activity. The latter shows strong preference for unpaired primer termini and presumably plays a proofreading role. The yeast gene encoding the catalytic subunit of DNA polymerase-γ (*MIP1*) has been cloned (Foury 1989). The amino acid sequence of the Mip1 protein is similar to both eukaryotic nuclear DNA polymerases and reverse transcriptases, but has no discernible resemblance to prokaryotic DNA polymerases.

Mitochondrial DBPs have also been identifed from several organisms (Pavco and Van Tuyle 1985; Mignotte et al. 1988; Van Dyck et al. 1992; Tiranti et al. 1993; Stroumbakis et al. 1994). Interestingly, these proteins are homologous to *Escherichia coli* SSB, and their physicochemical properties are quite similar to the prokaryotic enzyme as well. Yeast mutants lacking the *RIM1* gene, which encodes the mitochondrial SSB, are completely devoid of mitochondrial DNA, consistent with an essential role for the protein in mitochondrial DNA replication (Van Dyck et al. 1992). Presumably, the mitochondrial SSB facilitates the elongation of nascent DNA chains by the mitochondrial DNA polymerase. However, the functional interactions of the two proteins have not been extensively analyzed to date.

Another likely accessory protein for mitochondrial DNA synthesis is a DNA helicase, since H-strand synthesis requires melting of the parental strands at the replication fork. A helicase activity has been identified in highly purified mitochondria from bovine brain (Hehman and Hauswirth 1992). This activity, which translocates in the 3' to 5' direction on the single-stranded portion of partially duplex substrates, is a likely candidate for a replicative helicase in mitochondria, but this role has not been verified directly. The Pif1 helicase of budding yeast has been implicated in the repair, recombination, and replication of the mitochondrial genome (Lahaye et al. 1991). Genetic interactions have been observed between the *PIF1* gene and the *RIM1* gene, encoding the mitochondrial SSB (Van Dyck et al. 1992). Moreover, a yeast strain deficient in the Pif1 enzyme loses mitochondrial DNA at elevated temperatures.

Mitochondrial DNA replication remains the best example of a naturally occurring, completely continuous DNA replication system in eukaryotic cells. Although a number of the components required for DNA replication have been identified, and we have a general picture of the overall replication pathway, much remains to be learned about the biochemical mechanisms involved. Future in vitro studies of mitochondrial DNA synthesis with purified components will likely fill in many of the gaps in our knowledge.

SEMIDISCONTINUOUS DNA SYNTHESIS

Replication of Chromosomal DNA

In the continuous DNA replication systems that we have discussed so far, the two complementary DNA strands are synthesized relatively independently and asynchronously. In contrast, the replication of eukaryotic chromosomal DNA occurs by a semidiscontinuous mechanism in which the synthesis of the two strands is strongly coupled in space and time. One advantage of semidiscontinuous DNA replication over completely continuous DNA replication is that the generation of single-stranded DNA is much more localized and transient. This may help preserve the integrity of the genome, since breaks and other lesions in single-stranded DNA are difficult to repair.

Semidiscontinuous DNA replication is mediated by a complex protein machine assembled at each replication fork. The proteins comprising this replication machine act in concert to unwind the parental strands and carry out the simultaneous synthesis of the two progeny strands. Both progeny strands are synthesized in the 5' to 3' direction, but since the parental DNA strands are antiparallel, two distinct mechanisms of DNA synthesis are required. One of the two progeny strands (leading strand) is synthesized continuously in the direction of fork movement. The other strand (lagging strand) is synthesized discontinuously in the direction opposite to fork movement. Discontinuous DNA synthesis on the lagging-strand template involves the repeated synthesis of oligoribonucleotide primers, which are then elongated into short DNA chains (Okazaki fragments). Following their synthesis, Okazaki fragments are processed to remove the RNA primers and joined together to form an uninterrupted progeny strand.

The semidiscontinuous DNA replication mechanism appears to be quite ancient in origin, as its essential features have been conserved from bacteria to man. Although the basic biochemical processes that occur at eukaryotic and prokaryotic replication forks are similar, there are many differences in detail. For example, in *E. coli*, a single DNA polymerase is thought to form a dimeric complex that mediates DNA synthesis on both the leading and lagging strands (Maki et al. 1988; McHenry 1988), whereas eukaryotes contain at least three distinct DNA polymerase activities, all of which participate in DNA replication. The critical protein-protein interactions that are required for formation of an efficient replication machine also appear to differ between prokaryotes and eukaryotes. In *E. coli*, the primase activity responsible for initiating new DNA chains forms a specific complex with the helicase activity responsible for unwinding the parental strands as the fork moves (Kornberg and Baker 1992) whereas in eukaryotes, the primase is a subunit of one of the DNA polymerase molecules (Lehman and Kaguni 1989; Wang 1991). Some of the accessory proteins involved in DNA synthesis have more complex structures in eukaryotes than in prokaryotes. One example is the single-stranded DNA-binding protein, which is a single polypeptide chain in *E. coli* (Meyer and Laine 1990) but a large heterotrimeric protein in eukaryotes (Fairman and Stillman 1988; Wold and Kelly 1988). The greater complexity may be related to the more complex mechanisms required to regulate DNA replication in eukaryotes. Despite these apparent differences, there are many similarities between prokaryotic and eukaryotic DNA replication that reflect a common evolutionary origin (see Stillman, this volume). One particularly striking example is the remarkable structural similarity of the β subunit of *E. coli* DNA polymerase III and the eukaryotic protein PCNA, both of which function as "sliding clamps" to increase the processivity of DNA synthesis (Kong et al. 1992; Krishna et al. 1994).

A novel feature of eukaryotic chromosomes is the packaging of DNA into chromatin. This fact may account for the surprising disparity in the rate of fork movement in prokaryotes and eukaryotes. Whereas the *E. coli* replication machinery moves at the prodigious rate of 100 kb/minute, unwinding DNA at some 10,000 rpm, eukaryotic fork movement is much slower (0.5–5 kb/min) (Kornberg and Baker 1992). The presence of histones may limit the maximal rate at which DNA polymerization can occur. Such a constraint might have contributed to the evolution of intrinsically slower polymerases in eukaryotes. In addition, special mechanisms may have evolved to allow fork movement without disruption of either the replication complex or the nucleosomes. In this manner, the DNA remains packaged even while DNA synthesis is taking place. However, the ability of large protein assemblies to negotiate one another is not unique to eukaryotes, as in vitro experiments have shown that the phage T4 replication apparatus can bypass an RNA polymerase complex moving in either direction without complete dissociation of either set of proteins (Liu et al. 1993; Liu and Alberts 1995).

Viral Model Systems

Direct biochemical investigation into the mechanisms underlying cellular DNA replication has proven difficult. With the exception of special cases such as *Xenopus* eggs, it has not yet been possible to develop an in vitro cellular DNA replication system. In the absence of such a system, investigators have turned to viral models to study the mechanism of cellular DNA replication. Viruses present several advantages, including small genomes and well-defined origins of replication. The Papovaviridae, which include the SV40, polyoma, and papilloma viruses, have been particularly important to the study of chromosomal replication. Upon infection of suitable host cells, the genomes of these viruses are transported to the nucleus, where the double-stranded viral minichromosomes are replicated by mechanisms that closely resemble the cellular process. Initiation occurs at a single origin within the viral genome and proceeds bidirectionally in a semidiscontinuous manner. The great advantage of SV40 and other papovaviruses is that many of the important steps in DNA replication are performed by host proteins.

A significant advance in the study of eukaryotic DNA replication came with the development of a cell-free SV40 system in 1984 (Li and Kelly 1984). Primate cytoplasmic extract and a single viral protein, the large T antigen, carry out all the functions required for complete replication of SV40-origin-containing plasmid DNA. T antigen recognizes the origin and unwinds the duplex, providing access for the numerous host replication proteins that function coordinately to synthesize the daughter DNA strands (for review, see Stillman 1989; Borowiec et al. 1990; Kelly 1991). Fractionation of human extract and reconstitution of replication activity using an in vitro complementation assay have led to the identification of those cellular proteins necessary and sufficient for viral DNA replication (Ishimi et al. 1988; Wold et al. 1989; Tsurimoto et al. 1990; Weinberg et al. 1990; Eki et al. 1992; Waga et al. 1994). Given that the mechanisms of SV40 and cellular DNA replication appear to be very similar, there is considerable confidence that the same proteins have identical functions in the replication of the chromosomes. It should be noted that analysis of the viral system has not identified all of the proteins involved in eukaryotic DNA replication. For example, genetic studies in yeast strongly suggest that pol-ε is necessary for DNA replication in vivo (Araki et al. 1992; Budd and Campbell 1993), but this enzyme was not originally identified as necessary for SV40 DNA replication.

As discussed previously, the natural template for eukaryotic DNA replication is chromatin. Whereas the cell-free system was originally developed with naked DNA templates, a number of studies have attempted to more accurately reflect the in vivo replicative process by employing chromatin templates. Although in vitro replication of such minichromosomes is very inefficient relative to that of naked DNA, the repression of initiation due to tightly bound histones can be relieved by the presence of transcription factors bound near the origin (Cheng and Kelly 1989; Cheng et al. 1992). A similar stimulatory effect of transcription factors on replication efficiency has been observed in vivo. Several possible mechanisms could account for this derepression, including increased accessibility of the origin to the initiator protein or activation of the replication apparatus by contact with factors bound near the origin (see DePamphilis, this volume). Further studies have focused on the fate of parental nucleosomes during DNA replication and indicate that nucleosomes remain associated with the replicating DNA during DNA synthesis (Randall and Kelly 1992). In addition, the assembly of new chromatin appears to be coupled to SV40 DNA replication in vitro (Stillman 1986). Experiments such as these have added to our understanding of the mechanism by which the replication machinery negotiates other DNA-bound structures, a process that must occur continuously in the nucleus.

Enzymology of the Replication Fork

Replication Proteins

Studies employing the SV40 model system have resulted in the identification of many cellular replication proteins. Recent work employing both genetic and biochemical techniques has added to this list. A summary of known replication proteins is presented in Table 1 and in some detail below.

Helicases

The focal point of all replication forks is the helicase, which catalyzes the transition from double- to single-stranded DNA. In *E. coli,* the DnaB

Table 1 Cellular replication proteins

Protein	Subunit (kD)[a]	Replicative function
DNA polymerases		
pol-α:primase	180, 70, 58, 48	DNA polymerase, primase
pol-δ	125, 48	DNA polymerase, 3′ to 5′ exonuclease
pol-ε	258, 55	DNA polymerase, 3′ to 5′ exonuclease
Accessory proteins		
RP-A	70, 32, 14	single-stranded DNA binding
PCNA	36	pol-δ/ε processivity factor
RF-C	145, 40, 38, 37, 36.5	loads PCNA onto template
Nucleases		
ribonuclease H1	89	Okazaki fragment maturation
FEN-1 (MF-1)	44	Okazaki fragment maturation
Others		
DNA ligase I	102	joins Okazaki fragments
topoisomerase I	100	unlinks parental strands
topoisomerase II	172	unlinks parental strands and progeny duplexes

References include those cited within the text as well as Kesti et al. (1993), Syvaoja and Linn (1989), Eder and Walder (1991), and Miller et al. (1981).
[a] Approximiate molecular weights of human polypeptides.

helicase translocates in a 5′ to 3′ direction while unwinding DNA (LeBowitz and McMacken 1986) and therefore is bound to the lagging-strand template during DNA replication. The same mechanism is employed by the phages T4 and T7 (Matson et al. 1983; Richardson and Nossal 1989), which induce their own helicases to engage in semidiscontinuous replication. In direct contrast, SV40 T antigen is bound to the leading-strand template and moves in the 3′ to 5′ direction advancing the fork (Goetz et al. 1988; Wiekowski et al. 1988). Although the polarity of translocation is not conserved, the replicative helicases of *E. coli*, T4, T7, and SV40 are all hexamers, suggesting a common quaternary structure for enzymes with this function. To date, several eukaryotic helicases have been identified, but conclusive identification of the enzyme that acts at chromosomal replication forks must await further biochemical and genetic investigation.

Single-stranded DNA-binding Protein

As the replication fork advances, a helix-destabilizing protein is required to maintain the single-stranded DNA structure that serves as a template for RNA priming and DNA synthesis. In eukaryotic cells, replication protein A (RP-A; RF-A; HSSB) performs this function (Wobbe et al. 1987; Fairman and Stillman 1988; Wold and Kelly 1988). This phosphoprotein has three subunits with molecular weights of 70,000, 32,000, and 14,000. The large subunit contains the DNA-binding activity but cannot support SV40 DNA replication in vitro by itself (Kenny et al. 1990; Erdile et al. 1991); therefore, at least one of the two smaller subunits is likely to be required for replication. In support of this hypothesis, all three yeast genes encoding RP-A are essential for viability (Heyer et al. 1990; Brill and Stillman 1991), and antibodies directed against any of the three subunits inhibit SV40 DNA replication in vitro (Erdile et al. 1990, 1991; Kenny et al. 1990; Umbricht et al. 1993). Although the roles of the two smaller RP-A subunits remain unknown, the middle subunit is phosphorylated during S phase of the cell cycle (Din et al. 1990), suggesting that this subunit may play some role in regulating DNA replication.

The DNA-binding properties of RP-A have been investigated extensively, but there remains some disagreement about fundamental aspects of the protein-DNA interaction. RP-A binds relatively nonspecifically to single-stranded DNA but exhibits a modest preference for DNA sequences rich in pyrimidines (Kim et al. 1992). Several different binding site sizes have been reported, ranging from 8 to 30 nucleotides for human RP-A and up to 100 nucleotides for the yeast protein (Alani et al. 1992; Kim et al. 1992; Blackwell and Borowiec 1994). Some evidence has been presented that RP-A forms two different types of complexes with DNA, which may account for the variability in site size estimates. Further disagreement centers on the degree of cooperativity of the DNA-binding reaction. Most prokaryotic single-stranded DNA-binding proteins display a high level of cooperativity, allowing the rapid and complete binding of any exposed single-stranded regions in the genome. Steady-state fluorescence experiments indicate that this is also the case with yeast RP-A (Alani et al. 1992). However, a series of studies employing electrophoretic mobility shift of oligonucleotides in the presence of human RP-A have demonstrated little cooperativity (Kim et al. 1992; Kim and Wold 1995). Other studies employing a similar technique suggest that only one of two proposed binding modes is highly cooperative (Blackwell and Borowiec 1994). The lack of consensus with regard to the DNA-binding properties may result from differences in experimental technique, or possibly differences in the state of the protein itself, such as the extent of phosphorylation.

DNA Pol-α:Primase Complex

DNA synthesis is initiated by the bifunctional pol-α:primase complex, a heterotetrameric phosphoprotein (Wang 1991). The primase activity resides in the 48-kD D subunit and is tightly associated with the 58-kD C subunit, which is thought to tether the primase to the 180-kD polymerase A subunit (Copeland and Wang 1993; Santocanale et al. 1993; Bakkenist and Cotterill 1994; Stadlbauer et al. 1994). With the exception of the *Drosophila* enzyme (Cotterill et al. 1987), there is no proofreading exonuclease activity associated with the A subunit. The remaining 70-kD B subunit has no known catalytic function, but it may contribute to recruitment of pol-α:primase to the replication fork (Collins et al. 1993) (see below).

The main function of pol-α:primase is to serve as a priming enzyme. The primase catalyzes the synthesis of complementary oligoribonucleotides, which are then extended a short distance by the A subunit DNA polymerase activity. Although high concentrations of pol-α:primase can support the complete replication of SV40 origin-containing plasmid DNA in vitro (Ishimi et al. 1988), it is unlikely that cellular replication relies heavily on this polymerase during elongation. The low processivity of the enzyme and the lack of an associated proofreading exonuclease suggest that pol-α:primase serves exclusively to initiate DNA synthesis on the lagging strand. Dissociation of pol-α:primase from the DNA provides a primer terminus for the assembly of the PCNA/pol-δ or /pol-ε complexes, highly processive polymerases that can efficiently extend the RNA/DNA primers originally synthesized by pol-α:primase.

DNA Polymerases δ and ε

The heterodimeric DNA polymerases δ and ε are involved in the elongation stage of DNA replication. Unlike pol-α:primase, polymerases δ and ε do not act alone but require the action of two auxiliary factors. The multisubunit replication factor C (RF-C) (Tsurimoto and Stillman 1989; Lee et al. 1991a) binds to the primer terminus immediately after RNA/DNA primer synthesis has been completed by pol-α:primase, allowing the subsequent assembly of a functional pol-δ or pol-ε complex. Once bound to the primer-template junction, RF-C loads PCNA onto the DNA in an energy-dependent reaction. PCNA then functions as a processivity factor, binding to pol-δ or pol-ε and maintaining a stable interaction between polymerase and template (Tan et al. 1986; Prelich et al. 1987; Burgers 1991; Lee et al. 1991a; Tsurimoto and Stillman 1991; Podust and Hubscher 1993). Ultimately, one or both of these processive polymerases, with their intrinsic proofreading activities, probably synthesize all of the cellular DNA, thereby ensuring faithful duplication of the genome (see below).

DNA polymerases δ and ε have some similarities in catalytic function and are both essential for viability in yeast (for reviews, see Wang 1991; So and Downey 1992). However, it is not yet clear whether they have specialized roles in replication. pol-ε does not substitute well for pol-δ in the cell-free SV40 DNA replication system with purified proteins (Lee et al. 1991b), so it may contribute to DNA synthesis only when certain additional factors are present. It is also possible that these enzymes have important roles that are not limited to DNA replication. For example, recent experiments indicate that pol-ε is one member of the S-phase checkpoint pathway (Navas et al. 1995), a biochemical feedback mechanism that delays cell-cycle progression upon damage to the DNA or inhibition of DNA synthesis during the replicative phase. Thus, in addition to synthesizing DNA, pol-ε may be involved in monitoring the status of DNA replication.

Nucleases

Synthesis of the lagging strand results in the generation of DNA fragments with RNA primers at the 5′ ends. For lagging-strand synthesis to be completed, these primers must be removed and the resulting gap must be filled. In *E. coli,* DNA polymerase I, with its intrinsic 5′ to 3′ exonuclease activity, mediates both functions (Kornberg and Baker 1992). None of the replicative eukaryotic polymerases contains a 5′ to 3′ exonuclease, indicating that other factors are involved in the initial processing of Okazaki fragments. A 44-kD 5′ to 3′ exonuclease (FEN-1; MF-1) that is required for the formation of covalently closed circular DNA during in vitro SV40 DNA replication has been identified in human cells (Ishimi et al. 1988; Harrington and Lieber 1994; Waga et al. 1994). Recently, studies employing purified calf proteins and a model lagging-strand template have shown that FEN-1 and RNase H1 act together to process Okazaki fragments. RNase H1 nicks the primer on the 5′ side of the 3′ ribonucleotide, providing a suitable substrate for FEN-1, which removes the 3′-terminal ribonucleotide of the RNA primer. The oligoribonucleotide is displaced and the gap is filled by DNA polymerase, resulting in nicked double-stranded DNA (Turchi and Bambara 1993; Huang et al. 1994; Murante et al. 1994; Turchi et al. 1994).

DNA Ligase

After removal of the RNA primers and extension of the DNA chains through the resulting gaps, the nascent DNA fragments must be joined to complete the synthesis of the lagging strand. Of the three DNA ligases that have been identified in mammalian cells (Lindahl and Barnes 1992), DNA ligase I is the most likely candidate to carry out this function in vivo. In a cell-free SV40 system containing highly purified proteins, DNA ligase I, but not DNA ligase III, catalyzes the formation of covalently closed daughter molecules (Waga et al. 1994). Although DNA ligase II has not been tested, this enzyme is present at low abundance and is likely to be involved in DNA repair. Genetic studies support the conclusion that DNA ligase I is involved in cellular replication in vivo. It has been shown that the human DNA ligase I gene can complement a *Saccharomyces cerevisiae* cdc9 mutant, which is defective in DNA replication due to DNA ligase deficiency (Barnes et al. 1990).

Topoisomerases

As replication procceds and the parental strands are unwound, positive supercoils are potentially introduced ahead of the replication fork. The resulting accumulation of torsional strain could lead to inhibition of fork movement if not relieved by a DNA topoisomerase. In eukaryotic cells, two types of topoisomerases have been discovered (Wang 1985). The type I enzyme introduces a transient single-strand break in the DNA, thereby relaxing either negatively or positively supercoiled DNA. Type II topoisomerase introduces a transient double-strand DNA break through which duplex DNA is passed. In addition to relaxing DNA, this enzyme can decatenate intertwined molecules. Studies with the SV40 model system employing naked DNA have revealed that either type of topoisomerase is capable of removing the positive supercoils ahead of the fork, allowing rapid and efficient DNA synthesis (Yang et al. 1987). However, the progeny DNA molecules that are formed remain multiply intertwined because of failure to remove all of the links between the parental strands during DNA synthesis. Topoisomerase II is required to resolve this tangled structure into two separate progeny genomes, allowing subsequent segregation (Yang et al. 1987).

The topological problems accompanying SV40 DNA replication in vitro are likely to arise during cellular replication as well. The length of chromosomal DNA and its association with nuclear proteins to form chromatin probably precludes free rotation of the DNA during replication. Therefore, the two eukaryotic topoisomerases are likely to function in vivo as they do in the SV40 system. It has been suggested that topoisomerase II may have a special role in the completion of DNA synthesis where two adjacent forks converge (Ishimi et al. 1992). However, this conclusion is based on studies of a cell-free SV40 system containing only one DNA polymerase (pol-α:primase) and may not accurately reflect the normal cellular process.

Organization of the Replication Fork

Most of the studies aimed at defining the organization of the cellular replication fork have revolved around the SV40 system. Although the identity and mechanism of the cellular helicase remain unknown, it is likely that the known cellular proteins play similar roles in chromosomal replication as in the viral model. The current picture of the eukaryotic replication fork is summarized in Figure 5. Leading-strand synthesis is performed by a processive polymerase complex (PCNA/pol-δ or PCNA/pol-ε) and a single auxiliary protein (RP-A). Lagging-strand synthesis requires additional proteins. pol-α:primase synthesizes RNA/DNA primers that are extended by complexes similar to those operating on the leading strand. Primer removal and completion of DNA synthesis require FEN-1 nuclease, RNase H1, and DNA ligase I as described above.

Studies with both crude and purified systems have demonstrated that assembly of replication proteins at the replication fork is mediated by specific protein-protein contacts. These interactions are critical for the coordinated synthesis of both leading and lagging strands, and are likely to be important at all stages of the replicative process. Specific mechanisms involving multiple proteins at the replication fork are discussed below.

Protein-protein Interactions

In the SV40 system, the viral T antigen and the cellular proteins pol-α: primase and RP-A are required for initiation of DNA replication, and a variety of studies have provided evidence for significant molecular interactions among the three proteins. For example, biochemical analysis of model reactions has uncovered a number of instances in which one protein affects the activity of another (Kenny et al. 1989; Collins and Kelly 1991; Collins et al. 1993; Melendy and Stillman 1993; Murakami

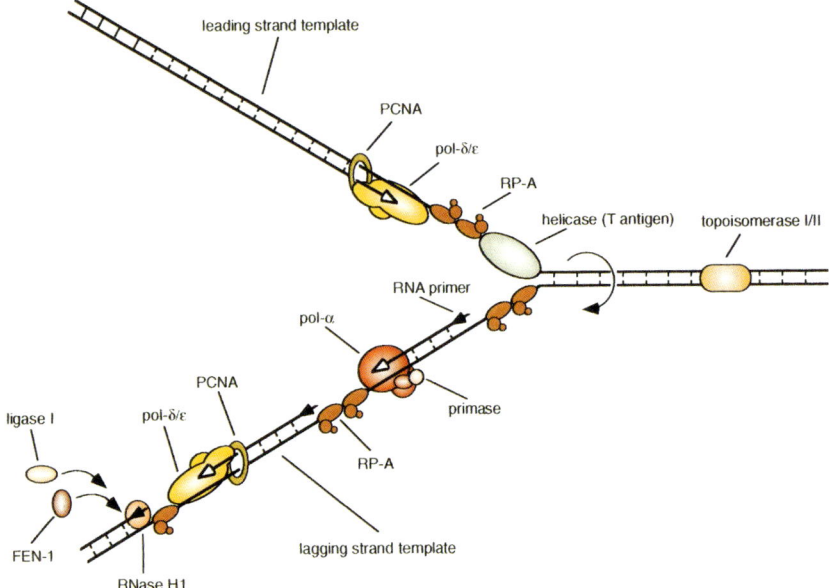

Figure 5 Semidiscontinuous DNA synthesis. This figure shows a diagrammatic view of the organization of the eukaryotic replication fork. The pol-δ/ε auxiliary factor RF-C is not included in this diagram because it is not known whether this protein stays associated with the replication apparatus after loading PCNA. See text for further details.

and Hurwitz 1993a). These biochemical data are supported by physical studies demonstrating direct association of each combination of two proteins. In particular, much information has been gathered on the pol-α/T-antigen interaction. This association appears complex, possibly involving three of the four pol-α subunits (A, B, and D) (Dornreiter et al. 1990, 1992; Gannon and Lane 1990; Collins et al. 1993; Bruckner et al. 1995).

It is very likely that the molecular interactions among initiation proteins are important determinants of efficiency and specificity in the initiation reaction. Unwinding of the SV40 origin requires T antigen and RP-A, and contact between these two proteins is likely to be important at this early stage of replication. Although the isolated large subunit of RP-A does not interact with T antigen, intact RP-A does (Dornreiter et al. 1992), suggesting that one or both of the smaller subunits could play a role in this interaction. Once the duplex is unwound, interaction between T antigen and pol-α may contribute to efficient priming on the RP-A-coated, single-stranded DNA template. It is possible that the different interaction domains on pol-α have different functional roles during replication. For example, the D subunit may help to coordinate initiation at the origin (Schneider et al. 1994), and the A and B subunits may be important for tethering pol-α:primase to the advancing helicase (see below). Priming is also likely to be influenced by an RP-A/primase interaction, which has been demonstrated to occur in vitro through the large subunit of RP-A (Dornreiter et al. 1992). The close association of the three proteins involved in initiation of SV40 DNA replication is quite striking and suggests that similar interactions may occur at cellular origins of replication. Confirmation of this possibility awaits identification of the cellular counterpart(s) of T antigen.

Protein-protein interactions are also critically important during elongation of the nascent DNA chains. As discussed above, efficient elongation requires a highly processive polymerase, and it is likely that pol-δ (or pol-ε), with its auxiliary factors RF-C and PCNA, carries out this function on both leading and lagging strands. The processivity of pol-δ is absolutely dependent on interaction with PCNA, which is topologically linked to the template DNA. There is no evidence for interaction between the processive polymerases and other replication proteins, although RP-A does stimulate pol-δ activity to some extent (Kenny et al. 1989).

Recent data suggest the possibility that protein-protein interactions may even be important during the final stage of DNA replication, when the Okazaki fragments are joined to form a continuous DNA chain. Studies with purified proteins have shown that DNA ligase I, but not DNA ligase III, can catalyze the formation of covalently closed daughter DNA molecules in the SV40 model system (Waga et al. 1994). Since both ligases are capable of sealing nicks in double-stranded DNA, the strict requirement for DNA ligase I may reflect specific contacts between the enzyme and other proteins involved in the maturation of nascent strands.

A number of interesting interactions between replication proteins and other cellular factors have been uncovered. Although the functions of these interactions are not entirely clear, some of these may be involved in regulating replication. RP-A interacts with several proteins, including the transcription factors p53, GAL4, and VP16 (He et al. 1993; Li and Botchan 1993) and the repair proteins XPA and XPG (He et al. 1995; Matsuda et al. 1995). Another example, discovered with the SV40 model system, is the coupling of chromatin assembly to DNA replication (Stillman 1986). This coupling is strongly suggestive of a direct interaction between the chromatin assembly apparatus and the replication machinery and probably evolved to ensure rapid and efficient packaging of the progeny DNA following synthesis.

Synthesis of Okazaki Fragments and Cycling of Pol-α:Primase

Because synthesis of the lagging strand proceeds in the direction opposite to fork movement, repeated priming events by the pol-α:primase are required. There are two general mechanisms by which this priming could occur. One possibility is that pol-α:primase completely dissociates from the template following completion of each RNA/DNA primer (distributive mechanism). In this scenario, synthesis of each RNA/DNA primer would involve association of a different molecule of pol-α:primase with the template. Alternatively, pol-α:primase might be tethered to the replication complex at the fork via specific protein-protein interactions. In this case, the enzyme would not leave the domain of the template following the completion of an RNA/DNA primer but would immediately reassociate with the template at a new site to initiate primer synthesis (processive mechanism).

At this point, the available data are not sufficient to choose between the two mechanisms. Evidence supporting a distributive model has come from dilution experiments in the cell-free SV40 DNA replication system (Murakami and Hurwitz 1993b). When the pol-α:primase concentration is decreased by dilution, the rate of DNA replication decreases proportionately. Although these data clearly indicate that pol-α:primase dissociates from the template at a measurable rate during DNA synthesis, it has not yet been possible to directly measure the number of priming events mediated by a given pol-α:primase prior to dissociation. Evidence for processive priming has been provided by the strong physical interaction between pol-α and T antigen (Collins et al. 1993). It is known that T antigen is capable of unwinding long segments of DNA without dissociation. Binding of pol-α:primase to T antigen could significantly increase the lifetime of pol-α:primase at the fork. Thus, a single pol-α:primase molecule could proceed through several priming cycles, allowing efficient synthesis of the lagging strand. In agreement with this model, the presence of T antigen increases the apparent processivity of pol-α:primase on model templates (Collins and Kelly 1991).

In addition to possibly increasing processivity of priming, interaction between pol-α:primase and T antigen may facilitate access of pol-α:primase to the template during replication. Experiments with model templates have shown that RP-A-coated single-stranded DNA is resistant to priming. In the presence of T antigen, this inhibition is relieved (Collins et al. 1993; Melendy and Stillman 1993). Therefore, interaction of pol-α:primase with T antigen may be required for primase to productively interact with the template DNA. During cellular DNA replication, a similar mechanism must be employed to allow efficient priming of the RP-A-coated DNA. It is possible that the cellular process is also mediated by a direct physical interaction between pol-α:primase and the replicative helicase.

Proofreading Mechanisms

The catalytic subunits of polymerases δ and ε contain 3′ to 5′ exonuclease activities, allowing high-fidelity DNA synthesis. Therefore, the majority of nuclear DNA synthesis employs the conventional proofreading mechanism first outlined in prokaryotic systems. However, pol-α:primase contains no obvious proofreading activity. It has been reported that a cryptic 3′ to 5′ exonuclease is uncovered when the *Drosophila* pol-α:primase complex is dissociated (Cotterill et al. 1987), but this phenomenon has not been observed with pol-α:primase complexes from other species. It is not clear how errors are corrected in DNA synthesized by pol-α:primase. One possibility is that a separate exonuclease proofreading activity exists. However, it is more likely that maturation of lagging-strand DNA fragments leads to removal of mismatches downstream from the RNA primer in pol-α-catalyzed regions. Prior to ligation of the nicked strand, FEN-1 nuclease and a proofreading polymerase could act in conjunction to accurately replace the DNA originally polymerized by pol-α. Since FEN-1 does not function on a gapped template but requires a nick, removal of DNA and resynthesis would have to occur concurrently. Alternating nucleolytic and synthetic steps by these two enzymes would closely resemble the "nick translation" reaction of *E. coli* DNA polymerase I (Murante et al. 1994).

Replication Centers

An interesting recent development has been the discovery that DNA replication may occur in relatively discrete foci in the nuclei of mammalian cells. When nascent DNA is labeled by exposure of cells to a short pulse of BrdU, the label is localized to a relatively small number of intranuclear sites. Enumeration of these sites or "replication centers" suggests that each may contain as many as 100 replication forks. Thus, DNA replication may be highly compartmentalized in eukaryotic cells (for review, see Cook 1991). Both PCNA and RP-A have been detected at these replication centers by immunofluorescence (Bravo and Macdonald-

Bravo 1987; Cardoso et al. 1993; Brenot-Bosc et al. 1995). Interestingly, the large subunit of RP-A has been detected at foci prior to the onset of replication and is a possible component of a prereplication complex at the origin (Adachi and Laemmli 1992).

The organization of replication centers is not well understood. One hypothesis invokes the nuclear matrix as an insoluble support that serves as a foundation for DNA replication and organizes the replication centers. However, a clear understanding of the structure and function of foci will have to await further biochemical investigation.

Control of Eukaryotic DNA Replication—Unanswered Questions

The basic enzymology of the eukaryotic replication fork has been uncovered through the identification and characterization of the replication proteins required for DNA synthesis in vitro. However, many fundamental questions remain regarding the control of replication in vivo. These problems focus on the coordination of DNA replication with other events in the cell cycle, including the timing of initiation, the prevention of multiple rounds of replication during a single replicative phase of the cell cycle, and the response to insults that may compromise the faithful duplication of the genetic material. Much of the current research on eukaryotic DNA replication is directed at characterizing these mechanisms.

Initiator proteins, which recognize origins of replication and unwind the double-stranded DNA, are likely to be central to the regulation of cellular DNA replication. The presence or the activation of these proteins may be responsible for controlling the timing of initiation during the cell cycle, and their subsequent removal or inactivation may be necessary to prevent re-replication during a single S phase. Identification of eukaryotic initiator proteins has proven to be quite difficult. However, a complex of six polypeptides that specifically recognizes yeast origin sequences has been purified recently (Bell and Stillman 1992). This origin recognition complex (ORC) is likely to function as an initiator, and thorough characterization of the protein should provide insight into the mechanisms that operate to regulate initiation. Studies have already shown that ORC protein level is constant through the cell cycle (Diffley and Cocker 1992). This observation suggests that the initiator protein is activated at the G_1/S transition, possibly by protein phosphorylation or by specific interactions with other cellular proteins.

Studies with the cell-free SV40 system have provided some evidence that protein phosphorylation may be involved in replication control.

SV40 DNA replication does not occur until the G_1/S transition has been reached in the host cell. Interestingly, extracts prepared from cells in G_1 are incompetent for in vitro SV40 DNA replication unless supplemented with protein phosphatase 2Ac (PP2Ac) or cdc2 kinase (Virshup et al. 1989; D'Urso et al. 1990). Although the key target of these G_1-activating enzymes is not yet clear, T antigen is one possible candidate. Both PP2Ac and cdc2 kinase can directly modify the phosphorylation state of T antigen in vitro, thereby regulating its ability to unwind the SV40 origin. It is possible that cellular initiator proteins may be activated at the G_1/S boundary by similar phosphorylation and/or dephosphorylation events.

Two cellular proteins involved in the initiation of SV40 DNA replication in vitro, pol-α:primase and RP-A, are phosphorylated in cell-cycle-dependent manners (Din et al. 1990; Nasheuer et al. 1991). There has been no clear demonstration that phosphorylation affects the replicative function of either protein, but this remains a reasonable possibility. It has also been suggested that RP-A phosphorylation might be involved in a signaling pathway that coordinates DNA replication with the cell cycle (Brush et al. 1994). In this case, RP-A would have both a replicative and a regulatory role. Such bifunctionality has recently been demonstrated with pol-ε, which appears to act as a replicative polymerase and as a member of the S-phase checkpoint pathway (Navas et al. 1995). This finding may help to explain the need for two rather redundant DNA polymerase activities in eukaryotic cells.

Observations such as those described above have provided some clues into the mechanism of DNA replication control, but it is clear that the majority of the process is not well understood. The characterization of the pathways and their components that allow communication between the replication machine and other cellular apparati will rely on both the biochemistry of the viral model systems and the genetics afforded by yeast. Since the basic mechanism of DNA replication has been conserved from yeast to humans, many of the regulatory mechanisms are likely to be conserved as well. As a result, these systems will continue to be used interchangeably to provide us with an understanding of how the cell efficiently and faithfully replicates its DNA in preparation for cell division.

REFERENCES

Adachi, Y. and U.K. Laemmli. 1992. Identification of nuclear pre-replication centers poised for DNA synthesis in *Xenopus* egg extracts: Immunolocalization study of replication protein A. *J. Cell Biol* **119:** 1–15.

Alani, E., R. Thresher, J.D. Griffith, and R.D. Kolodner. 1992. Characterization of DNA-binding and strand-exchange stimulation properties of y-RPA, a yeast single-strand-DNA-binding protein. *J. Mol. Biol.* **227:** 54–71.

Araki, H., P.A. Ropp, A.L. Johnson, L.H. Johnston, A. Morrison, and A. Sugino. 1992. DNA polymerase II, the probable homolog of mammalian DNA polymerase epsilon, replicates chromosomal DNA in the yeast *Saccharomyces cerevisiae*. *EMBO J.* **11:** 733–740.

Ashktorab, H. and A. Srivastava. 1989. Identification of nuclear proteins that specifically interact with adeno-associated virus type 2 inverted terminal repeat hairpin DNA. *J. Virol.* **63:** 3034–3039.

Bakkenist, C.J. and S. Cotterill. 1994. The 50-kDa primase subunit of *Drosophila melanogaster* DNA polymerase alpha: Molecular characterization of the gene and functional analysis of the overexpressed protein. *J. Biol. Chem.* **269:** 26759–26766.

Baldacci, G., B. Cherif-Zahar, and G. Bernardi. 1984. The initiation of DNA replication in the mitochondrial genome of yeast. *EMBO J.* **3:** 2115–2120.

Barnes, D.E., L.H. Johnston, K. Kodama, A.E. Tomkinson, D.D. Lasko, and T. Lindahl. 1990. Human DNA ligase I cDNA: Cloning and functional expression in *Saccharomyces cerevisiae*. *Proc. Natl. Acad. Sci.* **87:** 6679–6683.

Bell, S.P. and B. Stillman. 1992. ATP-dependent recognition of eukaryotic origins of DNA replication by a multiprotein complex [see comments]. *Nature* **357:** 128–134.

Berns, K.I. 1990a. Parvoviridae and their replication. In *Virology* (ed. B.N. Fields and D.M. Knipe), pp. 1743–1763. Raven Press, New York.

———. 1990b. Parvovirus replication. *Microbiol. Rev.* **54:** 316–329.

Berns, K.I. and W.W. Hauswirth. 1979. Adeno-associated viruses. *Adv. Virus Res.* **25:** 407–449.

Berns, K.I. and R.M. Linden. 1995. The cryptic life style of adeno-associated virus. *BioEssays* **17:** 237–245.

Blackwell, L.J. and J.A. Borowiec. 1994. Human replication protein A binds single-stranded DNA in two distinct complexes. *Mol. Cell. Biol.* **14:** 3993–4001.

Blanc, H. and B. Dujon. 1980. Replicator regions of the yeast mitochondrial DNA responsible for suppressiveness. *Proc. Natl. Acad. Sci.* **77:** 3942–3946.

Borowiec, J.A., F.B. Dean, P.A. Bullock, and J. Hurwitz. 1990. Binding and unwinding—How T antigen engages the SV40 origin of DNA replication. *Cell* **60:** 181–184.

Bosher, J., E.C. Robinson, and R.T. Hay. 1990. Interactions between the adenovirus type 2 DNA polymerase and the DNA binding domain of nuclear factor I. *New Biol.* **2:** 1083–1090.

Bravo, R. and H. Macdonald-Bravo. 1987. Existence of two populations of cyclin/proliferating cell nuclear antigen during the cell cycle: Association with DNA replication sites. *J. Cell Biol.* **105:** 1549–1554.

Brenot-Bosc, F., S. Gupta, R.L. Margolis, and R. Fotedar. 1995. Changes in the subcellular localization of replication initiation proteins and cell cycle proteins during G1- to S-phase transition in mammalian cells. *Chromosoma* **103:** 517–527.

Brill, S.J. and B. Stillman. 1991. Replication factor-A from *Saccharomyces cerevisiae* is encoded by three essential genes coordinately expressed at S phase. *Genes Dev.* **5:** 1589–1600.

Bruckner, A., F. Stadlbauer, L.A. Guarino, A. Brunahl, C. Schneider, C. Rehfuess, C. Prives, E. Fanning, and H.P. Nasheuer. 1995. The mouse DNA polymerase alpha-primase subunit p48 mediates species-specific replication of polyomavirus DNA in vitro. *Mol. Cell. Biol.* **15:** 1716–1724.

Brush, G.S., C.W. Anderson, and T.J. Kelly. 1994. The DNA-activated protein kinase is required for the phosphorylation of replication protein A during simian virus 40 DNA replication. *Proc. Natl. Acad. Sci.* **91**: 12520-12524.

Budd, M.E. and J.L. Campbell. 1993. DNA polymerases delta and epsilon are required for chromosomal replication in *Saccharomyces cerevisiae*. *Mol. Cell. Biol.* **13**: 496-505.

Burgers, P.M. 1991. *Saccharomyces cerevisiae* replication factor C. II. Formation and activity of complexes with the proliferating cell nuclear antigen and with DNA polymerases delta and epsilon. *J. Biol. Chem.* **266**: 22698-22706.

Cardoso, M.C., H. Leonhardt, and G.B. Nadal. 1993. Reversal of terminal differentiation and control of DNA replication: Cyclin A and Cdk2 specifically localize at subnuclear sites of DNA replication. *Cell* **74**: 979-992.

Challberg, M.D. and T.J. Kelly. 1979. Adenovirus DNA replication in vitro. *Proc. Natl. Acad. Sci.* **76**: 655-659.

―――. 1989. Animal virus DNA replication. *Annu. Rev. Biochem.* **58**: 671-717.

Challberg, M.D., S.V. Desiderio, and T.J. Kelly. 1980. Adenovirus DNA replication in vitro: Characterization of a protein covalently linked to nascent DNA strands. *Proc. Natl. Acad. Sci.* **77**: 5105-5109.

Chang, D.D. and D.A. Clayton. 1984. Precise identification of individual promoters for transcription of each strand of human mitochondrial DNA. *Cell* **36**: 635-643.

―――. 1985. Priming of human mitochondrial DNA replication occurs at the light-strand promoter. *Proc. Natl. Acad. Sci.* **82**: 351-355.

―――. 1987. A novel endoribonuclease cleaves at a priming site of mouse mitochondrial DNA replication. *EMBO J.* **6**: 409-417.

Chang, D.D., W.W. Hauswirth, and D.A. Clayton. 1985. Replication priming and transcription initiate from precisely the same site in mouse mitochondrial DNA. *EMBO J.* **4**: 1559-1567.

Chen, M., N. Mermod, and M.S. Horwitz. 1990. Protein-protein interactions between adenovirus DNA polymerase and nuclear factor I mediate formation of the DNA replication preinitiation complex. *J. Biol. Chem.* **265**: 18634-18642.

Cheng, L. and T.J. Kelly. 1989. Transcriptional activator nuclear factor I stimulates the replication of SV40 minichromosomes in vivo and in vitro. *Cell* **59**: 541-551.

Cheng, L.Z., J.L. Workman, R.E. Kingston, and T.J. Kelly. 1992. Regulation of DNA replication in vitro by the transcriptional activation domain of GAL4-VP16. *Proc. Natl. Acad. Sci.* **89**: 589-593.

Clayton, D.A. 1982. Replication of animal mitochondrial DNA. *Cell* **28**: 693-705.

―――. 1991. Replication and transcription of vertebrate mitochondrial DNA. *Annu. Rev. Cell Biol.* **7**: 453-478.

―――. 1992. Transcription and replication of animal mitochondrial DNAs. *Int. Rev. Cytol.* **141**: 217-232.

Collins, K.L. and T.J. Kelly. 1991. Effects of T antigen and replication protein A on the initiation of DNA synthesis by DNA polymerase alpha-primase. *Mol. Cell. Biol.* **11**: 2108-2115.

Collins, K.L., A.A. Russo, B.Y. Tseng, and T.J. Kelly. 1993. The role of the 70 kDa subunit of human DNA polymerase alpha in DNA replication. *EMBO J.* **12**: 4555-4566.

Cook, P.R. 1991. The nucleoskeleton and the topology of replication (Review). *Cell* **66**: 627-635.

Copeland, W.C. and T.S. Wang. 1993. Enzymatic characterization of the individual mammalian primase subunits reveals a biphasic mechanism for initiation of DNA replication. *J. Biol. Chem.* **268:** 26179–26189.

Cote, J. and A. Ruiz-Carrillo. 1993. Primers for mitochondrial DNA replication generated by endonuclease G. *Science* **261:** 765–769.

Cotmore, S.F. and P. Tattersall. 1987. The autonomously replicating parvoviruses of vertebrates. *Adv. Virus Res.* **33:** 91–174.

Cotterill, S.M., M.E. Reyland, L.A. Loeb, and I.R. Lehman. 1987. A cryptic proofreading $3'\rightarrow5'$ exonuclease associated with the polymerase subunit of the DNA polymerase-primase from *Drosophila melanogaster*. *Proc. Natl. Acad. Sci.* **84:** 5635–5639.

de Zamaroczy, M., G. Faugeron-Fonty, G. Baldacci, R. Goursot, and G. Bernardi. 1984. The ori sequences of the mitochondrial genome of a wild-type yeast strain: Number, location, orientation and structure. *Gene* **32:** 439–457.

Diffley, J.F.X. and J.H. Cocker. 1992. Protein-DNA interactions at a yeast replication origin. *Nature* **357:** 169–172.

Din, S., S.J. Brill, M.P. Fairman, and B. Stillman. 1990. Cell-cycle-regulated phosphorylation of DNA replication factor A from human and yeast cells. *Genes Dev.* **4:** 968–977.

Dornreiter, I., A. Hoss, A.K. Arthur, and E. Fanning. 1990. SV40 T antigen binds directly to the large subunit of DNA polymerase α. *EMBO J.* **9:** 3329–3336.

Dornreiter, I., L.F. Erdile, I.U. Gilbert, W.D. von Winkler, T.J. Kelly, and E. Fanning. 1992. Interaction of DNA polymerase alpha-primase with cellular replication protein A and SV40 T antigen. *EMBO J.* **11:** 769–776.

D'Urso, G., R.L. Marraccino, D.R. Marshak, and J.M. Roberts. 1990. Cell cycle control of DNA replication by a homologue from human cells of the p34cdc2 protein kinase. *Science* **250:** 786–791.

Eder, P.S. and J.A. Walder. 1991. Ribonuclease H from K562 human erythroleukemia cells: Purification, characterization, and substrate specificity. *J. Biol. Chem.* **266:** 6472–6479.

Eki, T., T. Matsumoto, Y. Murakami, and J. Hurwitz. 1992. The replication of DNA containing the simian virus 40 origin by the monopolymerase and dipolymerase systems. *J. Biol. Chem.* **267:** 7284–7294.

Erdile, L.F., M.S. Wold, and T.J. Kelly. 1990. The primary structure of the 32-kDa subunit of human replication protein A. *J. Biol. Chem.* **265:** 3177–3182.

Erdile, L.F., W.-D. Heyer, R. Kolodner, and T.J. Kelly. 1991. Characterization of cDNA encoding the 70-kDa subunit of human replication protein A (RP-A), a single-stranded DNA binding protein involved in DNA replication and recombination. *J. Biol. Chem.* **266:** 12090–12098.

Fairman, M.P. and B. Stillman. 1988. Cellular factors required for multiple stages of SV40 DNA replication in vitro. *EMBO J.* **7:** 1211–1218.

Field, J., R.M. Gronostajski, and J. Hurwitz. 1984. Properties of the adenovirus DNA polymerase. *J. Biol. Chem.* **259:** 9487–9495.

Fisher, R.P. and D.A. Clayton. 1985. A transcription factor required for promoter recognition by human mitochondrial RNA polymerase. Accurate initiation at the heavy- and light-strand promoters dissected and reconstituted in vitro. *J. Biol. Chem.* **260:** 11330–11338.

Foury, F. 1989. Cloning and sequencing of the nuclear gene MIP1 encoding the catalytic subunit of the yeast mitochondrial DNA polymerase. *J. Biol. Chem.* **264:** 20552–20560.

Gannon, J.V. and D.P. Lane. 1990. Interactions between SV40 T antigen and DNA polymerase α. *New Biol.* **2:** 84–92.

Gillum, A.M. and D.A. Clayton. 1979. Mechanism of mitochondrial DNA replication in mouse L-cells: RNA priming during the initiation of heavy-strand synthesis. *J. Mol. Biol.* **135:** 353–368.

Goetz, G.S., F.B. Dean, J. Hurwitz, and S.W. Matson. 1988. The unwinding of duplex regions in DNA by the simian virus 40 large tumor antigen-asssociated DNA helicase activity. *J. Biol. Chem.* **263:** 383–392.

Gray, H. and T.W. Wong. 1992. Purification and identification of subunit structure of the human mitochondrial DNA polymerase. *J. Biol. Chem.* **267:** 5835–5841.

Harrington, J.J. and M.R. Lieber. 1994. The characterization of a mammalian DNA structure-specific endonuclease. *EMBO J.* **13:** 1235–1246.

Hauswirth, W.W. and K.I. Berns. 1979. Adeno-associated virus DNA replication: Nonunit-length molecules. *Virology* **93:** 57–68.

Hay, R.T. and W.C. Russell. 1989. Recognition mechanisms in the synthesis of animal virus DNA. *Biochem J.* **258:** 3–16.

He, Z., L.A. Henricksen, M.S. Wold, and C.J. Ingles. 1995. RPA involvement in the damage-recognition and incision steps of nucleotide excision repair. *Nature* **374:** 566–569.

He, Z., B.T. Brinton, J. Greenblatt, J.A. Hassell, and C.J. Ingles. 1993. The transactivator proteins VP16 and GAL4 bind replication factor A. *Cell* **73:** 1223–1232.

Hehman, G.L. and W.W. Hauswirth. 1992. DNA helicase from mammalian mitochondria. *Proc. Natl. Acad. Sci.* **89:** 8562–8566.

Hermonat, P.L., M.A. Labow, R. Wright, K.I. Berns, and N. Muzyczka. 1984. Genetics of adeno-associated virus: Isolation and preliminary characterization of adeno-associated virus type 2 mutants. *J. Virol.* **51:** 329–339.

Heyer, W.-D., M.R.F. Rao, L.F. Erdile, T.J. Kelly, and R.D. Kolodner. 1990. An essential *Saccharomyces cerevisiae* single-stranded DNA binding protein is homologous to the large subunit of human RP-A. *EMBO J.* **9:** 2321–2329.

Hixson, J.E., T.W. Wong, and D.A. Clayton. 1986. Both the conserved stem-loop and divergent 5′-flanking sequences are required for initiation at the human mitochondrial origin of light-strand DNA replication. *J. Biol. Chem.* **261:** 2384–2390.

Hong, G., P. Ward, and K.I. Berns. 1992. In vitro replication of adeno-associated virus DNA. *Proc. Natl. Acad. Sci.* **89:** 4673–4677.

Huang, L., Y. Kim, J.J. Turchi, and R.A. Bambara. 1994. Structure-specific cleavage of the RNA primer from Okazaki fragments by calf thymus RNAse H1. *J. Biol. Chem.* **269:** 25922–25927.

Ikeda, J.E., T. Enomoto, and J. Hurwitz. 1982. Adenoviral protein-primed initiation of DNA chains in vitro. *Proc. Natl. Acad. Sci.* **79:** 2442–2446.

Im, D.S. and N. Muzyczka. 1990. The AAV origin binding protein Rep68 is an ATP-dependent site-specific endonuclease with DNA helicase activity. *Cell* **61:** 447–457.

Insdorf, N.F. and D.F. Bogenhagen. 1989a. DNA polymerase γ from *Xenopus laevis*. I. The identification of a high molecular weight catalytic subunit by a novel DNA polymerase photolabeling procedure. *J. Biol. Chem.* **264:** 21491–21497.

———. 1989b. DNA polymerase gamma from *Xenopus laevis*. II. A 3′→5′ exonuclease is tightly associated with the DNA polymerase activity. *J. Biol. Chem.* **264:** 21498–21503.

Ishimi, Y., A. Claude, P. Bullock, and J. Hurwitz. 1988. Complete enzymatic synthesis of

DNA containing the SV40 origin of replication. *J. Biol. Chem.* **263:** 19723–19733.
Ishimi, Y., K. Sugasawa, F. Hanaoka, T. Eki, and J. Hurwitz. 1992. Topoisomerase II plays an essential role as a swivelase in the late stage of SV40 chromosome replication in vitro. *J. Biol. Chem.* **267:** 462–466.
Kelly, T.J. 1991. DNA replication in mammalian cells: Insights from the SV40 model system. *Harvey Lect.* **85:** 173–188.
Kenny, M.K., S.H. Lee, and J. Hurwitz. 1989. Multiple functions of human single-stranded-DNA binding protein in simian virus 40 DNA replication: Single-strand stabilization and stimulation of DNA polymerases alpha and delta. *Proc. Natl. Acad. Sci.* **86:** 9757–9761.
Kenny, M.K., U. Schlegel, H. Furneaux, and J. Hurwitz. 1990. The role of human single-stranded DNA binding protein and its individual subunits in simian virus 40 DNA replication. *J. Biol. Chem.* **265:** 7693–7700.
Kesti, T., H. Frantti, and J.E. Syvaoja. 1993. Molecular cloning of the cDNA for the catalytic subunit of human DNA polymerase ε. *J. Biol. Chem.* **268:** 10238–10245.
Kim, C. and M.S. Wold. 1995. Recombinant human replication protein A binds to polynucleotides with low cooperativity. *Biochemistry* **34:** 2058–2064.
Kim, C., R.O. Snyder, and M.S. Wold. 1992. Binding properties of replication protein A from human and yeast cells. *Mol. Cell. Biol.* **12:** 3050–3059.
King, A.J. and P.C. Van der Vliet. 1994. A precursor terminal protein-trinucleotide intermediate during initiation of adenovirus DNA replication: Regeneration of molecular ends in vitro by a jumping back mechanism. *EMBO J.* **13:** 5786–5792.
Kong, X.P., R. Onrust, M. O'Donnell, and J. Kuriyan. 1992. Three-dimensional structure of the beta subunit of *E. coli* DNA polymerase III holoenzyme: A sliding DNA clamp. *Cell* **69:** 425–437.
Kornberg, A. and T.A. Baker. 1992. *DNA replication*, 2nd edition. Freeman, New York.
Krishna, T.S., X.P. Kong, S. Gary, P.M. Burgers, and J. Kuriyan. 1994. Crystal structure of the eukaryotic DNA polymerase processivity factor PCNA. *Cell* **79:** 1233–1243.
Lahaye, A., H. Stahl, S.D. Thines, and F. Foury. 1991. PIF1: A DNA helicase in yeast mitochondria. *EMBO J.* **10:** 997–1007.
LeBowitz, J.H. and R. McMacken. 1986. The *Escherichia coli* dnaB replication protein is a DNA helicase. *J. Biol. Chem.* **261:** 4738–4748.
Lechner, R.L. and T.J. Kelly. 1977. The structure of replicating adenovirus 2 DNA molecules. *Cell* **12:** 1007–1020.
Lee, S.H., A.D. Kwong, Z.Q. Pan, and J. Hurwitz. 1991a. Studies on the activator 1 protein complex, an accessory factor for proliferating cell nuclear antigen-dependent DNA polymerase delta. *J. Biol. Chem.* **266:** 594–602.
Lee, S.H., Z.Q. Pan, A.D. Kwong, P.M. Burgers, and J. Hurwitz. 1991b. Synthesis of DNA by DNA polymerase epsilon in vitro. *J. Biol. Chem.* **266:** 22707–22717.
Lehman, I.R. and L.S. Kaguni. 1989. DNA polymerase alpha. *J. Biol. Chem.* **264:** 4265–4268.
Li, J.J. and T.J. Kelly. 1984. Simian virus 40 DNA replication in vitro. *Proc. Natl. Acad. Sci.* **81:** 6973–6977.
Li, R. and M.R. Botchan. 1993. The acidic transcriptional activation domains of VP16 and p53 bind the cellular replication protein A and stimulate in vitro BPV-1 DNA replication. *Cell* **73:** 1207–1221.
Lindahl, T. and D.E. Barnes. 1992. Mammalian DNA ligases. *Annu. Rev. Biochem.* **61:** 251–281.

Lindenbaum, J.O., J. Field, and J. Hurwitz. 1986. The adenovirus DNA binding protein and adenovirus DNA polymerase interact to catalyze elongation of primed DNA templates. *J. Biol. Chem.* **261:** 10218–10227.

Liu, B. and B.M. Alberts. 1995. Head-on collision between a DNA replication apparatus and RNA polymerase transcription complex. *Science* **267:** 1131–1137.

Liu, B., M.L. Wong, R.L. Tinker, E.P. Geiduschek, and B.M. Alberts. 1993. The DNA replication fork can pass RNA polymerase without displacing the nascent transcript. *Nature* **366:** 33–39.

Lusby, E., R. Bohenzky, and K.I. Berns. 1981. Inverted terminal repetition in adeno-associated virus DNA: Independence of the orientation at either end of the genome. *J. Virol* **37:** 1083–1086.

Lusby, E., K.H. Fife, and K.I. Berns. 1980. Nucleotide sequence of the inverted terminal repetition in adeno-associated virus DNA. *J. Virol.* **34:** 402–409.

Madsen, C.S., S.C. Ghivizzani, and W.W. Hauswirth. 1993. Protein binding to a single termination-associated sequence in the mitochondrial DNA D-loop region. *Mol. Cell. Biol.* **13:** 2162–71.

Maki, H., S. Maki, and A. Kornberg. 1988. DNA polymerase III holoenzyme of *Escherichia coli*. IV. The holoenzyme is an asymmetric dimer with twin active sites. *J. Biol. Chem.* **263:** 6570–6578.

Matson, S.W., S. Tabor, and C.C. Richardson. 1983. The gene 4 protein of bacteriophage T7. Characterization of helicase activity. *J. Biol. Chem.* **258:** 14017–14024.

Matsuda, T., M. Saijo, I. Kuraoka, T. Kobayashi, Y. Nakatsu, A. Nagai, T. Enjoji, C. Masutani, K. Sugasawa, F. Hanaoka, et al. 1995. DNA repair protein XPA binds replication protein A (RPA). *J. Biol. Chem.* **270:** 4152–4157.

McCarty, D.M., J.H. Ryan, S. Zolotukhin, X. Zhou, and N. Muzyczka. 1994. Interaction of the adeno-associated virus Rep protein with a sequence within the A palindrome of the viral terminal repeat. *J. Virol.* **68:** 4998–5006.

McHenry, C.S. 1988. DNA polymerase III holoenzyme of *Escherichia coli. Annu. Rev. Biochem.* **57:** 519–550.

Melendy, T. and B. Stillman. 1993. An interaction between replication protein A and SV40 T antigen appears essential for primosome assembly during SV40 DNA replication. *J. Biol. Chem.* **268:** 3389–3395.

Meyer, R.R. and P.S. Laine. 1990. The single-stranded DNA-binding protein of *Escherichia coli. Microbiol. Rev.* **54:** 342–380.

Mignotte, B., J. Marsault, and M. Barat-Gueride. 1988. Effects of the *Xenopus laevis* mitochondrial single-stranded DNA-binding protein on the activity of DNA polymerase gamma. *Eur. J. Biochem.* **174:** 479–484.

Miller, K.G., L.F. Liu, and P.T. Englund. 1981. A homogeneous type II DNA topoisomerase from HeLa cell nuclei. *J. Biol. Chem.* **256:** 9334–9339.

Mul, Y.M. and P.C. Van der Vliet. 1992. Nuclear factor I enhances adenovirus DNA replication by increasing the stability of a preinitiation complex. *EMBO J.* **11:** 751–760.

Mul, Y.M., C.P. Verrijzer, and P.C. van der Vliet. 1990. Transcription factors NFI and NFIII/oct-1 function independently, employing different mechanisms to enhance adenovirus DNA replication. *J. Virol.* **64:** 5510–5518.

Murakami, Y. and J. Hurwitz. 1993a. DNA polymerase alpha stimulates the ATP-dependent binding of simian virus tumor T antigen to the SV40 origin of replication. *J. Biol. Chem.* **268:** 11018–11027.

———. 1993b. Functional interactions between SV40 T antigen and other replication proteins at the replication fork. *J. Biol. Chem.* **268:** 11008–11017.

Murante, R.S., L. Huang, J.J. Turchi, and R.A. Bambara. 1994. The calf 5′- to 3′-exonuclease is also an endonuclease with both activities dependent on primers annealed upstream of the point of cleavage. *J. Biol. Chem.* **269:** 1191–1196.

Muzyczka, N. 1992. Use of adeno-associated virus as a general transduction vector for mammalian cells. *Curr. Top. Microbiol. Immunol.* **158:** 97–129.

Nagata, K., R.A. Guggenheimer, and J. Hurwitz. 1983a. Adenovirus DNA replication in vitro: Synthesis of full-length DNA with purified proteins. *Proc. Natl. Acad. Sci.* **88:** 4266–4270.

———. 1983b. Specific binding of a cellular DNA replication protein to the origin of replication of adenovirus DNA. *Proc. Natl. Acad. Sci.* **80:** 6177–6181.

Nagata, K., R.A. Guggenheimer, T. Enomoto, J.H. Lichy, and J. Hurwitz. 1982. Adenovirus DNA replication in vitro: Identification of a host factor that stimulates synthesis of the preterminal protein-dCMP complex. *Proc. Natl. Acad. Sci.* **79:** 6438–6442.

Nasheuer, H.P., A. Moore, A.F. Wahl, and T.S. Wang. 1991. Cell cycle-dependent phosphorylation of human DNA polymerase alpha. *J. Biol. Chem.* **266:** 7893–7903.

Navas, T.A., Z. Zhou, and S.J. Elledge. 1995. DNA polymerase epsilon links the DNA replication machinery to the S phase checkpoint. *Cell* **80:** 29–39.

Ni, T.H., X. Zhou, D.M. McCarty, I. Zolotukhin, and N. Muzyczka. 1994. In vitro replication of adeno-associated virus DNA. *J. Virol.* **68:** 1128–1138.

Parisi, M.A., B. Xu, and D.A. Clayton. 1993. A human mitochondrial transcriptional activator can functionally replace a yeast mitochondrial HMG-box protein both in vivo and in vitro. *Mol. Cell. Biol.* **13:** 1951–1961.

Pavco, P.A. and G.C. Van Tuyle. 1985. Purification and general properties of the DNA-binding protein (P16) from rat liver mitochondria. *J. Cell. Biol.* **100:** 258–264.

Podust, V.N. and U. Hubscher. 1993. Lagging strand DNA synthesis by calf thymus DNA polymerases alpha, beta, delta and epsilon in the presence of auxiliary proteins. *Nucleic Acids Res.* **21:** 841–846.

Prelich, G., C.K. Tan, M. Kostura, M.B. Mathews, A.G. So, K.M. Downey, and B. Stillman. 1987. Functional identity of proliferating cell nuclear antigen and a DNA polymerase-delta auxiliary protein. *Nature* **326:** 517–520.

Pronk, R., W. Van Driel, and P.C. Van der Vliet. 1994. Replication of adenovirus DNA in vitro is ATP-independent. *FEBS Lett.* **337:** 33–38.

Pruijn, G.J., W. van Driel, and P.C. van der Vliet. 1986. Nuclear factor III, a novel sequence-specific DNA-binding protein from HeLa cells stimulating adenovirus DNA replication. *Nature* **322:** 656–659.

Randall, S.K. and T.J. Kelly. 1992. The fate of parental nucleosomes during SV40 DNA replication. *J. Biol. Chem.* **267:** 14259–14265.

Rawlins, D.R., P.J. Rosenfeld, R.J. Wides, M.D. Challberg, and T.J. Kelly. 1984. Structure and function of the adenovirus origin of replication. *Cell* **37:** 309–319.

Rekosh, D.M., W.C. Russell, A.J. Bellet, and A.J. Robinson. 1977. Identification of a protein linked to the ends of adenovirus DNA. *Cell* **11:** 283–295.

Richardson, R.W. and N.G. Nossal. 1989. Characterization of the bacteriophage T4 gene 41 DNA helicase. *J. Biol. Chem.* **264:** 4725–4731.

Rosenfeld, P.J. and T.J. Kelly. 1986. Purification of nuclear factor I by DNA recognition site affinity chromatography (published erratum appears in *J. Biol. Chem.* [1986] **261:** 10015) *J. Biol. Chem.* **261:** 1398–1408.

Rosenfeld, P.J., E.A. O'Neill, R.J. Wides, and T.J. Kelly. 1987. Sequence-specific interactions between cellular DNA-binding proteins and the adenovirus origin of DNA replication. *Mol. Cell. Biol.* **7**: 875-886.

Salas, M. 1991. Protein-priming of DNA replication. *Annu. Rev. Biochem.* **60**: 39-71.

Santocanale, C., M. Foiani, G. Lucchini, and P. Plevani. 1993. The isolated 48,000-dalton subunit of yeast DNA primase is sufficient for RNA primer synthesis. *J. Biol. Chem.* **268**: 1343-1348.

Schmitt, M.E. and D.A. Clayton. 1992. Yeast site-specific ribonucleoprotein endoribonuclease MRP contains an RNA component homologous to mammalian RNase MRP RNA and essential for cell viability. *Genes Dev.* **6**: 1975-1985.

―――. 1993. Conserved features of yeast and mammalian mitochondrial DNA replication. *Curr. Opin. Genet. Dev.* **3**: 769-774.

Schneider, C., K. Weisshart, L.A. Guarino, I. Dornreiter, and E. Fanning. 1994. Species-specific functional interactions of DNA polymerase alpha-primase with SV40 T antigen require SV40 origin DNA. *Mol. Cell. Biol.* **14**: 3176-3185.

Snyder, R.O., D.S. Im, and N. Muzyczka. 1990a. Evidence for covalent attachment of the adeno-associated virus (AAV) rep protein to the ends of the AAV genome. *J. Virol.* **64**: 6204-6213.

Snyder, R.O., R.J. Samulski, and N. Muzyczka. 1990b. In vitro resolution of covalently joined AAV chromosome ends. *Cell* **60**: 105-113.

So, A.G. and K.M. Downey. 1992. Eukaryotic DNA replication. *Crit. Rev. Biochem. Mol. Biol.* **27**: 129-155.

Srivastava, A., E.W. Lusby, and K.I. Berns. 1983. Nucleotide sequence and organization of the adeno-associated virus 2 genome. *J. Virol.* **45**: 555-564.

Stadlbauer, F., A. Brueckner, C. Rehfuess, C. Eckerskorn, F. Lottspeich, V. Forster, B.Y. Tseng, and H.P. Nasheur. 1994. DNA replication in vitro by recombinant DNA-polymerase-alpha-primase. *Eur. J. Biochem.* **222**: 781-793.

Stillman, B. 1986. Chromatin assembly during SV40 DNA replication in vitro. *Cell* **45**: 555-565.

―――. 1989. Initiation of eukaryotic DNA replication in vitro. *Annu. Rev. Cell. Biol.* **5**: 197-245.

Stohl, L.L. and D.A. Clayton. 1992. *Saccharomyces cerevisiae* contains an RNase MRP that cleaves at a conserved mitochondrial RNA sequence implicated in replication priming. *Mol. Cell. Biol.* **12**: 2561-2569.

Stroumbakis, N.D., Z. Li, and P.P. Tolias. 1994. RNA- and single-stranded DNA-binding (SSB) proteins expressed during *Drosophila melanogaster* oogenesis: A homolog of bacterial and eukaryotic mitochondrial SSBs. *Gene* **143**: 171-177.

Syvaoja, J. and S. Linn. 1989. Characterization of a large form of DNA polymerase δ from HeLa cells that is insensitive to proliferating cell nuclear antigen. *J. Biol. Chem.* **264**: 2489-2497.

Tan, C.K., C. Castillo, A.G. So, and K.M. Downey. 1986. An auxiliary protein for DNA polymerase-delta from fetal calf thymus. *J. Biol. Chem.* **261**: 12310-12316.

Tattersall, P. and D.C. Ward. 1976. Rolling hairpin model for replication of parvovirus and linear chromosomal DNA. *Nature* **263**: 106-109.

Temperley, S.M. and R.T. Hay. 1992. Recognition of the adenovirus type 2 origin of DNA replication by the virally encoded DNA polymerase and preterminal proteins. *EMBO J.* **11**: 761-768.

Tiranti, V., M. Rocchi, S. DiDonato, and M. Zeviani. 1993. Cloning of human and rat

cDNAs encoding the mitochondrial single-stranded DNA-binding protein (SSB). *Gene* **126:** 219-225.
Tratschin, J.D., I.L. Miller, and B.J. Carter. 1984. Genetic analysis of adeno-associated virus: Properties of deletion mutants constructed in vitro and evidence for an adeno-associated virus replication function. *J. Virol.* **51:** 611-619.
Tsurimoto, T. and B. Stillman. 1989. Purification of a cellular replication factor, RF-C, that is required for coordinated synthesis of leading and lagging strands during simian virus 40 DNA replication in vitro. *Mol. Cell. Biol.* **9:** 609-619.
―――. 1991. Replication factors required for SV40 DNA replication in vitro. I. DNA structure-specific recognition of a primer-template junction by eukaryotic DNA polymerases and their accessory proteins. *J. Biol. Chem.* **266:** 1950-1960.
Tsurimoto, T., T. Melendy, and B. Stillman. 1990. Sequential initiation of lagging and leading strand synthesis by two different polymerase complexes at the SV40 DNA replication origin. *Nature* **346:** 534-539.
Turchi, J.J. and R.A. Bambara. 1993. Completion of mammalian lagging strand DNA replication using purified proteins. *J. Biol. Chem.* **268:** 15136-15141.
Turchi, J.J., L. Huang, R.S. Murante, Y. Kim, and R.A. Bambara. 1994. Enzymatic completion of mammalian lagging-strand DNA replication. *Proc. Natl. Acad. Sci.* **91:** 9803-9807.
Umbricht, C.B., L.F. Erdile, E.W. Jabs, and T.J. Kelly. 1993. Cloning, overexpression, and genomic mapping of the 14-kDa subunit of human replication protein A. *J. Biol. Chem.* **268:** 6131-6138.
Van der Vliet, P.C. 1991. The role of cellular transcription factors in the enhancement of adenovirus DNA replication. *Semin. Virol.* **2:** 271-280.
Van Dyck, E., F. Foury, B. Stillman, and S.J. Brill. 1992. A single-stranded DNA binding protein required for mitochondrial DNA replication in *S. cerevisiae* is homologous to *E. coli* SSB. *EMBO J.* **11:** 3421-3430.
Virshup, D.M., M.G. Kauffman, and T.J. Kelly. 1989. Activation of SV40 DNA replication in vitro by cellular protein phosphatase 2A. *EMBO J.* **8:** 3891-3898.
Waga, S., G. Bauer, and B. Stillman. 1994. Reconstitution of complete SV40 DNA replication with purified replication factors. *J. Biol. Chem.* **269:** 10923-10934.
Wang, J.C. 1985. DNA topoisomerases. *Annu. Rev. Biochem.* **54:** 665-697.
Wang, T.S. 1991. Eukaryotic DNA polymerases. *Annu. Rev. Biochem.* **60:** 513-552.
Ward, P., E. Urcelay, R. Kotin, B. Safer, and K.I. Berns. 1994. Adeno-associated virus DNA replication in vitro: Activation by a maltose binding protein/Rep 68 fusion protein. *J. Virol.* **68:** 6029-6037.
Weinberg, D.H., K.L. Collins, P. Simancek, A. Russo, M.S. Wold, D.M. Virshup, and T.J. Kelly. 1990. Reconstitution of simian virus 40 DNA replication with purified proteins. *Proc. Natl. Acad. Sci.* **87:** 8692-8696.
Wernette, C.M. and L.S. Kaguni. 1986. A mitochondrial DNA polymerase from embryos of *Drosophila melanogaster*. Purification, subunit structure, and partial characterization. *J. Biol. Chem.* **261:** 14764-14770.
Wernette, C.M., M.C. Conway, and L.S. Kaguni. 1988. Mitochondrial DNA polymerase from *Drosophila melanogaster* embryos: Kinetics, processivity, and fidelity of DNA polymerization. *Biochemistry* **27:** 6046-6054.
Wides, R.J., M.D. Challberg, D.R. Rawlins, and T.J. Kelly. 1987. Adenovirus origin of DNA replication: Sequence requirements for replication in vitro. *Mol. Cell. Biol.* **7:** 864-874.

Wiekowski, M., M.W. Schwartz, and H. Stahl. 1988. Simian virus 40 large T antigen helicase: Characterization of the ATPase-dependent DNA unwinding activity and its substrate requirements. *J. Biol. Chem.* **263:** 436–442.

Wobbe, C.R., L. Weissbach, J.A. Borowiec, F.B. Dean, Y. Murakami, P. Bullock, and J. Hurwitz. 1987. Replication of SV40-origin-containing DNA with purified proteins. *Proc. Natl. Acad. Sci.* **84:** 1834–1838.

Wold, M.S. and T.J. Kelly. 1988. Purification and characterization of replication protein A, a cellular protein required for in vitro replication of simian virus 40 DNA. *Proc. Natl. Acad. Sci.* **85:** 2523–2527.

Wold, M.S., D.H. Weinberg, D.M. Virshup, J.J. Li, and T.J. Kelly. 1989. Identification of cellular proteins required for simian virus 40 DNA replication. *J. Biol. Chem.* **264:** 2801–2809.

Wong, T.W. and D.A. Clayton. 1985a. In vitro replication of human mitochondrial DNA: Accurate initiation at the origin of light-strand synthesis. *Cell* **42:** 951–958.

———. 1985b. Isolation and characterization of a DNA primase from human mitochondria. *J. Biol. Chem.* **260:** 11530–11535.

———. 1986. DNA primase of human mitochondria is associated with structural RNA that is essential for enzymatic activity. *Cell* **45:** 817–825.

Yang, L., M.S. Wold, J.J. Li, T.J. Kelly, and L.F. Liu. 1987. Roles of DNA topoisomerases in SV40 DNA replication in vitro. *Proc. Natl. Acad. Sci.* **84:** 950–954.

Yang, Q., A. Kadam, and J.P. Trempe. 1992. Mutational analysis of the adeno-associated virus *rep* gene. *J. Virol.* **66:** 6058–6069.

2
Origins of DNA Replication

M.L. DePamphilis
National Institute of Child Health and Human Development
National Institutes of Health
Bethesda, Maryland 20892-2753

Initiation of DNA replication involves three sequential steps. First, one or more specific *trans*-acting proteins bind to specific *cis*-acting DNA sequences referred to as origins of replication (DePamphilis 1993a,b; Kornberg and Baker 1992). Second, DNA unwinding of the two complementary templates begins. This step is usually carried out by a DNA helicase but can also be done by some DNA polymerases (e.g., adenovirus [Ad]). DNA unwinding is facilitated by single-strand-specific DNA-binding proteins such as replication protein A (RP-A), which coat the templates, and by topoisomerase I, which releases torsional stress generated by unwinding DNA. Third, DNA synthesis is initiated on one or both templates. In cellular chromosomes and DNA viruses that do not encode their own DNA polymerase (e.g., SV40, PyV, and PV), DNA polymerase-α:DNA primase complex synthesizes a short RNA-primed nascent DNA chain referred to as an Okazaki fragment. The first Okazaki fragment initiated on each template is extended continuously by DNA polymerase-δ and its accessory proteins to become the long nascent DNA strand on the forward arm of each of the two replication forks. The net result of these steps is bidirectional DNA replication employing bubble and fork structures such as those found in the chromosomes of prokaryotic and eukaryotic cells (Fig. 1). DNA replication is coupled to chromatin assembly, resulting in the random distribution of pre-fork histone octamers to both arms of the fork and rapid assembly of new histone octamers in the intervening regions of newly replicated DNA. Initiation of DNA replication can also occur in only one direction instead of both directions (geminiviruses, parvovirus, Ad, mtDNA) and can utilize preexisting DNA primers (parvovirus), RNA primers (retroviruses), and protein-nucleotide primers (Ad), instead of de novo synthesis of RNA primers. The sequences, proteins, and mechanisms referred to in each step are discussed in detail in various chapters of this book.

Concepts in Eukaryotic DNA Replication
© 1999 Cold Spring Harbor Laboratory Press 0-87969-557-9/99

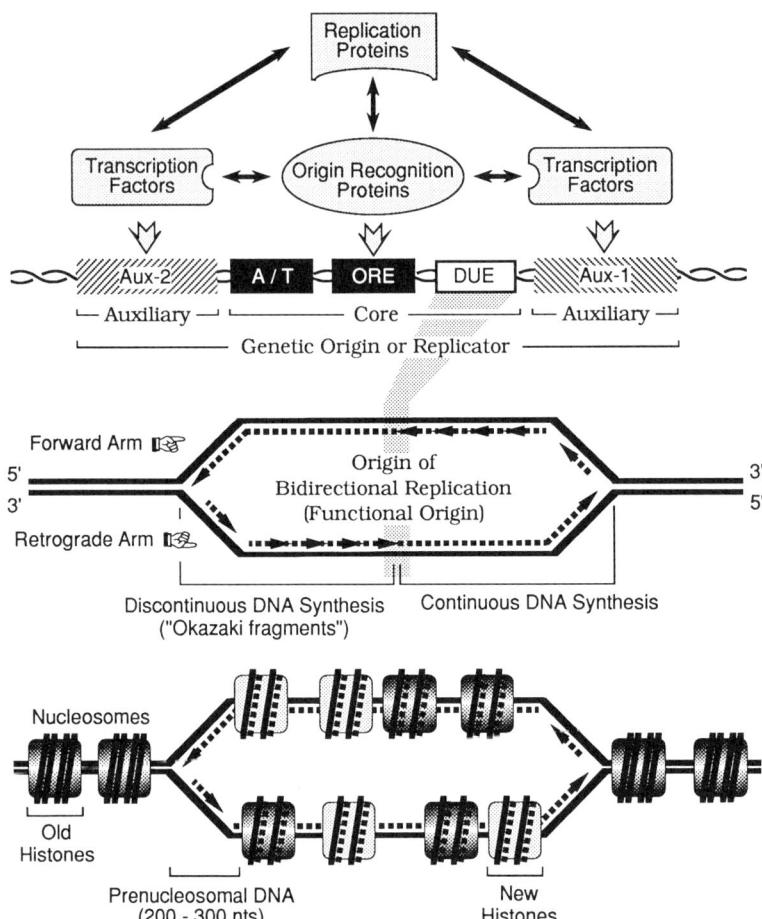

Figure 1 The modular concept of origins of DNA replication found in simple genomes (DePamphilis 1993a). Replication is initiated by a combination of replication proteins and transcription factors interacting with specific DNA sequences and with each other. Bidirectional DNA replication originates at the DUE, resulting in a replication bubble with two replication forks traveling in opposite directions. The structure of replication forks (DePamphilis and Wassarman 1980; Brush and Kelly; Stillman; both this volume), their organization into chromatin (Cusick et al. 1989; Wolffe, this volume), and the process of chromatin assembly (Gruss and Sogo 1992; Wolffe, this volume) have been described in detail previously.

Central to understanding how animal cells regulate DNA replication is understanding the nature of DNA sites where replication begins. Every genome analyzed so far contains at least one replication origin per chromosome, and the genomes of eukaryotic cells contain about one origin every 10–330 kb (Hand 1978). Replication origins play two important roles in DNA replication. First, they ensure that each time a cell divides, the entire genome is replicated efficiently. For example, a single mammalian cell contains about 1.8 meters of DNA that must be replicated in 6–8 hours, and early embryos of frogs, flies, and sea urchins replicate comparable amounts of DNA in 10–40 minutes. Second, replication origins provide a way in which to regulate when and where initiation events occur (Table 1). Replication of cellular chromosomes is restricted to one phase of the cell proliferation cycle (S phase), and initiation at each of several thousand replication origins is restricted to once per S phase (Blumenthal et al. 1974). Multiple initiation events at the same locus (gene amplification) can occur in tumors and transformed cell lines, but only rarely are genes amplified during normal animal development (Tlsty 1990). In contrast, replication of mtDNA and large viral DNA genomes such as herpes and vaccinia is not dependent on the cell entering S phase. mtDNA replicates randomly throughout the cell division cycle, and large viral genomes usurp the cell's machinery to provide their own replication components. Replication of small viral DNA genomes such as papovaviruses is restricted to S phase, but each genome copy may undergo two or more rounds of replication during a single S phase. Thus, it appears that mitochondrial and viral genomes in mammalian cells were designed to escape the very controls required for cellular DNA replication. This may not be true for simpler organisms such as flagellated protozoa, where replication of mtDNA (kinetoplast DNA) appears to follow the same rules as nuclear DNA.

Origins of DNA replication that function in eukaryotic cells can be divided into two groups: those found in "simple genomes" such as animal viruses (SV40, polyomavirus [PyV], PV, Ad, herpes simplex virus [HSV], Epstein-Barr virus [EBV]), mitochondria (human, mouse), protozoa (*Tetrahymena*), yeast (*Saccharomyces cerevisiae, Schizosaccharomyces pombe*), and slime mold (*Physarum*); and those found in "complex genomes" of metazoa such as flies (*Drosophila, Sciara*), frogs (*Xenopus*), and mammals (rodents, human). Simple origins that function in eukaryotes are similar to those that function in prokaryotes (Kornberg and Baker 1992). They have a modular anatomy composed of unique DNA sequence motifs and interactions with soluble proteins (DePamphilis 1993a,b). Whether or not metazoan chromosomes contain sequence-specific replication origins analogous to those found in simpler genomes is controversial, and therefore is discussed in more detail. What is clear is that replication begins at specific sites in metazoan chromosomes, that initiation of DNA replication requires nuclear structure, and that nuclear structure imposes site specificity.

Table 1 Characteristics of replication origins that function in eukaryotic cells

Genome	maps to a specific site	requires specific DNA sequences	Initiation of DNA replication			requires nuclear structure
			is restricted to S phase	occurs ?/origin/ cell cycle	exhibits ARS activity	
Mitochondria	~300 bp	30 and ~500 bp	no	>1	no	no
Animal viruses (nuclear)	1 to ~200 bp	18 to ~1000[b] bp	yes	>1	yes	no
Yeast	~300 bp	100 to 200 bp	yes	1	yes	
Protozoa (*Tetrahymena* macronucleus)	<1 kb	<2 kb	yes			
Slime mold	<1 kb		yes	1		
Flies (DNA amplification)	yes[a]	1 to 8 kb[c]	no	>1	no	
(S phase, cultured cells)	yes[a]		yes	1		
Mammals	yes[a]	1 to 8 kb[d]	yes	1	yes/no	yes

[a] See Table 4.
[b] EBV *oriP*, the largest origin, includes nonrequired sequences between *ori-core* and its enhancer.
[c] Estimated from genetic analyses of *Drosophila* chorion gene amplification locus.
[d] Estimated from four reported ARS activities and one natural deletion mutant (see Table 3).

CHARACTERISTICS OF SIMPLE ORIGINS

Replication origins in simple genomes are composed of modular units acting in concert to determine where and when DNA replication will occur (Table 1). In this sense, replication origins are equivalent to transcription promoters, but whereas each promoter is uniquely responsible for transcription of its associated gene, large chromosomes like yeast often contain more origins than are needed for their replication (Newlon et al. 1993). This flexibility in origin number is allowed because DNA regions that lack a replication origin can be replicated by replication forks that originate from origins located many kilobases away.

Simple origins exhibit four characteristics.

1. Simple origins are composed of unique, genetically required, sequences. These sequences occupy from 50 bp to about 1000 bp and are defined by *cis*-acting mutations that prevent DNA replication. They are anatomically similar to origins that function in prokaryotic cells and consist of at least four elements, three of which (origin recognition element [ORE], DNA unwinding element [DUE], A/T-rich element) are essential and therefore referred to as the origin core (*ori*-core) (Fig. 1) (DePamphilis 1993a,b). *ori*-cores occupy from 18 bp to about 120 bp. *ori*-core function is critically dependent on the spacing, orientation, and arrangement of these three elements (DePamphilis 1993a,b; Parsons and Tegtmeyer 1992; Harrison et al. 1994). In general, there is a lack of symmetry in both the organization of sequence elements and their specific functions in replication origins. As a result, DNA synthesis initiation events within *ori*-core may occur only on one DNA strand (e.g., Ad, mtDNA, SV40, PyV). Therefore, initiation of bidirectional replication (e.g., SV40, PyV) is unlikely to be a symmetrical event with two forks moving simultaneously out of the ORE. More likely is that forward-arm synthesis begins on one strand of *ori*-core and then progresses beyond *ori*-core before DNA synthesis is initiated in the opposite direction on the complementary strand (see "Initiation zone model" in DePamphilis et al. 1988).

   The fourth element consists of transcription-factor-binding sites that flank one or both sides of *ori*-core (DePamphilis 1993c): *aux-1* is proximal to the DUE element, and *aux-2* is proximal to the A/T element (Fig. 1). These transcription-factor-binding sites are referred to as *ori*-auxiliary components (*aux*), because they facilitate *ori*-core activity from 2- to 1000-fold, depending both on the origin and on experimental conditions, but are not required for replication under all conditions and do not affect the mechanism by which replication occurs (DePamphilis 1993a,b). Some yeast origins, for example, contain a binding site (element B3) for transcription factor ABF1, whereas others do not. In this sense, auxiliary components are analogous to transcription enhancers. However, although enhancers that stimulate transcription are generally independent of their distance and orientation relative to the promoter, orientation and spacing between auxiliary sequences and *ori*-core are critical in some origins (Ad, SV40, PyV), whereas in other origins (EBV, yeast), these parameters are flexible. These differences presumably reflect differences in the specificity and strength of interactions between transcription factors and origin recognition proteins, as well as differences in the mechanism by which *ori*-auxiliary components function.

2. The genetic origin is coincident with the functional origin. The functional origin is the site where DNA synthesis actually begins. It has been mapped to within ±300 bp of the genetic origin in yeast chromosomes and plasmids, and the transition from discontinuous to continuous DNA synthesis on each template (Fig. 1) mapped with nucleotide resolution to the genetic origins in SV40, PyV, Ad, mtDNA, *Escherichia coli* (Kohara et al. 1985; Rokeach and Zyskind 1986; Seufert and Messer 1987), and bacteriophage λ (Yoda et al. 1988). However, these transitions can lie outside *ori* when SV40 DNA replication is initiated in vitro (Bullock et al. 1991), suggesting that extensive DNA unwinding can precede initiation of DNA synthesis, thereby providing DNA polymerase-α:DNA primase with the opportunity to begin synthesis outside of *ori* (see "Initiation zone model" in DePamphilis et al. 1988). In vivo, the rate of DNA unwinding may be retarded by chromatin structure.

3. Simple origins can act as autonomously replicating sequences (ARSs). ARS elements confer on other DNA molecules the ability to replicate when transferred to either cells or cell extracts containing the required replication proteins. So far, ARS elements have been demonstrated only with viral and yeast origins. Functional origins in the chromosomes of *S. cerevisiae* have been shown to correspond to individual ARS elements that are genetically required for origin activity (Newlon, this volume). The same appears true for *S. pombe*, although sequence requirements for origins in *S. pombe* appear more diffuse and origin function less efficient than in *S. cerevisiae* (Caddle and Calos 1994; Dubey et al. 1994; Wohlgemuth et al. 1994).

4. Simple origins can function in a soluble cell-free DNA replication system. So far, this has been demonstrated only with viral origins, a

problem that may be overcome when large amounts of purified replication proteins become available for other systems.

ANATOMY OF SIMPLE REPLICATION ORIGINS

Origin Recognition Element

An ORE is the DNA-binding site for one or more origin recognition proteins (Table 2) that are required for initiation of DNA replication. This DNA-protein interaction can be regulated by posttranslational modifications, such as phosphorylation of specific amino acid residues in SV40 T antigen (see Weisshart and Fanning, this volume). Sequences within or flanking the ORE often exhibit the characteristics of bent DNA (Zahn and Blattner 1985; Deb et al. 1986; Williams et al. 1988), which may facilitate protein binding (Ryder et al. 1986). Moreover, although an ORE may exhibit a twofold symmetry, interaction with its recognition protein can be asymmetrical, leading to an asymmetrical opening of *ori-core* (SenGupta and Borowiec 1994).

Origin recognition proteins serve at least two functions. The first is to initiate DNA unwinding using either their own helicase activity (e.g., parvovirus, SV40, PyV, PV, HSV) or one that associates with it (EBV may be an example). The second function is to guide other replication proteins to the origin. For example, SV40 and PyV T antigen associate with DNA polymerase-α:DNA primase, the enzyme responsible for initiation of the first RNA-primed DNA chain (Okazaki fragment) at their respective origins, and with RP-A, a single-stranded DNA-binding protein that stimulates both helicase and polymerase action (Melendy and Stillman 1993; Schneider et al. 1994 and references therein). Similarly, Ad preterminal protein, the protein-dCTP primer that initiates Ad DNA synthesis, is complexed with Ad DNA polymerase, the enzyme responsible for Ad DNA synthesis (see Hay, this volume). In yeast, a six-protein origin recognition complex binds to an approximately 50-bp sequence that includes elements A and B1. All yeast origins require element A (containing the 11-bp conserved ARS consequence sequence) and at least two of the three B elements that lie to one side of element A (see Newlon, this volume). Origin recognition proteins such as EBV EBNA1 that do not have their own helicase activity (Frappier and O'Donnell 1991) presumably associate with a cellular helicase. These interactions between origin recognition proteins and replication proteins can determine the host-cell specificity of a replication origin (Melendy and Stillman 1993; Schneider et al. 1994).

52   M.L. DePamphilis

Table 2 Proteins that activate replication origins

	Origin recognition proteins[a]			Transcription factors[b]			
Origin	protein	enzymatic activities	associated rep. proteins	Aux-2	associated rep. proteins	Aux-1	associated rep. proteins
Gemini	Rep (AL1)	endonuclease, ATPase					
Parvo							
MVM	NS1	endonuclease, helicase		ATF, NS1	NS1		
AAV	Rep 68 and 78	endonuclease, helicase		Rep	Rep		
Papova							
SV40	T antigen	helicase	pol-α:primase RP-A	AP1, Sp1, NF1>T antigen (Gal4, VP16, c-Jun, GR)	RP-A	T antigen	T antigen
polyoma	T antigen	helicase	pol-α:primase RP-A	AP1>>Gal4, VP16, PEA3 c-Jun, v-Jun, E1A, E2, Sp1, Rel, E2F (CREB, pRB)	RP-A	T antigen	T antigen
Papilloma	E1	helicase		E2, VP16	E1, RP-A	E2, VP16	E1, RP-A
Herpes simplex	UL9	helicase[c]		several candidates			
Ad2	preterminal prot. Ad DNA pol	prime DNA synthesis DNA synthesis	Ad DNA pol preterminal prot.	NF1, Oct-1	Ad DNA pol		
Ad4	(same as Ad2)			(no Aux-2 factors)			
Epstein-Barr							
oriP	EBNA1	(no helicase)		EBNA1			
oriLyt	BZLF-1 ?			BZLF-1 (Jun, E2, Myc, VP16)			
Mitochondria							
oriH	CSB1 RNase MRP	endoribonuclease RNA primer synthesis		mtTFA, mtRNA pol			
oriL	mtDNA primase	binds ssDNA at ori					
Kinetoplast	p13.5						
S. cerevisiae	6-protein complex			ABF1, RAP1 Gal4			

[a] Adenovirus origin recognition protein is a complex of preterminal protein and Ad DNA polymerase. Pol-α:primase is DNA polymerase-α:DNA primase.
[b] Transcription factors in parentheses did not stimulate origin activity in vivo.
[c] Helicases encoded by papova and papilloma viruses can unwind DNA at the origin, but HSV helicase cannot. References are found in the appropriate chapters.

DNA Unwinding Element

DNA unwinding appears to begin at an easily unwound DNA sequence referred to as the DNA unwinding element (DUE, Fig. 1). A DUE is identified by *cis*-acting mutations that both increase the stability of the double helix (i.e., make DNA unwinding more difficult) and reduce DNA replication (Umek and Kowalski 1990b). Computer programs are available that determine DNA helical stability in known sequences (Rychlik and Rhoads 1989; Natale et al. 1992), and the effects of mutations on helical stability can be assessed within the context of a supercoiled plasmid using a single-strand-specific endonuclease, or two-dimensional gel electrophoresis of plasmid topoisomers. These approaches have identified DUEs in *E. coli oriC* (Kowalski and Eddy 1989), yeast ARS elements in plasmids (Natale et al. 1992; Huang and Kowalski 1993; Miller and Kowalski 1993), and replication origins in yeast chromosomes (Huang and Kowalski 1993) and SV40 (Lin and Kowalski 1994). In yeast replication origins the genetic and physical properties of element B2 are consistent with those of a DUE (Rao et al. 1994; Theis and Newlon 1994).

Although primary sequence is an important determinant of the energy required for DNA unwinding (because of the importance of base-stacking interactions), there is no unique consensus sequence for a DUE. DUE sequences in yeast are not conserved, and easily unwound DNA sequences that are not components of origins can substitute for the DUEs in both yeast and *E. coli* (Umek and Kowalski 1988; Kowalski and Eddy 1989). Therefore, DUEs are unlikely to be binding sites for specific replication proteins. Instead, it appears that one of the proteins binding to the ORE must interact with the DUE because the spatial relationship between these two core elements is critical. In yeast, the DUE always is located 3′ to the T-rich strand of the ARS consensus sequence (Natale et al. 1993), and reversal of the orientation of the ARS consensus sequence with respect to the DUE abolishes DNA replication (Holmes and Smith 1989). Binding of proteins to ORE results in DNA unwinding in the DUEs of *E. coli oriC* (Kowalski and Eddy 1989), SV40, and PyV. In SV40 and PyV, the DUE, the site where T antigen begins unwinding DNA, and the origin of bidirectional replication (OBR; defined by the transition between continuous and discontinuous DNA synthesis on each template [Fig. 1]) are all coincident (DePamphilis 1993a; Lin and Kowalski 1994). Similarly, in yeast ARS307, a majority of leading strands emanate from the proposed DUE (unpublished data cited in Theis and Newlon 1994). Therefore, the DUE appears to be the entry site for the replication machinery.

Not all easily unwound DNA sequences are part of replication origins, but the fact that such sequences can substitute for known DUEs qualifies them as "potential DUEs." Since the energetic, length, and spacing requirements of true DUEs remain to be defined, it is difficult to estimate the frequency at which potential DUEs occur in natural DNA. For purposes of comparison, a potential DUE can be defined as a 100-bp sequence whose helical stability is 20 or more kcal/mole below the average for the sequence analyzed. This definition is stringent enough to exclude some real DUEs. Such a potential DUE would be expected once every 3.2 kb in a random sequence of 60% A+T content (D. Natale, pers. comm.), suggesting that potential DUEs occur much more frequently than replication origins in yeast (1 origin/36 kb) and mammalian (1 origin/100 kb) DNA. Thus, the ability of an easily unwound sequence to function as a DUE likely depends on conditions such as its proximity to other origin elements (Fig. 1), the concentration of initiation factors, the influence of chromatin structure, and the amount of negative superhelical energy available (Umek and Kowalski 1990a).

A/T-rich Element

Most, but not all (e.g., EBV), replication origins contain an A/T rich sequence consisting of a T-rich and an A-rich strand. The length of the A/T-rich element is critical for origin function (Gerard and Gluzman 1986; Koff et al. 1991), a fact that may be related to its bent character (Deb et al. 1986). Bent DNA can interact more easily with proteins, which may account for the fact that binding of origin recognition proteins to OREs frequently distorts the A/T-rich element (Koff et al. 1991; Gillette et al. 1994; SenGupta and Borowiec 1994). Distortion of the A/T-rich element may facilitate either binding to the ORE or melting of the DUE.

Auxiliary Components

ori-auxiliary components stimulate replication only when they bind one or more transcription factors, and only when the transcription factor contains an activation domain that specifically interacts with the replication machinery (Table 2) (DePamphilis 1993c; van der Vliet, this volume). In some genomes (SV40, PyV, PV, EBV), the same sequence elements that function as promoters or enhancers in transcription also function as *aux* components in replication; *cis*-acting mutations that affect one process also affect the other. Auxiliary components could be used to regulate origin activity in two ways. First, the ability of a particular transcription factor to stimulate a particular origin may be limited to specific members of a transcription factor family, and to the availability of specific coactivator proteins (Guo and DePamphilis 1992). Thus, auxiliary components can stimulate the same origin to different extents in different cell types as the composition of available transcription factors changes during animal development (Rochford and Villarreal 1991). Second, just as transcription factors can initiate transcription of different genes at different times during the cell division cycle, they could regulate the temporal order of DNA replication during S phase, accounting for the observation that active genes are replicated early and inactive genes are replicated late (see Simon and Cedar, this volume).

There are four basic mechanisms by which transcription factors can stimulate *ori*-core (DePamphilis 1993c):

1. An upstream promoter can direct transcription through *ori*-core. The resulting mRNA is then cut by an endonuclease at specific sites to generate RNA primers for initiation of DNA synthesis. This mechanism occurs at mtDNA *oriH*, *E. coli* filamentous and T-odd bacteriophage, and *E. coli* plasmid ColE1 (Kornberg and Baker 1992).
2. Transcription factors can facilitate binding of origin recognition proteins to *ori*-core. For example, NFI binding to *aux-2* facilitates binding of subsaturating concentrations of the Ad2 preterminal protein/Ad DNA polymerase complex (pTP-pol) (see Hay, this volume). Binding of PV-encoded enhancer-specific activation protein, E2, to *aux-1* and possibly *aux-2* facilitates binding of E1 to *ori*-core (see Stenlund, this volume). Binding of EBV-encoded EBNA1 protein to the EBV enhancer (*aux-2*) stabilizes interaction of EBNA1 to the EBV *ori*-core (Frappier et al. 1994).
3. Transcription factors can facilitate the activity of an initiation complex after it has formed. For example, SV40 *ori*-auxiliary components stimulate SV40 *ori*-core by facilitating T-antigen-dependent DNA unwinding at *ori*-core (Gutierrez et al. 1990). This may occur by stabilizing the interaction of T antigen with ssDNA as it disrupts its own binding site by unwinding it (Gutierrez et al. 1990), and by recruiting RP-A, a single-strand DNA-binding protein, to stabilize ssDNA (He et al. 1993; Li and Botchan 1993). The T-antigen dimer that binds to *aux-1* appears to interact with the T-antigen hexamer bound to ORE (Guo et al. 1991), thus accounting for the observation that the need for *aux-1* to stimulate papovavirus origin activity is inversely related to the ability of T antigen to activate *ori*-core (Sock et al. 1993). In the chromosomes of *E. coli*, bacteriophage λ, and plasmid R6K, tran-

scription or association with RNA polymerase near *ori* can stimulate *ori* activity by removing torsional stress in DNA and thus facilitating DNA unwinding (Kornberg and Baker 1992).
4. Transcription factors can prevent chromatin structure from interfering with binding of replication factors to origins. Nucleosomes can repress replication origins in yeast (Simpson 1990), *Drosophila* (Karpen and Spradling 1990), and mammalian chromosomes (Forrester et al. 1990). Therefore, since transcriptionally active DNA sequences are not incorporated into nucleosomes (Morse et al. 1992), transcription through a replication origin may provide access to replication factors. Alternatively, binding of transcription factors to enhancers may allow interactions between the enhancer and *ori*-core, analogous to those between enhancer and promoter (Majumder and DePamphilis 1994), that prevent nucleosomes from repressing origins in the same way that it prevents nucleosomes from repressing promoters (Paranjape et al. 1994). This mechanism may apply to PyV, where the PyV enhancer (*aux-2*) is dispensable under conditions where a repressive chromatin structure appears to be absent (Prives et al. 1987; Martínez-Salas et al. 1988; Majumder et al. 1993), and to SV40 *aux-2* where prebinding some transcription factors (e.g., NFI, Gal4:VP16) can prevent chromatin assembly from interfering with SV40 DNA replication in vitro (see Hassell and Brinton, this volume). However, the facts that Gal4:VP16 does not stimulate the SV40 origin in vivo, and that prebinding T antigen alone can also prevent nucleosome repression, suggest that other mechanisms should be considered (DePamphilis 1993c).

VIRAL ORIGINS AS MODELS FOR CELLULAR ORIGINS

PV and EBV have been considered models for cellular DNA replication because their genomes replicate in the nucleus, their DNA replication is restricted to S phase, and they maintain a low number of genome copies per cell. Moreover, early studies on PV DNA replication concluded that a complex interaction between positive and negative controls restricted initiation of replication to once per origin per S phase (Roberts and Weintraub 1988). However, later studies (see Stenlund, this volume) revealed that PV origins are remarkably similar to those in papovaviruses, and that although *cis*-acting PV sequences can suppress the activity of lytic origins such as SV40 and PyV, they do not restrict them to one initiation event per S phase (Nallaseth and DePamphilis 1994). In fact, initiation of PV DNA replication is not restricted to once per S phase (Gilbert and Cohen 1987; Ravnan et al. 1992). EBV remains a candidate because its DNA replicates at the same rate as cellular chromosomes (Haase and Calos 1991; Yates and Guan 1991), but it remains to be determined whether or not EBV DNA replication, like cellular DNA replication, does not reinitiate when cells are limited to a single S phase in the presence of a mitotic inhibitor such as nocodazole (Nallaseth and DePamphilis 1994). This test should be applied to all putative mammalian ARS elements (Table 3) (Krysan et al. 1993; Masukata et al. 1993).

COMPLEX (METAZOAN) ORIGINS

In comparison with simple genomes, origins of replication in multicellular animals (the metazoa) often appear paradoxical. Early attempts to identify *ori* sequences in mammalian chromosomes by their ability to function as ARS elements were difficult to reproduce and therefore controversial (Gutierrez et al. 1988; Burhans and Huberman 1994), although some recent reports (Table 3) look promising. Nevertheless, most large (>10 kb) DNA fragments from mammalian chromosomes can provide some ARS activity in mammalian cells (Krysan et al. 1993; Masukata et al. 1993), suggesting that DNA length is more critical than DNA sequence. The same conclusion is reached when DNA is injected into the eggs of frogs, sea urchins, or fish, or when DNA is added to extracts of *Xenopus* eggs or *Drosophila* embryos (Coverley and Laskey 1994; see Laskey and Madine; Blow and Chong; both this volume). DNA replication is initiated at a single randomly chosen site within virtually any DNA molecule (Hines and Benbow 1982; Méchali and Kearsey 1984; Hyrien and Méchali 1992; Mahbubani et al. 1992). This lack of site-specific initiation also appears during chromosome replication in *Xenopus* (Hyrien and Méchali 1993) and *Drosophila* (Shinomiya and Ina 1991) embryos, suggesting that the lack of sequence requirements and site specificity observed when DNA is introduced into cultured cells, eggs, or egg extracts accurately reflects chromosome replication in situ.

Another approach to identify replication origins in metazoan chromosomes is in situ mapping of initiation sites for DNA replication in the hope that subsequent physical and genetic analysis of those loci will reveal the nature of a metazoan origin of DNA replication. This approach has revealed that DNA replication in metazoan chromosomes occurs at specific sites.

Table 3 Metazoan origins of replication are genetically determined

Organism	Location	Same origin in single copy and multicopy genomes[a]	Translocated to other sites	Deletion in OBR	ARS activity	APE activity
Hamster	DHFR gene (oriβ)	chromosomal (1)	active (2)			yes (3)
Mouse	rRNA gene		active (5)			yes (4)
Drosophila	chorion gene					
Human	hsp70 gene				yes (6)	
Mouse	IgH gene enhancer				yes (7)	
Mouse	ADA gene (late S phase)	episomal (8)			yes (8)	
Human	c-*myc* gene				yes (9)	
Human	cDNA 343				yes (10)	
Hamster	CAD gene group	episomal (11)				
Human	β-globin gene			inactive (12)		
Mouse	ADA gene (early S phase)	episomal (13)				

[a] Amplified gene copies are found in either chromosomal or episomal locations.
References (in parentheses): (1) Handeli et al. 1989; Burhans et al. 1990, 1991; Vassilev et al. 1990; Dijkwel and Hamlin 1992, 1995; DePamphilis 1993d; Tasheva and Roufa 1994a; Gilbert et al. 1995; M. Giacca, unpubl.; (2) Handeli et al. 1989; (3) Stolzenburg et al. 1994; (4) Hermann et al. 1994; (5) Orr-Weaver 1991; (6) Taira et al. 1994; (7) Ariizumi et al. 1993; Iguchi-Ariga et al. 1993; (8) Virta-Pearlman et al. 1993; (9) Berberich et al. 1995; (10) Wu et al. 1995; (11) Kelly et al. 1995; (12) Kitsberg et al. 1993; (13) Pearson et al. 1994.

MAPPING ORIGINS OF REPLICATION IN METAZOAN CHROMOSOMES

Methods for mapping DNA replication initiation sites fall into two categories: those that begin by labeling nascent DNA chains, and those that begin by fractionating DNA structures. The first category analyzes nascent DNA strands labeled by incorporation of radioactive or density-substituted deoxyribonucleotides during DNA replication, annealing them to sequence-specific probes in order to determine the amount and direction of synthesis that occurs at specific DNA sites (for discussion, see Vassilev and DePamphilis 1992). There are four basic methods.

The first method identifies the earliest labeled DNA fragments in cells that have been synchronized at their G_1/S border, permeabilized, and then released into S phase in the presence of a labeled deoxyribonucleotide. Labeled DNA should appear first at origins of DNA replication. Identification of the earliest labeled DNA fragment can be facilitated by cross-linking the DNA templates with psoralen prior to initiation of replication to prevent migration of replication forks away from the origin region. These methods are generally applied to cells containing amplified DNA sequences. Thus, replication sites identified under these conditions could either be unique to amplified DNA sequences or an artifact of the cell-synchronization conditions.

Examination of single-copy sequences in exponentially proliferating cells in the absence of metabolic inhibitors is possible with the help of the DNA polymerase chain reaction (PCR). Labeled nascent DNA strands are separated from unreplicated DNA and then fractionated according to their length. Replication origins can be localized either from the length (Vassilev et al. 1990) or abundance (Giacca et al. 1994; Yoon et al. 1995) of nascent DNA strands passing through a specific genomic sequence. An origin of bidirectional replication (OBR) (Fig. 1) lies halfway along the shortest nascent DNA strand passing through a specific sequence. In addition, the closer a specific sequence lies relative to an OBR, the greater its abundance in shorter nascent DNA chains relative to longer nascent DNA chains. By examining several specific sequences on either side of a putative OBR, bidirectional replication events can be distinguished from unidirectional events, and the resolution of an OBR increases. Quantitation of the number of nascent DNA chains that contain a specific sequence (and therefore OBR resolution) can be improved by competition between hybridization of the PCR primer to nascent DNA chains and a competitor DNA standard (Giacca et al. 1994).

The third group measures the distribution of Okazaki fragments between the two arms of a replication fork. If DNA replication occurs by the replication fork mechanism (Fig. 1), then DNA synthesis on the forward arm (leading strand) is a continuous process, but DNA synthesis on the retrograde arm (lagging strand) is a discontinuous process in which short (40–300 nucleotides) RNA-primed, nascent DNA chains (Okazaki fragments) are repeatedly formed and joined into longer chains (see Brush and Kelly, this volume). Therefore, the ratio of Okazaki fragments that anneal to unique DNA sequences representing the retrograde-arm template versus those representing the forward-arm template provides a minimum estimate of the fraction of replication forks traveling in a specified direction. By measuring the distribution of Okazaki fragments between the templates at several different genomic locations, the transition from continuous to discontinuous DNA synthesis that defines an OBR can be mapped (Fig. 1). With small circular genomes such as SV40 and PyV, sufficient Okazaki fragments can be isolated from infected cells to localize the SV40 and PyV OBRs to 2 bp and 20 bp, respectively, by digesting the DNA hybrids at a unique restriction site and sequencing the nascent DNA strands (DePamphilis et al. 1988). With metazoan chromosomes, it is necessary to synchronize cells at their G_1/S border in order to collect Okazaki fragments only from newly initiated replication origins and not from replication forks traveling through from upstream or downstream origins. This limited availability of labeled Okazaki fragments restricts resolution of OBRs in mammalian cells to the sizes of the sequence-specific probes used in the hybridization reaction.

A fourth method measures replication fork polarity in exponentially proliferating cells by inhibiting protein synthesis in vivo. Under these conditions, Okazaki fragment synthesis is preferentially inhibited, allowing accumulation of labeled long nascent DNA strands synthesized on the forward arms of replication forks. Again, the fraction of labeled strands that anneal to each of the two strands of a specific sequence identifies fork polarity. In this procedure, however, nascent DNA should preferentially anneal to the template complementary to that recognized by Okazaki fragments. Initial studies using this mapping protocol assumed that it depended on preferential segregation of pre-fork histone octamers to the forward arm of replication forks in the absence of histone synthesis, and therefore employed micrococcal nuclease to digest nascent DNA on retrograde arms (Handeli et al. 1989). Subsequent studies demonstrated that this mapping protocol did not depend on chromatin structure and therefore did not require micrococcal nuclease (Burhans et al. 1991; Kitsberg et al. 1993).

The second category of methods for identifying replication origins is based on fractionating DNA according to its size and shape using two-dimensional (2-D) gel electrophoresis (Fangman and Brewer 1991; Huberman 1994). Structures specifically associated with DNA replication, such as bubbles and forks (depicted in Fig. 1), can be released by digestion with restriction endonucleases. They are then recognized by their mobility patterns during gel electrophoresis, and their genomic locations are identified by blotting-hybridization with sequence-specific radiolabeled probes. This approach is particularly useful in systems such as yeast, where incorporation of labeled nucleotides into nascent DNA is difficult. The direction in which replication forks travel can be determined by first fractionating double-stranded DNA according to its size at neutral pH, and then fractionating it in a second dimension at alkaline pH to observe the lengths of single-stranded (nascent) DNA released from replication forks. The closer to an OBR, the greater the abundance of short nascent DNA strands, analogous to measuring the lengths of nascent (radiolabeled) DNA strands described above. 2-D neutral/neutral and 2-D neutral/alkaline gel electrophoresis fractionations can be run sequentially to determine more precisely the sizes of forks and bubbles at specific genomic locations (Liang and Gerbi 1994). These methods have localized replication origins in PV, yeast, and slime mold to specific sites of 0.2 to 1 kb (see Stenlund; Newlon; Pierron and Bénard; all this volume).

METAZOAN CHROMOSOMES INITIATE DNA REPLICATION AT SPECIFIC SITES

Mapping initiation sites for DNA replication at 18 different genomic locations (Table 4) has yielded the following characteristics of metazoan DNA replication origins:

1. DNA synthesis does not initiate randomly throughout cellular chromosomes, but at specific DNA sites. Therefore, at some point during animal development, specific replication origins are formed. These sites, however, appear to consist of a primary origin (OBR) flanked by many secondary origins (initiation zone).
2. From 80% to 95% of DNA synthesis occurs bidirectionally from specific genomic loci referred to as OBRs. This conclusion is based on the fraction of replication forks traveling in the same direction, as determined by 2-D neutral/alkaline gels (Liang et al. 1993; Liang and Gerbi 1994; Shinomiya and Ina 1994), the ratio of Okazaki fragments that hybridize to the two strands of a unique DNA probe (Burhans et

Table 4 Metazoan chromosomes initiate DNA replication at specific sites

Organism	Location	OBR[a] (kb)	Initiation zone[b] (kb)	Reference
Hamster	DHFR gene (*oriβ*)	0.5–3	55	1
Human	rRNA genes	8	31	2
Mouse	rRNA genes	3		3
Drosophila	DNA polymerase α gene	5	10	4
Drosophila	chorion gene	1[c]	12	5
Sciara	chromosome-2, locus 9	1[c]	6	6
Human	Hsp70 gene	0.4		7
Human	lamin B2 gene	0.5		8
Mouse	IgH gene enhancer	0.6		9
Mouse	ADA gene (late S phase)	2		10
Human	c-*myc* gene	2		11
Hamster	ribosomal protein S14 gene	2		12
Human	cDNA 343 (early S phase)	2		13
Human	β-globin gene	2		14
Mouse	CAD gene group	5		15
Hamster	rhodopsin gene	5		16
Hamster	DHFR gene (*oriγ*)	8		17
Mouse	ADA gene (early S phase)	11		18

[a] Mapped by labeling of nascent DNA strands.
[b] Mapped by 2-D gel analyses of replication bubbles.
[c] Origins of gene amplification that were mapped by 2-D gel analyses of replication fork direction.

References: (1) Handeli et al. 1989; Burhans et al. 1990, 1991; Vassilev et al. 1990; Dijkwel and Hamlin 1992, 1995; DePamphilis 1993d; Tasheva and Roufa 1994a; Gilbert et al. 1995; M. Giacca, unpubl.; (2) E. Gogel and F. Grummt, unpubl.; (3) Little et al. 1993; Yoon et al. 1995; (4) Shinomiya and Ina 1994; (5) Delidakis and Kafatos 1989; Heck and Spradling 1990; (6) Liang et al. 1993; Liang and Gerbi 1994; (7) Taira et al. 1994; (8) Giacca et al. 1994; (9) Ariizumi et al. 1993; Iguchi-Ariga et al. 1993; (10) Virta-Pearlman et al. 1993; (11) Vassilev and Johnson 1990; Berberich et al. 1995; (12) Tasheva and Roufa 1994a; (13) Wu et al. 1993; Pearson et al. 1994; (14) Kitsberg et al. 1993; (15) Kelly et al. 1995; (16) Gale et al. 1992; (17) Handeli et al. 1989; (18) Carroll et al. 1993.

al. 1990; Carroll et al. 1993; Tasheva and Roufa 1994a; Berberich et al. 1995; Kelly et al. 1995; Gilbert et al. 1995), and the ratio of long nascent DNA strands from forward arms of replication forks that hybridize to the two strands of a unique DNA probe (Handeli et al. 1989; Burhans et al. 1991; Kitsberg et al. 1993; Kelly et al. 1995). In addition, quantitative analysis of specific sequences (Gilbert et al. 1995), particularly those within long nascent DNA strands (Vassilev and Johnson 1990; Vassilev et al. 1990; Giacca et al. 1994; Yoon et al. 1995), reveals that most of them originate bidirectionally from a small chromosomal locus that resides within an initiation zone (see point 4).

3. An OBR is contained within 0.5 kb to 2 kb. This conclusion is based on 11 different OBRs that appear to lie within a 2-kb region and on the DHFR oriβ locus where five different nascent DNA strand methods have been applied with remarkable agreement (Table 4). Some OBRs lie within larger regions of 5-11 kb. All of these estimates are for the maximum size of an OBR; resolution is limited by the difficulty in preparing probes large enough to give a strong hybridization signal with radiolabeled nascent DNA chains, and devoid of repetitive sequences. Future refinements in origin-mapping techniques will likely resolve these OBRs to a smaller locus. Thus, metazoan replication origins appear to be 3-10 times larger than replication origins in simple genomes (0.05-1 kb). The fact that 18 OBRs have been identified by independent investigators using a variety of different methods gives confidence that site-specific initiation is not an artifact of the experimental conditions used to map them. Similar results were obtained with synchronized and unsynchronized cells, with cells containing single-copy sequences and with cells containing amplified multicopy sequences, with untreated cells and with cells treated with metabolic inhibitors, and with different methods for detecting specific DNA sequences.

Neutral/alkaline 2-D gel electrophoresis has been used to map an OBR to 1 kb at an amplification locus in *Sciara* (Liang et al. 1993; Liang and Gerbi 1994), giving credence to an earlier interpretation of 2-D gel electrophoresis mapping data that 80% of replication forks at the chorion gene amplification locus in *Drosophila* originate from a specific 1-kb site (Heck and Spradling 1990). Neutral/alkaline 2-D gel electrophoresis also identified an S-phase OBR 15-20 kb downstream from the *Drosophila* DNA polymerase α gene (Shinomiya and Ina 1994).

4. Replication bubbles are detected throughout a larger "initiation zone" of 6-55 kb that includes the OBR. This conclusion is based on analyses of DNA structures by neutral/neutral 2-D gel electrophoresis at five different genomic loci (Table 3) (Delidakis and Kafatos 1989; Heck and Spradling 1990; Dijkwel and Hamlin 1992, 1995; Liang et al. 1993; Little et al. 1993; Liang and Gerbi 1994), two of which (*Drosophila* chorion gene, *Sciara* locus 9) are developmentally programmed amplification origins. Four of these "initiation zones" encompass an OBR that was detected either by nascent strand analyses (rRNA genes, DHFR oriβ) or by measuring the direction of fork movement using neutral/alkaline 2-D gel electrophoresis (*Sciara* locus 9, *Drosophila* pol α gene). One study on the *Drosophila*

chorion gene locus (Heck and Spradling 1990) concluded that although multiple initiation sites may exist within a 12-kb locus, a model in which a single origin is preferred 70–80% of the time could explain their neutral/neutral 2-D gel electrophoresis data.

Results of neutral/neutral 2-D gel analyses are consistent with newly synthesized DNA analyses and most neutral/alkaline gel analyses if one assumes that the frequency of initiation events at the OBR is much greater than the frequency of initiation events outside the OBR. In fact, replication bubbles detected by neutral/neutral 2-D gel analyses appear more abundant in the 12-kb region containing the DHFR $ori\beta$ OBR (Dijkwel and Hamlin 1992), and in the 8-kb region at the 5'-end of the rRNA transcription unit (Little et al. 1993) where nascent DNA strand analyses revealed a >10-fold excess of newly synthesized DNA relative to other sites within the initiation zone (DePamphilis 1993d; Gilbert et al. 1995; Yoon et al. 1995). In practice, the relative number of initiation events in different DNA segments is difficult to quantify by 2-D gel analysis because of concerns over variable loss of replication bubbles and other technical problems (Dijkwel and Hamlin 1992; Krysan et al. 1993; Little et al. 1993), whereas analysis of labeled nascent DNA chains lends itself readily to quantification and thus reveals the preference for one site relative to another. For example, the ratio of DNA synthesis between the two templates of a specific DNA fragment automatically provides the minimum fraction of replication forks moving in the same direction through this region. Initiation events distributed randomly outside the OBR simply contribute to the background level in these mapping protocols.

METAZOAN ORIGINS OF REPLICATION ARE GENETICALLY DETERMINED

The simple fact that metazoan origins map to specific sites that replicate at specific times during S phase (see Simon and Cedar, this volume) demonstrates that origins of replication are inherited from one cell division to the next. This conclusion is reinforced by reports that the same OBR identified in cells containing two copies per diploid genome is also identified in cells containing 1,000 (hamster DHFR gene) to 30,000 (mouse ADA gene) tandem copies of either chromosomal or extrachromosomal (episomal) sequences (Table 3). Therefore, each copy of the amplified region that initiates replication must use the same OBR; otherwise, initiation would appear to occur at many different sites within the same DNA locus.

Direct evidence that metazoan replication origins are genetically determined comes from reports that DHFR *oriβ* (Handeli et al. 1989) and the chorion gene amplification origin (Orr-Weaver 1991) retain their activity when translocated to other chromosomal sites. Conversely, an 8-kb deletion between the human δ-globin and β-globin genes that includes the only OBR found within a 135-kb region eliminates bidirectional replication from this site; all replication forks now move in one direction through this 135-kb region (Kitsberg et al. 1993). These data demonstrate that metazoan origins of replication are determined by as yet undefined DNA sequences. Nevertheless, identification of genetically required DNA sequences that function as ARS elements has been difficult.

To date, five reports of ARS elements that function in mammalian cells and cell extracts have been documented in detail and shown to correspond to sites where replication occurs in mammalian chromosomes (Table 3). In other plasmid assays, replication appears to depend on the distribution of as yet undefined sequence signals over a large area (>10 kb), signals that are more prevalent in human DNA than in bacterial DNA (Krysan et al. 1993). Sequences have been identified in human DNA that stimulate plasmid replication about 3-fold and are present at a 2-kb OBR mapped in chromosomal DNA (Masukata et al. 1993). These results suggest that replication is stimulated by simple sequence features that occur frequently in mammalian DNA and therefore may promote initiation events throughout the initiation zone.

ARS activity in mammalian cells may depend on several variables. For example, some OBR regions may exhibit stronger ARS activity than others. Incubation of negatively supercoiled plasmid DNA with DNA polymerase-α:DNA primase, RP-A, T-antigen helicase, and DNA gyrase resulted in site-specific initiation of DNA replication at the strong yeast origin, ARS1, and at the c-*myc* OBR (Ishimi et al. 1994). These conditions employ the energy derived from negative superhelical turns to initiate DNA replication at DUEs that can be unwound by T antigen in the presence of RP-A. However, in the DHFR *oriβ* region where ARS activity has not been detected (Burhans et al. 1990), preference for the OBR region was observed, but it was less pronounced than with the other two origins.

Other studies on plasmid DNA replication in human cells (Caddle and Calos 1992) or in *Xenopus* eggs and egg extracts (Gilbert et al. 1995) failed to observe either preferential replication of plasmids containing the DHFR *oriβ* region or site-specific initiation within *oriβ* in those plasmids that contained this sequence. However, when nuclei from G_1-phase hamster cells were incubated in *Xenopus* egg extract, DNA replication was initiated specifically at or near the same *ori*β OBR utilized by hamster cells (Gilbert et al. 1995). Therefore, site-specific initiation of DNA replication in metazoan chromosomes involves nuclear structure, a requirement that may be difficult to fulfill with plasmid DNA. For example, matrix (scaffold) attachment regions (Schlake et al. 1994) or locus control regions (Bonifer et al. 1994) can increase transcription rates for integrated but not episomal templates, demonstrating that some potential components of replication origins function only in the context of cell chromosomes. Conversely, many sequences that can function as ARS elements in plasmids do not function as replication origins in chromosomes (Kipling and Kearsey 1990; Newlon et al. 1993). Therefore, plasmid DNA replication may not be an appropriate model for metazoan cellular DNA replication, because metazoan origins may function efficiently only within the context of a real chromosome.

Finally, the sequence context of an origin can strongly affect its activity. When two or more yeast ARS elements are in close proximity (~6 kb), the efficiency of each is reduced, and only one is activated in each cell cycle (Brewer and Fangman 1993; Marahrens and Stillman 1994). This phenomenon has been demonstrated in *S. pombe* chromosomes, where initiation zones have been shown to be composed of two or three independent origins (Dubey et al. 1994; Wohlgemuth et al. 1994). Other sequences in the neighborhood also can affect origin activity, making one yeast ARS element preferred over its neighbor (Newlon et al. 1993; Brewer and Fangman 1994). Thus, one could imagine that a metazoan initiation zone is composed of many "simple origins" of the type found in yeast, for example, and that the resulting interference patterns from neighboring origins and extraneous sequences would impose an OBR at one particular site. Moreover, the anatomical complexity observed for metazoan initiation zones could vary considerably as a function of the number and arrangement of the simple origins that comprise them.

The ability to detect ARS activity in mammalian cells thus may depend on a number of factors, among which are negative superhelical density in the extrachromosomal DNA, sequence context of the cellular OBR, number and proximity of initiation signals that comprise a replication origin, size of the extrachromosomal DNA, and the relative strengths of various OBRs. In addition, detection of ARS activity may require stringent selection conditions (Virta-Pearlman et al. 1993). Detection of ARS in *S. pombe*, for example, is more difficult than in *S. cerevisiae*, because virtually every DNA sequence, even vector DNA, is capable of replicating to a limited extent and therefore requires stringent selection conditions to identify "true" ARS elements (Caddle and Calos 1994;

Dubey et al. 1994; Wohlgemuth et al. 1994). Furthermore, if replication sites in nuclei are limited, only a small number of extrachromosomal origins will be accommodated, and detection may require a sensitive PCR-based assay (Taira et al. 1994).

An alternative assay for *cis*-acting sequences that initiate DNA replication is to look for an amplification promoting element (APE) that promotes formation of large numbers of integrated copies of a DNA sequence, rather than replication of extrachromosomal DNA sequences. A 370-bp APE has been identified in the nontranscribed spacer region of mouse rRNA gene (Hermann et al. 1994) and mediates a 40- to 800-fold amplification of the vector DNA in transformed cells. This DNA segment also contains an OBR that maps from 0.5 to −3.5 kb upstream of the transcription initiation site for mouse rRNA gene (E. Gogel and F. Grummt, unpubl.), in agreement with the OBR at the 5′ end of human rRNA genes (Little et al. 1993; Yoon et al. 1995). The 4.5-kb OBR region in the hamster DHFR *ori*β also acts as an APE and contains homologies with the APE found in rRNA genes (Stolzenburg et al. 1994). APE activity may provide a more reproducible assay for metazoan replication origin sequences, if these sequences function efficiently only in the context of a large chromosome.

DNA FEATURES OF A METAZOAN REPLICATION ORIGIN

Metazoan replication origins contain a number of structural features that may be related to their role in DNA replication, although none have so far been shown to be required for initiation of chromosomal DNA replication (see Heintz, this volume). Among these are binding sites for transcription factors, and those for c-Myc protein (Taira et al. 1994) and octamer-binding transcription factors (Iguchi-Ariga et al. 1993) have been reported to contribute to ARS activity in mammalian cells. Replication and transcription sites are colocalized in mammalian nuclei (Hassan et al. 1994), but transcription through cellular replication origins does not appear to be required for replication, since only 3 of the 18 OBRs in Table 4 are located within a transcribed region (cDNA 343, RPS14 gene, CAD gene). More likely, both processes are facilitated by an open chromatin structure that provides access to initiation factors and negative superhelical energy that facilitates DNA unwinding. These features may be provided by a particular nuclear structure.

One feature likely to be shared by all OBRs is the presence of one or more DUEs, since DNA unwinding must precede DNA synthesis and simple OBRs appear to initiate DNA unwinding at a DUE (see above).

Potential DUEs exist at or close to several OBRs, including DHFR *ori*β (Dobbs et al. 1994). A second feature is a densely methylated island (DMI) that consists of 127 bp (RPS14 gene OBR) to 512 bp (DHFR *ori*β) of DNA in which all dC residues are methylated on both strands, regardless of the adjacent nucleotide (Tasheva and Roufa 1994b). This unusual methylation pattern has been observed only in association with replication origins, and then only in proliferating cells. Intriguingly, DNA methyltransferase, the enzyme responsible for converting hemimethylated sites to methylated sites in nascent DNA, becomes associated with replication foci during S phase (Leonhardt et al. 1992). DMIs might act in a positive way by providing a binding site for a replication-specific factor, by altering DNA structure to promote unwinding at a neighboring DUE, or by altering chromatin structure to increase accessibility to initiation factors. In fact, the DMI overlapping the DHFR *ori*β OBR is flanked by binding sites for RIP60, a protein that can link the two sites to form a 736-bp DNA loop that encompasses the DMI and flanks a potential DUE (Mastrangelo et al. 1993). By analogy to *E. coli* DnaA, bacteriophage λ O protein (Kornberg and Baker 1992), and SV40 T-antigen (see Borowiec, this volume), binding of origin recognition proteins can impose superhelical tension that causes untwisting of DNA in nearby DUEs. Alternatively, by analogy to *E. coli oriC* (Herrick et al. 1994), the DMI may help to limit initiation to once per S phase. *oriC* is methylated on both strands at 11 sites. When these sites become hemimethylated as a result of replication, *oriC* associates with an outer membrane component, delaying rebinding of its origin recognition protein, DnaA, and thus delaying reinitiation.

A third feature is attachment sites for nuclear matrix or nuclear scaffold (Nakayasu and Berezny 1989; Hozák et al. 1993), components of nuclear structure that are commonly associated with newly replicated DNA and, in some cases, origins of cellular replication (see Laskey and Madine, this volume).

A fourth feature is palindromic sequences that can collapse into cruciform structures when sufficient negative superhelical energy is provided. These structures may promote initiation of DNA replication at specific sites. Cruciform extrusion at the origin of *E. coli* plasmid pT181 is promoted by the plasmid-encoded initiator protein RepC (Noirot et al. 1990), and antibodies directed against DNA cruciforms can stimulate overall DNA synthesis and copy number of specific genes in permeabilized mammalian cells (Zannis-Hadjopoulos et al. 1988). Moreover, staining mammalian cells with anti-cruciform antibodies suggests that cruciform structures accumulate as cells prepare to enter S phase (Ward et al. 1991). A role for cruciforms in replication origins must be considered cautiously, however, since cruciform structures can be a direct result of aberrant DNA replication involved in gene amplification (Cohen et al. 1994).

THE ROLE OF NUCLEAR STRUCTURE IN METAZOAN DNA REPLICATION

Prokaryotic genomes and animal virus genomes can all replicate in the presence of purified soluble proteins and cofactors; no requirement for a cellular structure has been observed, although there exists a transient interaction between *E. coli oriC* and the outer membrane that regulates the rate at which reinitiation can occur at *oriC* (Herrick et al. 1994). Whether or not the same is true for replication origins in simple eukaryotic organisms remains to be seen. However, one of the most striking requirements for initiation of DNA replication in metazoan chromosomes is that of nuclear structure (see Laskey and Madine, this volume).

DNA replication in metazoan chromosomes occurs at discrete nuclear foci. Clusters of replication origins initiate replication synchronously (Hand 1978), giving rise to discrete "replication complexes" that contain from 100 to 300 replication forks (see Laskey and Madine, this volume). Formation of these replication complexes accounts for the many observations that newly synthesized DNA is preferentially bound to components of nuclear structure generally referred to as nuclear matrix or nuclear scaffold (Nakayasu and Berezney 1989; Hozák et al. 1993). These replication complexes appear to be assembled in an energy-dependent process prior to S phase at the sites where replication begins (Fig. 2). RP-A, a heterotrimeric single-strand DNA-binding protein that is required for replication of metazoan chromosomes (Fang and Newport 1993), is bound at discrete foci in nuclei prior to DNA unwinding and DNA synthesis (Adachi and Laemmli 1994). High levels of cyclin B/ cdc2 protein kinase, an enzyme that is required for entrance into mitosis, prevents the appearance of these RP-A foci, consistent with their absence in mitotic chromosomes (Adachi and Laemmli 1994). Cyclin-A-dependent cdk2 protein kinase, an enzyme that is required for entrance into S phase (Fang and Newport 1991), colocalizes with RP-A (Cardoso et al. 1993). Proliferating cell nuclear antigen (PCNA), a cofactor for DNA polymerase-δ, and DNA polymerase-α, an enzyme required for synthesis of Okazaki fragments, are also found at replication foci in S-phase nuclei (Kill et al. 1991). Whether or not they are prebound to these foci before replication begins is not clear. Presumably, licensing factor (see Laskey and Madine; Blow and Chong; both this volume), a cytoplasmic initiation factor that gains access to replication origins only when the nuclei become permeable during mitosis, also binds to preinitiation complexes.

Initiation of DNA replication in metazoan chromosomes requires an intact nuclear structure. Replication of DNA introduced into either

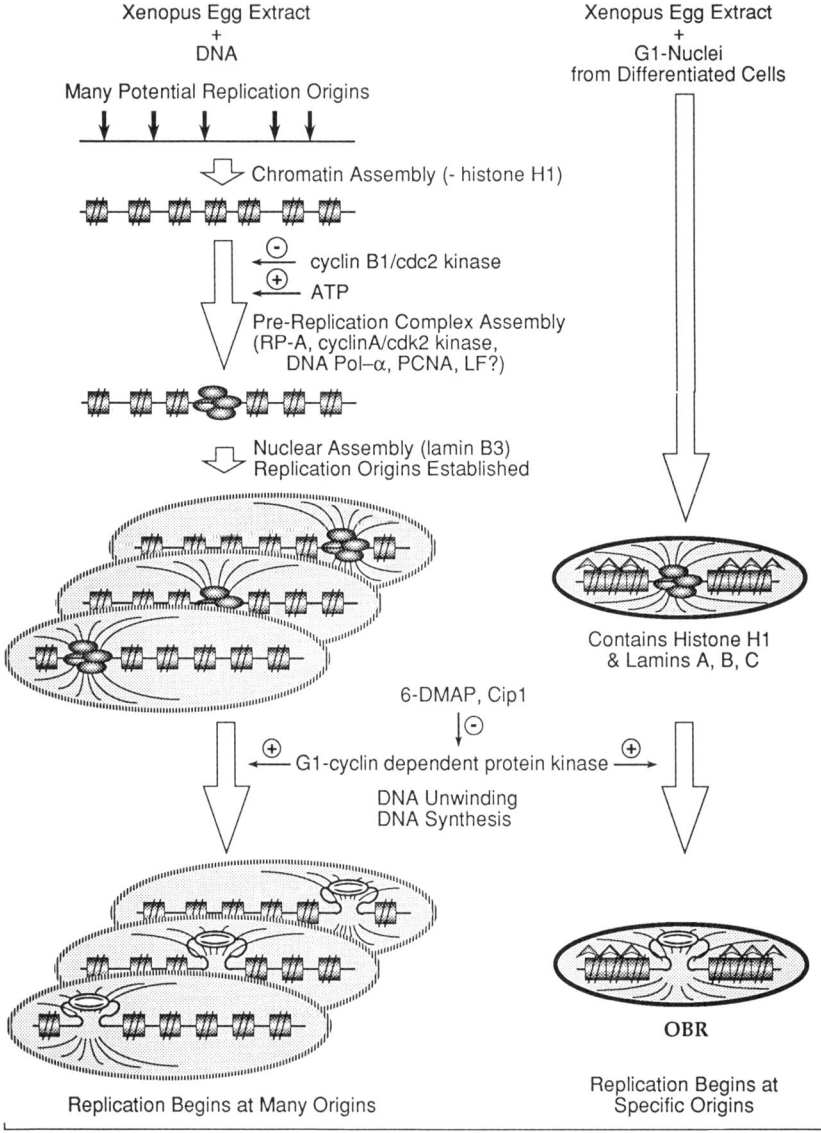

Figure 2 (See facing page for legend.)

Xenopus eggs or egg extracts does not occur unless DNA is first assembled into chromatin and then organized into a nuclear structure that includes lamin B3 and functional nuclear pores (see Laskey and Madine; Blow; both this volume). In addition, the nuclear envelope is instrumental in regulating the onset of S phase, apparently by regulating access of chromosomal DNA to one or more initiation factors (licensing factor) present in the cytoplasm (Coverley and Laskey 1994).

Site-specific initiation requires an intact nuclear structure. *Xenopus* egg extract can initiate DNA replication in purified DNA molecules only after the DNA is organized into a pseudo-nucleus, but under these conditions, DNA replication is independent of DNA sequence and begins at many sites distributed throughout the molecules. However, *Xenopus* egg extract can initiate DNA replication at specific sites in mammalian chromosomes, but only when the DNA is presented in the form of an intact nucleus from differentiated cells (Gilbert et al. 1995). Initiation of DNA synthesis in nuclei isolated from G_1-phase hamster cells is distinguished from continuation of DNA synthesis at preformed replication forks in S-phase nuclei by a delay that precedes DNA synthesis, a dependence on soluble *Xenopus* egg factors, sensitivity to the protein kinase inhibitor 6-dimethylaminopurine (DMAP), and complete labeling of nascent DNA chains. Initiation sites for DNA replication were mapped downstream from the amplified DHFR gene region by (1) identification of the earliest labeled DNA fragments (Gilbert et al. 1993), (2) quantitative hybridization of newly synthesized DNA to double-stranded DNA probes to reveal genomic loci where DNA synthesis began, and (3)

Figure 2 Acquisition of site-specific DNA replication. *Xenopus* eggs or egg extracts assemble bare DNA or sperm chromatin into a relaxed nuclear structure that permits initiation of DNA replication within many sequences, allowing the early-cleavage-stage amphibian embryo to rapidly replicate its genome. In contrast, preformed nuclei from G_1-phase differentiated mammalian cells initiate DNA replication under the same conditions at or near a site-specific OBR that was selected by the mammalian cell to be used as a replication origin during its subsequent S phase. Selection of initiation sites may be restricted by chromatin structure (nucleosome ▨ , histone H1 ⋀) masking some potential origins, and nuclear matrix (scaffold) associated regions (⇒⇐) stabilizing DNA unwinding at other potential origins. At some point during animal development, changes occur in nuclear organization that restrict the number of sites that can be used as origins of replication. In *Xenopus* (and presumably other animals whose embryos undergo rapid cell cleavages) this transition appears to occur after the mid-blastula transition (see text). (LF) Licensing factor.

quantitative hybridization of Okazaki fragments to single-stranded DNA probes to reveal the transition between continuous and discontinuous DNA synthesis on each template within this initiation locus. When bare DNA substrates are used, then *Xenopus* eggs or egg extracts do not distinguish between prokaryotic DNA, hamster DNA that does not contain a replication origin, and hamster DNA that does contain a replication origin. Moreover, initiation events were distributed equally throughout a 30-kb cosmid containing the DHFR *ori*β region. When nuclei are used, *Xenopus* egg extract continues DNA synthesis in S-phase nuclei at sites that had been initiated in hamster cells (e.g, DHFR *ori*β). When the integrity of the nuclear membrane is preserved, *Xenopus* egg extract initiates DNA replication in G_1-phase nuclei specifically at or near the OBR (*ori*β) utilized by hamster cells. When nuclear integrity is damaged, preference for initiation at *ori*β is significantly reduced or eliminated. Therefore, initiation sites for DNA replication in mammalian cells are established prior to S phase by some component of differentiated nuclear structure, and this replication origin can be recognized by soluble initiation factors present in *Xenopus* eggs.

Subsequent studies (Wu and Gilbert 1996) have revealed that *Xenopus* egg extract initiates replication at many sites throughout the DHFR gene region in nuclei isolated from early G_1-phase hamster cells, whereas the same extract initiates specifically at *ori*β in nuclei isolated from late G_1-phase hamster cells. Therefore, specific origins of replication in mammalian chromosomes are reestablished during each cell division cycle several hours after nuclear assembly occurs.

MODEL FOR METAZOAN ORIGINS OF DNA REPLICATION

There are three basic models for understanding the cumulative data on metazoan replication origins. The first is the "strand separation model" in which extensive DNA unwinding precedes initiation of RNA-primed DNA synthesis at many sites concurrently on both templates (Benbow et al. 1992). This model was based largely on reports of single-stranded DNA in *Xenopus* and *Drosophila* embryos, but the single-stranded DNA bubbles that one would expect as replication intermediates have not been detected by 2-D gel analyses of DNA replication in *Xenopus* eggs (Hyrien and Méchali 1992; Mahbubani et al. 1992) or *Drosophila* embryos (Liang and Gerbi 1994), or by electron microscopy of DNA replicating in mammalian cells (Hamlin et al. 1992; Gruss et al. 1994). On the contrary, the fact that replication forks and bubbles are detected by these methods, and the fact that replication fork polarity is observed by hybridization of either Okazaki fragments or long nascent DNA chains to complementary DNA templates (see above, Mapping Origins of Replication in Metazoan Chromosomes), provide compelling evidence that DNA synthesis occurs at replication forks in both simple and complex genomes.

A second model is one in which a replication complex assembles at the OBR, but then migrates to other positions on either side of the OBR before initiating DNA unwinding and DNA synthesis. This would generate a Gaussian distribution of initiation events with the OBR at its apex. However, one would expect the OBR sequence to strongly stimulate DNA replication under all conditions, a prediction that is difficult to reconcile with the lack of DNA sequence preference observed in the eggs and cleavage-stage embryos of frogs, flies, fish, and sea urchins.

A third model is suggested by the Jesuit dictum that "many are called, but few are chosen" and perhaps offers the simplest way of interpreting all the data (Fig. 3) (DePamphilis 1993a,b,d; Burhans and Huberman 1994). Whereas naked DNA contains many possible sites where replication can begin, assembly of DNA into chromatin can suppress initiation at some of these sites, and organization of chromatin into a nuclear structure can activate DNA replication at other sites by promoting DNA unwinding. Thus, initiation sites for DNA replication in metazoan chromosomes would consist of three parts:

1. *OBR*. Most initiation events occur bidirectionally within a 0.5- to 2-kb locus. In differentiated cells, where most potential replication origins are suppressed, favored sites may consist of an easily unwound DNA sequence in combination with other origin components such as DMIs, OREs, and transcription-factor-binding sites. These features, in addition to sequences that may attach to nuclear matrix (scaffold) in order to stabilize unwound DNA at the OBR (Bode et al. 1992), would comprise the site-specific, heritable replication origins that have been mapped in metazoan chromosomes (Table 1).

2. *Initiation zone*. Some initiation events are detected randomly distributed throughout a 6-kb to 55-kb region that encompasses the OBR. These sequences may remain accessible to replication factors so that the same cell can occasionally initiate replication at a nearby DUE. Alternatively, some cells within a population simply may have selected a different OBR in the same region, or are capable of utilizing more than one OBR in the same region, the choice of which can change during successive S phases. Secondary initiation sites also occur in simple genomes, although usually at a lower frequency. For example, papovavirus replication origins are at least 100-fold more effi-

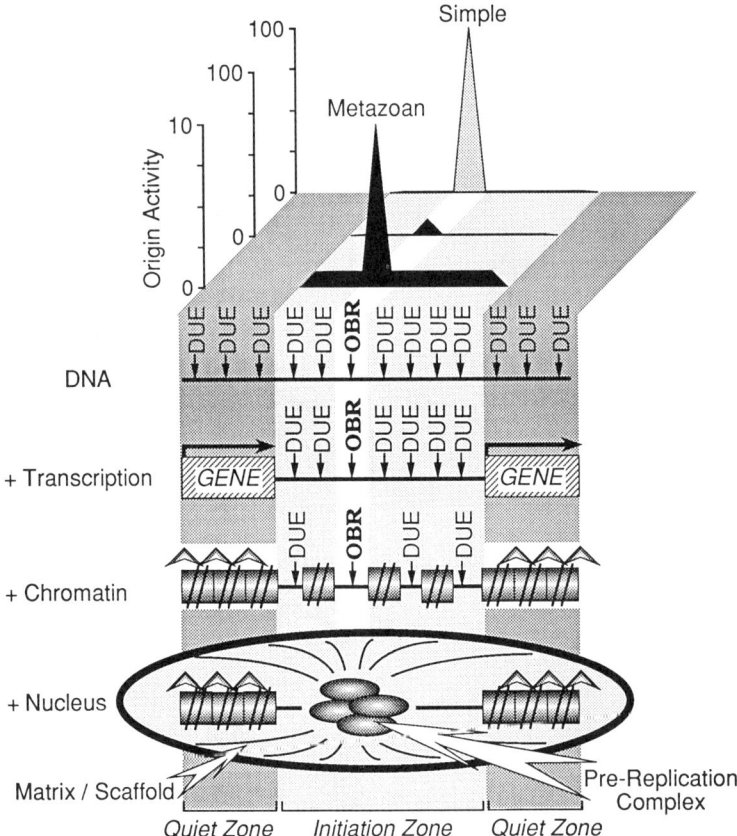

Figure 3 The "Jesuit Model" for metazoan replication origins and their relationship to simple replication origins. An OBR in simple genomes appears much more efficient than an OBR in metazoan genomes, thus reducing the relative intensity of secondary initiation events observed in the initiation zone surrounding the OBR. "Quiet zones" are sequences where no initiation events are detected. Eukaryotic DNA contains many potential origins; some may be simply DUEs whereas other, more efficient replication origins may be analogous to those in simple genomes (Fig. 1). The number of potential origins that can become active origins is restricted by actively transcribed DNA regions ("Genes") and chromatin structure (nucleosome ▨ , histone H1 ⋏), whereas nuclear structure, which is required for initiation of metazoan DNA replication, activates selected origins.

cient at initiating replication than neighboring DNA sequences, but initiation events still occur at other sites in the DNA molecule, although they are difficult to detect (Martin and Setlow 1980; Tack and Proctor 1987). Similarly, bacteriophages T4 and T7 contain both a primary replication origin and several secondary origins (Kornberg and Baker 1992). If we compare site specificity in a metazoan genome with that in a simple genome, initiation events outside the metazoan OBR would go undetected unless the scale is expanded to accommodate the lower activity of a metazoan replication origin (Fig. 3).

3. *Quiet zone.* In contrast to DNA replication in rapidly cleaving embryos where initiation events are detected throughout the genome, initiation events in differentiated cells are restricted to specific sites (OBR + initiation zone). Initiation events outside the initiation zone may be further suppressed by active transcription (Haase et al. 1994) and higher-order chromatin structure (Fig. 3) (Hand 1978; Forrester et al. 1990; Simpson 1990; Ferguson and Fangman 1992; Karpen and Spradling 1992), two features that are generally absent in rapidly cleaving embryos (Fig. 2) (Wolffe 1994).

DNA replication in *Xenopus* egg extracts provides a simple paradigm for the selection and assembly of initiation sites for DNA replication (Fig. 2). Naked DNA contains many potential sites where DNA replication can begin. These sites most likely correspond to easily unwound DNA sequences that are components of replication origins in simple (Natale et al. 1993) as well as metazoan (Dobbs et al. 1994) genomes, and that can promote site-specific replication in plasmid DNA (Ishimi et al. 1994). When naked DNA is introduced into *Xenopus* eggs or egg extract, chromatin is assembled in the absence of histone H1, which is required for compaction of DNA into 30-nm fibers (Wolffe 1994). In addition, nuclei assembled in *Xenopus* eggs are less compact than nuclei in differentiated cells. This may be due partially to the fact that the composition of nuclear lamin proteins, a component of nuclear structure required for DNA replication (Newport et al. 1990; Jenkins et al. 1993), is simpler in *Xenopus* embryos than in *Xenopus* differentiated cells (Benavante et al. 1985; Stick and Hausen 1985). This relaxed environment may allow assembly of replication complexes at any one of the many potential origins located within a large DNA region, thus facilitating rapid replication of the *Xenopus* embryonic genome. The fact that replication does not initiate at more than one site within a single plasmid molecule (Hines and Benbow 1982; Méchali and Kearsey 1984; Mahbubani et al. 1992; Hyrien and Méchali 1992) suggests that replication in embryos undergoing rapid nuclear division cycles (e.g., frogs, flies, fish, sea urchins) occurs at only one of many potential sites within a large DNA region. Formation of replication complexes does not require nuclear membrane formation (Adachi and Laemmli 1994), but initiation of DNA replication does, suggesting that some component of nuclear structure interacts directly with each replication origin. Moreover, activation of replication complexes appears to require the action of G_1-cyclin-dependent protein kinase cdk2. General inhibitors of protein kinases (e.g., DMAP) prevent initiation of DNA replication in either sperm chromatin or G_1 nuclei, but do not inhibit active replication forks (Blow 1993; Kubota and Takisawa 1993; Gilbert et al 1995). Cip1 protein, a specific inhibitor of cdk2 kinase, exhibits a similar effect, and this inhibition can be overcome with cyclin E (Strausfeld et al. 1994). Cip1 inhibition occurs after formation of replication complexes but before DNA unwinding has begun (Adachi and Laemmli 1994).

It has long been observed that the density of initiation sites (origin-to-origin distance) in metazoan chromosomes varies in different cell types and under different experimental conditions, leading to the conclusion that many more replication sites are available than are used during a normal S phase, and that the number of active replication origins depends on the ratio of initiation factors to DNA and accessibility of origins as determined by higher-order chromatin structure (Hand 1978). As animal development progresses and cells begin to undergo differentiation, a 5- to 10-fold reduction can occur in the frequency of initiation sites (Blumenthal et al. 1974; McKnight and Miller 1977; Buongiorno-Nardelli et al. 1982; Shinomiya and Ina 1991). In fact, although initiation events in *Xenopus* rRNA gene region are distributed randomly throughout these sequences during early development (prior to gastrulation), initiation events become confined to the intergenic regions as development progresses (Hyrien et al. 1995). Changes in chromatin and nuclear structure that occur at the mid-blastula transition in *Xenopus*, for example, may repress initiation of DNA replication at some loci while facilitating it at others (Fig. 3) (Gilbert et al. 1995). In yeast, for example, at least half of the ARS elements on chromosome III are not active as chromosomal replication origins (Newlon et al. 1993). Factors that likely contribute to these changes in metazoa include the onset of transcription and cell differentiation, and the appearance of histone H1 and lamins A and C. Thus, some of the initiation sites chosen in fertilized *Xenopus* eggs may remain as origins throughout development, whereas other sites may be eliminated.

SUMMARY AND PERSPECTIVE

DNA replication in differentiated eukaryotic cells is initiated at specific DNA sites, regardless of whether the genome is viral or mitochondrial, or whether the organism consists of one cell type or many cell types. Perhaps the most basic difference among origins is whether they are designed to initiate once per cell division cycle or more than once. However, whereas the anatomy of initiation sites in most simple genomes is fairly clear and their mechanism of action is understood in principle, if not in detail, initiation sites in metazoan chromosomes remain elusive: Their requirement for nuclear structure makes them more complex than those in simple genomes, and the sequences that determine them have not yet been identified. Metazoan replication origins may follow the same principles revealed by replication origins in simple genomes but solve the problems of determining where, when, and how to initiate DNA synthesis in different ways. Why such complexity appears in the chromosomes of metazoa but not single-cell organisms may involve differences in chromatin structure and nuclear organization that accompany the appearance of histone H1, nuclear lamins, and nuclear envelope breakdown during mitosis in metazoans.

The concept of a metazoan origin as a specific nucleoprotein complex that interacts with soluble replication factors, rather than simply a specific sequence that is recognized by soluble factors, can account for three important features of metazoan DNA replication. First, it provides a way to change the number of origins per chromosomal locus at different stages in animal development, thus accommodating the need for shorter S phases in the early developmental stages of some animals. Second, it provides a way to limit the number of origins per genome without simultaneously limiting the cell's capacity to sustain genetic alterations. A large number of less efficient origins that demand less sequence specificity allow more flexibility for rearranging genetic information than a small number of highly efficient origins that require a great deal of sequence specificity. Finally, it provides a way to restrict initiation at each origin to once per S phase. Initiation at each replication origin would be restricted to once per cell cycle if the nucleoprotein complex that defines a replication origin in G_1 nuclei was disrupted by the act of replication. In that event, reinitiation at the same site would be prevented, even in the presence of excess soluble initiation factors.

Metazoan replication origins in G_1 nuclei of differentiated mammalian cells may be analogous to the six-protein origin replication complex assembled at yeast ARS sites (see Newlon, this volume) or the EBNA1 protein/origin core DNA complex in EBV episomes (see Yates, this volume), both of which appear to be stable DNA-protein complexes that exist throughout the cell cycle. Thus, a licensing factor (see Laskey and Madine, this volume) may recognize this DNA/protein platform, permitting it to be activated by a G_1-cyclin-dependent protein kinase. Licensing factor is inactivated by the replication process and then replaced when the nuclear membrane becomes permeable to cytoplasmic factors during mitosis.

ACKNOWLEDGMENTS

I am indebted to Darren Natale, David Gilbert, William Burhans, and many other friends and colleagues for their intellectual contributions to this paper.

REFERENCES

Adachi. Y. and U.K. Laemmli. 1994. Study of the cell cycle-dependent assembly of the DNA pre-replication centres in *Xenopus* egg extract. *EMBO J.* **13:** 4153–4164.

Ariizumi, K., Z. Wang, and P.W. Tucker. 1993. Immunoglobulin heavy chain enhancer is located near or in an initiation zone of chromosomal DNA replication. *Proc. Natl. Acad. Sci.* **90:** 3695–3699.

Benavante, R., G. Krohne, and W.W. Franke. 1985. Cell type-specific expression of nuclear lamina proteins during development of *Xenopus laevis*. *Cell* **41:** 177–190.

Benbow, R.M., J. Zhao, and D.D. Larson. 1992. On the nature of origins of DNA replication in eukaryotes. *BioEssays* **14:** 661–670.

Berberich, S., A. Trivedi, D.C. Daniel, E.M. Johnson, and M. Leffak. 1995. In vitro replication of plasmids containing human c-*myc* DNA. *J. Mol. Biol.* **245:** 92–109.

Blow, J.J. 1993. Preventing re-replication of DNA in a single cell cycle: Evidence for a replication licensing factor. *J. Cell Biol.* **122:** 993–1002.

Blumenthal, A.B., H.J. Kriegstein, and D.S. Hogness. 1974. The units of DNA replication in *Drosophila melanogaster* chromosomes. *Cold Spring Harbor Symp. Quant. Biol.* **38:** 205–223.

Bode, J., Y. Kohwi, L. Dickinson, T. Joh, D. Klehr, C. Mielke, and T. Kohwi-Shigematsu. 1992. Biological significance of unwinding capability of nuclear matrix-associating DNAs. *Science* **55:** 195–197.

Bonifer, C., N. Yannoutsos, G. Krüger, F. Grosveld, and A.E. Sippel. 1994. Dissection of the locus control function located on the chicken lysozyme gene domain in transgenic mice. *Nucleic Acids Res.* **22:** 4202–4210.

Brewer, B.J. and W.L. Fangman. 1993. Initiation at closely spaced replication origins in a yeast chromosome. *Science* **262:** 1728–1731.

―――. 1994. Initiation preference at a yeast origin of replication. *Proc. Natl. Acad. Sci.* **91:** 3418–3422.

Bullock, P.A., Y.S. Seo, and J. Hurwitz. 1991. Initiation of SV40 DNA synthesis in vitro. *Mol. Cell. Biol.* **11:** 2350–2361.

Buongiorno-Nardelli, M., G. Micheli, M.T. Carri, and M. Marilley. 1982. A relationship between replicon size and supercoiled loop domains in the eukaryotic genome. *Nature* **298:** 100–102.

Burhans, W.C. and J.A. Huberman. 1994. DNA replication origins in animal cells: A question of context? *Science* **263:** 639–640.

Burhans, W.C., L.T. Vassilev, M.S. Caddle, N.H. Heintz, and M.L. DePamphilis. 1990. Identification of an origin of bidirectional DNA replication in mammalian chromosomes. *Cell* **62:** 955–965.

Burhans, W.C., L.T. Vassilev, J. Wu, J.M. Sogo, F.N. Nallaseth, and M.L. DePamphilis. 1991. Emetine allows identification of origins of mammalian DNA replication by imbalanced DNA synthesis, not through conservative nucleosome segregation. *EMBO J.* **10:** 4351–4360.

Caddle, M.S. and M.P. Calos. 1992. Analysis of the autonomous replication behavior in human cells of the DHFR putative origin of replication. *Nucleic Acids Res.* **20:** 5971–5978.

———. 1994. Specific initiation at an origin of replication from *Schizosaccharomyces pombe*. *Mol. Cell. Biol.* **14:** 1796–1805.

Cardoso, M.C., H. Leonhardt, and B. Nadal-Ginard. 1993. Reversal of terminal differentiation and control of DNA replication: Cyclin A and cdk2 specifically localize at subnuclear sites of DNA replication. *Cell* **74:** 979–992.

Carroll, S.M., M.L. DeRose, J.L. Kolman, G.H. Nonet, R.E. Kelly, and G.M. Wahl. 1993. Localization of a bidirectional DNA replication origin in the wild type and in episomally amplified murine ADA loci. *Mol. Cell. Biol.* **13:** 2971–2981.

Cohen, S., D. Hassin, S. Karby, and S. Lavi. 1994. Hairpin structures are the primary amplification products: A novel mechanism for generation of inverted repeats during gene amplification. *Mol. Cell. Biol.* **14:** 7782–7791.

Coverley, D. and R.A. Laskey. 1994. Regulation of eukaryotic DNA replication. *Annu. Rev. Biochem.* **63:** 745–776.

Cusick, M.E., P.M. Wassarman, and M.L. DePamphilis. 1989. Application of nucleases to visualizing chromatin organization at replication forks. *Methods Enzymol.* **170:** 290–316.

Deb, S.P., A.L. DeLucia, C. Bauer, A. Koff, and P. Tegtmeyer. 1986. Domain and structure of the SV40 core origin of replication. *Mol Cell. Biol.* **6:** 1663–1670.

Delidakis, C. and F.C. Kafatos. 1989. Amplification enhancers and replication origins in the autosomal chorion gene cluster of *Drosophila*. *EMBO J.* **8:** 891–901.

DePamphilis, M.L. 1993a. Eukaryotic DNA replication: Anatomy of an origin. *Annu. Rev. Biochem.* **62:** 29–63.

———. 1993b. Origins of DNA replication that function in eukaryotic cells. *Curr. Opin. Cell Biol.* **5:** 434–441.

———. 1993c. How transcription factors regulate origins of DNA replication in eukaryotic cells. *Trends Cell Biol.* **3:** 161–167.

———. 1993d. Origins of DNA replication in metazoan chromosomes. *J. Biol. Chem.* **268:** 1–4.

DePamphilis, M.L. and P.M. Wassarman. 1980. Replication of eukaryotic chromosomes: A close-up of the replication fork. *Annu. Rev. Biochem.* **49:** 627–666.

DePamphilis, M.L., E. Martínez-Salas, D.Y. Cupo, E.A. Hendrickson, C.E. Fritze, W.R. Folk, and U. Heine. 1988. Initiation of polyomavirus and SV40 DNA replication, and the requirements for DNA replication during mammalian development. *Cancer Cells* **6:** 165–175.

Dijkwel, P.A. and J.L. Hamlin. 1992. Initiation of DNA replication in the DHFR locus is confined to the early S-period in CHO cells synchronized with the plant amino acid mimosine. *Mol. Cell. Biol.* **12:** 3715–3722.

———. 1995. The Chinese hamster DHFR origin consists of multiple potential nascent strand start sites. *Mol. Cell. Biol.* **15**: 3023-3031.

Dobbs, D.L., W.-L. Shaiu, and R.M. Benbow. 1994. Modular sequence elements associated with origin regions in eukaryotic chromosomal DNA. *Nucleic Acids Res.* **22**: 2479-2489.

Dubey, D.D., J. Zhu, D.L. Carlson, K. Sharma, and J.A. Huberman. 1994. Three ARS elements contribute to the ura4 replication origin region in the fission yeast, *Schizosaccharomyces pombe*. *EMBO J.* **13**: 3638-3647.

Fang, F. and J. Newport. 1991. Evidence that the G1-S and G2-M transitions are controlled by different cdc2 proteins in higher eukaryotes. *Cell* **66**: 731-742.

―――. 1993. Distinct roles of cdk2 and cdc2 in RP-A phosphorylation during the cell cycle. *J. Cell Sci.* **106**: 983-994.

Fangman, W.L. and B.J. Brewer. 1991. Activation of replication origins within yeast chromosomes. *Annu. Rev. Cell Biol.* **7**: 375-402.

Ferguson, B.M. and W.L. Fangman. 1992. A position effect on the time of replication origin activation in yeast. *Cell* **68**: 333-339.

Forrester, W.C., E. Epner, M.C. Driscoll, T. Enver, M. Brice, T. Papayannopoulou, and M. Groudine. 1990. A deletion of the human β-globin locus activation region causes a major alteration in chromatin structure and replication across the entire β-globin region. *Genes Dev.* **4**: 1637-1649.

Frappier, L. and M. O'Donnell. 1991. Overproduction, purification and characterization of EBNA1, the origin binding protein of Epstein-Barr virus. *J. Biol. Chem.* **266**: 7819-7826.

Frappier, L., K. Goldsmith, and L. Bendell. 1994. Stabilization of the EBNA1 protein on the EBV latent origin of DNA replication by a DNA looping mechanism. *J. Biol. Chem.* **269**: 1057-1062.

Gale, J.M., R.A. Tobey, and J.A. D'Anna. 1992. Localization and DNA sequence of a replication origin in the rhodopsin gene locus of Chinese hamster cells. *J. Mol. Biol.* **224**: 343-352.

Gerard, R. and Y. Gluzman. 1986. Functional analysis of the role of the A/T rich region and upstream flanking sequences in SV40 DNA replication. *Mol. Cell. Biol.* **6**: 4570-4577.

Giacca, M., L. Zentilin, P. Norio, S. Diviacco, D. Dimitrova, G. Contreas, G. Biamonti, G. Perini, F. Weighardt, S. Riva, and A. Falaschi. 1994. Fine mapping of a replication origin of human DNA. *Proc. Natl. Acad. Sci.* **91**: 7119-7124.

Gilbert, D.M. and S.N. Cohen. 1987. BPV plasmids replicate randomly in mouse fibroblasts throughout S-phase of the cell cycle. *Cell* **50**: 59-68.

Gilbert, D.M., H. Miyazawa, and M.L. DePamphilis. 1995. Site-specific initiation of DNA replication in *Xenopus* egg extract requires nuclear structure. *Mol. Cell. Biol.*. **15**: 2942-2954.

Gilbert, D.M., H. Miyazawa, F.S. Nallaseth, J.M. Ortega, J.J. Blow, and M.L. DePamphilis. 1993. Site-specific initiation of DNA replication in metazoan chromosomes and the role of nuclear organization. *Cold Spring Harbor Symp. Quant. Biol.* **58**: 475-485.

Gillette, T.G., M. Lusky, and J. Borowiec. 1994. Induction of structural changes in the BPV-1 origin of replication by the viral E1 and E2 proteins. *Proc. Natl. Acad. Sci.* **91**: 8846-8850.

Gruss, C. and J.M. Sogo. 1992. Chromatin replication. *BioEssays* **14**: 1-8.

Gruss, C., J. Wu, and JM. Sogo. 1994. Disruption of nucleosomes at the replication fork.

EMBO J. **12:** 4533-4545.
Guo, Z.-S. and M.L. DePamphilis. 1992. Specific transcription factors stimulate both simian virus 40 and polyomavirus origins of DNA replication. *Mol. Cell. Biol.* **12:** 2514-2524.
Guo, Z.-S., U. Heine, and M.L. DePamphilis. 1991. T-antigen binding to site I facilitates initiation of SV40 DNA replication but does not affect bidirectionality. *Nucleic Acids Res.* **19:** 7081-7088.
Gutierrez, C., Z-S. Guo, J. Roberts, M.L. DePamphilis. 1990. Simian virus origin auxiliary sequences weakly facilitate binding of T-antigen, but strongly facilitate initiation of DNA unwinding. *Mol. Cell. Biol.* **10:** 1719-1728.
Gutierrez, C., Z.-S. Guo, W. Burhans, M.L. DePamphilis, J. Farrell-Towt, and G. Ju. 1988. Is c-myc protein directly involved in DNA replication? *Science* **240:** 1202-1203.
Haase, S.B. and M.P. Calos. 1991. Replication control of autonomously replicating human sequences. *Nucleic Acids Res.* **19:** 5053-5058.
Haase, S.B., S.S. Heinzel, and M.P. Calos. 1994. Transcription inhibits the replication of autonomously replicating plasmids in human cells. *Mol. Cell. Biol.* **14:** 2516-2526.
Hamlin, J.L., P.A. Dijkwel, and J.P. Vaughn. 1992. Initiation of replication in the Chinese hamster DHFR domain. *Chromosoma* (suppl. 1) **102:** 17-23.
Hand, R. 1978. Eucaryotic DNA: Organization of the genome for replication. *Cell* **15:** 317-325.
Handeli, S., A. Klar, M. Meuth, and H. Cedar. 1989. Mapping replication units in animal cells. *Cell* **57:** 909-920.
Harrison, S., K. Fisenne, and J. Hearing. 1994. Sequence requirements of the EBV latent origin of DNA replication. *J. Virol.* **68:** 1913-1925.
Hassan, A.B., R.J. Errington, N.S. White, D.A. Jackson, and P.R. Cook. 1994. Replication and transcription sites are colocalized in human cells. *J. Cell Sci.* **107:** 425-434.
He, Z., B.T. Brinton, J. Greenblatt, J.A. Hassell, and C.J. Ingles. 1993. Transactivator proteins VP16 and GAL4 bind replication factor A. *Cell* **73:** 1223-1232.
Heck, M.M.S. and A.C. Spradling. 1990. Multiple replication origins are used during *Drosophila* chorion gene amplification. *J. Cell Biol.* **4:** 903-914.
Hermann, C., E. Gartner, U.H. Weidle, and F. Grummt. 1994. High copy expression vector based on amplification promoting sequences. *DNA Cell Biol.* **13:** 437-445.
Herrick, J., R. Kern, S. Guha, A. Landulsi, O. Fayet, A. Malki, and M. Kohiyama. 1994. Parental strand recognition of the DNA replication origin by the outer membrane in *E. coli. EMBO J.* **13:** 4695-4703.
Hines, P.J. and R.M. Benbow. 1982. Initiation of replication at specific origins in DNA molecules microinjected into unfertilized of the frog *Xenopus laevis. Cell* **30:** 459-468.
Holmes, S.G. and M.M. Smith. 1989. Interaction of the H4 autonomously replicating sequence core consensus sequence and its 3'-flanking domain. *Mol. Cell. Biol.* **9:** 5464-5472.
Hozák P., A.B. Hassan, D.A. Jackson, and P.R. Cook. 1993. Visualization of replication factories attached to nucleoskeleton. *Cell* **73:** 361-373.
Huang, R.-Y. and D. Kowalski. 1993. A DNA unwinding element and an ARS consensus comprise a replication origin within a yeast chromosome. *EMBO J.* **12:** 4521-4531.
Huberman, J.A. 1994. Analysis of DNA replication origins and directions by two-dimensional gel electrophoresis. In *The cell cycle: A practical approach* (ed. P. Fantes and R.F. Brooks), pp. 213-234. Oxford University Press, United Kingdom.
Hyrien, O. and M. Méchali. 1992. Plasmid replication in *Xenopus* eggs and egg extracts:

A 2D gel electrophoretic analysis. *Nucleic Acids Res.* **20:** 1463-1469.
―――. 1993. Chromosomal replication initiates and terminates at random sequences but at regular intervals in the ribosomal DNA of *Xenopus* early embryos. *EMBO J.* **12:** 4511-4520.
Hyrien, O., C. Maric, and M. Méchali. 1995. Transition in specification of embryonic metazoan DNA replication origins. *Science* **270:** 994-997.
Iguchi-Ariga, S.M.M., N. Ogawa, and H. Ariga. 1993. Identification of the initiation region of DNA replication in the murine immunoglobulin heavy chain gene and possible function of the octamer motif as a putative DNA replication origin in mammalian cells. *Biochim. Biophys. Acta* **1172:** 73-81.
Ishimi, Y., K. Matsumoto, and R. Ohba. 1994. DNA replication from initiation zones of mammalian cells in a model system. *Mol. Cell. Biol.* **14:** 6489-6496.
Jenkins, H., T. Holman, C. Lyon, B. Lane, R. Stick, and C. Hutchison. 1993. Nuclei that lack a lamina accumulate karyophilic proteins and assemble a nuclear matrix. *J. Cell Sci.* **106:** 275-285.
Karpen, G.H. and A.C. Spradling. 1992. Reduced DNA polytenization of a minichromosome region undergoing position effect variegation in *Drosophila*. *Cell* **63:** 97-107.
Kelly, R.E., M.L. DeRose, B.W. Draper, and G.M. Wahl. 1995. Identification of an origin of bidirectional replication within the coding region of the ubiquitously expressed CAD gene. *Mol. Cell. Biol.* **15:** 4136-4148.
Kill, I.R., J.M. Bridges, K.H.S. Campbell, G. Maldonado-Codina, and C.J. Hutchison. 1991. The timing of the formation and usage of replicase clusters in S-nuclei of human diploid fibroblasts. *J. Cell Sci.* **100:** 869-876.
Kipling, D. and S.E. Kearsey. 1990. Reversion of autonomously replicating sequence mutations in *S. cerevisiae*: Creation of a eucaryotic replication origin within procaryotic vector DNA. *Mol. Cell. Biol.* **10:** 265-272.
Kitsberg, D., S. Selig, I. Keshet, and H. Cedar. 1993. Replication structure of the human β-globin gene domain. *Nature* **366:** 588-590.
Koff, A., J.F. Schwedes, and P. Tegtmeyer. 1991. HSV origin-binding protein (UL9) loops and distorts the viral replication origin. *J. Virol.* **65:** 3284-3292.
Kohara, Y., N. Tohdoh, X.-W. Jiang, and T. Okazaki. 1985. The distribution and properties of RNA primed initiation sites of DNA synthesis at the replication origin of *Escherichia coli* chromosome. *Nucleic Acids Res.* **13:** 6847-6866.
Kornberg, A. and T. Baker. 1992. *DNA replication*. W.H. Freeman, New York.
Kowalski, D. and M.J. Eddy. 1989. The DNA unwinding element: A novel, *cis*-acting component that facilitates opening of the *Escherichia coli* replication origin. *EMBO J.* **8:** 4335-4344.
Krysan, P.J., J.G. Smith, and M.P. Calos. 1993. Autonomous replication in human cells of multimers of specific human and bacterial DNA sequences. *Mol. Cell. Biol.* **13:** 2688-2696.
Kubota, Y. and H. Takisawa. 1993. Determination of initiation of DNA replication before and after nuclear formation in *Xenopus* egg cell free extracts. *J. Cell Biol.* **123:** 1321-1331.
Leonhardt, H, A.W. Page, H.U. Weier, and T.H. Bestor. 1992. A targeting sequence directs DNA methyltransferase to sites of DNA replication in mammalian nuclei. *Cell* **71:** 865-873.
Li, R. and M.R. Botchan. 1993. Acidic transcription activation domains of VP16 and p53

bind cellular replication protein A and stimulate in vitro BPV-1 DNA replication. *Cell* **73:** 1207–1221.
Liang, C. and S.A. Gerbi. 1994. Analysis of an origin of DNA amplification in *Sciara coprophila* by a novel three dimensional gel method. *Mol. Cell. Biol.* **14:** 1520–1530.
Liang, C., J.D. Spitzer, H.S. Smith, and S.A. Gerbi. 1993. Replication initiates at a confined region during DNA amplification in *Sciara* DNA puff II/9A. *Genes Dev.* **7:** 1072–1084.
Lin, S. and D. Kowalski. 1994. DNA helical instability facilitates initiation at the SV40 replication origin. *J. Mol. Biol.* **235:** 496–507.
Little, R.D., T.H.K. Platt, and C.L. Schildkraut. 1993. Initiation and termination of DNA replication in human rRNA genes. *Mol. Cell. Biol.* **13:** 6600–6610.
Mahbubani, H.M., T. Paull, J.K. Elder, and J.J. Blow. 1992. DNA replication initiates at multiple sites on plasmid DNA in *Xenopus* egg extracts. *Nucleic Acids Res.* **22:** 1457–1462.
Majumder, S. and M.L. DePamphilis. 1994. TATA-dependent enhancer stimulation of promoter activity in mice is developmentally acquired. *Mol. Cell. Biol.* **14:** 4258–4268.
Majumder, S., M. Miranda, and M.L. DePamphilis. 1993. Analysis of gene expression in mouse preimplantation embryos demonstrates that the primary role of enhancers is to relieve repression of promoters. *EMBO J.* **12:** 1131–1140.
Marahrens, Y. and B. Stillman. 1994. Replicator dominance in a eukaryotic chromosome. *EMBO J.* **13:** 3395–3400.
Martin, R.G. and V.P. Setlow. 1980. Initiation of SV40 DNA synthesis is not unique to the replication origin. *Cell* **20:** 381–391.
Martínez-Salas, E., D.Y. Cupo, and M.L. DePamphilis. 1988. The need for enhancers is acquired upon formation of a diploid nucleus during early mouse development. *Genes Dev.* **2:** 1115–1126.
Mastrangelo, I.A., P.G. Held, L. Dailey, J.S. Wall, P.V.C. Hough, N. Heintz, and N.H. Heintz. 1993. RIP60 dimers assemble link structures at an origin of bidirectional replication in the DHFR amplicon of CHO cells. *J. Mol. Biol.* **232:** 766–778.
Masukata, H., H. Satoh, C. Obuse, and T. Okazaki. 1993. Autonomous replication of human chromosomal DNA fragments in human cells. *Mol. Biol. Cell* **4:** 1121–1132.
McKnight, S.L. and O.L. Miller, Jr. 1977. Electron microscopic analysis of chromatin replication in the cellular blastoderm *Drosophila melanogaster* embryo. *Cell* **12:** 795–804.
Méchali, M. and S. Kearsey. 1984. Lack of specific sequence requirement for DNA replication in *Xenopus* eggs compared with high sequence specificity in yeast. *Cell* **38:** 55–64.
Melendy, T. and B. Stillman. 1993. An interaction between replication protein A and T-antigen appears essential for primosome assembly during SV40 DNA replication. *J. Biol. Chem.* **268:** 3389–3395.
Miller, C.A. and D. Kowalski. 1993. *cis*-Acting components in the replication origin from ribosomal DNA of *Saccharomyces cerevisiae*. *Mol. Cell. Biol.* **13:** 5360–5369.
Morse, R.H., S.Y. Roth, and R.T. Simpson. 1992. A transcriptionally active tRNA gene interferes with nucleosome positioning in vivo. *Mol. Cell. Biol.* **12:** 4015–4025.
Nakayasu, H. and R. Berezney. 1989. Mapping replication sites in the eukaryotic cell nucleus. *J. Cell Biol.* **108:** 1–11.
Nallaseth, F.S. and M.L. DePamphilis. 1994. Papillomavirus contains *cis*-acting sequences that can suppress but not regulate origins of DNA replication. *J. Virol.* **68:**

3051-3064.

Natale, D.A., A.E. Schubert, and D. Kowalski. 1992. DNA helical stability accounts for mutational defects in a yeast replication origin. *Proc. Natl. Acad. Sci.* **89:** 2654-2658.

Natale, D.A., R.M. Umek, and D. Kowalski. 1993. Ease of DNA unwinding is a conserved property of yeast replication origins. *Nucleic Acids Res.* **21:** 555-560.

Newlon, C., I. Collins, A. Dershowitz, A.M. Deshpande, S.A. Greenfeder, L.Y. Ong, and J.F. Theis. 1993. Analysis of replication origin function on chromosome III of *Saccharomyces cerevisiae*. *Cold Spring Harbor Symp. Quant. Biol.* **58:** 415-423.

Newport, J.W, K.L. Wilson, and W.G. Dunphy. 1990. A lamin-independent pathway for nuclear envelope assembly. *J. Cell Biol.* **111:** 2247-2259.

Noirot, P., J. Bargonetti, and R.P. Novick. 1990. Initiation of rolling circle replication in pT181 plasmid: Initiator protein enhances cruciform extrusion at the origin. *Proc. Natl. Acad. Sci.* **87:** 8560-8564.

Orr-Weaver, T.L. 1991. *Drosophila* chorion genes: Cracking the eggshell's secrets. *BioEssays* **13:** 97-105.

Paranjape, S.M., R.T. Kamakaka, and J.T. Kadonaga. 1994. Role of chromatin structure in the regulation of transcription by RNA polymerase II. *Annu Rev. Biochem.* **63:** 265-297.

Parsons, R. and P. Tegtmeyer. 1992. Spacing is crucial for coordination of domain functions within the SV40 core origin. *J. Virol.* **66:** 1933-1942.

Pearson, C.E., A. Shihab-el-Deen, G.B. Price, and M. Zannis-Hadjopoulos. 1994. Electron microscopic analysis of *in vitro* replication products of ors 8, a mammalian origin enriched sequence. *Somatic Cell Mol. Genet.* **20:** 147-152.

Prives, C., Y. Murakami, F.G. Kern, W. Folk, C. Basilico, and J. Hurwitz. 1987. DNA sequence requirements for replication of polyomavirus DNA in vivo and in vitro. *Mol. Cell. Biol.* **7:** 3694-3704.

Rao, H., Y. Marahrens, and B. Stillman. 1994. Functional conservation of multiple elements in yeast chromosomal replicators. *Mol. Cell. Biol.* **14:** 7643-7651.

Ravnan, J.-B., D.M. Gilbert, G. Kelly, T. Hagen, and S.N. Cohen. 1992. Random-choice replication of extrachromosomal BPV molecules in heterogeneous, clonally derived BPV-infected cell lines. *J. Virol.* **66:** 6946-6952.

Roberts, J.M. and H. Weintraub. 1988. *cis*-Acting negative control of DNA replication in eukaryotic cells. *Cell* **52:** 397-404.

Rochford, R. and L.P. Villarreal. 1991. Polyomavirus DNA replication in the pancreas and in a transformed pancreas cell line has distinct enhancer requirements. *J. Virol.* **65:** 2108-2112.

Rokeach, L.A. and J.W. Zyskind. 1986. RNA termination within the *E. coli* origin of replication: Stringent regulation and control by DnaA protein. *Cell* **46:** 763-771.

Rychlik, W. and R.E. Rhoads. 1989. A computer program for choosing optimal oligonucleotides for filter hybridization, sequencing and in vitro amplification of DNA. *Nucleic Acids Res.* **17:** 8543-8551.

Ryder, K., S. Silver, A.L. DeLucia, E. Fanning, and P. Tegtmeyer. 1986. An altered DNA conformation in origin region I is a determinant for the binding of SV40 large T-antigen. *Cell* **44:** 719-725.

Schlake, T., D. Klehr-Wirth, M. Yoshida, T. Beppu, and J. Bode. 1994. Gene expression within a chromatin domain: The role of core histone hyperacetylation. *Biochemistry* **33:** 4197-4206.

Schneider, C., K. Weisshart, L.A. Guarino, I. Dornreiter, and E. Fanning. 1994. Species specific functional interactions of DNA polymerase-α-primase with SV40 T-antigen require SV40 origin DNA. *Mol. Cell. Biol.* **14:** 3176-3185.

SenGupta, D.J. and J.A. Borowiec. 1994. Strand and face: Topography of interactions between the SV40 origin of replication and T-antigen during initiation of replication. *EMBO J.* **13:** 982-992.

Seufert, W., and W. Messer. 1987. Start sites for bidirectional in vitro DNA replication inside the replication origin, oriC, of *E. coli. EMBO J.* **6:** 2469-2472.

Shinomiya, T. and S. Ina. 1991. Analysis of chromosomal replicons in early embryos of *Drosophila melanogaster* by two-dimensional gel electrophoresis. *Nucleic Acids Res.* **19:** 3935-3941.

———. 1994. Mapping an initiation region of DNA replication at a single-copy chromosomal locus in *Drosophila melanogaster* cells by two-dimensional gel methods and PCR-mediated nascent-strand analysis: Multiple replication origins in a broad zone. *Mol. Cell. Biol.* **14:** 7394-7403.

Simpson, R.T. 1990. Nucleosome positioning can affect the function of a *cis*-acting DNA element in vivo. *Nature* **343:** 387-389.

Sock, E., M. Wegner, E.A. Fortunato, and F. Grummt. 1993. Large T-antigen and sequences within the regulatory region of JC virus both contribute to the features of JC virus DNA replication. *Virology* **197:** 537-548.

Stick, R. and P. Hausen. 1985. Changes in the nuclear lamina composition during early developmnet of *Xenopus laevis. Cell* **42:** 191-200.

Stolzenburg, F., R. Gerwig., E. Dinkl, and F. Grummt. 1994. Structural homologies and functional similarities between mammalian origins of replication and amplification promoting sequences. *Chromosoma* **103:** 209-214.

Strausfeld, U.P., M. Howell, R. Rempel, J.L. Maller, T. Hunt, and J.J. Blow. 1994. Cip1 blocks the initiation of DNA replication in *Xenopus* extracts by inhibition of cyclin-dependent kinases. *Curr. Biol.* **4:** 876-883.

Tack, L.C. and G.N. Proctor. 1987. Two major replicating SV40 chromosome classes: Synchronous replication fork movement is associated with bound large T antigen during elongation. *J. Biol. Chem.* **262:** 6339-6349.

Taira, T., S.M.M. Iguchi-Ariga, and H. Ariga. 1994. A novel DNA replication origin identified in the human heat shock protein 70 gene promoter. *Mol. Cell. Biol.* **14:** 6386-6397.

Tasheva, E.S. and D.J. Roufa. 1994a. A mammalian origin of bidirectional DNA replication within the Chinese hamster RPS14 locus. *Mol. Cell. Biol.* **14:** 5628-5635.

———. 1994b. Densely methylated DNA islands in mammalian chromosomal replication origins. *Mol. Cell. Biol.* **14:** 5636-5644.

Theis, J.F. and C.S. Newlon. 1994. Domain B of ARS307 contains two functional elements and contributes to chromosomal replication origin function. *Mol. Cell. Biol.* **14:** 7652-7659.

Tlsty, T.D. 1990. Normal diploid human and rodent cells lack a detectable frequency of gene amplification. *Proc. Natl. Acad. Sci.* **87:** 3132-3136.

Umek, R.M. and D. Kowalski. 1988. The ease of DNA unwinding as a determinant of initiation at yeast replication origins. *Cell* **52:** 559-567.

———. 1990a. The DNA unwinding element in a yeast replication origin functions independently of easily unwound sequences present elsewhere on a plasmid. *Nucleic Acids Res.* **18:** 6601-6605.

———. 1990b. Thermal energy suppresses mutational defects in DNA unwinding at a

yeast replication origin. *Proc. Natl. Acad. Sci.* **87:** 2486-2490.
Vassilev, L.T. and M.L. DePamphilis. 1992. Guide to identification of origins of DNA replication in eukaryotic cell chromosomes. *Crit. Rev. Biochem. Mol. Biol.* **27:** 445-472.
Vassilev, L.T. and E.M. Johnson. 1990. An initiation zone of chromosomal DNA replication located upstream of the c-*myc* gene in proliferating HeLa cells. *Mol. Cell Biol.* **10:** 4899-4904.
Vassilev, L.T., W.C. Burhans, and M.L. DePamphilis. 1990. Mapping an origin of DNA replication at a single copy locus in exponentially proliferating mammalian cells. *Mol. Cell. Biol.* **10:** 4685-4689.
Virta-Pearlman, V.J., P.H. Gunaratne, and A.C. Chinault. 1993. Analysis of a replication initiation sequence from the adenosine deaminase region of the mouse genome. *Mol. Cell. Biol.* **13:** 5931-5942.
Ward, G.K., A. Shihab-el-Deen, and M. Zannis-Hadjopoulos. 1991. DNA cruciforms and the nuclear supporting stucture. *Exp. Cell. Res.* **195:** 92-98.
Williams, J.S., T.T. Eckdahl, and J.N. Anderson. 1988. Bent DNA as a replication enhancer in *S. cerevisiae. Mol. Cell. Biol.* **8:** 2763-2769.
Wohlgemuth, J.G., G.H. Bulboaca, M. Moghadam, M.S. Caddle, and M.P. Calos. 1994. Physical mapping of origins of replication in the fission yeast *Schizosaccharomyces pombe. Mol. Biol. Cell.* **5:** 839-849.
Wolffe, A.P. 1994. The role of transcription factors, chromatin structure and DNA replication in 5S RNA gene regulation. *J. Cell Sci.* **107:** 2055-2063.
Wu, C., M. Zannis-Hadjopoulos, and G.B. Price. 1993. In vivo activity for initiation of DNA replication resides in a transcribed region of the human genome. *Biochim. Biophys. Acta* **1174:** 258-268.
Wu, J.-R. and D.M. Gilbert. 1996. A distinct G1-phase step required to specify a mammalian replication origin. *Science* (in press).
Yates, J.L. and N. Guan. 1991. Epstein-Barr virus-derived plasmids replicate only once per cell cycle and are not amplified after entry into cells. *J. Virol.* **65:** 483-488.
Yoda, K.-Y., H. Yasuda, X.-W. Jiang, and T. Okazaki. 1988. *Nucleic Acids Res.* **16:** 6531-6546.
Yoon, Y., J.A. Sanchez, C. Brun, and J.A. Huberman. 1995. Mapping of replication initiation sites in human ribosomal DNA by nascent strand abundance analysis. *Mol. Cell. Biol.* **15:** 2482-2489.
Zahn, K. and F.R. Blattner. 1987. Direct evidence for DNA bending at the λ-replication origin. *Science* **236:** 416-422.
Zannis-Hadjopoulos, M., L. Frappier, M. Khoury, and G.B.Price. 1988. Effect of anticruciform DNA monoclonal antibodies on DNA replication. *EMBO J.* **7:** 1837-1844.

3
Roles of Transcription Factors in DNA Replication

Peter C. van der Vliet
Laboratory for Physiological Chemistry
Utrecht University
3508 TA Utrecht
The Netherlands

Sequence-specific recognition of DNA by proteins is essential for both transcription and DNA replication. Transcription factors form a growing family of regulatory proteins that can positively or negatively influence transcription by binding to regulatory elements in DNA contacting components of the basal transcription machinery. In recent years, it has become clear that several transcription factors are multifunctional and also directly influence initiation of DNA replication. This was first detected during studies of the replication of eukaryotic viruses, such as adenovirus and papovaviruses, but it may well be a more general phenomenon. Many viral origins consist of a core origin and auxiliary regions that contribute to the initiation of replication and are required for optimal viral growth (for review, see DePamphilis 1988, 1993a; Challberg and Kelly 1989; Stillman 1989). These auxiliary regions contain transcription factor-binding sites. In polyomavirus, presence of the transcriptional enhancer stimulated DNA replication up to 1000-fold, and presence of the enhancer increased SV40 replication approximately 100-fold. The auxiliary region of adenovirus stimulates initiation of replication up to 200-fold, although this region is not directly involved in transcription. The transcription factors used for enhancement of replication can be cellular as well as viral. Cellular transcription factors are involved in adenovirus, SV40, and polyomavirus replication. For papillomavirus and Epstein-Barr virus, the viral proteins (BPV-E2, EBNA1) combine functions in transcription and replication, as is the case for T antigen of SV40 and polyomavirus.

The availability of initiation systems that can be reconstituted with purified proteins for adenovirus and papovaviruses, combined with efficient transfection assays with origin-containing plasmids, has facilitated studies of the mechanism by which transcription factors enhance initiation. A number of distinct modes of action have been discovered. As a general rule, transcription factors enhance replication by facilitating rate-limiting steps in the initiation process. As summarized in Table 1 and Figure 1, these steps include:

1. Recruiting initiation proteins such as adenovirus DNA polymerase, the single-strand binding protein RP-A, or BPV-E1 helicase to the origin. By binding these proteins and targeting them to the origin employing their DNA-binding domains, transcription factors can facilitate the assembly of a multiprotein preinitiation complex and also stabilize such a complex. This method of action requires, understandably, a strict positioning of the transcription factors relative to the core origin.
2. Changing the activity of initiation proteins. Several examples exist in which the helicase activities of SV40 T antigen or BPV-E1 are influenced by transcription factors.
3. Induction of structural changes in origin DNA. This could lead to facilitated formation of a preinitiation complex by bringing initiation proteins closer together or by winding DNA on the surface of a multiprotein complex. Many transcription factors such as ABF-1, NFI, and Oct-1 can bend DNA. Bending could also be instrumental in origin opening.
4. Global changes in the chromatin structure surrounding the origin. Opening up the nucleosome structure could prevent the inhibitory effect of histones and nucleosomes on initiation (antirepressor effect) and could lead to a destabilization of the origin region, which could facilitate binding of initiation proteins. For this mode of action, the position of the transcription factor-binding site relative to the core origin is less fixed.

Of course, these various mechanisms are not mutually exclusive and may operate together to optimize the initiation process. The contribution of each individual mechanism depends on the system and the conditions. No evidence has been obtained so far that transcription factors also are involved in elongation or even origin clearance. On the contrary, several transcription factors seem to contact their targets only during a limited time span, being released soon after initiation or even before the actual initiation reaction has taken place.

Whether transcription factors employ the same mechanisms to enhance both processes, DNA replication and transcription, is not yet clear.

Table 1 How can transcription factors enhance replication?

	System	Transcription factor	Target protein
Recruitment of replication proteins (leading to assembly and stabilization of a multimeric initiation complex)	Ad	NFI Oct-1	pTP-polymerase pTP-polymerase
	polyoma	T-ag AP1	T-ag, pol-α, RP-A RP-A
	SV40	T-ag SP1	T-ag, pol-α, RP-A ?
	BPV	E2	E1
	EBV oriP oriLyt	EBNA1 BLTF-1	EBNA1 ?
Changing the activity of an initiator protein	SV40 BPV	T-ag E2	T-ag helicase E2 helicase
Changing the global architecture or nucleosome structure of the origin, e.g., by bending or looping	Ad BPV *S. cerevisiae* EBV	Oct-1 E2 ABF-1 EBNA1	

Only naturally bound transcription factors are mentioned. For substitution of binding sites by synthetic ones, see Table 2. (T-ag) Large T antigen.

Global changes of chromatin structure could be instrumental in both processes. Such changes require, in general, the transcription activation domains of the transcription factors. On the contrary, interactions with initiation proteins are specific and use protein interfaces that are different from the ones involved in contacting the transcription machinery.

In most cases, the role of transcription factors in DNA replication is independent of the transcription process itself. Bound transcription factors do not appear to activate DNA replication by increasing transcription from, or through, the origin of DNA replication. An exception is mitochondrial DNA replication (Clayton 1991). Similar to prokaryotic systems (Baker and Kornberg 1992), the role of transcription factors is to regulate the synthesis of transcripts that act as a primer or a regulator. This is beyond the scope of this chapter; mitochondrial DNA replication is discussed in detail in Clayton (this volume). Below, I focus on recent results obtained in several viral systems as well as in yeast. Several reviews on the role of transcription factors in DNA replication have appeared previously (DePamphilis 1988, 1993b; Van der Vliet 1991; Heintz 1992).

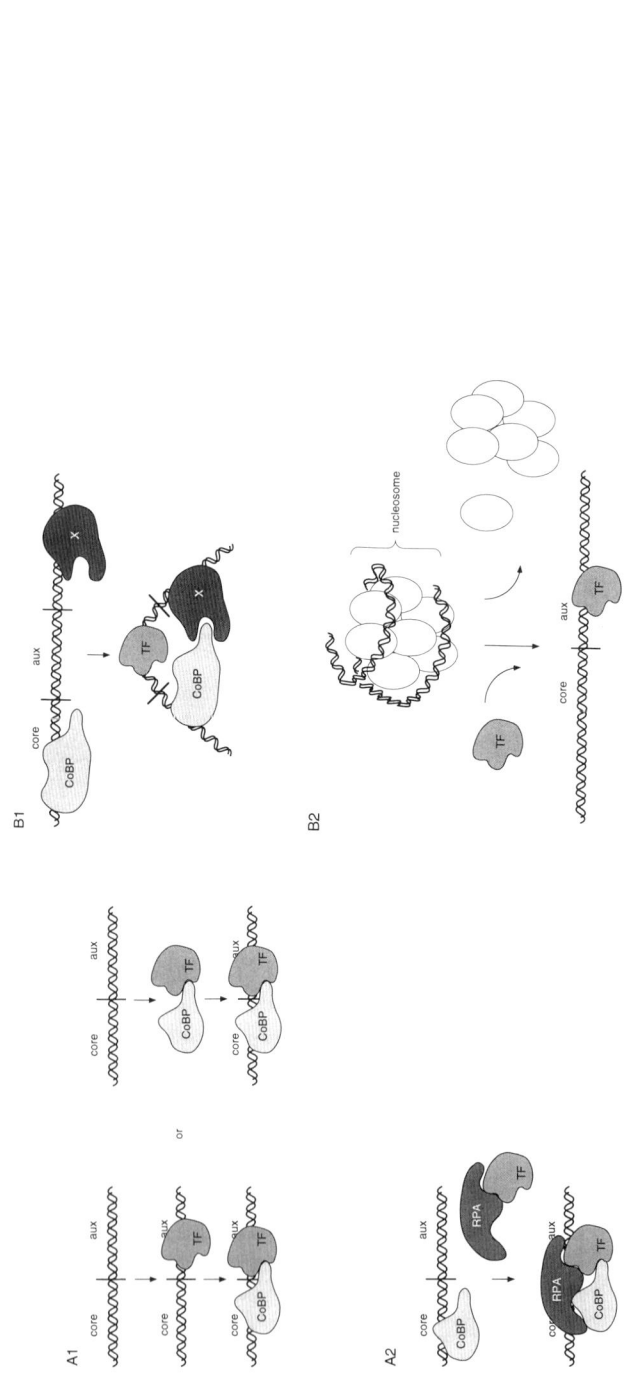

Figure 1 Ways in which transcription factors can stimulate initiation of DNA replication. (*A*) Direct contacts: (1) A core origin binding protein (CoBP) is recruited by a transcription factor (TF) binding to the auxiliary origin region (aux). Formation of a CoBP-TF complex may occur in solution or on the DNA. In this way a transcription factor stabilizes binding of CoBP to the core origin and may also influence the activity of CoBP. (2) A transcription factor may also recruit auxiliary initiation proteins, e.g., RP-A, that could help unwind the origin. (*B*) Structural changes in DNA or chromatin. (1) Bending of DNA by a transcription factor could bring two essential components, CoBP and X, another protein of the initiation complex, in close proximity. (2) Transcription factors disrupt the nucleosome structure, thereby enabling the initiation proteins to bind to the core origin (antirepressor effect).

VIRAL SYSTEMS

Adenovirus

The 36-kbp linear genome of human adenoviruses replicates via protein-primed initiation followed by elongation via a displacement mechanism (see Hay, this volume). The development of an in vitro initiation system has led to the purification of novel viral and cellular proteins that could be used to reconstitute efficient replication in vitro. Genetic and biochemical analysis indicates that three viral proteins are required for replication. These are the precursor terminal protein (pTP) and the viral DNA polymerase (pol), forming a pTP-pol heterodimer, and the DNA-binding protein (DBP).

The viral proteins can only sustain a very limited level of initiation, and two nuclear transcription factors stimulate initiation considerably. These are nuclear factor 1 (NFI, Nagata et al. 1982; Leegwater et al. 1985; Rosenfeld and Kelly 1986) and nuclear factor III (Pruijn et al. 1986; O'Neill and Kelly 1988). Nuclear factor III is identical to the octamer-binding transcription factor Oct-1 (O'Neill et al. 1988; Pruijn et al. 1989). NFI and Oct-1 function by binding to the origins of replication located at each molecular end. These consist, for Ad2 and Ad5, of a conserved core region, located at nucleotides 9–18 from the termini, two less well conserved space regions (1–9 and 18–24), and an auxiliary region (25–55). The core region binds pTP-pol, and the auxiliary region represents the binding site for NFI and Oct-1. The various domains within the origin are closely spaced, and enhancement of replication by NFI and Oct-1 depends critically on the spatial arrangement of their binding sites (Adhya et al. 1986; Wides et al. 1987; Bosher et al. 1990; Coenjaerts et al. 1991), suggesting essential protein-protein interactions.

How Does NFI Stimulate Initiation?

Human NFI is the collective name of a family of transcription factors (52–66 kD) generated by differential splicing from a single gene. They contain a highly conserved DNA-binding and dimerization domain (NFI-BD) located within the 220–240 amino-terminal amino acids (Paonessa et al. 1988; Santoro et al. 1988; Meisterernst et al. 1989). This region suffices for stimulation of replication (Mermod et al. 1989; Gounari et al. 1990). The NFI-BD dimer binds to positions 25–39 by contacting two successive regions in the major groove of the DNA, with a consensus TGGCA-N5-GCCAA (Nagata et al. 1983; Gronostajski 1986; De Vries et al. 1987). Interestingly, binding of NFI or NFI-BD to the origin is enhanced by the viral DBP (Cleat and Hay 1989; Stuiver and Van der Vliet 1990; Bosher et al. 1991), presumably through structural changes that DBP induces in origin DNA (Stuiver et al. 1992).

The level of stimulation by NFI is strongly dependent on the pTP-pol concentration and drops from 60-fold at low concentrations to approximately 2-fold at high concentrations of pTP-pol (Mul et al. 1990). This suggests that NFI recruits pTP-pol to the origin. In agreement with this, a direct DNA-independent interaction between NFI and pTP-pol has been reported (Bosher et al. 1990; Chen et al. 1990; Mul et al. 1990). The domain interacting with polymerase is located in NFI-BD between amino acids 68 and 150 (Chen et al. 1990) and is genetically distinguishable from the DNA-binding domain. Mutations in NFI that fail to bind pTP-pol are also defective for replication stimulation (Armentero et al. 1994). NFI not only recruits pTP-pol, but also stabilizes its interaction with the origin, as shown by template competition experiments and gel retardation (Mul and Van der Vliet 1992). Kinetic considerations are in accord with an increase in the amount of active preinitiation complex rather than an effect on the activity of pTP-pol, since NFI increases the V_{max} without changing the K_m (Mul and Van der Vliet 1993). The stabilizing effect of NFI, as measured by the half-life of a pTP-pol-DNA complex, was only approximately 10-fold compared to a 60-fold enhancement of initiation. This may indicate that NFI acts in multiple ways, possibly involving also structural changes within the origin (Zorbas et al. 1989). A role for NFI in origin unwinding is less likely, however, since NFI is still required when the template strand in the core origin is already unwound (Kenny and Hurwitz 1988).

Since NFI does not bind efficiently to single-stranded DNA, it can be anticipated that unwinding of the NFI site by the passing DNA polymerase leads to dissociation of NFI from DNA. Unexpectedly, dissociation of NFI occurs very early in initiation as soon as the DNA polymerase encounters dNTPs (Coenjaerts and Van der Vliet 1994). This result may be explained by conformational changes in the DNA polymerase and in origin DNA during initiation that could lead to disruption of both the NFI-pTP-pol interaction and the interaction between NFI and DNA. Apparently NFI delivers pTP-pol to the template but is released and recycled after initiation. This is reminiscent of similar effects observed in the BPV E1-E2 complex, where E2 dissociates as soon as E1 starts its helicase action (Lusky et al. 1994) and to the recruitment of DnaB helicase by DnaC or by λP protein (Baker and Kornberg 1992). In the latter case, λP protein must be removed by heat shock proteins before replication can start.

How Does Oct-1 Stimulate Initiation?

Oct-1 is a 91-kD ubiquitous transcription factor belonging to the POU-protein family (Herr et al. 1988; Verrijzer and Van der Vliet 1993). It binds to the sequence ^{40}ATGATAATGA^{49} in the Ad2 origin and stimulates initiation approximately 7-fold, independently of NFI (Mul et al. 1990; Coenjaerts et al. 1994). Like the NFI site (Hay 1985), the Oct-1 site is important for optimal replication in vivo. If other compensating DNA sequences are present, removal of the Oct-1 site is without consequence (Hatfield and Hearing 1993).

Deletion analysis indicated that the DNA-binding POU domain suffices for stimulation of replication (Verrijzer et al. 1990). This domain is highly conserved, and POU domain transcription factors from different subclasses all enhance replication, albeit to slightly different extents (Verrijzer et al. 1992).

In several aspects, similarities between enhancement of replication by Oct-1 and NFI exist. Like NFI, Oct-1 enhances the V_{max} of the initiation reaction and still stimulates initiation when the origin is partially unwound. Stimulation by Oct-1 is also dependent on the pTP-pol concentration, although less sharply than observed for NFI (Mul and Van der Vliet 1992, 1993; Coenjaerts et al. 1994). In contrast to NFI, direct contacts between pTP-pol and Oct-1 or the POU domain were not observed in solution, but recently an interaction could be found employing a GST-POU fusion protein immobilized on glutathione-agarose beads (Coenjaerts et al. 1994). Previous attempts to detect this interaction in solution might have been hampered by a low affinity. Binding was also observed with the POU$_{HD}$, although more weakly than with the intact POU domain, indicating that additional contacts with POUs cannot be excluded. Interestingly, the POU-pTP-pol interaction could not be competed away by NFI, suggesting that the regions contacted on pTP-pol are different for NFI and Oct-1 (Coenjaerts et al. 1994). This also explains the independent stimulatory effects observed with these two transcription factors.

DNA bending by the POU domain (Verrijzer et al. 1991; Johansen et al. 1993) may contribute to the stimulation, but the way in which this occurs is not well understood. Bending may facilitate the interaction between the various proteins and thus have an architectural role in the formation and assembly of a multiprotein initiation complex. It is not likely that bending per se is sufficient for enhancement, since substitution of the Oct-1-binding site by an AP1 site did not lead to stimulation in the presence of Fos-Jun combinations that are able to bend the DNA to a similar extent as Oct-1 (Coenjaerts et al. 1994). This stresses the importance of specific protein-protein interactions between Oct-1 and pTP-pol and explains the strict positioning of its binding site.

In conclusion, NFI and Oct-1 stimulate initiation mainly by a direct interaction with different, so far undefined, domains in pTP-pol. Small distortions of the origin structure induced by these transcription factors may further facilitate the assembly of a preinitiation complex. POU_{HD} and the NFI dimer bind to the same side of the DNA helix and, together with pTP-pol, the preinitiation complex spans five helical turns requiring considerable bending of the DNA to allow the necessary surface contacts. In addition to NFI and Oct-1, the terminal protein and DBP are required for optimal initiation. The presence of the terminal protein in the viral DNA leads to a modest increase in binding of pTP-pol to the origin (Pronk and Van der Vliet 1993), whereas bound DBP enhances binding of NFI. Together, these five proteins optimize formation and stability of a preinitiation complex. NFI and Oct-1 seem to function only in assembly of the preinitiation complex. Evidence for a function of NFI or Oct-1 in elongation or any other event during initiation is lacking. Unwinding of the origin or changing the kinetics of initiation appears to be mainly a function of the viral DBP (Mul and Van der Vliet 1993; Zijderveld and Van der Vliet 1994; Monaghan et al. 1994). A nucleosome antirepression effect, as suggested for papovaviruses, is unlikely since optimal stimulation is detected in the absence of nucleosomes, and adenovirus DNA is not constrained in a regular nucleosome structure in infected cells. Apparently, the viral origin and pTP-pol complex have evolved to allow interactions with conserved DNA-binding domains of transcription factors. Why these particular transcription factors are employed is not clear, and the adaptation process varies between serotypes, since Ad4 does not have an NFI-binding site. Presumably, the Ad4 pTP-pol complex has a stronger interaction with the core origin, thereby bypassing the need for NFI recruitment (Harris and Hay 1988). This is reminiscent of the T antigens of polyomavirus and SV40, which also differ in their affinity for the core origin, leading to an increased need for transcription factors in polyomavirus, which encodes a T antigen that binds more weakly to the origin than SV40 T antigen.

SV40

The 5243-bp circular SV40 DNA replicates bidirectionally from a fixed origin composed of a 64-bp core and adjacent auxiliary regions, Aux-1 and Aux-2, which each span approximately 40 bp. The core origin contains a 24-bp central region that forms the binding site for the SV40 T-antigen double hexamer flanked on one side by a 17-bp AT-rich element and on the other by a 10-bp DNA unwinding element in which the origin of bidirectional replication is located (see also Hassell and Brinton, this volume).

Although the core origin suffices for initiation, the flanking auxiliary sequences function synergistically to increase the frequency of initiation up to 100-fold, depending on the conditions and methods used to measure DNA replication (Guo et al. 1989; Gutierrez et al. 1990). Aux-1 contains binding sites for T antigen, and precise positioning of Aux-1 relative to the core origin is required for its enhancing effect. Deletion of Aux-1 had only a very slight (1.6-fold) effect on binding of T antigen to the core origin or on the bidirectionality of initiation, even under conditions of limiting T antigen. In contrast, a strong effect on T-antigen-dependent DNA unwinding was observed (Guo et al. 1991). This suggests that any interaction between T-antigen molecules bound at Aux-1 and at the core origin influences the helicase activity rather than increasing the binding of T antigen. The level of stimulation by Aux-1 depends on the type of papovavirus. JC virus relies on Aux-1 to the greatest extent, then SV40, and BK virus the least (Sock et al. 1993). This may be related to small differences in the interaction between the various T antigens, leading to different strengths of the intramolecular interactions.

Aux-2 also contains binding sites for T antigen, albeit weak ones, and is recognized by several cellular transcription factors, including SP1, LSF, AP1, AP2, AP4, GT-1B, and p53 (Table 2). Due to the complexity of this region, it has been difficult to establish, by mutational analysis, which transcription factor-binding sites are essential for stimulation of initiation. Therefore, another approach was taken. Aux-2 was replaced by multimers of one site or by nonnatural binding sites for various transcription factors, and their effects were studied by transfection (see Table 2). Substitution of Aux-2 by two synthetic SP1-binding sites restored replication to wild-type levels (Guo and DePamphilis 1992; Lednicky and Folk 1992), whereas one copy was less effective. Such a copy-number-dependent stimulation was also found upon substitution with NFI or ATF sites (Hoang et al. 1992) and is reminiscent of the results obtained with transcriptional enhancers.

In a more extensive study using a number of different transcription factors, AP1 and, to a lesser extent, T antigen, were shown to be active. However, other DNA-binding proteins like Gal4 could not substitute, even when fused to the strong *trans*-activation domains of VP16 or c-Jun (Guo and DePamphilis 1992). The same was true for binding of *Escherichia coli lac* repressor adjacent to the origin (Cheng et al. 1992). This shows that binding of proteins to Aux-2 is insufficient and indicates

Table 2 Natural auxiliary regions can sometimes be substituted by synthetic auxiliary regions

System	Natural sites	Substituting sites and proteins	Activity in vivo	Activity in vitro	References
Adenovirus	Oct-1	AP1		–	1
SV40 Aux-2	SP1, T-ag, LSF,	(SP1)3	+	–	2,3
	p53, AP1, AP2,	(AP1)3	+	–	2
	AP4, GI-IB	(Tag)3	+	+	2
		VP16	–	+	2,4,5
		c-Jun	–		2
		Gal4	–		2
		NFI	+		4,6
Polyoma Aux-2	AP1 (PEA1),	(SP)3	+		2
	AP3 (EBP-20),	(AP1)3	+		2
	T-ag, cETS	(Tag)3	+		2
	(PEA3), PEA2	VP16	+		2,8
		c-Jun	+		2,8,10,11
		v-Jun	+		8,10,11
		Gal4	+		2,7,8
		CREB	–		10
		Rel	+		12
		E1A	+		8
		Fos	+		8
S. cerevisiae	ABF-1	RAP-1	+		13
ARS-1 B3		Gal4	+		13

Transcription activation domains were bound to different DNA-binding domains, with comparable results. The positions of the binding sites vary. The levels of activity differ considerably, but this is not indicated in this table. For details, see the references. Since the conditions for testing are different, a direct comparison of the results is not always possible. PEA3, PEA1, and EBP-20 are the murine homologs of human c-ETS, AP1, and AP3, respectively. References: (1) Coenjaerts et al. 1994; (2) Guo and DePamphilis 1992; (3) Lednicky and Folk 1992; (4) Hoang et al. 1992; (5) Cheng et al. 1992; (6) Cheng and Kelly 1989; (7) Bennett Cook and Hassell 1991; (8) Baru et al. 1991; (9) Morgan et al. 1993; (10) Murakami et al. 1991; (11) Wasylyk et al. 1990; (12) Ishikawa et al. 1993; (13) Marahrens and Stillman 1992.

specificity, presumably related to the possibility to interact with other replication proteins such as T antigen or RP-A.

Binding of the active transcription factors to T antigen has not been reported, but a direct interaction was observed between VP16 and the large subunit of RP-A, RPA-1 (He et al. 1993; Li and Botchan 1993). This has led to the interesting hypothesis that recruitment of RP-A could stabilize single-stranded DNA in the origin or assist in RP-A-dependent unwinding by the T-antigen helicase. RP-A can also specifically interact with DNA polymerase-α and SV40 T antigen (Dornreiter et al. 1992; Melendy and Stillman 1993), and these protein-protein interactions may facilitate primosome assembly. Therefore, RP-A may function to influence two rate-limiting steps, the formation of an initiation complex and the subsequent helicase activity. For recruitment of RP-A to be important, the intracellular or local concentrations of RP-A should be limiting. RP-A is a rather abundant protein involved in many different processes; within the nucleus, competition for the protein may be strong. Thus, it is of advantage for the virus to recruit as much RP-A as possible for its own replication. Nevertheless, some caution in the interpretation is needed. The VP16 *trans*-activation domain is notoriously promiscuous in its interactions and, in addition to RP-A, has been reported to bind to many other proteins, including TBP, TFIIB, TFIIH, TAF40, and Oct-1. Moreover, Gal4-VP16 does not stimulate initiation in vivo (Guo and DePamphilis 1992; Hoang et al. 1992) despite its interaction with RP-A. This may be caused by sufficiently high concentrations of RP-A present in the cells used, or it may indicate that in vivo this mechanism is not important.

Another explanation of the effect of transcription factors on Aux-2 is based on replacement of Aux-2 by NFI-binding sites. In vivo, strong stimulation was observed, but the effect of adding NFI in a standard SV40 cell-free system was negligible. However, NFI specifically prevented the repression of DNA replication occurring when the template was preassembled into chromatin, suggesting a role as antirepressor (Cheng and Kelly 1989). A similar antirepressing effect was observed for Gal4-VP16 (Cheng et al. 1992). This agrees with results obtained with the naturally bound SP1, which can counteract histone H1-mediated inhibition of RNA pol II transcription (Croston et al. 1991; Laybourn and Kadonaga 1991). Several observations, however, argue against a model in which perturbation of the local nucleosome structure is the primary mechanism by which transcription factors stimulate SV40 replication. As mentioned above, Gal4-VP16 does not stimulate in vivo. Furthermore, a number of transcription factors, including T antigen itself, can inhibit chromatin assembly and prevent nucleosomes from repressing initiation (Ishimi 1992; Gruss et al. 1993). Finally, it has been reported that enhancement by SP1 is maintained when nucleosomes have been stripped (Lednicky and Folk 1992). Thus, although antirepression can be observed, it is not necessarily the primary reason for enhancement but, rather, could be a consequence of transcription factor binding. This explanation has also been put forward for similar effects observed upon binding of the BPV E1-E2 complex (see below).

Although binding of transcription factors to Aux-2 is the most likely way in which these proteins function, it may not be the only way. Recently, Oct-1 was shown to bind to the AT-rich element of the core origin, and binding inhibits the T-antigen-dependent helicase activity (Kilwinski et al. 1995). This may provide another level of control by transcription factors.

Polyomavirus

The polyomavirus (Py) genome is closely related to SV40 but replicates only in murine cells. The origin consists also of a core region flanked by two auxiliary regions. Like SV40, Py encodes a large T antigen essential for replication that binds to the core origin as a hexamer, has DNA-dependent ATPase and helicase activities, and binds DNA polymerase-α:DNA primase. However, Py T antigen binds more weakly to DNA and thus may rely more heavily on auxiliary factors (see Hassell and Brinton, this volume).

Binding of T antigen to Aux-1 activates replication 5- to 10-fold and presumably serves the same role as in SV40. The 200-bp Aux-2 region, coinciding with the enhancer element (de Villiers et al. 1984), stimulates 200- to 1000-fold, which is more than in SV40. Within Aux-2, two elements have been recognized (α and β) that are functionally redundant. Proteins binding to these sites are known as polyoma enhancer binding proteins (PEBP) or polyoma enhancer activators (PEA) and are homologs of the AP1 and ETS families of transcription factors, respectively. Several experiments indicate that binding of these transcription factors is important for replication and that overexpression of AP1 leads to stimulation of DNA replication (Murakami et al. 1991). Fusion proteins of various DNA-binding domains and the *trans*-activation domains of Fos-Jun were active in a DNA-binding-dependent manner, indicating the need for *trans*-activation domains (Wasylyk et al. 1990; Morgan et al. 1993). Interestingly, stimulation of replication was under control of growth-promoting agents such as the phorbol ester TPA. This suggests that protein kinase C-dependent signal transduction pathways are coupled directly to DNA replication, possibly through phosphorylation of the *trans*-activation domain. This effect seemed to be specific for AP1, since overexpression of CREB, also recognizing the AP1 site, stimulated only transcription but not replication (Murakami et al. 1991). So far, specific interactions between AP1 and other replication proteins have not been reported.

To uncouple effects on transcription and replication and to determine if other transcription factors could substitute for AP1 or ETS, a similar approach as for SV40 was employed. Natural Aux-2 sequences were replaced by synthetic binding sites capable of binding just one of the transcription factors, either individually or in tandem. Also tested were sites for well-characterized transcription factors such as AP1 and NF-κB (Guo and DePamphilis 1992; Ishikawa et al. 1993), and factors without a mammalian counterpart, such as Gal4 or fusion proteins between the Gal4 DNA-binding domain and the *trans*-activation domains of VP16, E1A, c-Jun, or BPV-E2 (Table 2) (Wasylyk et al. 1990; Baru et al. 1991; Murakami et al. 1991; Guo and DePamphilis 1992; Morgan et al. 1991). These binding sites, in the presence of appropriate proteins, could all substitute for Aux-2, but to a different extent and dependent on the positions of the sites.

In contrast to SV40, stimulation by VP16 was observed, although natural AP1 stimulates better than Gal4-VP16, Gal4, and c-Jun. Except for Rel oncoproteins (Ishikawa et al. 1993), *trans*-activation domains were essential. Synergistic activation was achieved by multimerization of the Gal4-binding sites (Baru et al. 1991; Bennett Cook and Hassell 1991), which should be juxtaposed to the core origin for optimal functioning. Gal4-VP16 stimulated replication of polyomavirus but not SV40 DNA in vivo, which may be related to the weak binding of Py T antigen and correspondingly higher requirement for auxiliary factors. Alternatively, there may be different levels of coactivators in the cells used for transfection.

The models put forward to explain the mechanisms by which these *trans*-activation domains work are the same as described above for SV40. A chromatin interference model is not excluded but is difficult to reconcile with the finding that one cluster of five identical Gal4-binding sites stimulates replication, whereas two such clusters are inhibitory (Baru et al. 1991). A strong argument for the importance of an interaction with RP-A comes from mutagenesis studies. Mutations in VP16 that reduce the ability of Gal4-VP16 to stimulate replication are also defective in RP-A binding (He et al. 1993). An extensive analysis (B.T. Brinton et al., in prep.; Hassell and Brinton, this volume) revealed that mutation of 6 of the 46 amino acids that make up the amino-terminal domain of VP16 coordinately reduced DNA replication, as well as binding to RPA-1. This provides strong evidence in favor of RP-A recruitment as the main mechanism for stimulation of DNA replication. Whether RP-A is a limiting factor is not known, but even if RP-A is present in excess, binding of VP16 could be essential to activate RP-A or to induce structural changes.

Papillomaviruses

The papillomavirus genome is slightly larger than that of papovaviruses and encodes two, rather than one, proteins required for replication (E1 and E2). E2 is a viral transcription factor. Much of the information on viral DNA replication has been derived from bovine papillomavirus type-1 (BPV-1) and has been extended recently for several human serotypes (HPV) (see Stenlund, this volume).

BPV-1 contains an origin of bidirectional replication spanning approximately 60 bp. A core origin consisting of an AT-rich region and an imperfect inverted repeat is flanked by two auxiliary regions (BS11 and BS12) that stimulate replication considerably (Ustav et al. 1991). The 68-kD E1 phosphoprotein (Sun et al. 1990) binds specifically to the core origin (Ustav et al. 1991). It possesses DNA-dependent ATPase and helicase activities (Seo et al. 1993a; Thorner et al. 1993; Yang et al. 1993) and distorts the origin structure severely upon binding in an ATP-dependent fashion, as shown by $KMnO_4$ sensitivity assays (Gillette et al. 1994). Interestingly, E2 can also bind the p180 subunit of DNA polymerase-α:DNA primase (Park et al. 1994). Thus, in many respects E1 resembles T antigen and is the essential initiator protein, although only limited sequence homology with T antigen exists. Under in vitro conditions, E1 is sufficient for initiation if added in high amounts to the reaction (Li and Botchan 1993). In vivo the minimal origin includes an E2-binding site (Ustav et al. 1993).

E2 binds as a dimer to a 12-bp palindromic sequence, ACC N6 GGT, present in BS11 and BS12, and binding leads to severe bending of the DNA. The E2 open reading frame can encode three proteins by differential splicing: E2, E2-TR, and E8/E2. All three contain the 85-amino-acid carboxy-terminal DNA-binding domain, but only the full-length 48-kD E2 enhances replication and transcription. The other two proteins function as repressors of transcription and possibly also DNA replication. The regions of E2 required for enhancement of replication and transcription are not identical (Winokur and McBride 1992).

Enhancement of initiation by E2 is directly related to its capacity to bind the E1 helicase. The interaction between E1 and E2 is cooperative and leads to an increase in the affinity of E1 for its binding site (Mohr et al. 1990; Blitz and Laimins 1991; Lusky et al. 1993; Seo et al. 1993b). E2 can even rescue an E1 mutant defective in DNA binding (Spalholz et al. 1993; Thorner et al. 1993). Moreover, the amount of E1 required to generate structural changes in the origin is lowered by the presence of E2 (Gillette et al. 1994). E2 also enhances origin-dependent unwinding by E1 (Seo et al. 1993b). Thus, E2 stimulates replication by enhancing several properties of E1; in agreement with this, the highest level of stimulation by E2 in vitro occurs at low E1 concentrations (Müller et al. 1994). Interestingly, an intact DNA-binding domain is not a strict requirement for E2 to enhance replication (Winokur and McBride 1992). This has led to the hypothesis that the cooperative nature of the interaction, accompanied by structural changes in E1 upon complex formation with E2, suffices for enhancement of E1 binding to DNA. This targeting role of E2 may be short-lived, as indicated by band-shift experiments (Lusky et al. 1994). The E1-E2 complex binds to the origin but is still replication-inactive. It may function as a focal point for attracting other replication proteins, leading to a replication-active complex that can start unwinding. As soon as this occurs, E2 is released, possibly through structural changes induced in E1 and the origin. This is reminiscent of the early release of NFI during initiation of adenovirus DNA replication (Coenjaerts and Van der Vliet 1994) and of the recruitment of DnaB helicase by DnaC or λP protein (Baker and Kornberg 1992).

The mechanism of action by E2 might well involve more than just interacting with E1. E2 can also recruit RP-A, and mutants defective in RP-A binding are less effective in enhancing BPV replication in vitro (Li and Botchan 1993). Moreover, E2 interferes with nucleosome binding, but this antirepressor action is not specific for E1 and E2 and may be a consequence of E1/E2 binding rather than a primary effect (Li and Botchan 1994).

The results obtained with BPV were largely confirmed for several HPV serotypes, although some differences were observed (Frattini and Laimins 1994; Gopalakrishnan and Khan 1994). Interestingly, mutational analysis of HPV-11 showed that the presence of an E1-binding site is less important than the presence of an E2-binding site (Russell and Botchan 1995), stressing the function of E2 as the major targeting protein for the E1 helicase, which binds with lower specificity. Moreover, E2 also suppressed nonspecific initiation by E1 in vitro (Kuo et al. 1994).

In conclusion, the combination of E1 and E2 has many similarities to T antigen. Not only does it have similar enzymatic activities, but it also attracts both RP-A and DNA polymerase-α:DNA primase. However, the binding of E1 to the core origin is not always strong enough, and the most important role of E2 appears to be to facilitate binding of limiting amounts of E1. The intracellular ratio of E1 to E2 may be an important factor in determining the replication efficiency and thus the copy number. Once they have been delivered by E2, possibly assisted by E2-induced DNA bending and by the cooperative nature of the E1-E2 interaction, E1 multimers may attract other replication proteins, and a preinitiation complex will be formed. The cooperative binding of E1 and E2 leads to ATP-induced structural changes that may release E2 and activate the helicase activity.

Herpesvirus

DNA replication has been most extensively studied in the α-herpesvirus HSV and the γ-herpesvirus Epstein-Barr virus (EBV). In the 150-kbp linear HSV-DNA, three distinct origins have been described, ori_L and two copies of ori_S, located in the inverted repeats of the S component of the DNA. Seven viral proteins required for replication have been described (see Challberg, this volume).

As in other DNA viruses, both ori_S and ori_L are flanked by *cis*-acting sequences containing transcriptional regulatory elements that have been postulated to function in DNA replication as shown by deletion analysis (Stow and McMonagle 1983; Wong and Schaffer 1991). Whether binding of transcription factors to these elements is essential and how these transcription factors contribute to replication efficiency will be difficult to establish in view of the lack of an origin-dependent in vitro system. Besides the mechanisms proposed for papovaviruses, even a role of transcription itself is not excluded since the elements are still active in transcription during replication. This transcription process could influence ori function by modifying the local DNA structure in a way similar to that described for *E. coli oriC* (Baker and Kornberg 1988).

EBV oriP Is Activated by Transcription Factor-induced DNA Looping

The 192-kbp linear genome of EBV is maintained as multiple copies of a plasmid in latently infected B lymphocytes. Latent replication originates from *oriP*, the plasmid origin of replication, and requires only one viral protein, EBNA1, which also functions as a transcription factor.

oriP contains two noncontiguous elements, DS (dyad symmetry) and FR (family of repeats), separated by approximately 1 kbp. DS forms the functional origin of replication and contains four imperfect binding sites for EBNA1, two within a 65-bp region of dyad symmetry and two flanking this sequence. FR contains 20 tandemly repeated copies of a 30-bp sequence, each containing a 12-bp palindrome core that forms the EBNA1-binding site. This region functions as an enhancer both for transcription and DNA replication and also governs stable segregation of *oriP*-containing plasmids.

Upon initiation, the EBNA1 dimers bind first to the sites in FR, then to sites in DS. EBNA1 complexes formed on the two elements interact, and this leads to looping-out of the DNA (Frappier and Donnell 1991; Su et al. 1991). Binding of EBNA1 to FR or DS alone is much weaker than to both elements, indicating that the intramolecular interactions between

EBNA1 bound to FR and DS stabilize the DNA/protein complex, as shown by competition with nonspecific DNA (Frappier and Donnell 1991; Su et al. 1991). Thus, looping enables site saturation at lower concentrations via cooperativity. These interactions take place in a part of the EBNA1 molecule separate from the normal dimerization domain or the DNA-binding domain (Chen et al. 1993) and are located in a small region between amino acids 350 and 361 (Frappier et al. 1994). Contact with FR-bound EBNA1 leads to stabilization of EBNA1 at the DS element and could well be a trigger to activate initiation of DNA synthesis, either by recruitment of other replication factors or by a change in DNA structure. In particular, the protein-mediated looping might facilitate unwinding by introducing topological stress. Just DS-DS-bound EBNA1 does not stabilize binding. Interestingly, looping is also an important mechanism for bacteriophage R6k replication, where origin-bound Rep protein stabilizes the other Rep proteins bound weakly to *oriα*. This cooperativity leads also to efficient replication. In these cases, apparently specific interacting surfaces can only make optimal contacts when positioned on the DNA such that a loop is occurring. Only this position leads to optimal protein-protein interactions, thus imposing torsional strain on the DNA that can be used for origin opening.

Lytic Origin of Replication

EBV has both a latent state and a lytic replication cycle. The lytic phase can be induced by treating cells with phorbol esters or by introducing a vector expressing the *BZLF1* gene encoding a viral transcription factor. Two copies of *oriLyt* separated from *oriP* have been found, one in DS-L and one in DS-R (Hammerschmidt and Sugden 1988). These origins contain multiple regions required for replication and additional sequences that increase replication and can be functionally substituted by a transcriptional enhancer. Similar to HSV-1, at least seven viral genes required for replication have been found (Fixman et al. 1992). The DS-L origin is quite large and complex and consists of a 225-bp AT-rich region that is presumed to be the site of initiation, flanked by two elements, 320 and 370 bp long, respectively (Schepers et al. 1993b). These flanking elements contain promoters of two divergently transcribed genes, *BHLF1* and *BHRF1*, and also act as enhancers for replication. The BHLF1 promoter/replication enhancer is controlled by the viral transcription factor BZLF1 (synonyms: EBI, zta, ZEBRA, z) which belongs to the b-Zip class and is related to the AP1 family. Four binding sites are found in the promoter of *BHLF1* and three in *BHRF1*, to which BZLF1 binds with different affinities. Another *trans*-activator that binds to the *BHRF1* promoter is Rta, which may also be involved in enhancement of replication. Employing a transient replication and cotransfection protocol, it could be shown that for replication enhancement, the amino-terminal transcription activation domains of *BZLF1* are required (Schepers et al. 1993a). In contrast to results in the papovavirus system, other transcription factors (c-Jun, BPV-E2, Myc, or VP16) could not substitute for BZLF1, suggesting the requirement for specific interactions, although the target proteins have not been identified.

Other Viral Systems

The need for transcription factors to optimize DNA replication is likely not confined to the systems described above. Evidence is also present for Geminivirus (see Bisaro, this volume) and parvoviruses such as minute virus of mice (MVM). Here the active origin (50 bp) contains an ATF-binding site (Cotmore and Tattersall 1994). However, in these cases, the evidence for a function of these transcription factors is still circumstantial.

Cellular Origins

The use of transcription factors by viruses for the enhancement of their replication seems to be a rule rather than an exception. This is likely caused by adaptation of the viral genome to the presence of these proteins in their host cells. The need for economic use of genetic information due to packaging constraints may have forced the use of sequence information in *cis*-acting elements such as enhancers for dual purposes. This may also hold for the employment of viral transcription factors such as BPV-E2 and EBNA1 for multiple functions. Such sequence constraints do not exist in eukaryotic cells, and therefore an extrapolation of the need for transcription factors to cellular DNA synthesis is not a matter of fact. Nevertheless, recent results have indicated a strong link between transcription and replication also in cellular origins, and at least two proteins required for the initiation of DNA replication in *Saccharomyces cerevisiae* are also essential for control of transcription.

Origins in Yeast

In *S. cerevisiae*, replication origins have been well characterized, and although an in vitro system for initiation is not yet established, information on the replication proteins has recently been obtained both by biochemical and by genetic means. Yeast origins coincide with ARS elements and are located in AT-rich regions of the genome with an average spacing of approximately 100 kbp (see Newlon, this volume). ARS elements are approximately 100 bp in length and contain two essential elements, A and B. The A element contains an 11-bp consensus sequence (ACS), which is the only region common to all *S. cerevisiae* origins. The B element is composed of three functional elements, B1, B2, and B3, which are also important for origin functions but are less conserved. The A element is bound in vitro by the multiprotein origin recognition complex (ORC) (Bell and Stillman 1992), which can also be observed in vivo (Diffley and Cocker 1992). ORC binds the ACS in an ATP-dependent manner and also covers part of the B1 element. The complex is composed of six different polypeptides, ORC1–ORC6, which are essential for yeast viability. Interestingly, the 72-kD ORC2 protein is also implicated in transcriptional silencing of the HMR-E locus (Rivier and Rine 1992; Bell et al. 1993; Foss et al. 1993; Micklem et al. 1993). In agreement with this, the *cis*-acting sequences required for silencing can function as ARS elements. These results provide strong evidence for a role in vivo of ORC in replication as well as in transcriptional control, in particular in silencing. Possibly the ORC2 protein can be present in different complexes, dependent on its function. Alternatively, DNA replication could be involved directly in transcription silencing by some form of mechanistic coupling in which ORC is involved (Rivier and Rine 1992). Due to the lack of an in vitro system, the exact role of ORC2 in replication remains obscure.

The ORC complex remains bound to the ACS throughout the cell cycle but can occur in two states differing in footprint pattern at different parts of the cycle (Diffley et al. 1994). Near the end of mitosis and through G_1, an additional region of protection overlapping the ORC footprint is observed that is lost after entry into S phase. Diffley and coworkers suggest that this postreplicative state, which resembles the footprint pattern obtained in vitro with ORC and ABF-1, represents a ground state laid down immediately after S phase and remaining throughout the cell cycle even into G_0. Thus, ORC and ABF-1, a transcription factor described below, may have to be activated to create a prereplicative state.

The B3 element contains the binding site for the multifunctional transcription factor ABF1 (Buchman et al. 1988; Diffley and Stillman 1988). ABF1-binding sites have been identified in several ARSs at variable positions up to 1.2 kbp from the A element, indicating that they can function in a position- and orientation-independent manner to stimulate replication, thereby resembling transcriptional enhancers. ABF-1 is an abundant 81-kD phosphoprotein. It binds in a sequence-specific manner to a consensus sequence $GCAN_4Y_2RCTR$ (R=purine, Y=pyrimidine). A detailed contact point analysis has indicated that ABF-1 is mainly oriented along one face of the helix (McBroom and Sadowski 1994a). Considerable bending (120º) of the DNA has been inferred from circular permutation and phasing analysis (McBroom and Sadowski 1994b). ABF-1 is involved in positive control of a number of genes, in silencing of the mating-type loci HML and HMR, and in DNA replication.

To understand its role in DNA replication in more detail, the B3 element from ARS-1 has been substituted by the binding sites of several other transcription factors including RAP-1, Gal4, and a LexA-Gal4 fusion protein (Marahrens and Stillman 1992). These binding sites substituted efficiently, in both orientations and in a DNA-binding-dependent manner in the presence of the appropriate transcription factors, indicating that the B3 element may be a rather nonspecific enhancer.

Two hypotheses, not mutually exclusive, have been put forward to explain the role of ABF-1. One is a direct contact between the ABF-1 activation domain and replication proteins, present at the core origin. A candidate is RP-A, which can be bound directly by Gal4 (He et al. 1993). Recruitment of RP-A followed by binding to the B1-B2 unwinding element and stabilizing single-stranded DNA or assisting in unwinding is an attractive hypothesis. However, a direct interaction between ABF-1 and RP-A has not been described, and the ABF-1 site can vary in position and orientation without loss of activity. Moreover, such a recruitment model is difficult to reconcile with the continuous presence of ABF-1 in vivo at the core origin (Diffley et al. 1994). As shown by in vivo footprinting of an ARS element in which the ABF-1 site is mutated, ABF-1 is not involved in recruitment of ORC (J. Cocker and J. Diffley, pers. comm.) or the ORC-bound factor Dbf4 (Dowell et al. 1994). An alternative is that the *trans*-activation domain of ABF-1 is an architectural component facilitating the interaction between multiprotein complexes and DNA or influencing the local chromatin structure, employing its capacity to bend DNA. Interestingly, RAP-1, which can substitute for ABF-1, can also strongly bend DNA (Gilson et al. 1993).

Another transcription factor that has been implicated in DNA replication is the TATA-box-binding protein TBP (Lue and Kornberg 1993). TBP can bind to the ACS, and TBP-binding sites can activate replication origins in vivo. Moreover, the ARS-1 B domain can act as a weak upstream activating sequence in transcription. However, this could just be fortuitous; no example of an origin within a transcription unit in yeast chromosomes has been found. There is also no evidence that ARS elements have promoter activity (Fangman and Brewer 1991).

On the basis of genetic evidence, another class of proteins, Mcm (*m*inichromosome *m*aintenance), have been suggested to influence DNA replication (Christ and Tye 1991). One of these, Mcm-1, is a transcription factor that bridges the $\alpha 1$ and $\alpha 2$ regulatory homeobox proteins involved in mating-type control (Smith and Johnson 1992) and also interacts with other transcription factors. Its potential role in DNA replication is based on a plasmid maintenance defect in Mcm-1 mutants. As yet there is no evidence that Mcm-1 interacts directly with ARSs, and its role may be limited to transcriptional enhancement of genes coding for essential replication proteins.

In summary, the evidence for a link between DNA replication enhancement and transcriptional control, in particular silencing of HML-E, is compelling, but the mechanisms by which ABF-1 and ORC2 exert their multiple functions remain unclear. Most evidence points to an architectural role of ABF-1 rather than a direct interaction with other replication proteins.

Origins in Higher Eukaryotes

A number of potential eukaryotic origins in metazoa have been described recently (see DePamphilis, this volume). In those cases where sequence analysis of the potential origins is available, a large number of transcription factor-binding sites have been found. Since no mutational analysis has been performed, a functional analysis is not possible, and the significance of these sites is not clear. Although it is tempting to assume that they may act as enhancers of replication in a similar fashion as in yeast and the viral systems, they may just be localized close to origins by chance and represent promoter regions without any functional relationship to initiation of DNA replication. (Details of the binding sites in cellular origins can be found in DePamphilis, this volume.)

SUMMARIZING REMARKS

Why Are Transcription Factors Used for Replication?

Although at first sight the lack of a separate set of replication-enhancing proteins seems strange, it is understandable from an evolutionary point of view. Only a limited number of basic principles to recognize DNA have been found to date. This number may expand upon further structural analysis, but nevertheless, it seems that Nature uses variations on successful binding modes such as helix-turn-helix, zinc fingers, and helix-loop-helix above development of completely novel modes. These conserved modes of DNA recognition can apparently be used for multiple purposes, in both DNA replication and transcription. A similar situation is found for RP-A, also multifunctional, and for TFIIH, active in transcription and repair (Schaeffer et al. 1993; Drapkin et al. 1994). An advantage of the use of transcription factors is that, in general, they open up chromatin structure. This may explain why transcriptionally active regions of chromosomal DNA replicate early in S phase. But why should transcription activation domains not have been adapted specifically to interact with replication proteins? Presumably, the flexible nature of activation domains increases the potential for different, specific interactions. Moreover, the specificity might be increased by the use of different coactivators. Transcription activation domains can be mimicked by reiteration of short peptide segments, revealing their modular organization (Tanaka et al. 1994). Their flexibility has so far hampered detailed structural analysis, but they may adopt a more fixed structure upon binding their target proteins functioning in the transcriptional machinery or in initiation of DNA replication. Such a flexible interaction exists also for major histocompatibility complex molecules in binding peptides (Young et al. 1994) and in the interaction between molecular chaperones and unfolded proteins (Braig et al. 1994). Thus, transcriptional activation domains may be very suitable to permit interactions with different molecules of the replication machinery.

A further advantage of the use of the transcription factors is that modification, in particular phosphorylation, could enable a subtle regulation of initiation. This has been described for transcription control by Fos-Jun (AP1), and in that respect it is interesting that a natural AP1 site in polyomaviruses plays such a prominent role and may be involved in replication timing.

Why would a virus not circumvent the need for transcription factors by enhancing the levels of rate-limiting factors directly? One reason may be that, when viral replication proteins and DNA are in low concentration early in infection, cellular proteins could enhance replication, thus enabling a quick start. However, this would just give a small advantage in time. More likely, for limiting factors such as E1 and pTP-pol, the ability to bind in a sequence-specific manner would be counterproductive. This holds in particular for DNA polymerase, which must clear the origin to enable elongation. Transcription factor guiding can also prevent nonproductive action outside the origin, which might occur if high levels of the initiator are present (Kuo et al. 1994). Consistent with this notion, the BPV-E2 protein stimulated the origin-specific unwinding reaction catalyzed by E1, but had no stimulatory effect on nonspecific helicase activity (Seo et al. 1993b).

Are Similar Mechanisms Used to Activate Transcription and Replication?

As summarized in Table 3, the various mechanisms proposed for enhancement of replication apply to a great extent also for transcription. For the latter, various target proteins have been detected, like TBP, TAFs, or TFIIB, and even TFIIH or RNA polymerase itself. In one instance (TFIIB), a change in the structure is observed upon interaction with VP16 (Roberts and Green 1994). Transcription factors are involved in modulation of the global chromatin structure required for transcription, as exemplified by the MMTV-LTR promoter and NFI. These effects on chromatin structure might be similar in transcription and replication and therefore require the same domains of the transcription factors.

In most cases where direct protein-protein interactions have been observed, however, a high specificity is found, and the regions required for transcription activation and replication enhancement are different. This is most conspicuous when just a DNA-binding domain is required for repli-

Table 3 Transcription factor action in replication and transcription compared

Action	Transcription	Replication
Recruitment of basal proteins leading to stable preinitiation complex	TBP	Ad pTP-pol
	TFIIB	BPV-E1
	TFIIF	RP-A
	TFIIH	pol-α:primase
Influencing helicase activity	TFIIH?	BPV-E1
		T-ag (SV40)
Interaction with coactivators	TAF 40, 110, 150	?
	CREB-binding protein	
Multimerization	many	EBNA1
		BPV-E1
		T-ag
Influencing target conformation	TFIIB	?
Effect on chromatin structure	SWI/SNF	yes
Promoter/origin clearance	TFIIE	?
	TFIIH	

cation (adenovirus: NFI, Oct-1; polyoma: Rel oncoprotein). Despite their name, these DNA-binding domains can interact not only with DNA, but with other proteins as well. For instance, the Oct-1 POU homeodomain can interact with several other proteins besides the Ad DNA polymerase and employs different amino acids to contact these different targets (Coenjaerts et al. 1994).

Future experiments will doubtless reveal the details of interacting domains and the consequences for these interactions, employing reconstituted replication systems as well as structural studies. A major question will be whether these transcription factors only optimize viral replication systems, or whether they are also involved in cellular replication, in particular, timing and control of initiation of cell division. Potential candidates are transcription factors binding to the locus control region of the β-globin gene, which influences the time in S phase when the locus is replicated (Forrester et al. 1990; Kitsberg et al. 1993).

ACKNOWLEDGMENTS

I thank Brad Brinton and Jackie Russell for sending submitted manuscripts or preprints. Experiments performed in the author's laboratory were supported by the Netherlands Foundation for Chemical Research (SON) with financial support from the Netherlands Organization for Scientific Research (NWO). Critical reading of the manuscript by H.T.M. Timmers and J. Dekker is gratefully acknowledged.

REFERENCES

Adhya, S., P.S. Shneidman, and J. Hurwitz. 1986. Reconstruction of adenovirus replication origins with a human nuclear factor I binding site. *J. Biol. Chem.* **261:** 3339–3346.
Armentero, M.-T., M. Horwitz, and N. Mermod. 1994. Targeting of DNA polymerase to the adenovirus origin of DNA replication by interaction with nuclear factor I. *Proc. Natl. Acad. Sci.* **91:** 11537–11541.
Baker, T.A. and A. Kornberg. 1988. Transcriptional activation of initiation of replication from the *E. coli* chromosomal origin: An RNA-DNA hybrid near oriC. *Cell* **55:** 113–123.
———. 1992. *DNA replication.* W.H. Freeman, San Francisco.
Baru, M., M. Shlissel, and H. Manor. 1991. The yeast GAL4 protein transactivates the polyomavirus origin of DNA replication in mouse cells. *J. Virol.* **65:** 3496–3503.
Bell, S.P. and B. Stillman. 1992. ATP-dependent recognition of eukaryotic origins of DNA replication by a multiprotein complex. *Nature* **357:** 128–134.
Bell, S.P., R. Kobayashi, and B. Stillman. 1993. Yeast origin recognition complex functions in transcription silencing and DNA replication. *Science* **262:** 1844–1849.
Bennett Cook, E.R. and J.A. Hassell. 1991. Activation of polyomavirus DNA replication

by yeast GAL4 is dependent on its transcriptional activation domains. *EMBO. J.* **10:** 959-969.

Blitz, I.L. and L.A. Laimins. 1991. The 68-kilodalton E1 protein of bovine papillomavirus is a DNA binding phosphoprotein which associates with the E2 transcriptional activator in vitro. *J. Virol.* **65:** 649-656.

Bosher, J., E.C. Robinson, and R.T. Hay. 1990. Interactions between the adenovirus type 2 DNA polymerase and the DNA binding domain of nuclear factor I. *New Biol.* **2:** 1083-1090.

Bosher, J., I.R. Leith, S.M. Temperley, M. Wells, and R.T. Hay. 1991. The DNA-binding domain of nuclear factor I is sufficient to cooperate with the adenovirus type 2 DNA binding protein in viral DNA replication. *J. Gen. Virol.* **72:** 2975-2980.

Braig, K., Z. Otwinowski, R. Hedge, D.C. Boisvert, A. Joachimiak, A.L. Horwich, and P.B. Sigler. 1994. The crystal structure of the bacterial chaperonin GroEL at 2.8 Å. *Nature* **371:** 578-580.

Buchman, A.R., N.F. Lue, and R.D. Kornberg. 1988. Connections between transcriptional activators upstream activation sequences, autonomously replicating sequences, and telomeres in *Saccharomyces cerevisiae*. *Mol. Cell. Biol.* **8:** 210-225.

Challberg, M.D. and T.J. Kelly. 1989. Animal virus DNA replication. *Annu. Rev. Biochem.* **58:** 671-717.

Chen, M., N. Mermod, and M.S. Horwitz. 1990. Protein-protein interactions between adenovirus DNA polymerase and nuclear factor 1 mediate formation of the DNA replication preinitiation complex. *J. Biol. Chem.* **265:** 18634-18642.

Chen, M.R., J.M. Middeldorp, and S.D. Hayward. 1993. Separation of the complex DNA binding domain of EBNA-1 into DNA recognition and dimerization subdomains of novel structure. *J. Virol.* **67:** 4875-4885.

Cheng, L. and T.J. Kelly. 1989. Transcriptional activator nuclear factor 1 stimulates the replication of SV40 minichromosomes in vivo and in vitro. *Cell* **59:** 541-551.

Cheng, L.Z., J.L. Workman, R.E. Kingston, and T.J. Kelly. 1992. Regulation of DNA replication in vitro by the transcriptional activation domain of GAL4-VP16. *Proc. Natl. Acad. Sci.* **89:** 589-593.

Christ, C. and B.-K. Tye. 1991. Functional domains of the yeast transcription/replication factor MCM1. *Genes Dev.* **5:** 751-764.

Clayton, D.A. 1991. Replication and transcription of vertebrate mitochondrial DNA. *Annu. Rev. Cell Biol.* **7:** 453-478.

Cleat, P.H. and R.T. Hay. 1989. Co-operative interactions between NFI and the adenovirus DNA binding protein at the adenovirus origin of replication. *EMBO J.* **8:** 1841-1848.

Coenjaerts, F.E.J. and P.C. Van der Vliet. 1994. Early dissociation of nuclear factor I from the origin during initiation of adenovirus DNA replication studied by origin immobilization. *Nucleic Acids Res.* **22:** 5235-5240.

Coenjaerts, F.E.J., J.A.W.M. Van Oosterhout, and P.C. Van der Vliet. 1994. The Oct-1 POU domain stimulates adenovirus DNA replication by a direct interaction between the viral precursor terminal protein-DNA polymerase complex and the POU homeodomain. *EMBO J.* **13:** 5401-5409.

Coenjaerts, F.E.J., E. De Vries, G.J.M. Pruijn, W. Van Driel, S.M. Bloemers, N.M.T. van der Lugt, and P.C. Van der Vliet. 1991. Enhancement of DNA replication by transcription factors NFI and NFIII/Oct-1 depends critically on the positions of their binding sites in the adenovirus origin of replication. *Biochim. Biophys. Acta* **1090:** 61-69.

Cotmore, S.F. and P. Tattersall. 1994. An asymmetric nucleotide in the parvoviral 3' hairpin directs segregation of a single active origin of DNA replication. *EMBO J.* **13**: 4145-4152.

Croston, G.E., L.A. Kerrigan, L.M. Lira, D.R. Marshak, and J.T. Kadonaga. 1991. Sequence-specific antirepression of histone H1-mediated inhibition of basal RNA polymerase II transcription. *Science* **251**: 643-649.

DePamphilis, M.L. 1988. Transcriptional elements as components of eukaryotic origins of DNA replication. *Cell* **52**: 635-638.

―――. 1993a. Eukaryotic DNA replication: Anatomy of an origin. *Annu. Rev. Biochem.* **62**: 29-62.

―――. 1993b. How transcription factors regulate origins of DNA replication in eukaryotic cells. *Trends Cell Biol.* **3**: 161-167.

de Villiers, J., W. Schaffner, C. Tyndall, S. Lupton, and R. Kamen. 1984. Polyoma virus DNA replication requires an enhancer. *Nature* **312**: 242-246.

De Vries, E., W. Van Driel, S.J. van den Heuvel, and P.C. Van der Vliet. 1987. Contact point analysis of the nuclear factor I recognition site reveals symmetric binding at one side of the DNA helix. *EMBO J.* **6**: 161-168.

Diffley, J.F.X. and J.H. Cocker. 1992. Protein-DNA interactions at a yeast replication origin. *Nature* **357**: 169-172.

Diffley, J.F.X. and B. Stillman. 1988. Purification of a yeast protein that binds to origins of DNA replication and a transcriptional silencer. *Proc. Natl. Acad. Sci.* **85**: 2120-2124.

Diffley, J.F.X., J.H. Cocker, S.J. Dowell, and A. Rowley. 1994. Two steps in the assembly of complexes at yeast replication origins in vivo. *Cell* **78**: 303-316.

Dornreiter, I., L.F. Erdile, I.U. Gilbert, D. Von Winkler, T.J. Kelly, and E. Fanning. 1992. Interaction of DNA polymerase α-primase with cellular replication protein A and SV40 T antigen. *EMBO J.* **11**: 769-776.

Dowell, S.J., P. Romanowski, and J.F.X. Diffley. 1994. Interaction of Dbf4, the Cdc7 protein kinase regulatory subunit, with yeast replication origins in vivo. *Science* **265**: 1243-1246.

Drapkin, R., J.T. Reardon, A. Ansari, J.-C. Huang, L. Zawel, K. Ahn, A. Sancar, and D. Reinberg. 1994. Dual role of TFIIH in DNA excision repair and in transcription by RNA polymerase II. *Nature* **368**: 769-771.

Fangman, W.L. and B.J. Brewer. 1991. Activation of replication origins within yeast chromosomes. *Annu. Rev. Cell Biol.* **7**: 375-402.

Fixman, E.D., G.S. Hayward, and S.D. Hayward. 1992. *trans*-Acting requirements for replication of Epstein-Barr virus ori-Lyt. *J. Virol.* **66**: 5030-5039.

Forrester, W.C., E. Epner, M.C. Driscoll, T. Enver, M. Brice, and M. Papaya Groudine. 1990. A deletion of the human β-globin locus activation region causes a major alteration in chromatin structure and replication across the entire β-globin locus. *Genes Dev.* **4**: 1637-1649.

Foss, M., F.J. McNally, P. Laurenson, and J. Rine. 1993. Origin recognition complex (ORC) in transcriptional silencing and DNA replication in *S. cerevisiae*. *Science* **262**: 1838-1845.

Frappier, L. and M. Donnell. 1991. Epstein-Barr virus nuclear antigen 1 mediates a DNA loop within the latent replication origin of Epstein-Barr virus. *Proc. Natl. Acad. Sci.* **88**: 10875-10879.

Frappier, L., K. Goldsmith, and L. Bendell. 1994. Stabilization of the EBNA1 protein on

the Epstein-Barr virus latent origin of DNA replication by a DNA looping mechanism. *J. Biol. Chem.* **269:** 1057-1062.
Frattini, M.G. and L.A. Laimins. 1994. The role of the E1 and E2 proteins in the replication of human papillomavirus type 31b. *Virology* **204:** 799-804.
Gillette, T.G., M. Lusky, and J.A. Borowiec. 1994. Induction of structural changes in the bovine papillomavirus type 1 origin of replication by the viral E1 and E2 proteins. *Proc. Natl. Acad. Sci.* **91:** 8846-8850.
Gilson, E., M. Roberge, R. Giraldo, D. Rhodes, and S.M. Gasser. 1993. Distortion of the DNA double helix by RAP1 at silencers and multiple telomeric binding sites. *J. Mol. Biol.* **231:** 293-310.
Gopalakrishnan, V. and S.A. Khan. 1994. E1 protein of human papillomavirus type 1α is sufficient for initiation of viral DNA replication. *Proc. Natl. Acad. Sci.* **91:** 9597-9601.
Gounari, F., R. De Francesco, J. Schmidt, P.C. Van der Vliet, R. Cortese, and H.G. Stunnenberg. 1990. Amino terminal domain of NFI binds to DNA as a dimer and activates adenovirus DNA replication. *EMBO J.* **9:** 559-566.
Gronostajski, R.M. 1986. Analysis of nuclear factor I binding to DNA using degenerate oligonucleotides. *Nucleic Acids Res.* **14:** 9117-9132.
Gruss, C., J. Wu, T. Koller, and J.M. Sogo. 1993. Disruption of the nucleosomes at the replicational fork. *EMBO J.* **12:** 4533-4545.
Guo, Z.S. and M.L. DePamphilis. 1992. Specific transcription factors stimulate simian virus 40 and polyomavirus origins of DNA replication. *Mol. Cell. Biol.* **12:** 2514-2524.
Guo, Z.S., U. Heine, and M.L. DePamphilis. 1991. T antigen binding to site I facilitates initiation of SV40 DNA replication but does not affect bidirectionality. *Nucleic Acids Res.* **19:** 7081-7088.
Guo, Z.S., C. Gutierrez, U. Heine, J.M. Sogo, and M.L. DePamphilis. 1989. Origin auxiliary sequences can facilitate initiation of simian virus 40 DNA replication in vitro as they do in vivo. *Mol. Cell. Biol.* **9:** 3593-3602.
Gutierrez, C., Z.S. Guo, J. Roberts, and M.L. DePamphilis. 1990. Simian virus 40 origin auxiliary sequences weakly facilitate T antigen binding but strongly facilitate DNA unwinding. *Mol. Cell. Biol.* **10:** 1719-1728.
Hammerschmidt, W. and B. Sugden. 1988. Identification and characterization of oriLyt, a lytic origin of DNA replication of Epstein-Barr virus. *Cell* **55:** 427-433.
Harris, M.P.G. and R.T. Hay. 1988. DNA sequences required for the initiation of adenovirus type 4 DNA replication in vitro. *J. Mol. Biol.* **201:** 57-68.
Hatfield, L. and P. Hearing. 1993. The NFIII/Oct-1 binding site stimulates adenovirus DNA replication in vivo and is functionally redundant with adjacent sequences. *J. Virol.* **67:** 3931-3939.
Hay, R.T. 1985. Origin of adenovirus DNA replication. Role of nuclear factor I binding site in vivo. *J. Mol. Biol.* **186:** 129-136.
He, Z., B.T. Brinton, J. Greenblatt, J.A. Hassell, and C.J. Ingles. 1993. The transactivator proteins VP16 and GAL4 bind replication factor A. *Cell* **73:** 1223-1232.
Heintz, N.H. 1992. Transcription factors and the control of DNA replication. *Curr. Opin. Cell Biol.* **4:** 459-467.
Herr, W., R.A. Sturm, R.G. Clerc, L.M. Corcoran, D. Baltimore, P.A. Sharp, H.A. Ingraham, M.G. Rosenfeld, M. Finney, G. Ruvkun, and H.R. Horvitz. 1988. The POU domain: A large conserved region in the mammalian pit-1, oct-1, oct-2 and *Caenorhabditis elegans* unc-86 gene products. *Genes Dev.* **2:** 1513-1516.
Hoang, A.T., W. Wang, and J.D. Gralla. 1992. The replication activation potential of

selected RNA polymerase II promoter elements at the simian virus 40 origin. *Mol. Cell. Biol.* **12:** 3087-3093.
Ishikawa, H., M. Asano, T. Kanda, S. Kumar, C. Gelinas, and Y. Ito. 1993. Two novel functions associated with the Rel oncoproteins: DNA replication and cell-specific transcriptional activation. *Oncogene* **8:** 2889-2896.
Ishimi, Y. 1992. Preincubation of T antigen with DNA overcomes repression of SV40 DNA replication by nucleosome assembly. *J. Biol. Chem.* **267:** 10910-10913.
Johansen, T., U. Moens, T. Holm, A. Fjose, and S. Krauss. 1993. Zebrafish pou[c]: A divergent POU family gene ubiquitously expressed during embryogenesis. *Nucleic Acids Res.* **21:** 475-483.
Kenny, M. and J. Hurwitz. 1988. Initiation of adenovirus DNA replication. II. Structural requirements using synthetic oligonucleotide adenovirus templates. *J. Biol. Chem.* **263:** 9809-9817.
Kilwinski, J., M. Baack, S. Heiland, and R. Knippers. 1995. Transcription factor Oct1 binds to the AT-rich segment of the simian virus 40 replication origin. *J. Virol.* **69:** 575-578.
Kitsberg, D., S. Selig, I. Keshet, and H. Cedar. 1993. Replication structure of the human β-globin gene domain. *Nature* **366:** 588-590.
Kuo, S.R., J.S. Liu, T.R. Broker, and L.T. Chow. 1994. Cell-free replication of the human papillomavirus DNA with homologous viral E1 and E2 proteins and human cell extracts. *J. Biol. Chem.* **269:** 24058-24065.
Laybourn, P.J. and J.T. Kadonaga. 1991. Role of nucleosomal cores and histone H1 in regulation of transcription by RNA polymerase II. *Science* **254:** 238-245.
Lednicky, J. and W.R. Folk. 1992. Two synthetic Sp1-binding sites functionally substitute for the 21-base-pair repeat region to activate simian virus 40 growth in CV-1 cells. *J. Virol.* **66:** 6379-6390.
Leegwater, P.A.J., W. Van Driel, and P.C. Van der Vliet. 1985. Recognition site of nuclear factor I, a sequence specific DNA-binding protein from HeLa cells that stimulates adenovirus DNA replication. *EMBO J.* **4:** 1515-1521.
Li, R. and M.R. Botchan. 1993. The acidic transcriptional activation domains of VP16 and p53 bind the cellular replication protein A and stimulate in vitro BPV-1 DNA replication. *Cell* **73:** 1207-1221.
———. 1994. Acidic transcription factors alleviate nucleosome-mediated repression of DNA replication of bovine papillomavirus type 1. *Proc. Natl. Acad. Sci.* **91:** 7051-7055.
Lue, N.F. and R.D. Kornberg. 1993. A possible role for the yeast TATA-element-binding protein in DNA replication. *Proc. Natl. Acad. Sci.* **90:** 8018-8022.
Lusky, M., J. Hurwitz, and Y.S. Seo. 1993. Cooperative assembly of the bovine papilloma virus E1 and E2 proteins on the replication origin requires an intact E2 binding site. *J. Biol. Chem.* **268:** 15795-15803.
———. 1994. The bovine papillomavirus E2 protein modulates the assembly of but is not stably maintained in a replication-competent multimeric E1-replication origin complex. *Proc. Natl. Acad. Sci.* **91:** 8895-8899.
Marahrens, Y. and B. Stillman. 1992. A yeast chromosomal origin of DNA replication defined by multiple functional elements. *Science* **255:** 817-823.
McBroom, L.D. and P.D. Sadowski. 1994a. Contacts of the ABF1 protein of *Saccharomyces cerevisiae* with a DNA binding site at MATα. *J. Biol. Chem.* **269:** 16455-16460.

──────. 1994b. DNA bending by *Saccharomyces cerevisiae* ABF1 and its proteolytic fragments. *J. Biol. Chem.* **269:** 16461-16468.

Meisterernst, M., L. Rogge, R. Foechler, M. Karaghiosoff, and E.L. Winnacker. 1989. Structural and functional organization of a porcine gene coding for nuclear factor I. *Biochemistry* **28:** 8191-8200.

Melendy, T. and B. Stillman. 1993. An interaction between replication protein A and SV40 T antigen appears essential for primosome assembly during SV40 DNA replication. *J. Biol. Chem.* **268:** 3389-3395.

Mermod, N., E.A. O'Neill, T.J. Kelly, and R. Tjian. 1989. The proline-rich transcriptional activator of CTF/NFI is distinct from the replication and DNA binding domain. *Cell* **58:** 741-753.

Micklem, G., A. Rowley, J. Harwood, K. Nasmyth, and J.F.X. Diffley. 1993. Yeast origin recognition complex is involved in DNA replication and transcriptional silencing. *Nature* **366:** 87-89.

Mohr, I.J., R. Clark, S. Sun, E.J. Androphy, P. MacPherson, and M.R. Botchan. 1990. Targeting the E1 replication protein to the papillomavirus origin of replication by complex formation with the E2 transactivator. *Science* **250:** 1694-1699.

Monaghan, A., A. Webster, and R.T. Hay. 1994. Adenovirus DNA binding protein: Helix destabilising properties. *Nucleic Acids Res.* **22:** 742-748.

Morgan, I.M., M. Asano, L.S. Havarstein, H. Ishikawa, T. Hiiragi, Y. Ito, and P.K. Vogt. 1993. Amino acid substitutions modulate the effect of Jun on transformation, transcriptional activation and DNA replication. *Oncogene* **8:** 1135-1140.

Mul, Y.M. and P.C. Van der Vliet. 1992. Nuclear factor I enhances adenovirus DNA replication by increasing the stability of a preinitiation complex. *EMBO J.* **11:** 751-760.

──────. 1993. The adenovirus DNA binding protein effects the kinetics of DNA replication by a mechanism distinct from NFI or Oct-1. *Nucleic Acids Res.* **21:** 641-647.

Mul, Y.M., C.P. Verrijzer, and P.C. Van der Vliet. 1990. Transcription factors NFI and NFIII/Oct-1 function independently, employing different mechanisms to enhance adenovirus DNA replication. *J. Virol.* **64:** 5510-5518.

Müller, F., Y.-S. Seo, and J. Hurwitz. 1994. Replication of bovine papillomavirus type 1 origin-containing DNA in crude extracts and with purified proteins. *J. Biol. Chem.* **269:** 17086-17094.

Murakami, Y., M. Satake, Y. Yamaguchi Iwai, M. Sakai, M. Muramatsu, and Y. Ito. 1991. The nuclear protooncogenes c-*jun* and c-*fos* as regulators of DNA replication. *Proc. Natl. Acad. Sci.* **88:** 3947-3951.

Nagata, K., R.A. Guggenheimer, and J. Hurwitz. 1983. Specific binding of a cellular DNA replication protein to the origin of replication of adenovirus DNA. *Proc. Natl. Acad. Sci.* **80:** 6177-6181.

Nagata, K., R.A. Guggenheimer, T. Enomoto, J.H. Lichy, and J. Hurwitz. 1982. Adenovirus DNA replication in vitro: Identification of a host factor that stimulates synthesis of the preterminal protein-dCMP complex. *Proc. Natl. Acad. Sci.* **79:** 6438-6442.

O'Neill, E.A. and T.J. Kelly. 1988. Purification and characterization of nuclear factor III (origin recognition protein C), a sequence-specific DNA binding protein required for efficient initiation of adenovirus DNA replication. *J. Biol. Chem.* **263:** 931-937.

O'Neill, E.A., C. Fletcher, C.R. Burrow, N. Heintz, R.G. Roeder, and T.J. Kelly. 1988. Transcriptional factor OTF-1 is functionally identical to the DNA replication factor NFIII. *Science* **241:** 1210-1213.

Paonessa, G., F. Gounari, R. Frank, and R. Cortese. 1988. Purification of a NFI-like DNA-binding protein from rat liver and cloning of the corresponding cDNA. *EMBO J.* **7:** 3115–3123.

Park, P., W. Copeland, L. Yang, T. Wang, M.R. Botchan, and I.J. Mohr. 1994. The cellular DNA polymerase α-primase is required for papillomavirus DNA replication and associates with the viral E1 helicase. *Proc. Natl. Acad. Sci.* **91:** 8700–8704.

Pronk, R. and P.C. Van der Vliet. 1993. The adenovirus terminal protein influences binding of replication proteins and changes the origin structure. *Nucleic Acids Res.* **21:** 2293–2300.

Pruijn, G.J.M., W. Van Driel, and P.C. Van der Vliet. 1986. Nuclear factor III, a novel sequence-specific DNA-binding protein from HeLa cells stimulating adenovirus DNA replication. *Nature* **322:** 656–659.

Pruijn, G.J.M., P.C. Van der Vliet, N.A. Dathan, and I.W. Mattaj. 1989. Anti-OTF-1 antibodies inhibit NFIII stimulation of in vitro adenovirus DNA replication. *Nucleic Acids Res.* **17:** 1845–1863.

Rivier, D.H. and J. Rine. 1992. An origin of DNA replication and a transcription silencer require a common element. *Science* **256:** 659–663.

Roberts, S.G.E. and M.R. Green. 1994. Activator-induced conformational change in general transcription factor TFIIB. *Nature* **371:** 717–720.

Rosenfeld, P.J. and T.J. Kelly. 1986. Purification of nuclear factor I by DNA recognition site affinity chromatography. *J. Biol. Chem.* **261:** 1398–1408.

Russell, J. and M.R. Botchan. 1995. *cis*-Acting components of human papillomavirus (HPV) DNA replication: Linker substitution analysis of the HPV type 11 origin. *J. Virol.* **69:** 651–660.

Santoro, C., N. Mermod, P.C. Andrews, and R. Tjian. 1988. A family of human CAAT-box-binding proteins active in transcription and DNA replication: Cloning and expression of multiple cDNAs. *Nature* **334:** 218–224.

Schaeffer, L., R. Roy, S. Humbert, V. Moncollin, W. Vermeulen, J.H.J. Hoeijmakers, P. Chambon, and J.-M. Egly. 1993. DNA repair helicase: A component of BTF2 (TFIIH) basic transcription factor. *Science* **260:** 58–63.

Schepers, A., D. Pich, and W. Hammerschmidt. 1993a. A transcription factor with homology to the AP-1 family links RNA transcription and DNA replication in the lytic cycle of Epstein-Barr virus. *EMBO. J.* **12:** 3921–3929.

Schepers, A., D. Pich, J. Mankertz, and W. Hammerschmidt. 1993b. *cis*-Acting elements in the lytic origin of DNA replication of Epstein-Barr virus. *J. Virol.* **67:** 4237–4245.

Seo, Y.S., F. Muller, M. Lusky, and J. Hurwitz. 1993a. Bovine papilloma virus (BPV)-encoded E1 protein contains multiple activities required for BPV DNA replication. *Proc. Natl. Acad. Sci.* **90:** 702–706.

Seo, Y.S., F. Muller, M. Lusky, E. Gibbs, H.Y. Kim, B. Phillips, and J. Hurwitz. 1993b. Bovine papilloma virus (BPV)-encoded E2 protein enhances binding of E1 protein to the BPV replication origin. *Proc. Natl. Acad. Sci.* **90:** 2865–2869.

Smith, D.L. and A.D. Johnson. 1992. A molecular mechanism for combinatorial control in yeast: MCM1 protein sets the spacing and orientation of the homeodomains of an α2 dimer. *Cell* **68:** 133–142.

Sock, E., M. Wegner, E.A. Fortunato, and F. Grummt. 1993. Large T antigen and sequences within the regulatory region of JC virus both contribute to the features of JC virus DNA replication. *Virology* **197:** 537–548.

Spalholz, B.A., A.A. McBride, T. Sarafi, and J. Quintero. 1993. Binding of bovine papillomavirus E1 to the origin is not sufficient for DNA replication. *Virology* **193**: 201-212.

Stillman, B.M. 1989. Initiation of eukaryotic DNA replication in vitro. *Annu. Rev. Cell Biol.* **5**: 197-245.

Stow, N.D. and E.C. McMonagle. 1983. Characterization of the TRs/IRs origin of DNA replication of herpes simplex virus type 1. *Virology* **130**: 427-438.

Stuiver, M.H. and P.C. Van der Vliet. 1990. The adenovirus DNA binding protein forms a multimeric protein complex with double-stranded DNA and enhances binding of nuclear factor I. *J. Virol.* **64**: 379-386.

Stuiver, M.H., W.G. Bergsma, A.C. Arnberg, H. Van Amerongen, R. Van Grondelle, and P.C. Van der Vliet. 1992. Structural alterations of double-stranded DNA in complex with the adenovirus DNA-binding protein. Implications for its function in DNA replication. *J. Mol. Biol.* **225**: 999-1011.

Su, W., T. Middleton, B. Sugden, and H. Echols. 1991. DNA looping between the origin of replication of Epstein-Barr virus and its enhancer site: Stabilization of an origin complex with Epstein-Barr nuclear antigen 1. *Proc. Natl. Acad. Sci.* **88**: 10870-10874.

Sun, S., L. Thorner, M. Lentz, P. McPherson, and M. Botchan. 1990. Identification of a 68-kilodalton nuclear ATP-binding phosphoprotein encoded by bovine papillomavirus type 1. *J. Virol.* **64**: 5093-5105.

Tanaka, M., W.M. Clouston, and W. Herr. 1994. The Oct-2 glutamine-rich and proline-rich activation domains can synergize with each other or duplicates of themselves to activate transcription. *Mol. Cell. Biol.* **14**: 6046-6055.

Thorner, L.K., D.A. Lim, and M.R. Botchan. 1993. DNA-binding domain of bovine papillomavirus type 1 E1 helicase: Structural and functional aspects. *J. Virol.* **67**: 6000-6014.

Ustav, M., E. Ustav, P. Szymanski, and A. Stenlund. 1991. Identification of the origin of replication of bovine papillomavirus and characterization of the viral origin recognition factor E1. *EMBO J.* **10**: 4321-4329.

―――. 1993. The bovine papillomavirus origin of replication requires a binding site for the E2 transcriptional activator. *Proc. Natl. Acad. Sci.* **90**: 898-902.

Van der Vliet, P.C. 1991. The role of cellular transcription factors in the enhancement of adenovirus DNA replication. *Semin. Virol.* **2**: 271-280.

Verrijzer, C.P. and P.C. Van der Vliet. 1993. POU domain transcription factors. *Biochim. Biophys. Acta* **1173**: 1-21.

Verrijzer, C.P., A.J. Kal, and P.C. Van der Vliet. 1990. The DNA binding domain (POU domain) of transcription factor Oct-1 suffices for stimulation of DNA replication. *EMBO J.* **9**: 1883-1888.

Verrijzer, C.P., M. Strating, Y.M. Mul, and P.C. Van der Vliet. 1992. POU domain transcription factors from different subclasses stimulate adenovirus DNA replication. *Nucleic Acids Res.* **20**: 6369-6375.

Verrijzer, C.P., J.A.W.M. Van Oosterhout, W. Van Weperen, and P.C. Van der Vliet. 1991. POU proteins bend DNA via the POU-specific domain. *EMBO J.* **10**: 3007-3014.

Wasylyk, C., J. Schneikert, and B. Wasylyk. 1990. Oncogene v-*jun* modulates DNA replication. *Oncogene* **5**: 1055-1058.

Wides, R.J., M.D. Challberg, D.R. Rawlins, and T.J. Kelly. 1987. Adenovirus origin of DNA replication: Sequence requirements for replication in vitro. *Mol. Cell. Biol.* **7**: 864-874.

Winokur, P.L. and A.A. McBride. 1992. Separation of the transcriptional activation and replication functions of the bovine papillomavirus-1 E2 protein. *EMBO. J.* **11**: 4111-4118.

Wong, S.W. and P.A. Schaffer. 1991. Elements in the transcriptional regulatory region flanking herpes simplex virus type 1 oriS stimulate origin function. *J. Virol.* **65**: 2601-2611.

Yang, L., I. Mohr, E. Fouts, D.A. Lim, M. Nohaile, and M.R. Botchan. 1993. The E1 protein of bovine papilloma virus 1 is an ATP-dependent DNA helicase. *Proc. Natl. Acad. Sci.* **90**: 5086-5090.

Young, A.C.M., W. Zhang, J.C. Sacchettini, and S.G. Nathenson. 1994. The three-dimensional structure of H-2Db at 2.4 Å resolution: Implications for antigen-determinant selection. *Cell* **76**: 39-50.

Zijderveld, D.C. and P.C. Van der Vliet. 1994. Helix-destabilizing properties of the adenovirus DNA binding protein. *J. Virol.* **68**: 1158-1164.

Zorbas, H., L. Rogge, M. Meisterernst, and E.L. Winnacker. 1989. Hydroxyl radical footprints reveal novel structural features around the NFI binding site in adenovirus DNA. *Nucleic Acids Res.* **17**: 7735-7748.

4

Roles of Nuclear Structure in DNA Replication

Ronald Laskey and Mark Madine
Wellcome/CRC Institute
Cambridge CB2 1QR
United Kingdom

Department of Zoology
University of Cambridge
Cambridge CB2 3EJ
United Kingdom

The cell nucleus is the defining feature of eukaryotes. It is bounded by a nuclear envelope consisting of two concentric layers of membrane perforated by nuclear pores that serve as channels of communication between the nucleus and the cytoplasm. DNA is not randomly packed into the nucleus but packaged precisely in such a way that all regions are accessible for replication each cell cycle. Partial access and therefore partial replication would result in chromosome breakage or nondisjunction at mitosis, with disastrous consequences. The packing hierarchy involves radial loop organization from an axial scaffold, as well as the compaction resulting from coiling DNA twice around the nucleosome subunits of chromatin (Schedl and Grosveld 1995; Van Holde et al. 1995).

A crucial feature of eukaryotic chromosomal DNA replication is that it always occurs within a nucleus. In lower eukaryotes such as fungi or *Physarum*, the nuclear membrane remains intact throughout the cell cycle, whereas it breaks down during mitosis of higher eukaryotes. Nevertheless, replication is constrained to interphase when the nuclear membrane is intact. We argue that this constraint has important regulatory consequences.

Further key features of eukaryotic DNA replication are that multiple initiations occur within a single chromosome and that these initiations are coordinated so that each region of the chromosome replicates, but replicates only once in any cell cycle. We argue that nuclear structure has essential roles to play in coordinating multiple initiations to replicate the chromosome exactly once.

Concepts in Eukaryotic DNA Replication
© 1999 Cold Spring Harbor Laboratory Press 0-87969-557-9/99

NUCLEAR STRUCTURE IS REQUIRED FOR CELLULAR DNA REPLICATION

Studies of eukaryotic DNA replication have been held back by a shortage of cell-free systems that initiate chromosomal DNA replication efficiently in vitro. Efficient replication systems have been developed for several viruses (see other chapters in this volume), but the only systems that clearly initiate efficiently on cellular DNA are derived from animal eggs. Initiating cell-free replication systems have been developed from unfertilized eggs of *Xenopus* and *Drosophila* (Blow and Laskey 1986; Newport 1987; Crevel and Cotterill 1991). In both cases, the egg contains a prefabricated stockpile of materials for exceptionally rapid DNA replication and nuclear assembly. *Xenopus* and *Drosophila* embryos reach 10,000 cells faster than a proliferating mammalian cell divides once.

These features make animal eggs attractive systems to study cell proliferation (see Blow and Chong, this volume), but they have some compensating disadvantages. For example, they lack G_0, G_1, and G_2 phases of the cell cycle, making them unsuitable for studies of transition between normal cell-cycle phases. A cell-free system that initiates cellular chromosomal replication efficiently in extracts of mammalian cells or yeast would have important applications.

A conspicuous feature of both the replication systems from *Xenopus* eggs is their dependence on nuclear structure to initiate DNA replication. Initiation is only observed when the template DNA is enclosed within a complete nuclear membrane (Newport 1987; Sheehan et al. 1988; Blow and Sleeman 1990). Nuclei whose membranes have been permeabilized by nonionic detergents or lysolecithin must be repaired by membrane vesicles before initiation is observed (Leno et al. 1992; Coverley et al. 1993).

Two further lines of evidence emphasize the importance of nuclear structure in initiation of DNA replication. First, purified DNA is replicated reasonably efficiently when added to *Xenopus* egg extracts (Blow and Laskey 1986; Newport 1987; Blow and Sleeman 1990; Newport et al. 1990; Cox and Laskey 1991), but replication is observed only when the DNA is assembled into pseudonuclei. Furthermore, the efficiency of replication depends on the efficiency of nuclear assembly, and only the DNA that is enclosed within the pseudonuclei replicates (Blow and Sleeman 1990).

The second line of evidence that nuclear structure is crucial for DNA replication comes from studies of nuclear membrane assembly in vitro. These indicate that the nuclear membrane defines the DNA it encloses as an integral unit of replication. Thus, Blow and Watson (1987) found that individual sperm nuclei in a common egg-extract environment act as integrated and independent units of replication. All the DNA in a nucleus replicates roughly synchronously, but not synchronously with its neighbors. This conclusion was strengthened by Leno and Laskey (1991), who exploited an unusual behavior of chicken erythrocyte nuclei in egg extract. These nuclei tend to clump to form aggregates. The response of the extract is to assemble a single nuclear membrane around the entire clump, not around the individual nuclei it contains. In this circumstance, all the nuclei enclosed within a single nuclear membrane replicated in precise synchrony, even though different clumps replicated at different times, and even though individual nuclei outside the clumps replicated asynchronously. Therefore, the nuclear membrane defines the DNA it encloses as a unit for initiation of replication.

One conspicuous way in which the nuclear membrane could regulate initiation of DNA replication is by concentrating nuclear proteins in the nucleus from the cytoplasm. The importance of this process for initiation of replication has been demonstrated by use of the inhibitor wheat germ agglutinin (WGA). This inhibits protein import and also prevents initiation of replication (Cox 1992).

A second way in which replication depends on an aspect of nuclear structure concerns the nuclear lamina. When lamins are removed from egg extract by immunodepletion, nuclear assembly and protein import are not inhibited, but the nuclei formed are unable to initiate replication (Newport et al. 1990; Meier et al. 1991; Jenkins et al. 1993).

The conclusion we offer from this section is that nuclear structure is essential for initiation of DNA replication in eukaryotes. This in turn suggests that attempts to obtain cell-free initiation of DNA replication from mammalian cells should focus on using nuclei as the template.

REPLICATION OCCURS AT DISCRETE SITES WITHIN THE NUCLEUS

Replication forks are not distributed diffusely throughout the nucleus. Instead, they are clustered in replication foci or factories (Fig. 1) (Nakamura et al. 1986; Mills et al. 1989; Nakayasu and Berezney 1989; Cox and Laskey 1991; Fox et al. 1991; O'Keefe et al. 1992; Hozak et al. 1993). These may contain tens to several hundreds of replication forks, depending on the type of cell. They are revealed by pulse-labeling sites of DNA replication with precursors that can be detected fluorescently.

The significance of replication fork clustering is not clear. It might facilitate coordination between polymerases and accessory proteins on the

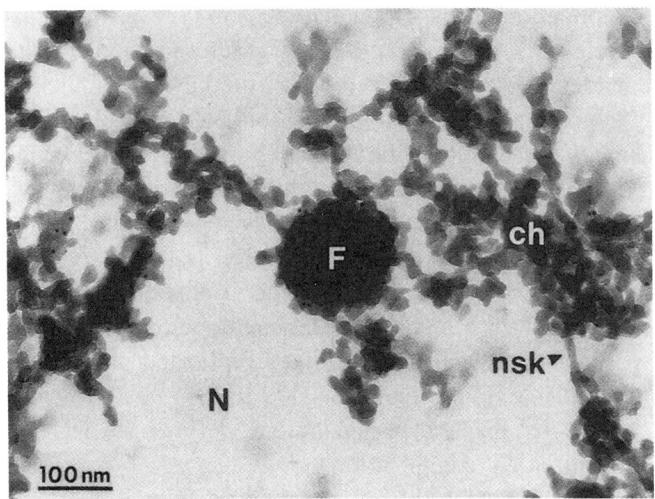

Figure 1 A replication factory. Resinless electron microscopy of HeLa nuclei following chromatin digestion reveals residual chromatin (ch) attached to an underlying nucleoskeleton (nsk). In S-phase cells, dense "factories" (F) are sites of DNA synthesis. Bar, 100 nm. (By courtesy of P. Hozak and P. Cook.)

two sides of each fork and on the two forks resulting from each initiation. More interestingly, it might provide a structural framework for the task of ensuring that all DNA is completely replicated.

The clustering of replication forks has added fuel to the debate of whether mobile DNA polymerases move along a fixed DNA track, or alternatively, whether mobile DNA is spooled through immobilized replication machinery (Hozak et al. 1993; Jackson 1995). Opinions remain divided on this issue, although it appears to us that evidence for spooling through fixed replication sites is slowly growing stronger. Electron microscopy of the initial DNA unwinding reaction for SV40 DNA replication clearly illustrates two T-antigen complexes at the replication forks which remain together while loops of single-stranded DNA are extruded (Wessel et al. 1992). If spooling is the correct interpretation, then one particular problem arises from the need to replicate the last regions of DNA between two replication foci or factories. Are they pulled in two directions? Are they completed by mobile polymerases detaching from the cluster? Alternatively, does DNA move in the same direction through consecutive clusters? Answers to these questions and the underlying issue of spooling would help us to understand the structural basics of replication.

Patterns of replication fork clusters remain the same throughout S phase of *Xenopus* sperm nuclei replicating in *Xenopus* egg extracts (Mills et al. 1989), but in somatic mammalian cells they change as S phase progresses (Nakayasu and Berezney 1989; Fox et al. 1991; O'Keefe et al. 1992). Some foci appear late, and further clustering together of entire foci occurs late in replication. Perhaps this aggregation of foci contributes to solving the problem referred to above of how the last stretch of DNA between adjacent foci is replicated.

What specifies the pattern of foci? This question has been addressed by two extreme examples seen in Figure 2. At one extreme, purified DNA from bacteriophage has been used as a template for reassembly into pseudonuclei and replication in egg extracts. Even phage DNA is replicated under strict cell-cycle control in *Xenopus* eggs (Harland and Laskey 1980; Méchali et al. 1983; Newport 1987). As shown in Figure 2, phage DNA is also replicated under the egg's spatial control. Even though it lacks eukaryotic sequences, the pattern of clustered replication forks strikingly resembles that of *Xenopus* sperm nuclei, indicating that specific eukaryotic DNA sequences are not required for this level of spatial regulation.

At the opposite extreme, polytene nuclei from salivary glands of *Drosophila* larvae also have a similar pattern of clustered replication forks superimposed on them by *Xenopus* egg extract (Fig. 2). The polytene chromatin is remodeled and decondensed by the extract during this process, losing its banded appearance. From these examples, we conclude that the pattern of clustered replication forks can arise de novo even on prokaryotic DNA that has never been subjected to such patterns, or on highly organized polytene chromosomes. The pattern is specified by the egg extract, not by the incoming nuclei, and it must be imposed by a structural measuring mechanism that is independent of DNA sequence.

One series of experiments demonstrates that *Xenopus* eggs can recognize and use preexisting nuclear structure, at least transitionally, for initiation of replication. Gilbert et al. (1995) showed that nuclei from Chinese hamster ovary cells with highly amplified genes for dihydrofolate reductase initiate replication nonrandomly at preferred sites. These sites are the same as those used by cells in vivo. However, when the nuclear membrane was disrupted or naked DNA was used as the substrate, specificity of initiation was lost.

REPLICATION LICENSING AND THE NUCLEAR MEMBRANE

Above, we argued that the nuclear membrane plays a regulatory role by coordinating the replication of all the DNA it contains. Here we argue

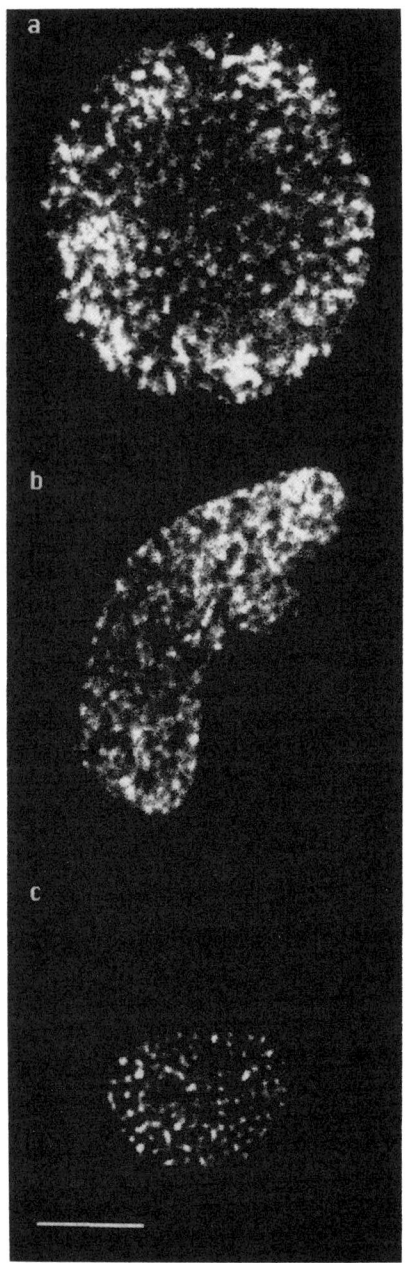

Figure 2 Discrete sites of replication seen as fluorescent foci in three types of nuclei replicating in *Xenopus* egg extracts: (*a*) *Drosophila* polytene nuclei, (*b*) *Xenopus* sperm nuclei, and (*c*) bacteriophage λ DNA pseudonuclei. Bar, 5 μm. (Original photographs provided by A.D. Mills and L.S. Cox.)

that it plays a further regulatory role by preventing reinitiation of DNA replication within a single cell cycle. The evidence for this assertion comes from experiments in which replicated, G_2 nuclei are transferred to fresh egg extract. They do not reinitiate replication if their nuclear membranes are intact. If, however, their nuclear membranes are permeabilized by nonionic detergents or lysolecithin before they are transferred, they reinitiate replication efficiently (Fig. 3). These experiments have been performed using *Xenopus* sperm nuclei as the templates or using synchronized HeLa nuclei (Blow and Laskey 1988; Leno et al. 1992; Coverley et al. 1993; Madine et al. 1995). These observations can be explained by a licensing model (Fig. 4A) (Blow and Laskey 1988). This model postulates an essential initiation factor (originally called "licensing factor," but see below) that is necessary, but not sufficient, for initiation of replication and that is unable to cross the nuclear envelope because it lacks a nuclear localization signal. Therefore, it could only bind to chromatin when the nuclear membrane breaks down in mitosis.

The model postulates that the factor would be inactivated by the act of replication so, in this way, each round of replication would need to be individually licensed by nuclear envelope breakdown at mitosis. Permeabilizing the nuclear membrane artificially would simply mimic this effect of mitosis, allowing the factor to act on chromatin to generate

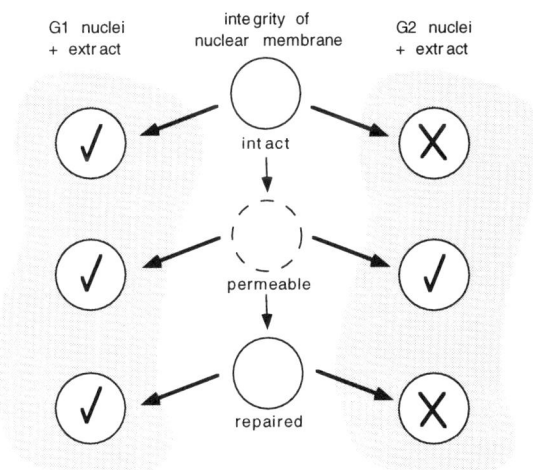

Figure 3 Summary of replication capacity of synchronized HeLa nuclei in *Xenopus* egg extract. G_1 nuclei replicate whether or not their nuclear membranes are intact, whereas G_2 nuclei are only able to replicate when their nuclear membrane is permeable (for references, see text).

A Licensing Factor Model

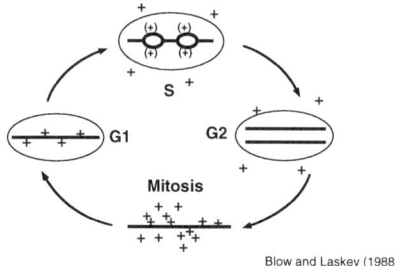

B *S. cerevisiae* CDC46/MCM5
subcellular localisation

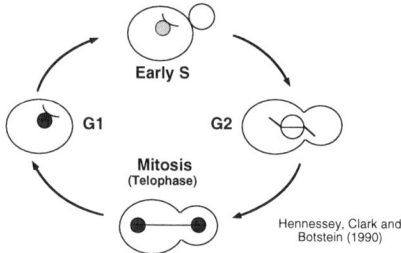

Figure 4 (*A*) "Licensing factor" model for mechanism limiting DNA replication to once per cell cycle. Licensing factor binds to DNA during mitosis but is unable to enter the nuclear envelope during interphase. It is inactivated by initiation or passage of a replication fork so that replication cannot reinitiate until the nuclear envelope breaks down again in the next mitosis. (Modified from Blow and Laskey 1988.) (*B*) Cartoon depicting nuclear localization of CDC46/MCM5 during the cell cycle of *S. cerevisiae* (Hennessey et al. 1990). Dark shading represents nuclear localization.

a new license. Evidence supporting a positive factor of this type has come from three laboratories (Blow 1993; Coverley et al. 1993; Kubota and Takisawa 1993) and has been extended recently (Chong et al. 1995; Kubota et al. 1995; Madine et al. 1995). These studies suggest, however, that the activity can be resolved into at least two components.

A family of proteins that were discovered in yeast have become excellent candidates for one of the components required for replication licensing. These proteins are becoming a growth industry. They are called MCM or *mini chromosome maintenance* proteins. They have been discovered independently in *Saccharomyces cerevisiae*, *Schizosaccharomyces pombe*, mouse, *Pleurodeles*, and *Notophthalmus*, and have also been cloned from several other species, including human and *Xenopus*. In yeast, they are required for DNA replication, and their localization is strikingly similar to that predicted for a licensing factor. Not only are they required for replication, but they are cytoplasmic until mitosis, when they enter the nucleus at anaphase and remain there until S phase, when they are removed (Fig. 4B).

This coincidence was investigated by cloning, expressing, and raising antibodies against *Xenopus* MCM3 (Madine et al. 1995). Affinity-purified antibodies precipitated a complex containing several MCM proteins. Removal of this complex by immunodepletion inhibits replication of *Xenopus* sperm nuclei (Fig. 5). Replication is restored by re-addition of the MCM complex. When synchronized HeLa cells were used as templates, permeable G_2 nuclei replicated in mock-depleted extract as expected, but not in extract depleted of the MCM complex. In contrast, G_1 HeLa nuclei replicated whether or not the *Xenopus* MCM complex was present, implicating the MCM complex in the mechanism that distinguishes G_1 from G_2 nuclei and preventing further replication of G_2 nuclei (Fig. 6). Similar conclusions were reached independently by Kubota et al. (1995) and Chong et al. (1995) using different approaches but also implicating the MCM complexes.

An obvious question is, Can the MCM proteins cross the nuclear membrane? Experiments with an MCM3 fusion protein indicated that it cannot cross an interphase nuclear membrane (Kubota et al. 1995). How-

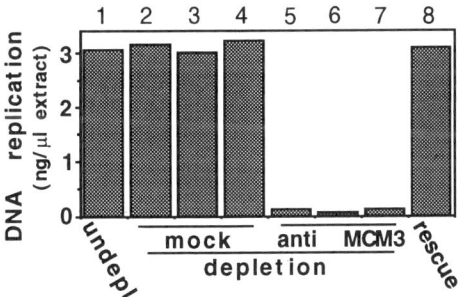

Figure 5 Immunodepletion by anti-XMCM3 antibodies inhibits DNA replication of *Xenopus* sperm nuclei in *Xenopus* egg extracts. Sperm nuclei replicate efficiently in undepleted (track *1*) extract or mock immunodepleted extracts (tracks *2–4*). In contrast, immunodepletion of the MCM protein complex with any one of three anti-XMCM3 antibodies abolishes replication (tracks *5–7*). Readdition of MCM complex purified from a fourth antibody restores replication (for details, see Madine et al. 1995). (Reprinted, with permission, from Madine et al. 1995 [copyright Macmillan].)

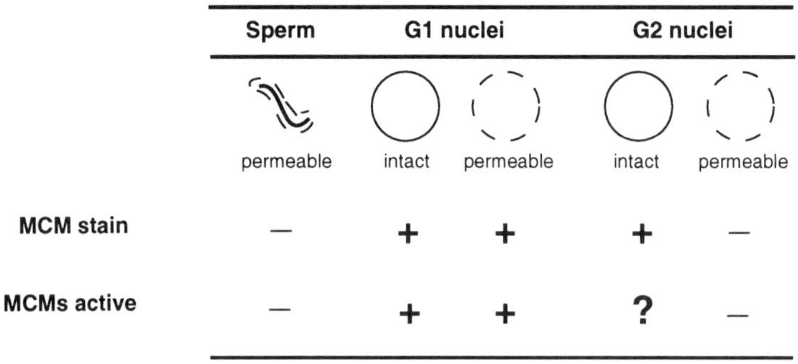

	Sperm	G1 nuclei		G2 nuclei	
	permeable	intact	permeable	intact	permeable
MCM stain	−	+	+	+	−
MCMs active	−	+	+	?	−

Figure 6 Summary of the MCM status of nuclei from *Xenopus* sperm or synchronized G_1 and G_2 HeLa cells. "MCM stain" refers to immunofluorescence with antibodies against MCM3. "MCMs active" refers to the ability of nuclei to replicate in an egg extract from which the MCM complex has been depleted (for details, see Madine et al. 1995).

ever, when transport of the native MCM complex is examined, MCM3 does cross an intact nuclear membrane (Madine et al. 1995). One possible explanation of this difference is a folding problem of the recombinant fusion protein, but a more interesting possibility is that another member of the MCM family is responsible for targeting the whole MCM complex to the nucleus.

If the MCM complex can cross an intact nuclear membrane, then how does it relate to the requirement for nuclear membrane breakdown or permeability to reinitiate DNA replication? Clues may come from two directions. First, M.A. Madine et al. (in prep.) have shown that nuclear envelope permeability is needed to allow binding of MCM3 to chromatin, rather than for entry into the nucleus. Second, Chong et al. (1995) have shown by fractionation that a second activity is required for replication licensing. Thus, it appears that two separate activities are required for replication licensing. It is possible to argue semantically about which of these should be called "licensing factor." We suggest that this argument is unhelpful and that the term licensing factor should be replaced by referring instead to "replication licensing," since at least two classes of factors are involved.

ACKNOWLEDGMENTS

We thank A.D. Mills and L.S. Cox for providing original photographs for Figure 2; special thanks to A.D. Mills for help with preparation of figures and to Christine Fox for preparation of the manuscript. The work in this laboratory was supported by the Cancer Research Campaign.

REFERENCES

Blow, J.J. 1993. Preventing re-replication of DNA in a single cell cycle: Evidence for a replication licensing factor. *J. Cell Biol.* **122:** 993-1002.

Blow, J.J. and R.A. Laskey. 1986. Initiation of DNA replication in nuclei and purified DNA by a cell-free extract of *Xenopus* eggs. *Cell* **47:** 577-587.

―――. 1988. A role for the nuclear envelope in controlling DNA replication within the cell cycle. *Nature* **332:** 706-709.

Blow, J.J. and A.M. Sleeman. 1990. Replication of purified DNA in *Xenopus* egg extracts is dependent on nuclear assembly. *J. Cell Sci.* **95:** 383-391.

Blow, J.J. and J.V. Watson. 1987. Nuclei act as independent and integrated units of replication in a *Xenopus* cell-free system. *EMBO J.* **6:** 1997-2002.

Chong, J.P.J., H.M. Mahbubani, C.-Y. Khoo, and J.J. Blow. 1995. Purification of an Mcm-containing complex as a component of the DNA replication licensing factor. *Nature* **374:** 418-421.

Coverley, D., C.S. Downes, P. Romanowski, and R.A. Laskey. 1993. Reversible effects of nuclear membrane permeabilisation on DNA replication: Evidence for a positive licensing factor. *J. Cell Biol.* **122:** 985-992.

Cox, L.S. 1992. DNA replication in cell-free extracts from *Xenopus* eggs is prevented by disrupting nuclear envelope function. *J. Cell Sci.* **101:** 43-53.

Cox, L.S. and R.A. Laskey. 1991. DNA replication occurs at discrete sites in pseudonuclei assembled from purified DNA in vitro. *Cell* **66:** 271-275.

Crevel, G. and S. Cotterill. 1991. DNA replication in cell-free extracts from *Drosophila melanogaster*. *EMBO J.* **10:** 4361-4369.

Fox, M.H., D.J. Arndt-Jovin, T.M. Jovin, P.H. Baumann, and M. Robert-Nicoud. 1991. Spatial and temporal distribution of DNA replication sites localised by immunofluorescence and confocal microscopy in mouse fibroblasts. *J. Cell Sci.* **99:** 247-53.

Gilbert, D.M., H. Miyazawa, and M. DePamphilis. 1995. Site-specific initiation of DNA replication in *Xenopus* egg extracts requires nuclear structure. *Mol. Cell. Biol.* **15:** 2942-2954.

Harland, R. and R.A. Laskey. 1980. Regulated replication of DNA microinjected into eggs of *Xenopus laevis*. *Cell* **21:** 761-771.

Hennessy, K.M., C.D. Clark, and D. Botstein. 1990. Subcellular localization of yeast CDC46 varies with the cell cycle. *Genes Dev.* **4:** 2252-2263.

Hozak, P., A.B. Hassan, D.A. Jackson, and P.R. Cook. 1993. Visualization of replication factories attached to a nucleoskeleton. *Cell* **73:** 361-373.

Jackson, D.A. 1995. Nuclear organization: Uniting replication foci, chromatin domains and chromosome structure. *BioEssays* **17:** 587-591.

Jenkins, H., T. Holman, C. Lyon, B. Lane, R. Stick, and C. Hutchison. 1993. Nuclei that lack a lamina accumulate karyophilic proteins and assemble a nuclear matrix. *J. Cell Sci.* **106:** 275-285.

Kubota, Y. and H. Takisawa. 1993. Determination of initiation of DNA replication before and after nuclear formation in *Xenopus* egg cell free extracts. *J. Cell Biol.* **123:** 1321-1331.

Kubota, Y., S. Mimura, S. Nishimoto, H. Takisawa, and H. Nojima. 1995. Identification of the yeast MCM3 related protein as a component of *Xenopus* DNA replication licensing factor. *Cell* **81:** 601-609.

Leno, G.H. and R.A. Laskey. 1991. The nuclear membrane determines the timing of

DNA replication in *Xenopus* egg extracts. *J. Cell Biol.* **112:** 557–566.

Leno, G.H., C.S. Downes, and R.A. Laskey. 1992. The nuclear membrane prevents replication of human G2 nuclei but not G1 nuclei in *Xenopus* egg extract. *Cell* **69:** 151–158.

Madine, M.A., C.-Y. Khoo, A.D. Mills, and R.A. Laskey. 1995. An MCM3 complex is required for cell cycle regulation of DNA replication in vertebrate cells. *Nature* **375:** 421–424.

Méchali, M., F. Méchali, and R.A. Laskey. 1983. Tumour promoter TPA increases initiation of replication on DNA injected into *Xenopus* eggs. *Cell* **35:** 63–69.

Meier, J., K.H.S. Campbell, C.C. Ford, R. Stick, and C.J. Hutchison. 1991. The role of lamin L substructure in nuclear assembly and DNA replication, in cell free extracts of *Xenopus* eggs. *J. Cell Sci.* **98:** 271–278.

Mills, A.D., J.J. Blow, J.G. White, W.B. Amos, D. Wilcock, and R.A. Laskey. 1989. Replication occurs at discrete foci spaced throughout nuclei replicating in vitro. *J. Cell Sci.* **94:** 471–477.

Nakamura, H., T. Morita, and C. Sato. 1986. Structural organisations of replicon domains during DNA synthetic phase in the mammalian nucleus. *Exp. Cell Res.* **165:** 291–297.

Nakayasu, H. and R. Berezney. 1989. Mapping replicational sites in the eukaryotic cell nucleus *J. Cell Biol.* **108:** 1–11.

Newport, J. 1987. Nuclear reconstitution in vitro: Stages of assembly around protein-free DNA. *Cell* **48:** 205–217.

Newport, J., K.L. Wilson, and W.G. Dunphy. 1990. A lamin independent pathway for nuclear envelope assembly. *J. Cell Biol.* **111:** 2247–2259.

O' Keefe, R.T., S.C. Henderson, and D.L. Spector. 1992. Dynamic organisation of DNA replication in mammalian cell nuclei: Spatially and temporally defined replication of chromosome specific α-satellite DNA sequences. *J. Cell Biol.* **116:** 1095–1110.

Schedl, P. and F. Grosveld. 1995. Domains and boundaries. In *Chromatin structure and gene expression* (ed. S.C.R. Elgin), pp. 172–191. IRL/Oxford University Press, Oxford, United Kingdom.

Sheehan, M.A., A.D. Mills, A.M. Sleeman, R.A. Laskey, and J.J. Blow. 1988. Steps in the assembly of replication-competent nuclei in a cell-free system from *Xenopus* eggs. *J. Cell Biol.* **106:** 1–12.

Van Holde, K., J. Zlatanova, G. Arents, and E. Moudrianakis. 1995. Elements of chromatin structure: Histones, nucleosomes and fibres. In *Chromatin structure and gene expression* (ed. S.C.R. Elgin), pp. 1–21. IRL/Oxford University Press, Oxford, United Kingdom.

Wessel, R., J. Schweizer, and H. Stahl. 1992. Simian virus 40 T-antigen DNA helicase is a hexamer which forms a binary complex during bidirectional unwinding from the viral origin of DNA replication. *J. Virol.* **66:** 804–815.

5
Mechanisms for Priming DNA Synthesis

Margarita Salas
Centro de Biología Molecular "Severo Ochoa" (CSIC-UAM)
Universidad Autónoma, Canto Blanco
28049 Madrid, Spain

Jennifer T. Miller and Jonathan Leis
Department of Biochemistry
Case Western Reserve University School of Medicine
Cleveland, Ohio 44106-4935

Melvin L. DePamphilis
National Institute of Child Health and Human Development
National Institutes of Health
Bethesda, Maryland 20892-2753

DNA replication is a semiconservative process in which a DNA polymerase uses one DNA strand as a template for the synthesis of a second, complementary, DNA strand. However, in contrast to RNA polymerases, which can initiate RNA synthesis on a DNA template de novo, all DNA polymerases require a preexisting primer on which to initiate DNA synthesis (Kornberg and Baker 1992). One apparent exception to this rule is a mitochondrial DNA (mtDNA)-encoded reverse transcriptase (RT) in *Neurospora* (Wang and Lambowitz 1993). Preexisting primers can be classified into four groups. The simplest primer consists of the 3'-hydroxyl (3'-OH) termini of DNA chains that are complementary to the DNA template and thereby form a stable duplex structure at the site where DNA synthesis begins. This primer is used for DNA repair (Friedberg and Wood, this volume), parvovirus DNA replication (Brush and Kelly; Cotmore and Tattersall; both this volume), some RTs. The second type of primer consists of a deoxyribonucleoside monophosphate that is covalently attached to a specific serine, threonine, or tyrosine residue of a protein. Examples are bacteriophage, plasmids, and animal viruses that replicate as a linear DNA genome, and animal viruses such as hepadnaviruses whose genome is partially double-stranded and partially single-stranded. The third type of primer consists of tRNA molecules that anneal to specific sequences in the RNA genomes of retroviruses where their 3′-OH termini are utilized by RT. The fourth class of primers consists of nascent RNA chains. These comprise nascent RNA transcripts that are processed to create a primer at a specific site in the template and short nascent oligoribonucleotides (initiator RNA) that are synthesized at many sites in the template and rapidly extended into short RNA-DNA primers by DNA polymerase-α:DNA primase (pol-α:primase). Nascent RNA transcripts are used during initiation of mtDNA replication, whereas replication forks in cellular chromosomes and double-stranded DNA (dsDNA) viral genomes that replicate within the nucleus use the initiator RNA mechanism.

INITIATION OF DNA SYNTHESIS ON PROTEIN PRIMERS

From prokaryotes to eukaryotes, linear DNAs exist in nature that cannot form circular structures through cohesive ends, hairpin structures at their termini, or concatemers during replication. In such cases, initiation of replication cannot take place by either RNA or DNA priming. However, most of these linear DNAs contain a protein covalently linked to their 5′ ends that acts as primer during initiation of DNA replication. This protein is called terminal protein (TP), and its role in DNA replication is summarized in Figure 1 and has been previously reviewed by Salas (1991).

The first evidence for the existence of protein attached at the ends of a linear dsDNA was the finding that the 19.3-kb virion DNA of *Bacillus subtilis* phage φ29 could be isolated as circular molecules and concatemers that were converted into unit-length linear DNA by treatment with proteolytic enzymes (Ortín et al. 1971). Such a treatment greatly reduced the capacity of φ29 DNA to transfect competent *B. subtilis* cells (Hirokawa 1972), and DNA isolated from a φ29 temperature-sensitive mutant in gene 3 was thermolabile for transfection (Yanofsky et al. 1976). Later it was shown that the 28-kD protein product of viral gene 3 was covalently linked at each 5′ end of the viral DNA (Salas et al. 1978). Other *B. subtilis* phages that contain linear dsDNA and TP of similar size fall into three serological classes: (1) φ15, PZA, and PZE in the group of φ29; (2) Nf, M2, and B103; and (3) GA-1. All phages of the φ29 family have a short inverted terminal repeat (ITR, Fig. 2), which is six nucleotides long for φ29, PZA, φ15, and B103 DNAs; eight nucleotides long for Nf and M2 DNAs; and seven nucleotides long for GA-1 DNA (Salas 1991). The linkage between φ29 TP and DNA is a phosphoester bond between the OH group of Ser-232 in TP and 5′-

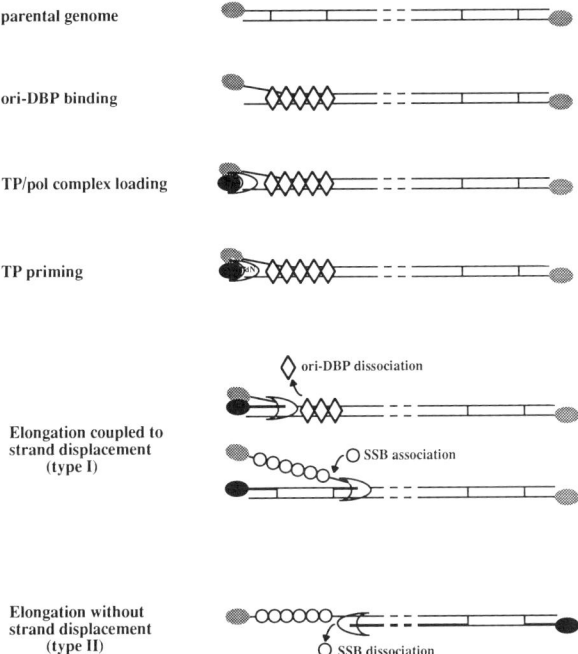

Figure 1 Model for protein priming of DNA replication. Only initiation starting at one DNA end is shown.

dAMP. Prediction of secondary structure in this region suggests that Ser-232 is located in a β turn, probably in the external part of the molecule, preceded by an α helix (Hermoso et al. 1985).

Replication of linear genomes starts at either DNA end, non-simultaneously, and proceeds by strand displacement toward the other end (Fig. 1). A dsDNA-binding protein (ori-DBP) binds to sequences close to the DNA ends, forming a nucleoprotein complex that helps to open the DNA ends to facilitate the interaction of a heterodimer that forms between TP and the DNA polymerase encoded by the linear genome. Initiation of replication occurs at either DNA terminus by the covalent linkage of a specific dNMP to the OH group of a specific serine, threonine, or tyrosine residue on TP, in a reaction catalyzed by the DNA polymerase and directed by an internal nucleotide at the 3' end of the template. After sliding-back (or jumping-back) (see below), elongation occurs coupled to strand displacement giving rise to type I molecules, with the likely dissociation of ori-DBP and the association of single-stranded DNA-binding protein (SSB) to the displaced ssDNA. Type I molecules are unit-length linear dsDNA with one or more single-

B. subtilis phages	φ29 φ15 PZA Nf M2 B103 GA1	3' TTTCATCCCATG 3' TTTCATCCCATG 3' TTTCATCCCATG 3' TTTCATTCCAAG 3' TTTCATTCCAAG 3' TTTCATCCCATG 3' TTTATCTAAGGG
E. coli phage	PRD1	3' CCCCTATGCACG
S. pneumoniae phage	Cp-1	3' TTTCGTACATGA
Eukaryotic virus	Adeno 2	3' GTAGTAGTTATT
Linear plasmids	S1 S2 pSKL pGKL1 pGKL2 pMC3-2 Kalilo Maranhar pSCP1 pSCL pSLP2 pSLA2	3' TTTTCATATGTT 3' TTTTCATATGTT 3' TTTTCCATATCT 3' TGTGTATTGTAT 3' TTTTCCATATAT 3' CAACACATATGG 3' CACATTCCCCGT 3' XXCCCCGTAGCC 3' (CC)CCCGCCTCT 3' GGGCGCCTCGCC 3' GGGCGCCTCGCC 3' GGGCGCCTCGCC
Bacterial chromosome	S. lividans	3' GGGCGCCTCGCC

Figure 2 The 3'-terminal nucleotide sequences of TP-containing DNA genomes.

stranded tails from the same or a different DNA end. Type II molecules are then formed in which elongation occurs without strand displacement, with the concomitant dissociation of the SSB protein. Type II molecules are unit-length linear molecules partially double-stranded and partially single-stranded. Thus, two fully replicated molecules result from initiation at each DNA end.

Requirements for Initiation of DNA Replication in the Phage φ29 Genome

To test whether a fully displaced parental DNA strand can be replicated in vivo, recombinant φ29 DNA molecules containing parental TP at only one DNA end were constructed. No replication in *B. subtilis* protoplasts was obtained, suggesting that the fully displaced DNA strand is not a template for replication (Escarmís et al. 1989). When the above recombinant φ29 DNA molecules were used as templates in the in vitro system (see below), no type II molecules were found (Gutiérrez et al. 1991a). Moreover, when replication of φ29 TP-DNA was studied in the in vitro system, a significant amount of type II replicative intermediates was found at an incubation time at which no synthesis of full-length φ29 DNA was detected. These results indicate that the appearance of type II replicative intermediates does not require synthesis of full-length DNA and full displacement of the parental strand; rather, the results support a model in which initiation of replication can occur from both DNA ends, and type II molecules are produced by separation of the two displacement forks when they meet.

When extracts from φ29-infected *B. subtilis* were incubated with [α-^{32}P]dATP in the presence of φ29 TP-DNA, a ^{32}P-labeled protein with the electrophoretic mobility of the TP, product of the viral gene 3, was found (Peñalva and Salas 1982; Shih et al. 1982). Incubation of the labeled protein with piperidine released 5'-dAMP, indicating the formation of a TP-dAMP covalent complex (Peñalva and Salas 1982). A covalent complex between the M2 TP and 5'-dAMP was also found when extracts from phage M2-infected *B. subtilis* were used (Matsumoto et al. 1983).

Phage φ29 genes 2 and 3 and phage M2 genes G and E were shown to be essential for the in vitro initiation reaction, that is, the formation of the TP-dAMP covalent complex using the corresponding TP-DNA as template (Blanco et al. 1983; Matsumoto et al. 1983). Phage φ29 genes 2 and 3 were cloned, and both proteins, p2 (Blanco and Salas 1984) and p3 (TP) (Prieto et al. 1984; Watabe et al. 1984a), were overproduced and purified in a functional form. Purified p2, in addition to catalyzing the initiation reaction, has DNA polymerase activity (Blanco and Salas 1984; Watabe et al. 1984a) and 3'→5' exonuclease activity, believed to be involved in proofreading (Watabe et al. 1984b; Blanco and Salas 1985a), which is about 10-fold more active on single- than on double-stranded DNA (Garmendia et al. 1992). φ29 DNA polymerase can also catalyze TP-deoxynucleotidylation in the absence of DNA template (Blanco et al. 1992).

The in vitro initiation reaction is greatly stimulated by NH_4^+ ions (Blanco and Salas 1985b), which stabilize the formation of a heterodimer between DNA polymerase and TP (Blanco et al. 1987), required for the initiation of replication. In fact, the two proteins are purified as a complex when extracts from φ29-infected cells are used (Watabe et al. 1983; Matsumoto et al. 1984). In addition, Mn^{++} ions are better activators of the initiation reaction than Mg^{++} due to a 50-fold decrease in the K_m for dATP (Esteban et al. 1992).

Once the viral DNA polymerase catalyzes the covalent linkage of dAMP to the OH group of Ser-232 in the TP, the same DNA polymerase processively elongates the nascent DNA strand in vitro to give full-length φ29 DNA (Blanco and Salas 1985b; Blanco et al. 1989). Further studies using primed M13 DNA as template indicated that the φ29 DNA polymerase is a highly processive enzyme (over 70 kb) that can catalyze strand displacement without need of accessory proteins (Blanco et al. 1989). Other viral proteins required for φ29 DNA replication are the origin-binding protein p6 and SSB p5 (see below).

Using the four φ29 replication proteins: TP, DNA polymerase, origin-binding protein p6, and SSB p5, and limiting amounts of φ29 TP-DNA, it is possible to amplify in vitro φ29 DNA by a factor of 1000-fold (Blanco et al. 1994).

Interaction of φ29 DNA Polymerase and Terminal Protein

Mutants obtained in each putative active-site residue at the Exo I, Exo II, or Exo III motifs of φ29 DNA polymerase, located at the amino end, did not impair protein-primed initiation and DNA polymerization; however, exonuclease activity was strongly reduced in all cases (Bernad et al. 1989; Soengas et al. 1992; Esteban et al. 1994). Unexpectedly, these mutant proteins were almost inactive when assayed for φ29 TP-DNA replication. This defect was shown to be due mainly to a 10- to 50-fold decrease in the rate of DNA synthesis coupled to strand displacement. Therefore, the strand-displacement activity of the φ29 DNA polymerase resides in the amino-terminal domain, probably overlapping with the $3' \rightarrow 5'$ exonuclease active site (Soengas et al. 1992). Site-directed mutagenesis of the conserved motifs in the carboxy-terminal portion of φ29 DNA polymerase indicated that this domain of the φ29 DNA polymerase contains the protein-primed initiation and DNA-polymerization activities of this enzyme (Blanco and Salas 1995). Assuming that the φ29 DNA polymerase structure is similar to that of the Klenow fragment of *Escherichia coli* DNA polymerase I (Ollis et al. 1985; Blanco et al. 1991), the polymerase DNA-binding cleft is proposed to be also the TP-binding site (Salas et al. 1993). Thus, the TP molecule bound to the DNA polymerase cleft could place the specific priming residue (Ser-232), acting as OH donor, next to the dNTP-binding site.

Deletion mutagenesis studies in the φ29 TP indicated the existence of two DNA polymerase-binding regions, located at positions 72–80 and 241–261, and three DNA-binding regions, at positions 13–18, 30–51, and 56–71 (Salas 1991). Site-directed mutagenesis showed that changing Ser-232 to a threonine inactivated the protein, whereas changing it to cysteine reduced the priming activity to about 0.7% of wild-type TP.

Changing Leu-220, Ser-223, and Ser-226 independently into proline resulted in mutant proteins with 3%, 140%, and 1%, respectively, of wild-type TP priming activity. All of the mutant TPs could interact with DNA polymerase and DNA, suggesting that Leu-220 and Ser-226, in addition to Ser-232, form part of a functional domain involved in initiation of DNA replication (Salas 1991).

The amino acid motif RGD, characteristic of cell adhesion proteins, is present in the TP of phage φ29, and it was proposed to be involved in the interaction of the TP/DNA polymerase heterodimer with the parental TP bound to the DNA (Kobayashi et al. 1991a,b). In addition, a sequence very similar to KKGCPPDD, found in the β subunit of the fibronectin receptor protein, is also present in φ29 TP. Analysis of synthetic peptides suggested that this sequence acts as a receptor in the parental TP, whereas the RGD sequence acts as an effector in the new TP primer.

φ29 Origin of DNA Replication

Treatment of φ29 TP-DNA with proteinase K inactivates it as a template for DNA replication. However, when the residual peptide that remains attached to the 5' end is removed with piperidine, template activity is restored, although it is 5- to 10-fold less than that obtained with φ29 TP-DNA. The minimal φ29 replication origins are located within the terminal 12 bp at each DNA end (Salas 1991). Template activity is also obtained with single-stranded oligonucleotides 12 bases long corresponding to the 3'-terminal sequence at the right or left φ29 DNA ends, giving rise to formation of a TP-dAMP initiation complex that can be fully elongated (Méndez et al. 1992). Interestingly, deoxynucleotidylation of the TP can also be obtained in the absence of any template, although in this case, any of the four dNTPs work, and the affinity for the nucleotide is greatly decreased (Blanco et al. 1992). Therefore, the template provides nucleotide specificity and increases the affinity of TP for its DNA polymerase.

φ29 Protein p6 and Formation of a Nucleoprotein Complex

Protein p6 is a 123-amino-acid protein that stimulates initiation of φ29 DNA replication by reducing the K_m for dATP and facilitating the transition from initiation to elongation (Blanco et al. 1986, 1988). Stimulation by p6 is due to formation of a nucleoprotein complex that spans 200–300 bp from each DNA end (Prieto et al. 1988; Serrano et al. 1989). p6 binds preferentially the φ29 replication origins at nucleotides 46–68 at the left end of φ29 DNA and nucleotides 62–125 at the right end (Serrano et al. 1989). These regions do not show any sequence similarity, but they contain DNA sequences predicted to be bendable every 12 bp, suggesting that bendability may be the major determinant for protein p6 recognition (Serrano et al. 1989). In agreement with this, tandem repetitions of a 24-bp bendable sequence, present in one of the main p6 recognition regions, bind protein p6 in the same positions and with higher affinity than the φ29 DNA replication origins (Serrano et al. 1993).

An α-helical structure located in the amino-terminal region of protein p6 is involved in binding DNA through its minor groove (Freire et al. 1994). A p6 dimer binds 24 bp, bending or kinking the DNA every 12 bp. Binding is highly cooperative, giving rise to a large multimeric complex in which a right-handed superhelix wraps around a protein core (Serrano et al. 1990), restraining positive supercoils when formed on a covalently closed plasmid (Prieto et al. 1988). p6 binding results in a 4.2-fold compaction of the DNA in which one superhelical turn has 63 bp (2.6 protein p6 dimers) with a pitch of 5.1 nm and a diameter of 6.6 nm. Therefore, the DNA should be bent 66° every 12 bp and underwound 11.5 bp per turn (Serrano et al. 1993). These features could facilitate the initial unwinding of DNA required to start replication by the DNA polymerase. In agreement with this, activation by p6 is greatest at lower temperatures (Salas et al. 1993).

φ29 Protein p5

Protein p5 is a 13-kD SSB that stimulates φ29 DNA replication (Salas 1991). Stimulation does not result from an increase either in formation of the TP-dAMP initiation complex (Martín et al. 1989) or in DNA elongation rate (Gutiérrez et al. 1991b). p5 binds nonspecifically to ssDNAs, including the ssDNA portions of replicative intermediates produced during φ29 DNA replication in vitro (Gutiérrez et al. 1991b). Each p5 monomer covers 3–4 nucleotides with a binding constant of 10^5 M^{-1} and an unlimited cooperativity parameter of 50–80 (Soengas et al. 1994). p5 can facilitate removal of secondary structure in the displaced ssDNA in replicative intermediates and displace oligonucleotides annealed to ssDNA (Soengas et al. 1995). Therefore, in addition to protecting the ssDNA produced during φ29 DNA replication from nuclease degradation (Martín et al. 1989) and preventing unproductive binding of φ29 DNA polymerase to ssDNA (Gutiérrez et al. 1991b), φ29 SSB could help to unwind secondary structure that may form in the displaced ssDNA during φ29 DNA replication. Consistent with this hypothesis, the rate of φ29

DNA elongation by φ29 DNA polymerase mutants defective in strand displacement is stimulated about 5-fold by addition of φ29 SSB (Soengas et al. 1995).

The interaction of the proteins and replication origin described above for φ29 DNA replication is summarized in Figure 1. In many respects, φ29 DNA replication is a paradigm for replication of other linear DNA genomes.

Protein-primed DNA Replication in Other Genomes

Phage PRD1

Phage PRD1 infects gram-negative bacteria including *E. coli* and *Salmonella typhimurium*. The 5' termini of the 14.9-kb PRD1 linear genome are linked to a 28-kd TP by a phosphoester bond between dGMP and tyrosine residue 190 (Bamford et al. 1983; Bamford and Mindich 1984; Shiue et al. 1991). DNA of PRD1 and closely related phages PR3, PR4, PR5, PR722, and L17 all have a 110-bp-long ITR (Mindich and Bamford 1988). TPs from PRD1, φ29, PZA, Nf, adenovirus, and the core antigen of duck hepatitis B virus all contain the conserved motif Tyr-Ser-Arg-Leu-Arg-Thr (Hsieh et al. 1990).

PRD1 DNA replication appears quite similar to that of φ29. It requires viral genes I (DNA polymerase) and VIII (TP) (Hsieh et al. 1987; Jung et al. 1987; Savilahti and Bamford 1987) in addition to some host components. Purified DNA polymerase catalyzes the formation of TP-dGMP covalent complex, has processive DNA chain-elongation activity, can catalyze strand displacement, and contains 3'→5' exonuclease activity (Savilahti et al. 1991; Zhu and Ito 1994). As with φ29 DNA polymerase, initiation is strongly activated by Mn^{++} (Caldentey et al. 1992). PRD1 DNA polymerase can also catalyze TP-deoxynucleotidylation in the absence of DNA template (Caldentey et al. 1992). Mutagenesis of conserved Lys-340 in the Kx$_3$NSxYG motif of PRD1 DNA polymerase generated a protein that had lost protein-priming and polymerization activities without affecting the 3'→5' exonuclease activity (Zhu et al. 1994). Mutagenesis in residue Arg-174 of the PRD1 TP, corresponding to the conserved motif YSRLRT, resulted in an inactive TP that was unable to form an initiation complex (Hsieh et al. 1990). As with φ29, linear duplex, protein-free, DNA molecules containing the 20 bp from the PRD1 DNA ends can undergo replication by protein priming in vitro. Similarly, a 27-mer single-stranded oligonucleotide containing the 20 bases of the 3' end of the PRD1 genome supported the formation of the TP-dGMP complex (Yoo and Ito 1991).

Phage Cp-1

The 5' termini of the 19.3-kb DNA of *Streptococcus pneumoniae* phages Cp-1, Cp-5, and Cp-7 are covalently linked to a 28-kD TP and contain ITRs of 236 bp, 343 bp, and 347 bp, respectively (García et al. 1983; Salas 1991). The amino acid sequence of Cp-1 TP is 71% homologous to that of φ29 (Martín et al. 1995) and contains a phosphoester bond between threonine and 5'-dAMP that can be elongated in vitro (García et al. 1986). Phage Cp-1 DNA polymerase is 96% homologous to that of phage φ29 (Martín et al. 1995). The 40-kb DNA of HB3 and related phages HB-623 and HB-746 (Romero et al. 1990) carries a 23-kD TP whose role in DNA replication remains to be determined, although the invertron model has also proposed a role in integration (Sakaguchi 1990).

Linear Plasmids

A variety of linear plasmids have been isolated from bacteria, yeast, fungi, and higher plants. In most cases, long ITRs have been characterized, and in many of them, evidence for the existence of a TP linked at the 5' ends of the DNA has been reported (Meinhardt et al. 1990; Salas 1991; Rohe et al. 1992). Linear DNA of yeast killer plasmids pGKL1 and pGKL2 replicates by a strand-displacement mechanism similar to that described for phage φ29 (Fujimura et al. 1988). Extracts from cells carrying pGKL1 contain an activity called terminal region recognition factor 1 (TRFl) that recognizes the termini of both pGKL1 (bp 107–183 within the ITR) and pGKL2 (bp 126–179 within the ITR) (McNeel and Tamanoi 1991). This binding protein could be similar in function to protein p6 and factors NFI and NFIII that bind to the replication origins of phage φ29 and Ad, respectively, and stimulate the initiation of replication. Furthermore, ORF1 in plasmid pGKL1 and ORF2 in pGKL2 contain regions of amino acid homology located at the carboxyl region of eukaryotic-type DNA polymerases, including the protein-primed DNA polymerases. By analogy with linear viral genomes, it is likely that the TPs of pGKL1 and pGKL2 are plasmid-encoded (Salas 1991).

Replication of plasmid pAI2 from *Ascobolus immersus* also starts at the DNA ends. In addition, a large open reading frame (ORF) spanning 1202 amino acids shares homology with eukaryotic-type DNA polymerases and with putative DNA polymerases from linear plasmids (Kempken et al. 1989). ORF1 from the *Claviceps purpurea* plasmid pClK1 encodes a protein of 1097 amino acids, which is also likely to be a DNA polymerase according to the above criteria (Oeser and Tudzynski 1989).

The gene encoding TP is unknown. However, taking into account that the amino acid involved in the linkage of φ29 and Ad TPs to DNA is located in a β turn preceded by an α helix (Hermoso et al. 1985), it is possible that the TP of pClK1 is encoded by 400 amino acids at the amino-terminal part of ORF1. In fact, serine residue 327 of this ORF was reported to meet the criteria for a nucleotide-linking site (Oeser and Tudzynski 1989). ORF3 of plasmid S1 also codes for a protein with amino acid homology with eukaryotic-type DNA polymerases (Paillard et al. 1985). A similar case is that of ORFs of plasmids pEM from *Agaricus bitorquis*, pMC3-2 from *Morchella conica*, and the *kalilo* and *maranhar* plasmids from *Neurospora* (for review, see Rohe et al. 1992). All the above data suggest that the linear plasmids replicate by a protein-priming mechanism.

Linear Chromosomes

Two copies of a DNA sequence similar or identical to the right end of the linear plasmid pSLP2 were found at the ends of the 8-Mb *Streptomyces lividans* linear chromosome. The telomeres contain a 25-kb ITR and carry covalently bound protein. The TP is removed by piperidine treatment, suggesting that, like phage φ29, the linkage to DNA occurs through a serine residue in the *S. lividans* TP (Lin et al. 1993). Chromosomal DNA of six other *Streptomyces* species also behave as linear molecules of about 8 Mb, suggesting that chromosomal linearity may be common among the streptomycetes (Lin et al. 1993).

A functional *oriC* also has been located at the center of the *S. lividans* chromosome (Zakrzewska-Czerwinska and Schrempf 1992). Since this chromosome can be circularized by joining the two ends by artificial targeted recombination or by spontaneous deletions spanning both telomeres, it appears to exist as either a linear or a circular molecule (Lin et al. 1993). The linear chromosome appears to replicate bidirectionally from its center. Presumably, replication of the ends is completed by protein-primed DNA replication. It has been postulated that the more primitive bacterial chromosomes replicated from the telomeres, and that the complex machinery for internal initiation came later. It is an open question whether bacterial chromosomes that replicate exclusively from the telomeres still exist in bacteria (Chen et al. 1995).

Adenovirus

The human adenovirus (Ad) genome consists of linear, dsDNA of 36 kb with ITRs of about 100 bp (Steenbergh et al. 1977) and a protein of 55 kD covalently linked at the 5' ends of the DNA (Carusi 1977; Rekosh et al. 1977). The Ad TP is synthesized as a precursor, the preTP (pTP) (Challberg et al. 1980), and pTP forms a phosphoester bond between serine residue 580 and 5'-dCMP (Challberg et al. 1980; Desiderio and Kelly 1981; Smart and Stillman 1982).

Initiation of Ad DNA replication occurs at either DNA end and proceeds by a strand-displacement mechanism giving rise to type I and type II molecules, as described above for phage φ29 DNA replication (Brush and Kelly; Hay; both this volume). A model first proposed by Rekosh et al. (1977) suggested that a free molecule of the TP could act as a primer for the initiation of replication by formation of a covalent linkage with dCMP, the 5'-terminal nucleotide, that would provide the 3'-OH group needed for elongation by the DNA polymerase. The 5' end of each nascent daughter strand is covalently linked to a protein of 80 kD (Challberg et al. 1980; Lichy et al. 1981), structurally related to the 55-kD protein covalently linked to the 5' ends of Ad DNA. The 80-kD pTP is processed to the 55-kD TP during virus maturation (Challberg and Kelly 1981). As with φ29 and PRD1 DNA replication, pTP and a 140-kD Ad DNA polymerase are required for Ad DNA replication. pTP and Ad DNA polymerase form a heterodimer that binds to the core origin comprising the 20 terminal nucleotides (Temperley and Hay 1992). Initiation of replication occurs by DNA polymerase-catalyzed transfer of dCMP onto the OH group of Ser-580 in pTP. Ad DNA polymerase, like φ29 and PRD1 DNA polymerases, has, in addition to its initiation and polymerization activities, 3'→5' exonuclease activity presumably involved in proofreading (Lindenbaum et al. 1986). In most of the mutants obtained, both initiation and polymerization activities were lost, indicating that these functions might be closely linked, as is the case with the φ29 DNA polymerase. The amino terminus of pTP is essential for priming activity and DNA binding (Pettit et al. 1989).

Another viral protein, the DNA-binding protein (DBP), binds ssDNA and dsDNA and stimulates the initiation of replication. Elongation of Ad DNA replication catalyzed by Ad DNA polymerase proceeds by strand displacement and also requires DBP. Ad DBP has properties that it shares with both p6 (dsDNA-binding protein) and p5 (SSB) of phage φ29. Like p6, DBP binds to the replication origins forming a nucleoprotein complex and stimulates the initiation of replication by decreasing the K_m for the initiating dNTP. Like p5, DBP binds to ssDNA and has unwinding activity. NFI and NFIII/Oct-1 are transcription factors that bind to the auxiliary region of the Ad replication origin, where they enhance formation of a stabilized nucleoprotein structure with the three viral proteins (van der Vliet, this volume).

Hepadnaviruses: Reverse Transcription

Hepadnaviruses contain a 3-kb circular, dsDNA genome in which the minus (uncoding) strand is complete with unique 5' and 3' termini. The plus strand is incomplete with a unique 5' end and a heterogeneous 3' end (Seeger and Mason, this volume). Three members of this virus family (human hepatitis B, ground squirrel hepatitis, and duck hepatitis B) contain a protein covalently linked to the 5' end of the complete minus strand (Gerlich and Robinson 1980; Molnar-Kimber et al. 1983; Weiser et al. 1983).

The mechanism by which hepadnavirus DNA replicates involves reverse transcription (Seeger and Mason, this volume). The RNA template for reverse transcription, the pre-genome, is produced by copying the minus-strand DNA in the closed circular dsDNA molecule formed from the virion DNA (Summers and Mason 1982; Ganem and Varmus 1987; Seeger et al. 1991). The polymerase gene product of hepadnaviruses encodes the TP at the amino-terminal quarter of this 785-amino-acid polypeptide, the DNA polymerase/RT in the central part, and the RNase H at the carboxy-terminal region. Furthermore, TP and RT domains are linked by a tether region that tolerates amino acid substitutions and deletions (Bartenschlager and Schaller 1988; Chang et al. 1990; Radziwill et al. 1990). Tyr-96 donates the OH group for the formation of the covalent bond between the polymerase gene product and viral DNA (Zoulim and Seeger 1994). Interestingly, Tyr-96 is part of a Gly-x-Tyr motif present in the TP of picornavirus at the site where the protein is linked to the RNA (Khudyakov and Makhov 1989). Thus, unlike other RTs that use tRNA as primers, hepatitis B virus RT uses a protein as primer. However, in contrast to other genomes that use proteins to prime DNA synthesis, DNA priming and polymerase activities in hepadnaviruses reside on the same polypeptide.

Sliding-back (Jumping-back) Mechanism for the Transition from Initiation to Elongation and for the Maintenance of the DNA Ends

Use of single-stranded homopolymers in the in vitro $\phi29$ replication system led to the conclusion that TP priming is a template-directed event (Méndez et al. 1992). Therefore, a mutational analysis of the $\phi29$ DNA right replication origin was carried out using as templates 12-mer single-stranded oligonucleotides containing the 3' end of the natural $\phi29$ DNA sequence (TTTCAT.......) or mutant derivatives with single changes in the first, second, or third T, to determine which T residue directs the formation of the TP-dAMP initiation complex. The results obtained clearly indicate that the second nucleotide of the template directs the linkage of dAMP to the primer protein (Méndez et al. 1992). This unexpected result was confirmed using a variety of single-stranded oligonucleotides and also dsDNA fragments containing the natural φ29 DNA left replication origin, or with a mutation at the second nucleotide from the end. Addition of the origin-binding protein p6 to the dsDNA fragments gives similar results, indicating that the initiation site of φ29 DNA replication is the second 3'-terminal nucleotide. The physiological role of this internal initiation event is supported by the fact that all the nucleotides in the template, including the 3'-terminal one, are replicated. Moreover, a terminal repetition of at least two nucleotides is required for efficient elongation of the initiation complex.

A sliding-back mechanism was proposed for the transition from initiation to elongation (Fig. 3) (Méndez et al. 1992). Once the TP-dAMP initiation complex has been formed, directed by the second nucleotide (T) at the 3' end, the TP-dAMP complex slides backward, locating the dAMP in front of the first nucleotide of the template (asymmetric translocation). Then the next nucleotide (A) is incorporated to the TP-dAMP complex, using again the second T of the template as a director. Further nucleotide addition involves normal translocation of both template and DNA primer terminus (normal elongation).

This strategy for maintaining the integrity of the φ29 DNA ends could also increase the fidelity of the TP-primed initiation reaction (Esteban et al. 1993). If a mismatched initiation event occurs, transition to elongation would not be efficient, and the incorrect initiation complex could dissociate from the DNA. Concomitant removal of both TP and the mismatched nucleotide may be the only possibility for editing, because the exonucleolytic activity of φ29 DNA polymerase, which acts as a proofreading enzyme during elongation (Garmendia et al. 1992), cannot excise the first dNMP linked to the TP (Esteban et al. 1993). Nonetheless, if a mutation is established in the first nucleotide of the DNA, the second 3' nucleotide of the template would be used again, restoring the terminal repetition at the end of the molecule. Therefore, according to the sliding-back model, errors in the first replication event would be lost because the 3'-terminal nucleotide is not used as the initiation site.

A similar strategy was employed to show that in PRD1 DNA, the fourth base from the 3' end of the template directs, by base complementarity, the dNMP to be linked to the TP in the initiation reaction (Fig. 3) (Caldentey et al. 1993). Thus, phage PRD1 maintains its 3'-end DNA sequences via a sliding-back mechanism. Unlike TP-DNA, the

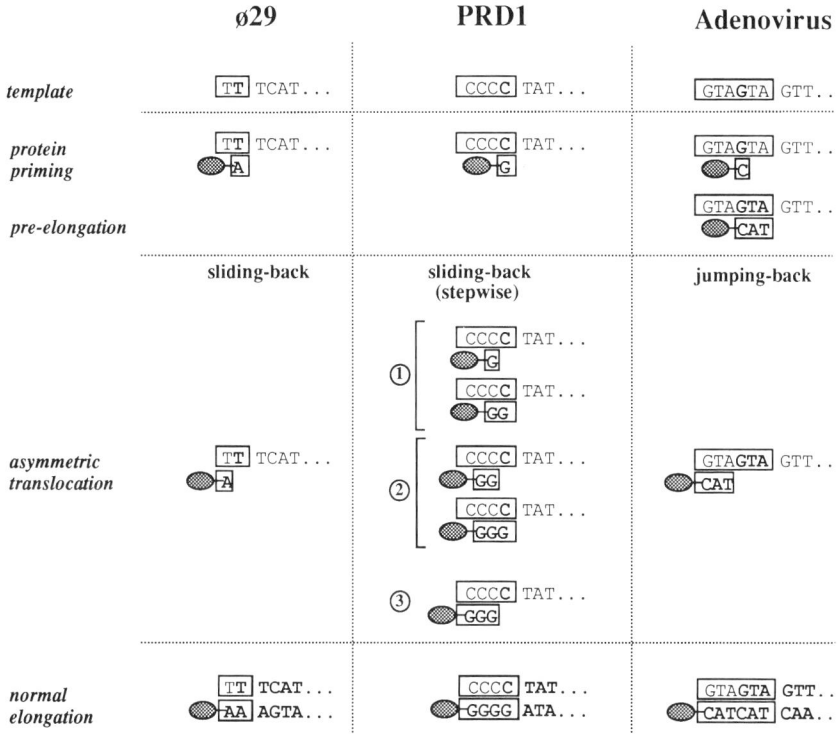

Figure 3 Sliding-back (jumping-back) model for the transition from initiation to elongation.

ssDNA templates could not be elongated by TP and DNA polymerase in vitro. Nevertheless, analysis of the transition products obtained with TP-DNA and origin-containing oligonucleotides suggest that sliding-back occurs stepwise, the fourth base being the directing position during the entire process.

In the Ad genome, the terminal sequence of the template strand has the more complex reiteration 3'-GTAGTA. Changes in the first G residue do not affect the formation of the pTP-dCMP complex, whereas additional substitution of the G residue at position four drastically impairs this reaction (Dobbs et al. 1990). Therefore, the G residue at position four could act as the template nucleotide during formation of the pTP-dCMP complex (Méndez et al. 1992). Recently, it was reported that a kinetic barrier to further elongation of the Ad pTP-CAT product is relieved by a high dCTP concentration (Mul and van der Vliet 1993). This is consistent with a rate-limiting, sliding-back step in which the pTP-CAT product, initiated from the fourth template residue, slides back in a single jump to be paired with the first three bases of the template, thus regenerating the Ad DNA ends (Caldentey et al. 1993). Recent results (King and van der Vliet 1994) support the model in which the pTP-CAT intermediate, synthesized opposite to positions 4–6, jumps back to positions 1–3 of the template to start elongation (Fig. 3). This jumping-back mechanism ensures the integrity of terminal sequences during replication of the linear genome.

Other linear genomes that contain TP have some sequence repetition at the DNA ends (Fig. 2), suggesting that internal initiation followed by sliding-back either in one step (phage φ29), stepwise (phage PRD1), or in a single jump (Ad) could be applicable to other systems that use a protein primer. This mechanism may also apply to RNA replication. It has been suggested that initiation of Tacaribe arenavirus RNA replication occurs at the second 3'-terminal nucleotide, and that the initiation complex slips backward before elongation can continue (Garcin and Kolakofsky 1992). Another example may be hepatitis B virus, where the polymerase binds to an RNA hairpin that serves as a template for formation of a short DNA primer covalently linked to protein. Following its synthesis, the nascent DNA strand apparently dissociates from its template and reanneals with complementary sequences at the 3' end of the RNA genome where DNA synthesis continues (Wang and Seeger 1993).

INITIATION OF DNA SYNTHESIS ON PREFORMED RNA PRIMERS

Retroviruses, as well as retroelements, recruit preexisting host-encoded tRNAs to prime DNA synthesis by RT (Weiss et al. 1985). Various tRNA species are utilized, depending on the retrovirus (Leis et al. 1993); tRNA^{Trp} is used by avian sarcoma leukosis viruses and tRNA^{Pro}, tRNA^{Lys,1,2}, and tRNA^{Lys,3} are used by a variety of murine, feline, simian, or human viruses.

Encapsidation of tRNA Primers by Retroviruses

Specific tRNA primers are encapsidated into virions through interactions with RT (Panet et al. 1975; Levin and Siedman 1979, 1981; Peters and Hu 1980). These interactions involve the TψC and the DHU arms of the tRNA (Hu and Dahlberg 1983; Barat et al. 1991). Furthermore, the recognition of the primer occurs at the level of the virus-encoded *gag-pol* precursor rather than that of free RT, since viral protease activation is not necessary for selective encapsulation of primers into virions (Crawford and Goff 1985; Stewart et al. 1990; Mak et al. 1994). The presence of modified bases in the tRNA also does not appear to be important either for binding or functionality as a primer (Aiyar et al. 1992; Wohrl et al. 1993; Huang et al. 1994).

Minus-strand Priming during Retrovirus Replication

Annealing of the primer to viral RNA is catalyzed by RT (Barat et al. 1989; Aiyar et al. 1992). The viral nucleocapsid protein may stimulate this process, but it is not necessary (Aiyar et al. 1994). The site to which the primer binds is in the 5′-untranslated region of viral RNA referred to as the primer binding site (PBS) (Weiss et al. 1985). The acceptor stem of the tRNA is unwound, and between 14 and 22 nucleotides (Weiss et al. 1985) form a base-paired duplex with the PBS (Fig. 4). The viral RNA 5′ of the PBS is termed U5, and the RNA 3′ of the PBS is termed leader. Complete base-pairing between the PBS and the primer tRNA is not necessary for priming of DNA synthesis. As many as 9 of the 18 base pairs between the human immunodeficiency virus (HIV-1) PBS and tRNA ^{Lys3} (Wakefield et al. 1994) and 3 of the 18 base pairs between the Rous sarcoma virus (RSV) PBS and tRNA^{Trp} (Aiyar et al. 1994) can be mismatched without abrogating primer function, provided that the mismatches do not interfere with the base-pairing at the 3′ end of the primer. Li et al. (1994) have shown that mutations that replace the HIV-1 PBS with sequences complementary to the closely related tRNA^{Lys1,2} or tRNA^{Phe} are biologically functional, although they produce viruses with delayed growth properties. When one examines the complement of tRNAs packaged into these mutant virions, one finds that they do not differ from wild type. This indicates that the PBS is not involved in selective packaging of tRNAs into virions. Subsequent rounds of infection by these mutant viruses result in reversion of the PBS to wild-type sequences.

RNA Secondary Structure Influences Initiation of Minus-strand DNA Synthesis during Reverse Transcription

Although there is significant base-pairing between the primer tRNA and the PBS sequence in viral RNA, the ability to utilize the primer efficiently involves an additional set of viral RNA secondary structures. Such structures are best defined in avian retroviruses (Cobrinik et al. 1988, 1991; Aiyar et al. 1992, 1994). The potential secondary structures of nucleotides 56–130 around the PBS of avian sarcoma/leukosis virus RNA is depicted in Figure 4. As shown, there is a potential to form base

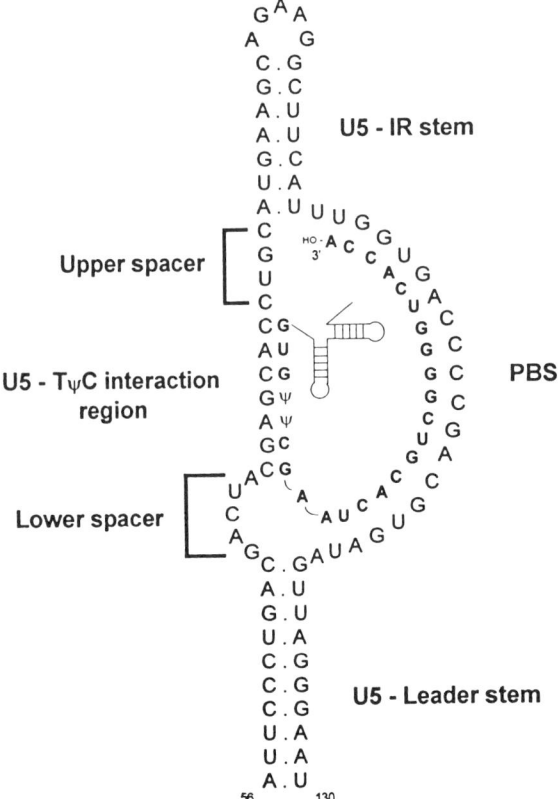

Figure 4 A linear representation of the 5' terminus of avian sarcoma/leukosis virus RNA. The U5 and leader regions are as indicated. A tRNA^{Trp} primer is shown schematically annealed to the primer binding site (PBS). Two sets of inverted repeat sequences that flank the PBS are represented by base-paired regions termed the U5-IR stem and the U5-leader stem. An interaction with the TψC arm of the primer and U5 RNA is also shown. Non-base-paired regions between the various duplex elements are shown by the square brackets.

pairs in the viral RNA between inverted repeat sequences near the PBS called the U5-leader and U5-IR stems. If either duplex is disrupted by deletion or base substitutions, partial defects to initiation of reverse transcription result and viruses containing these mutations grow more slowly (Cobrinik et al. 1988, 1991; Aiyar et al. 1992, 1994). Mutations to the U5-IR stem also result in viral integration defects, since these sequences at the level of DNA form part of the 5' long terminal repeat (LTR) integrase recognition site (Cobrinik et al. 1991). In addition to the U5-IR and U5-leader stems, an interaction between 7 nucleotides of the TψC arm of the tRNA^{Trp} primer and sequences in U5 lying between the stem structures is also necessary for efficient initiation of reverse transcription (Fig. 4) (Aiyar et al. 1992). Interactions of this type were originally proposed by Haseltine et al. (1977) and suggested by the data of Cordell et al. (1979).

Each of the above RSV duplex elements is separated by short non-base-paired spacers (Fig. 4). Small insertions or deletions, but not base substitutions, within these regions cause decreases in initiation of reverse transcription in vitro (Aiyar et al. 1994). These data suggest a need to maintain a specific orientation of these RNA structures with respect to one another, and this is further supported by the studies of Olson et al. (1992), who placed a 4-nucleotide insertion into U5 RNA between the U5-leader stem and the U5-TψC interaction site in a reticuloendotheliosis virus-based plasmid vector. After one round of replication, the majority of replication-competent viruses contained viral RNA with wild-type U5 sequences. The biological reason for these structures and their specific spatial orientation is not fully understood, but they may be required by RT to properly recognize and form the initiation complex. For a more detailed review of these secondary structure interactions and the effect of mutations that disrupt their potential base-pairing, the reader is referred to Leis et al. (1993).

Several other retroviruses are proposed by sequence analysis to form similar secondary structures in viral RNA. In HIV-1 (Baudin et al. 1993), HIV-2 (Berkhout and Schoneveld 1993), and Moloney murine leukemia virus (Mo-MLV) (Mougel et al. 1993), the existence of the U5-IR and U5-leader stems has been substantiated by direct chemical and enzymatic probing. Additionally, Murphy and Goff (1989) demonstrated for Mo-MLV that a deletion mutant in the region corresponding to the U5-IR stem was defective in initiation of cDNA synthesis. For HIV-1 replication, RNA structural interactions with sequences downstream from the PBS (Kohlstaedt and Steitz 1992) or further upstream of the U5-leader stem (Isel et al. 1993) have been reported to be required for initiation of reverse transcription. Bacterial retrons also require specific secondary structures at the 5' end of the template RNA for the cDNA-priming reaction (Shimamoto et al. 1993). However, the priming in this case probably involves a 2' rather than a 3'-hydroxyl group (Shimada et al. 1994).

Plus-strand Priming during Retrovirus Replication

The presence of the PBS near the 5' terminus of viral RNA was originally unexpected, since RT extends DNA synthesis from the 3'-OH end of the primer in the 5' to 3' direction. Thus, to synthesize full-length minus-strand DNA, the primer and RT must be repositioned at the 3' end of the viral RNA. A mechanism to accomplish this has been described (Fig. 5) and involves multiple steps dependent on the various enzymatic activities associated with RT (RNA- and DNA-dependent DNA polymerase, RNase H, unwinding activity) (Weiss et al. 1985; Coffin 1990). In contrast to the recruitment of cellular tRNAs as primers for cDNA synthesis, primers for plus-strand DNA synthesis are enzymatically produced through the action of RT-associated RNase H on the viral RNA. The initial plus-strand primer(s) is created in a polypurine tract at the 3' end of viral RNA after reverse transcription has produced an RNA-DNA hybrid substrate (Fig. 5) (Smith et al. 1984; Rattray and Champoux 1987; Charneau and Clavel 1991). An analysis of viral DNA in cells shortly after infection indicates that, in contrast to minus-strand cDNA synthesis, plus-strand DNA synthesis is discontinuous. This implies that there must be multiple plus-strand initiation events. Consistent with this is the finding of a second plus-strand initiation site in the HIV-1 *pol* gene by Charneau et al. (1992). Full-length plus-strand DNA is sub sequently assembled by the action of DNA ligase. The final product of reverse transcription is a linear viral DNA flanked by two LTRs, which then integrates into the host chromosome in a reaction that is dependent on the viral integrase (Brown et al. 1987; Craigie et al. 1990; Katz et al. 1990).

Retrotransposons: Initiation of Reverse Transcription

Transposable genetic elements of yeast and higher eukaryotes replicate via a mechanism closely related to that of retroviruses. Like retroviruses, these elements possess a PBS sequence near their 5' ends and recruit a host tRNA to prime cDNA synthesis. For the yeast (*S. cerevisiae*) transposable elements TY1, TY2, and TY3, the cellular initiator $tRNA_i^{Met}$ is used as primer (Warmington et al. 1985; Hansen and Sandmeyer 1990; Chapman et al. 1992; Von Pawel-Rammingen et al. 1992). For TY4, $tRNA^{Asn}$ may be used (Janetzky and Lehle 1992; Stucka et al. 1992). Several higher eukaryotic elements such as gypsy, 297, and 412 contain PBS sequences complementary to tRNAs other than $tRNA_i^{Met}$ (Britten and Springer 1993). As with retroviruses, the primer is selectively packaged into virus-like particles (Pochart et al. 1993). However, unlike retroviruses, the base-pairing between the PBS and the tRNA is not tolerant of mismatching. Transposition in *S. cerevisiae* cannot occur if heterologous tRNAs are provided from the *Schizosaccharomyces pombe* strain of yeast (Voytas and Boeke 1993). More recently, Lauermann and

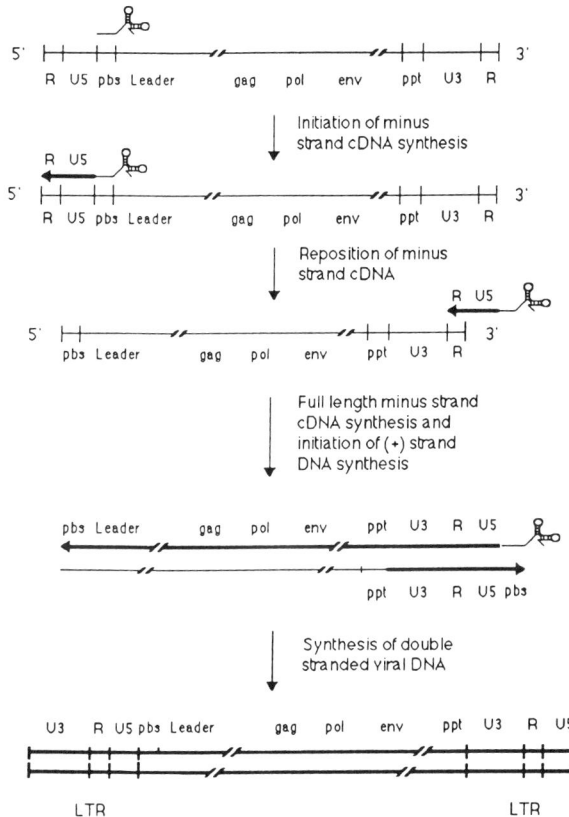

Figure 5 A diagrammatic representation of early reverse transcription of retroviral RNA. Major steps in this process are outlined to indicate key features. Thick lines with arrows indicate the direction of DNA synthesis. Features of the viral RNA (*thin line*) are as follows: (R) repeat sequence; (U5) unique 5' RNA; (pbs) primer binding site; (Leader) 5'-untranslated RNA; (*gag, pol, env*) coding sequences; (ppt) polypurine tract; (U3) unique 3'-RNA. The primer tRNA is shown annealed to the PBS.

Boeke (1994) reported that a single base change introduced into the TY1 PBS is not "healed" during the reverse transcription process. This latter result suggests that the primer sequence is not copied into the PBS during reverse transcription and thus may represent a fundamental difference between reverse transcription catalyzed by retroviruses and retrotransposons.

Several retrotransposable elements, including copia from *Drosophila melanogaster*, TY5 of *S. cerevisiae*, and elements in *Physarum polycephalum* and *Volvox carteri* (Kikuchi et al. 1986; Rothnie et al.

1991; Voytas and Boeke 1992; Lindauer et al. 1993), contain PBS sequences that are not homologous to the 3′ end of any natural tRNA molecule. Instead, the PBS is complementary to 15 nucleotides of the 3′ end of a 39-nucleotide fragment derived from the 5′ end of tRNA$_i^{Met}$. The nuclease that is responsible for the cleavage of the tRNA$_i^{Met}$ is not known. However, it has been shown that the *E. coli* RNA-processing enzyme, RNase P, is capable of cleaving tRNA$_i^{Met}$ at several sites, including one that would produce the above primer fragment (Kikuchi and Sasaki 1992). More recently, Hayashi and Stark (1994) have identified an endoribonuclease (RNase Zma) from *Zea mays* that may be the eukaryotic equivalent of the *E. coli* RNase P.

INITIATION OF DNA SYNTHESIS ON NASCENT RNA PRIMERS

Initiation of mtDNA Synthesis

Initiation of mtDNA replication in mammalian systems requires an RNA molecule to prime DNA synthesis but differs from the retroelements in several aspects. The most salient difference is that the primer is not a host cellular tRNA, but a nascent RNA transcript derived from its own genome which is then cleaved to create the primer. Replication of mtDNA initiates in a 150-bp region that controls both transcription and DNA replication (Clayton, this volume). A site-specific endoribonuclease, termed RNase mitochondrial RNA processing (RNase MRP), has been described that cleaves RNA polymerase transcripts near the site of transition from RNA to DNA synthesis (Chang and Clayton 1987). This enzyme is related to *E. coli* RNase P in that it shares a common secondary structure for its associated RNA (Forster and Altman 1990; Schmitt et al. 1993) and an antigenic determinant (Liu et al. 1994).

The Mauriceville and closely related Varkud mitochondrial plasmids of *Neurospora crassa* probably initiate DNA synthesis via multiple mechanisms, one of which may involve self-priming utilizing the 3′-OH end of a 3′-terminal tRNA-like structure as the primer (Wang and Lambowitz 1993; Kennell et al. 1994).

Initiation of Okazaki Fragment Synthesis

In those genomes in which DNA replication initiates internally and DNA synthesis occurs concomitantly on both templates, DNA synthesis is continuous on one arm but discontinuous on the other (Brush and Kelly, this volume). DNA synthesis in the direction of fork movement (forward arm of fork or leading-strand template) involves the continuous incorporation of dNTPs by DNA polymerase-δ to form long nascent DNA strands, whereas DNA synthesis in the direction opposite fork movement (retrograde arm of fork or lagging-strand template) is carried out discontinuously by repeated initiation, elongation, and, finally, joining of short nascent DNA chains (Okazaki fragments) to the 5' end of the long nascent DNA strand. Examples of this mechanism are found in prokaryotic and eukaryotic cell chromosomes, plasmids that can replicate in prokaryotic cells or in the nuclei of eukaryotic cells, and bacteriophage and nuclear animal viruses with dsDNA genomes (Kornberg and Baker 1992).

The first Okazaki fragment initiated on each template of the replication origin becomes the continuously synthesized nascent DNA strand on the forward arm of the newly created replication fork (DePamphilis et al. 1988). Once replication forks are established, Okazaki fragments originate predominantly, perhaps exclusively, from the retrograde arm. This has been shown for Okazaki fragments synthesized during DNA replication in vivo (SV40, Perlman and Huberman 1977; Kaufmann et al. 1978; Kaufmann 1981; Hay and DePamphilis 1982; Hay et al. 1984; polyomavirus, Hendrickson et al. 1987a,b; mammalian chromosomes, Burhans et al. 1990; Carroll et al. 1993; Tasheva and Roufa 1994; Berberich et al. 1995; Kelly et al. 1995), as well as in vitro in the absence of DNA ligase (Ishimi et al. 1988). Initiation of DNA synthesis on either arm of replication forks in eukaryotic cells involves a series of intermediates (Fig. 6). Okazaki fragments are transient intermediates in DNA replication ($t_{1/2}$ ~ 1 min at 30°C in mammalian cells) that consist of nascent DNA chains from 40 to 300 nucleotides long with an oligoribonucleotide covalently linked to their 5' end (DePamphilis and Wassarman 1980). Synthesis of this oligoribonucleotide is the first step in Okazaki fragment metabolism,

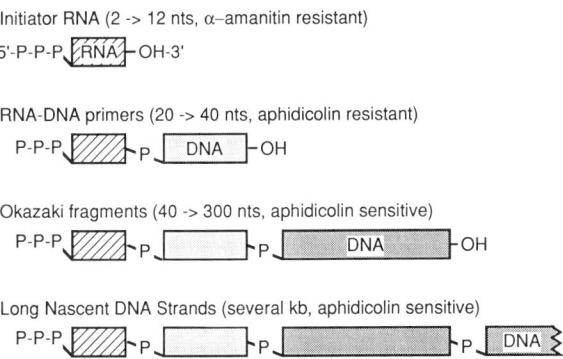

Figure 6 Intermediates in DNA synthesis found at replication forks.

providing an RNA primer on which to initiate DNA synthesis. This RNA primer has been designated initiator RNA (iRNA), because of its unique role in DNA replication (Reichard et al. 1974).

iRNA

Characteristics of iRNA on Newly Replicated DNA

The data documenting the structure, size, and sequence composition of iRNA (Fig. 7) synthesized at the sites where DNA synthesis begins come from extensive studies on SV40, polyomavirus, and mammalian nuclear DNA replication in whole cells or in isolated nuclei incubated with cytoplasm. These results have been reviewed previously in detail (DePamphilis and Wassarman 1980, 1982; DePamphilis and Bradley 1986; DePamphilis 1987). The following is an updated summary of the key points.

Transient, covalent phosphodiester linkages between RNA and DNA have been demonstrated by radiolabeling nascent DNA chains during their synthesis with a specific $[\alpha^{-32}P]dNTP$ substrate and then incubating them in alkali in order to transfer $^{32}PO_4$ from the 5'-terminal deoxyribonucleotide to its neighboring ribonucleotide. The frequency of each of the four possible rN-p-dN linkages is revealed by the amount of each of the four $[2'(3')^{-32}P]rNMPs$ produced. With replicating SV40 (Anderson et al. 1977), polyomavirus (Reichard et al. 1974; Hunter and Francke 1974), and mammalian cell (Waqar and Huberman 1975; Tseng and Goulian 1975) DNA, all 16 possible rN-p-dN linkages are present at frequencies consistent with a random distribution along the DNA template, and rN-p-dN linkages are excised at the same rate that Okazaki fragments are joined to nascent DNA (Anderson et al. 1977).

These rN-p-dN linkages result from iRNA that can be radiolabeled internally by incorporation of $[\alpha^{-32}P]rNTPs$ during DNA replication (Eliasson and Reichard 1978; Tseng and Goulian 1980; Kaufmann 1981). iRNA synthesis is insensitive to α-amanitin, a specific inhibitor of RNA polymerases II and III, revealing that these enzymes are not required for iRNA synthesis. The enzyme that is required (DNA primase) can incorporate dNTPs in place of their corresponding rNTPs when low concentrations of rNTPs are used (Eliasson and Reichard 1979; Tseng and Goulian 1980). Nascent DNA can be removed from the $[^{32}P]iRNA$ by digesting with DNase I, leaving 1-3 dNMPs at the 3' end of the iRNA. This material migrated during gel electrophoresis as oligonucleotides of 10 ± 2 residues. iRNA present on isolated replicating intermediates of SV40 and polyomavirus DNA also has been radiolabeled at its 5'

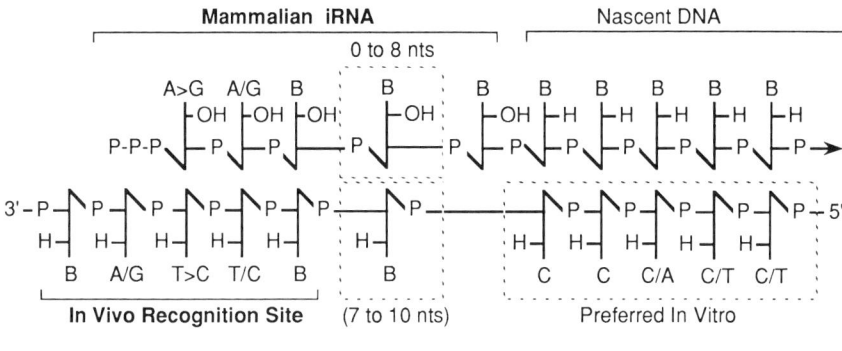

Figure 7 DNA primase recognition sequences. (*Top* panel) Typical initiation site for RNA-primed DNA nascent DNA synthesized in whole cells or in isolated nuclei in cell lysates is diagrammed with specific bases indicated by A, G, C, or T. B indicates any one of the four bases. Purified pol-α:primase prefers pyrimidine-rich sites in vitro, sometimes upstream of a CC(C/A) consensus sequence. (*Bottom* panel) Typical initiation sites for purified DNA primase alone or in the company of other proteins. Template sequences are 3'-5'.

terminus using [γ-^{32}P]ATP and T4 polynucleotide kinase and then digested with T4 exonuclease to remove all but one 3'-deoxyribonucleotide (Hay et al. 1984; Hendrickson et al. 1987b). This iRNA contained 6-9 rNMPs that began primarily with A (60%) or G (25%).

To distinguish full-length iRNA from degraded RNA primers or RNA fragments from other sources, the 5'-terminal ribonucleotide has been radiolabeled exclusively by incorporation of [β-^{32}P]rATP or [β-^{32}P]rGTP (Eliasson and Reichard 1978; Tseng et al. 1979; Kaufmann 1981). The length of this iRNA was the same as described above. Newly replicated DNA has also been isolated first and then radiolabeled with [α-^{32}P]rGTP using vaccinia virus guanylyltransferase, the enzyme that "caps" mRNA (Hendrickson et al. 1987b). This procedure labeled only iRNA with a 5′-end terminal ribonucleoside di- or triphosphate. Following digestion of nascent DNA, most of this iRNA contained 6–9 rNMPs, although iRNA as short as 2 and as long as 12 residues was detected. Release of the GTP-capped terminal nucleotide with P1 nuclease revealed that 70% of the primers began with A and 30% with G.

The presence of a short oligoribonucleotide at one end of nascent DNA chains has been confirmed by degrading RNA with either alkali or ribonuclease, and then measuring the change in the size of nascent DNA chains. This strategy revealed that 50% of the Okazaki fragments synthesized in vivo by SV40 contained iRNA 6–10 residues long (Kaufmann et al. 1977; Hay et al. 1984), and 90% of the Okazaki fragments synthesized by mammalian chromosomes contained iRNA 8–12 residues long (Burhans et al. 1991).

Site Specificity of iRNA Synthesis In Vivo

Evolution would be at a distinct disadvantage if DNA replication required specific DNA sequence signals distributed throughout the genome every 100 or so base pairs. Thus, it is not surprising that initiation of DNA synthesis at replication forks does not require specific template sequences, although it does show preference for certain sequence motifs. Nascent DNA chains in replicating SV40 (Hay and DePamphilis 1982; Hay et al. 1984) and polyomavirus (Hendrickson et al. 1987a; DePamphilis et al. 1988) DNA isolated from virus-infected cells have been selectively radiolabeled at their iRNA-p-DNA junction by removing the iRNA with alkali to unmask a 5′-terminal OH group on the nascent DNA chain. These 5′ termini were then labeled using [γ-^{32}P]rATP and T4 polynucleotide kinase. The resulting ^{32}P-labeled DNA chains were annealed to their template, cut at a unique restriction site, and then fractionated by gel electrophoresis alongside their own DNA sequence ladder in order to identify the position of their 5′-terminal nucleotide (i.e., the iRNA-p-DNA junction). The 5′ terminus of iRNA-DNA chains (i.e., the starting nucleotide location for iRNA) was determined by mapping the locations of end-labeled nascent DNA chains before and after removing their iRNA.

Results from these experiments revealed that >99% of initiation events occurred on the retrograde template and that at least 88% of these initiation events occurred at either 3′-purine-T-pyrimidine-5′ (PuTPy) or 3′-purine-C-pyrimidine-5′ (PuCPy) sites in the template. iRNA synthesis began at the underlined nucleotide with either ATP or GTP. PuTPy sites comprised 78% of initiation sites, consistent with a strong preference for initiation with ATP, and iRNA-p-DNA linkages revealed no sequence preferences, as expected from the nearest-neighbor analyses described above. Moreover, iRNA initiation sites encoded 8 ± 2 ribonucleotides (range = 2–12). This mapping strategy also revealed the transition from discontinuous (presence of initiation sites) to continuous (absence of initiation sites) that marks the origin of bidirectional replication (DePamphilis, this volume).

DNA Polymerase-α:DNA Primase

Pol-α:primase consists of four subunits whose structure is well conserved from yeast to mammals (Wang, this volume). Judged by its sensitivities to a variety of DNA replication inhibitors and its requirement for SV40, polyomavirus, and cellular DNA replication in vitro, pol-α:primase is the enzyme responsible for initiation of Okazaki fragment synthesis (DePamphilis and Bradley 1986; Brush and Kelly; Wang; Hassell and Brinton; all this volume). This enzyme contains a 49-kD and a 58-kD polypeptide, both of which are involved in iRNA synthesis in higher eukaryotes (Lehman and Kaguni 1989), and both of which are essential for DNA replication in yeast (Foiani et al. 1989a). The 49-kD subunit contains the rNTP-binding site and DNA primase catalytic activity (Nasheuer and Grosse 1988; Foiani et al. 1989b; Bakkenist and Cotterill 1994), whereas the 58-kD subunit may help to stabilize NTP binding to the 49-kD subunit (Santocanale et al. 1993).

The properties of DNA primase are consistent with the characteristics of iRNA synthesis. DNA primase is insensitive to α-amanitin and template-dependent, although it readily substitutes dNTPs in place of the corresponding rNTP (Sheaff and Kuchta 1994). Initiation of RNA primer synthesis is the rate-limiting step in DNA synthesis in vitro; RNA synthesis is about 100-fold slower than DNA synthesis (Grosse and Krauss 1985; Sheaff and Kuchta 1993). Pol-α:primase appears to slide along its template looking for primase recognition sites whose K_m values range from 7 pM to 100 pM (Davey and Faust 1990). Initiation depends on formation of an enzyme-DNA-NTP_1-NTP_2 complex complementary to the first two nucleotides in the template initiation site (Sheaff and Kuchta 1993), emphasizing the importance of the 3′-PuTPy-5′ template sequence as a recognition signal. This importance of the second or third nucleotide in primase site selection is reminiscent of site selection by protein primers (Fig. 3).

Pol-α:primase does not synthesize an RNA primer of unique length, but a family of oligoribonucleotides ranging from 2 to 14 residues, most of which are 7 to 13 residues long. Their size is strongly influenced by template sequence, the ratio of ATP to GTP (Yamaguchi et al. 1985a,b; Suzuki et al. 1993), and their proximity to secondary DNA structures such as cruciforms (Tseng and Prussak 1989). In the absence of polymerase, multimeric RNA primers are made, demonstrating that primase is responsible for determining RNA primer length (Tseng and Ahlem 1983; Singh et al. 1986; Cotterill et al. 1987). Pol-α prefers elongating primers at least 7 residues long (Kuchta et al. 1990). However, two accessory proteins (C_1C_2) are frequently associated with pol-α and stimulate its activity on ssDNA templates from 180- to 1800-fold by reducing the K_m of its primer (Pritchard et al. 1983; Kawasaki et al. 1986). C_1C_2 appear to eliminate nonproductive binding of this enzyme to ssDNA, allowing the polymerase to slide along the template until it recognizes a primer. C_1C_2 also decrease the average length of RNA primers synthesized by pol-α:primase by an average of 3–4 bases (Viswanatha et al. 1986). Since DNA replication appears dependent on these proteins (Kumble et al. 1992), they may facilitate initiation of DNA synthesis at replication forks.

Site Specificity of Pol-α:Primase In Vitro

As with iRNA initiation sites selected on viral chromosomes replicating in nuclei, no specific sequences emerge as pol-α:primase start sites on ssDNA templates in vitro. However, in vitro, this enzyme prefers to initiate RNA-primed DNA synthesis at or near pyrimidine-rich sequences (Fig. 7). The minimum size of the recognition site for mammalian pol-α:primase is 3′-purine-[pyrimidine]$_6$-5′ (Davey and Faust 1990; Suzuki et al. 1993). Selection of RNA primer initiation sites is strongly affected by the ratio of ATP to GTP and therefore affects the nucleotide composition of RNA primers and the frequency with which they initiate with either A or G (Yamaguchi et al. 1985b; Suzuki et al. 1993). The K_m for ATP is 100 μM (Kuchta et al. 1990). Mapping of initiation sites for purified pol-α:primase from monkey cells on unique segments of SV40 ssDNA revealed a preference for 3′-(Py)$_n$C$\underline{T}$TT(Py)$_n$ (80%) and 3′-(Py)$_n$C$\underline{C}$C(Py)$_n$ (20%), where the underlined nucleotide is complementary to the first nucleotide in the RNA primer (Yamaguchi et al. 1985a; Vishwanatha et al. 1986). These sites were less frequent (1/16 bases) than those for iRNA synthesis in vivo (1/7 bases), the RNA primers were shorter (7 ± 1 bases in vitro, 10 ± 1 bases in vivo), and the sites chosen in vitro were not the same as those chosen in vivo. Similar results were obtained from mapping initiation sites for RNA-primed DNA synthesis on plasmid DNA undergoing replication in a HeLa cell extract (Bullock et al. 1994). RNA primers most frequently began at 3′-NTT sites located, on average, once every 19 bases. However, the authors' conclusion that RNA primers had no preferred 5′-terminal nucleotide is contrary to all other studies and probably resulted from a 1-base mapping error. Other sequence preferences have also been reported. RNA primers that begin with pppAG or pppGG are favored, and 3′-dCT is strongly favored over 3′-dAC in the template (Sheaff and Kuchta 1993). Initiation sites in which 3′-CC(C/A)-5′ occurs 10–13 nucleotides downstream from the RNA primer start site are strongly preferred (Davey and Faust 1990), since point mutations that disrupt this motif decrease the K_m for DNA approximately 7-fold. One explanation for the variation observed in identifying initiation sites in vitro could be the protein composition of the enzyme preparation tested. For example, DNA primase alone (Tseng and Prussak 1989) appears to select far fewer sites than when it is associated with DNA polymerase-α. Phage T4 DNA-binding protein confines synthesis of RNA primers by the phage T4 DNA primase/DNA helicase complex to those sites where Okazaki fragment synthesis begins (Cha and Alberts 1990).

RNA-DNA Primers

iRNA is first extended into pppRNA-p-DNA chains of ≤40 nucleotides with a mean length of about 35 nucleotides. These RNA-DNA chains were originally referred to as "DNA primers" (Nethanel et al. 1988), but are referred to here as RNA-DNA primers to distinguish them from DNA primers that appear in parvovirus replication and DNA repair. RNA-DNA primers can be labeled during SV40-driven DNA replication in isolated nuclei or cell extracts by briefly incubating them with [α-^{32}P]NTPs or [α-^{32}P]dNTPs and then fractionating the labeled DNA by gel electrophoresis (Nethanel et al. 1988, 1992; Nethanel and Kaufmann 1990; Bullock et al. 1994). Like Okazaki fragments (Anderson and DePamphilis 1979), DNA primers contain iRNA, originate predominantly from the lagging-strand template, are transient intermediates in DNA replication, and are separated from the 5′ ends of nascent downstream DNA chain by a short gap. Synthesis of RNA-DNA primers is insensitive to aphidicolin, a specific inhibitor of DNA pols α, δ, and ε (Nethanel et al. 1988), and so is the first 30–40 nucleotides polymerized by pol-α:primase (Decker et al. 1986), suggesting that RNA-DNA primers are synthesized by this enzyme. Moreover, butylphenyl-dGTP, a specific inhibitor of pol-α, and neutralizing antibodies against pol-α, inhibit synthesis of RNA-DNA primers. Assembly of RNA-DNA primers into longer DNA chains requires proliferating cell nuclear antigen (PCNA) (Bullock et al. 1991) and ATP (Nethanel et al. 1992), suggesting that this step requires pol-δ or -ε and an ATP-requiring protein such as DNA ligase. The limited processivity of pol-α:primase in DNA synthesis (10–15 dNMPs [Yamaguchi et al. 1985b]) should facilitate the switch to another, more processive enzyme, such as pol-δ.

So far, RNA-DNA primers have been reported only in SV40 replicating intermediates. However, with the exception of T antigen, SV40 DNA replication depends entirely on cellular proteins. Therefore, it is likely that replication of cellular chromosomes, as well as host-dependent viral chromosomes such as other papovaviruses and papillomaviruses, will also involve RNA-DNA primers.

Summary of Priming by Nascent RNA

Nascent iRNA usually begins with pppA and sometimes pppG, but not with pppU or pppC. Most iRNA consists of 6–9 ribonucleotides, although iRNA as short as 2 and as long as 12 ribonucleotides has been detected. iRNA initiation sites show a preference for PuTPy and PuCPy where T and C, respectively, are complementary to the first ribonucleotide. Other than this, neither the RNA moeity nor the RNA-p-DNA junction exhibits any sequence specificity. Therefore, the transition from RNA to DNA synthesis depends on primer length rather than on template sequence signal. However, since primer length appears to vary from one initiation site to another, template sequence must indirectly determine the nucleotide site where DNA synthesis begins. For example, the transition from iRNA to DNA synthesis, as well as from RNA-DNA primers to Okazaki fragment synthesis, may result from changes in secondary structure that result from the fact that RNA:DNA hybrids adopt the A form whereas DNA:DNA hybrids adopt the B form (Selsing and Wells 1979; Arnott et al. 1986).

Initiation of RNA-primed DNA synthesis in eukaryotes and prokaryotes shares certain features, although differences exist in their preferred template initiation sites (Fig. 7). All DNA primases initiate synthesis only with a purine ribonucleoside triphosphate (ATP>GTP) and can substitute dNTPs for rNTPs at a low frequency during iRNA synthesis. All DNA primase template recognition sequences include a preferred nucleotide at position −1 relative to the start site. Herpes simplex virus (HSV) DNA primase, like mammalian primase, prefers a pyrimidine-rich template sequence (Tenney et al. 1995; Challberg, this volume). *E. coli* DNA primase most frequently begins with pppAG at the template sequence 3'-GTC-5' (Swart and Griep 1993), and the consensus recognition sequence for *E. coli* DNA primase-DnaB helicase is 3'-PuPyPy-5' (Yoda and Okazaki 1991). This is strikingly similar to iRNA synthesis in mammalian nuclei. DNA primases from bacteriophages T4 and T7 prefer to begin with pppAC at a template sequence that contains a pyrimidine at position −1 (Cha and Alberts 1990; Mendelman and Richardson 1991). On the whole, these results suggest a conservation of function between prokaryotic and eukaryotic DNA primases.

Models for Okazaki Fragment Synthesis

There are two models for synthesis of Okazaki fragments (Fig. 8). The first is the "initiation zone" model, which proposes that one iRNA initiation site is selected stochastically among the many potential sites located within a defined template region (initiation zone). Formation of an Okazaki fragment would then follow an ordered sequential pathway: (1) synthesis of iRNA by DNA primase, (2) extension of iRNA into a DNA primer by pol-α, (3) synthesis of Okazaki fragment by pol-α or -δ, (4) excision of the iRNA in front of the Okazaki fragment (a two-step process [Anderson and DePamphilis 1979] involving two different nucleases [Brush and Kelly, this volume]), (5) filling in the resulting gap by pol-δ or -ε, and (6) ligation of the 3' end of the Okazaki fragment to the 5' end of the long nascent DNA strand. Thus, the size of an Okazaki fragment would vary from the length of a single RNA-DNA primer (~40 bases) to the maximum length of the initiation zone (~300 bases), and only a small subset of potential iRNA initiation sites would be used in a single DNA molecule during each round of DNA replication. In order for Okazaki fragments of different sizes to accumulate waiting for iRNA excision, gap filling, and ligation to occur (Anderson and DePamphilis 1979), initiation of RNA-primed DNA synthesis must be slow relative to the rate of DNA synthesis that follows. This is consistent with the properties of pol-α:primase (Sheaff and Kuchta 1993) and pol-δ (Wang, this volume).

What determines the size of the initiation zone? One possibility is the periodic structure of chromatin in front of the replication fork (DePamphilis and Wassarman 1980; DePamphilis et al. 1988). If the rate-limiting step in replication is the ability of the fork to move from

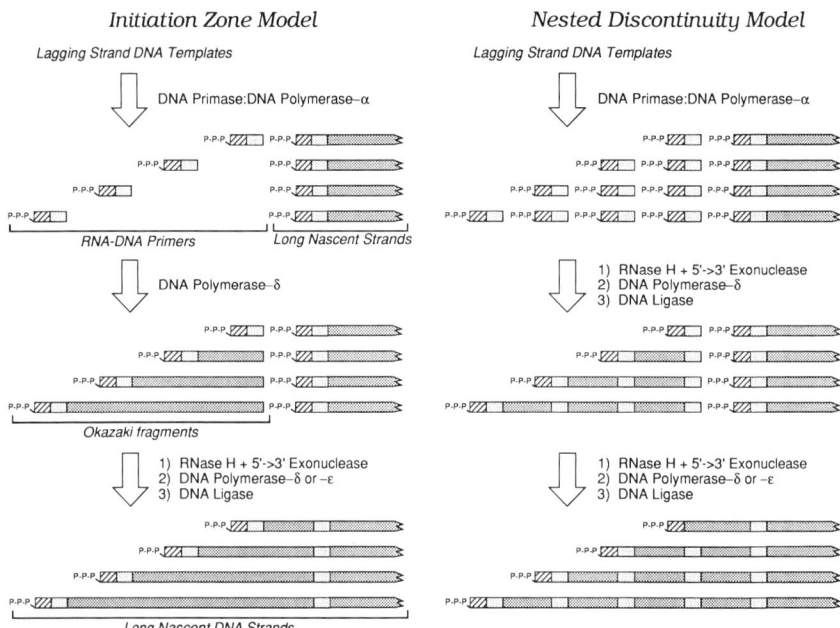

Figure 8 Two models for Okazaki fragment synthesis at DNA replication forks. Newly synthesized DNA is represented on four individual lagging-strand templates. Symbols are the same as in Fig. 6. See text for description.

one nucleosome to the next (i.e., nucleosome disassembly), then the amount of template DNA exposed on the retrograde arm of the fork (i.e., lagging-strand DNA template) will be the same as the average length of internucleosomal DNA. This would produce lagging-strand template on the retrograde arm with the mean length and size range of Okazaki fragments. In prokaryotic cells where nucleosomes are absent, Okazaki fragments average about 1500 nucleotides in length (Kornberg and Baker 1992). A second possibility is the organization of a replication fork. The average length of an Okazaki fragment may be determined by the length of retrograde DNA arm that is wrapped around the DNA replication complex (Stillman, this volume). This model is supported by the fact that Okazaki fragments are produced on newly initiated DNA molecules in vitro with a mean size and length distribution that is similar to those made in vivo (Ishimi et al. 1988). However, the added DNA substrate is rapidly assembled into nucleosomes under these conditions, resulting in the appearance of nucleosomes both in front of and behind replication forks (Gruss et al. 1990).

The second model proposes that synthesis of an Okazaki fragment itself is a discontinuous process in which Okazaki fragments result from assembly of a nested series of RNA-DNA primers (nested discontinuity model, Fig. 8; Nethanel et al. 1988, 1992). Instead of a stochastic selection of iRNA initiation sites, the first RNA-DNA primer is synthesized close to the 5' end of the long nascent DNA strand. This is followed in rapid sucession by synthesis of additional RNA-DNA primers as the pol-α:primase complex travels in a 5' → 3' direction on the lagging-strand template. iRNA excision occurs concomitantly, and the resulting gaps between RNA-DNA primers are filled in by pol-δ or -ε to form an Okazaki fragment. The final iRNA excision/gap-filling event then occurs in front of the Okazaki fragment, and it is ligated to the long nascent DNA strand. What delays the first RNA-DNA primer from joining to the long nascent DNA strand is not clear, although it could be an inherent property of the replication complex.

ACKNOWLEDGMENTS

M.S. thanks her colleagues for many helpful discussions, L. Blanco for his help in designing figures, those who sent preprints of their work, and Ms. M. Angeles Martínez for typing the manuscript. Work in the laboratory of M.S. was supported by grants from the National Institutes of Health (5 RO1 GM-27242-15), Dirección General de Investigación Científica y Técnica (PB93-0173), European Economic Community (CHRX-CT93-0248), and Fundación Ramón Areces.

REFERENCES

Aiyar, A., Z. Ge, and J. Leis. 1994. A specific orientation of RNA secondary structures is required for initiation of reverse transcription. *J. Virol.* **68:** 611–618.

Aiyar, A., D. Cobrinik, Z. Ge, H.-J. Kung, and J. Leis. 1992. Interaction between retroviral U5 RNA and the TΨC loop of the tRNA^{Trp} primer is required for efficient initiation of reverse transcription. *J. Virol.* **66:** 2464–2472.

Anderson, S. and M.L. DePamphilis. 1979. Metabolism of Okazaki fragments during simian virus 40 DNA replication. *J. Biol. Chem.* **254:** 11495–11504.

Anderson, S., G. Kaufmann, and M.L. DePamphilis. 1977. RNA primers in simian virus 40 DNA replication: Identification of transient RNA-DNA covalent linkages in replication DNA. *Biochemistry* **16:** 4990–4998.

Arnott, S., R. Chandrasekaran, R.P. Millane, and H-S Park. 1986. DNA-RNA hybrid secondary structures. *J. Mol. Biol.* **188:** 631–640.

Bakkenist, C.J. and S. Cotterill. 1994. The 50 kd primase subunit of *Drosophila melanogaster* DNA polymerase-α. *J. Biol. Chem.* **269:** 26759–26766.

Bamford, D.H. and L. Mindich. 1984. Characterization of the DNA-protein complex at the termini of the bacteriophage PRD1 genome. *J. Virol.* **50:** 309–315.

Bamford, D.H., T. McGraw, G. MacKenzie, and L. Mindich. 1983. Identification of a protein bound to the termini of bacteriophage PRD1 DNA. *J. Virol.* **47:** 311–316.

Barat, C., S. LeGrice, and J.-L. Darlix. 1991. Interaction of the HIV-1 reverse transcriptase with a synthetic form of its replication primer, tRNA^{Lys3}. *Nucleic Acids Res.* **19:** 751-757.

Barat, C., V. Lullien, O. Shatz, G. Keith, M.T. Nugeyre, F. Gruninger-Leitch, F. Barre-Sinoussi, S.F.J. LeGrice, and J.L. Darlix. 1989. HIV-1 reverse transcriptase specifically interacts with the anticodon domain of its cognate primer tRNA. *EMBO J.* **8:** 3279-3285.

Bartenschlager, R. and H. Schaller. 1988. The amino-terminal domain of the hepadnaviral P-gene encodes the terminal protein (genome-linked protein) believed to prime reverse transcription. *EMBO J.* **7:** 4185-4192.

Baudin, F., R. Marquet, C. Isel, J.-L. Darlix, B. Ehresmann, and C. Ehresmann. 1993. Functional sites in the 5' region of human immunodeficiency virus type 1 RNA form defined structural domains. *J. Mol. Biol.* **229:** 382-397.

Berberich, S., A. Trivedi, D.C. Daniel, E.M. Johnson, and M. Leffak. 1995. In vitro replication of plasmids containing human c-*myc* DNA. *J. Mol. Biol.* **245:** 92-109.

Berkhout, B. and I. Schoneveld. 1993. Secondary structure of the HIV-2 leader RNA comprising the tRNA-primer binding site. *Nucleic Acids Res.* **21:** 1171-1178.

Bernad, A., L. Blanco, J.M. Lázaro, G. Martín, and M. Salas. 1989. A conserved 3'→5' exonuclease active site in prokaryotic and eukaryotic DNA polymerases. *Cell* **59:** 219-228.

Blanco, L. and M. Salas. 1984. Characterization and purification of a phage φ29 coded DNA polymerase required for the initiation of replication. *Proc. Natl. Acad. Sci.* **81:** 5325-5329.

———. 1985a. Characterization of a 3'→5' exonuclease activity in the phage φ29-encoded DNA polymerase. *Nucleic Acids Res.* **13:** 1239-1249.

———. 1985b. Replication of φ29 DNA with purified terminal protein and DNA polymerase: Synthesis of full-length φ29 DNA. *Proc. Natl. Acad. Sci.* **82:** 6404-6408.

———. 1995. Mutational analysis of bacteriophage φ29 DNA polymerase. *Methods Enzymol.* **262:** 283-294.

Blanco, L., A. Bernad, and M. Salas. 1988. Transition from initiation to elongation in the protein-primed φ29 DNA replication. Salt-dependent stimulation by the viral protein p6. *J. Virol.* **62:** 4167-4172.

Blanco, L., A. Bernad, M.A. Blasco, and M. Salas. 1991. A general structure for DNA-dependent DNA polymerases. *Gene* **100:** 27-38.

Blanco, L., A. Bernad, J.A. Esteban, and M. Salas. 1992. DNA-independent deoxynucleotidylation of the φ29 terminal protein by the φ29 DNA polymerase. *J. Biol. Chem.* **267:** 1225-1230.

Blanco, L., J.A. García, M.A. Peñalva, and M. Salas. 1983. Factors involved in the initiation of phage φ29 DNA replication *in vitro*: Requirement of the gene 2 product for the formation of the protein p3-dAMP complex. *Nucleic Acids Res.* **11:** 1309-1323.

Blanco, L., J. Gutiérrez, J.M. Lázaro, A. Bernad, and M. Salas. 1986. Replication of phage φ29 DNA *in vitro*: Role of the viral protein p6 in initiation and elongation. *Nucleic Acids Res.* **14:** 4923-4937.

Blanco, L., J.M. Lázaro, M. de Vega, A. Bonnin, and M. Salas. 1994. Terminal protein-primed DNA amplification. *Proc. Natl. Acad. Sci.* **91:** 12198-12202.

Blanco, L., A. Bernad, J.M. Lázaro, G. Martín, C. Garmendia, and M. Salas. 1989. Highly efficient DNA synthesis by the phage φ29 DNA polymerase. Symmetrical mode of DNA replication. *J. Biol. Chem.* **264:** 8935-8940.

Blanco, L., I. Prieto, J. Gutiérrez, A. Bernad, J.M. Lázaro, J.M. Hermoso, and M. Salas. 1987. Effect of NH_4^+ ions on φ29 DNA-protein p3 replication: Formation of a complex between the terminal protein and the DNA polymerase. *J. Virol.* **61:** 3983–3991.

Britten, R.J. and M.S. Springer. 1993. Phylogenetic relationships of reverse transcriptase and RNase H sequences and aspects of genome structure in the gypsy group of retrotransposons. *Mol. Biol. Evol.* **10:** 1370–1379.

Brown, B., B. Bowerman, H.E. Varmus, and J.M. Bishop. 1987. Correct integration of retroviral DNA *in vitro*. *Cell* **49:** 347–356.

Bullock, P.A., Y.S. Seo, and J. Hurwitz. 1991. Initiation of SV40 DNA synthesis in vitro. *Mol. Cell. Biol.* **11:** 2350–2361.

Bullock, P.A., S. Tevosian, C. Jones, and D. Denis. 1994. Mapping initiation sites for SV40 DNA synhsis events in vitro. *Mol. Cell. Biol.* **14:** 5043–5055.

Burhans, W.C., L.T. Vassilev, M.S. Caddle, N.H. Heintz, and M.L. DePamphilis. 1990. Identification of an origin of bidirectional DNA replication in mammalian chromosomes. *Cell* **62:** 955–965.

Burhans, W.C., L.T. Vassilev, J. Wu, J.M. Sogo, F. Nallaseth, and M.L. DePamphilis. 1991. Emetine allows identification of origins of mammalian DNA replication by imbalanced DNA synthesis, not through conservative nucleosome segregation. *EMBO J.* **10:** 4351–4360.

Caldentey, J., L. Blanco, D.H. Bamford, and M. Salas. 1993. In vitro replication of bacteriophage PRD1 DNA. Characterization of the protein-primed initiation site. *Nucleic Acids Res.* **21:** 3725–3730.

Caldentey, J., L. Blanco, H. Savilahti, D.H. Bamford, and M. Salas. 1992. *In vitro* replication of bacteriophage PRD1 DNA. Metal activation of protein-primed initiation and DNA elongation. *Nucleic Acids Res.* **20:** 3971–3976.

Carroll, S.M., M.L. DeRose, J.L. Kolman, G.H. Nonet, R.E. Kelly, and G.M. Wahl. 1993. Localization of a bidirectional DNA replication origin in the wild type and in episomally amplified murine ADA loci. *Mol. Cell. Biol.* **13:** 2971.

Carusi, E.A. 1977. Evidence for blocked 5′ termini in human adenovirus. *Virology* **76:** 389–394.

Cha, T.-A. and B.M. Alberts. 1990. Effects of the bacteriophage T4 gene 41 and gene 32 proteins on RNA primer synthesis: Coupling of leading and lagging strand DNA synthesis at a replication fork. *Biochemistry* **29:** 1791–1798.

Challberg, M.D. and T.J. Kelly. 1981. Processing of the adenovirus terminal protein. *J. Virol.* **38:** 272–277.

Challberg, M.D., S.V. Desiderio, and T.J. Kelly. 1980. Adenovirus DNA replication *in vitro*: Characterization of a protein covalently linked to nascent DNA strands. *Proc. Natl. Acad. Sci.* **77:** 5105–5109.

Chang, D.D. and D.A. Clayton. 1987. A novel endoribonuclease cleaves at a priming site of mouse mitochondrial DNA replication. *EMBO J.* **6:** 409–417.

Chang, L.-J., R.C. Hirsch, D. Ganem, and H.E. Varmus. 1990. Effects of insertional and point mutations on the functions of the duck hepatitis B virus polymerase. *J. Virol.* **64:** 5553–5558.

Chapman, K.B., A.S. Bystrom, and J.D. Boeke. 1992. Initiator methionine tRNA is essential for Ty1 transposition in yeast. *Proc. Natl. Acad. Sci.* **89:** 3236–3240.

Charneau, P. and F. Clavel. 1991. A single-stranded gap in human immunodeficiency virus unintegrated linear DNA defined by a central copy of the polypurine tract. *J. Virol.* **65:** 2415–2421.

Charneau, P., M. Alizon, and F. Clavel. 1992. Second origin of DNA plus-strand synthesis is required for optimal human immunodeficiency virus replication. *J. Virol.* **66:** 2814-2820.

Chen, C.W., Y.S. Lin, Y.L. Yang, M.F. Tsou, H.M. Chang, H.M. Kieser, and D.A. Hopwood. 1995. The linear chromosomes of *Streptomyces*: Structure and dynamics. *Actinomycetologia* (in press).

Cobrinik, D., L. Soskey, and J. Leis. 1988. A retroviral RNA secondary structure required for efficient initiation of reverse transcription. *J. Virol.* **62:** 3622-3630.

Cobrinik, D., A. Aiyar, Z. Ge, M. Katzman, H. Huang, and J. Leis. 1991. Overlapping retrovirus U5 sequence elements are required for efficient integration and initiation of reverse transcription. *J. Virol.* **65:** 3864-3872.

Coffin, J.M. 1990. Retroviridae and their replication. In *Virology*, 2nd edition (ed. B.N. Fields et al.), pp. 1437-1500. Raven Press, New York.

Cordell, B., R. Swanstrom, H.M. Goodman, and J.M. Bishop. 1979. tRNA^{Trp} as primer for RNA-directed DNA polymerase: Structural determinants of function. *J. Biol. Chem.* **254:** 1866-1874.

Cotterill, S., G. Chui, and I.R. Lehman. 1987. DNA polymerase-primase from embryos of *Drosophila melanogaster* DNA primase subunits. *J. Biol. Chem.* **262:** 16105-16108.

Craigie, R., T. Fujiwara, and F. Bushman. 1990. The IN protein of Moloney murine leukemia virus processes the viral DNA ends and accomplishes their integration in vitro. *Cell* **62:** 829-837.

Crawford, S. and S.P. Goff. 1985. A deletion mutation in the 5' part of the *pol* gene of Moloney murine leukemia virus blocks proteolytic processing of the Gag and Pol proteins. *J. Virol.* **53:** 899-907.

Davey, S.K. and E.A. Faust. 1990. Murine DNA polymerase-α:primase initiates RNA primed DNA synthesis preferentially upstream of a 3'-CC(C/A)-5' motif. *J. Biol. Chem.* **265:** 3611-3614.

Decker, R.S., M. Yamaguchi, R. Possenti, and M.L. DePamphilis. 1986. Initiation of SV40 DNA replication *in vitro*: Aphidicolin causes accumulation of early replicating intermediates and allows determination of the initial direction of DNA synthesis. *Mol. Cell. Biol.* **6:** 3815-3825.

DePamphilis, M.L. 1987. Replication of simian virus 40 and polyoma virus chromosomes. In *Molecular aspects of papovaviruses* (ed. Y. Aloni), pp. 1-40. Martinus Nijhoff, Boston.

DePamphilis, M.L. and M.K. Bradley. 1986. Replication of simian virus 40 and polyoma virus chromosomes. In *The Papovaviridae* (ed. N.P. Salzman), vol. 1, pp. 99-246. Plenum Press, New York.

DePamphilis, M.L. and P.M. Wassarman. 1980. Replication of eukaryotic chromosomes: A close-up of the replication fork. *Annu. Rev. Biochem.* **49:** 627-666.

———. 1982. Organization and replication of papovavirus DNA. In *Organization and replication of viral DNA* (ed. A.S. Kaplan), pp. 37-114, CRC Press, Boca Raton, Florida.

DePamphilis, M.L., E. Martínez-Salas, D.Y. Cupo, E.A. Hendrickson, C.E. Fritze, W.R. Folk, and U. Heine. 1988. Initiation of polyomavirus and SV40 DNA replication, and the requirements for DNA replication during mammalian development. *Cancer Cells* **6:** 165-175.

Desiderio, S.V. and T.J. Kelly. 1981. Structure of the linkage between adenovirus DNA and the 55,000 molecular weight terminal protein. *J. Mol. Biol.* **145:** 319-337.

Dobbs, L., L.J. Zhao, G. Sripad, and R. Padmanabhan. 1990. Mutational analysis of single-stranded DNA templates active in the *in vitro* initiation assay for adenovirus DNA replication. *Virology* **178:** 43–51.
Eliasson, R. and P. Reichard. 1978. Replication of polyoma DNA in isolated nuclei. Synthesis and distribution of initator RNA. *J. Biol. Chem.* **253:** 7469–7475.
―――. 1979. Replication of polyoma DNA in isolated nuclei. VII. Initiator RNA synthesis during nucleotide depletion. *J. Mol. Biol.* **129:** 393–409.
Escarmís, C., D. Guirao, and M. Salas. 1989. Replication of recombinant φ29 DNA molecules in *Bacillus subtilis* protoplasts. *Virology* **169:** 152–160.
Esteban, J.A., M. Salas, and L. Blanco. 1993. Fidelity of φ29 DNA polymerase. Comparison between protein-primed initiation and DNA polymerization. *J. Biol. Chem.* **268:** 2719–2726.
Esteban, J.A., A. Bernad, M. Salas, and L. Blanco. 1992. Metal activation of synthetic and degradative activities of φ29 DNA polymerase, a model enzyme for protein-primed DNA replication. *Biochemistry* **31:** 350–359.
Esteban, J.A., M.S. Soengas, M. Salas, and L. Blanco. 1994. 3'→5' exonuclease active site of φ29 DNA polymerase. Evidence favoring a metal-ion-assisted reaction mechanism. *J. Biol. Chem.* **269:** 31946–31954.
Foiani, M., C. Santocanale, P. Plevani, and G. Lucchini. 1989a. A single essential gene, PRI2, encodes the large subunit of DNA primase in *Saccharomyces cerevisiae*. *Mol. Cell. Biol.* **9:** 3081–3087.
Foiani, M., A.G. Linder, G.R. Hartmann, G. Lucchini, and P. Plevani. 1989b. Affinity labeling of the active center and ribonucleoside triphosphate binding site of yeast DNA primase. *J. Biol. Chem.* **264:** 2189–2194.
Forster, A.C. and S. Altman. 1990. Similar cage-shaped structures for the RNA components of all ribonuclease P and ribonuclease MRP enzymes. *Cell* **62:** 407–409.
Freire, R., M. Salas, and J.M. Hermoso. 1994. A new protein domain for binding to DNA through the minor groove. *EMBO J.* **13:** 4353–4360.
Fujimura, H., T. Yamada, F. Hishinuma, and N. Gunge. 1988. DNA replication *in vivo* of linear DNA killer plasmids pGKL1 and pGKL2 in *Saccharomyces cerevisiae*. *FEMS Microbiol. Lett.* **49:** 441–444.
Ganem, D. and H.E. Varmus. 1987. The molecular biology of hepatitis B viruses. *Annu. Rev. Biochem.* **67:** 651–693.
García, E., A. Gómez, C. Ronda, C. Escarmís, and R. López. 1983. Pneumococcal bacteriophage Cp-1 contains a protein tightly bound to the 5' termini of its DNA. *Virology* **128:** 92–104.
García, P., J.M. Hermoso, J.A. García, E. García, R. López, and M. Salas. 1986. Formation of a covalent complex between the terminal protein of pneumococcal bacteriophage Cp-1 and 5'-dAMP. *J. Virol.* **58:** 31–35.
Garcin, A. and D. Kolakofsky. 1992. Tacaribe arenavirus RNA synthesis *in vitro* is primer dependent and suggests an unusual model for the initiation of genome replication. *J. Virol.* **66:** 1370–1376.
Garmendia, C., A. Bernad, J.A. Esteban, L. Blanco, and M. Salas. 1992. The bacteriophage φ29 DNA polymerase, a proofreading enzyme. *J. Biol. Chem.* **267:** 2594–2599.
Gerlich, W.H. and W.S. Robinson. 1980. Hepatitis B virus contains protein attached to the 5' terminus of its complete DNA strand. *Cell* **21:** 801–809.
Grosse, F. and G. Krauss. 1985. The primase activity of DNA polymerase-α from calf

thymus. *J. Biol. Chem.* **260**: 1881-1888.

Gruss, C., C. Gutiérrez, W.C. Burhans, M.L. DePamphilis, T. Koller, and J.M. Sogo. 1990. Nucleosome assembly in mammalian cell extracts before and after DNA replication. *EMBO J.* **9**: 2911-2922.

Gutiérrez, C., J.M. Sogo, and M. Salas. 1991a. Analysis of replicative intermediates produced during bacteriophage φ29 DNA replication *in vitro. J. Mol. Biol.* **222**: 983-994.

Gutiérrez, C., G. Martín, J.M. Sogo, and M. Salas. 1991b. Mechanism of stimulation of DNA replication by bacteriophage φ29 SSB protein p5. *J. Biol. Chem.* **266**: 2104-2111.

Hansen, L.J. and S.B. Sandmeyer. 1990. Characterization of a transpositionally active Ty3 element and identification of the Ty3 integrase protein. *J. Virol.* **64**: 2599-2607.

Haseltine, W.A., A.M. Maxam, and W. Gilbert. 1977. Rous sarcoma virus genome is terminally redundant: The 5' sequence. *Proc. Natl. Acad. Sci.* **74**: 989-993.

Hay, R.T. and M.L. DePamphilis. 1982. Initiation of simian virus 40 DNA replication in vivo: Location and structure of 5'-ends of DNA synthesized in the *Ori* region. *Cell* **28**: 767-779.

Hay, R.T., E.A. Hendrickson, and M.L. DePamphilis. 1984. Sequence specificity for the initiation of RNA primed-SV40 DNA synthesis in vivo. *J. Mol. Biol.* **175**: 131-157.

Hayashi, D.K. and B.C. Stark. 1994. A novel tRNA precursor cleaving endoribonuclease from *Zea mays. Arch. Biochem. Biophys.* **309**: 123-128.

Hendrickson, E.A., C.E. Fritze, W.R. Folk, and M.L. DePamphilis. 1987a. The origin of bidirectional DNA replication in polyoma virus. *EMBO J.* **6**: 2011-2018.

―――.1987b. Polyoma virus DNA replication is semi-discontinuous. *Nucleic Acids. Res.* **15**: 6369-6385.

Hermoso, J.M., E. Méndez, F. Soriano, and M. Salas. 1985. Location of the serine residue involved in the linkage between the terminal protein and the DNA of phage φ29. *Nucleic Acids Res.* **13**: 7715-7728.

Hirokawa, H. 1972. Transfecting deoxyribonucleic acid of *Bacillus* bacteriophage φ29 that is protease sensitive. *Proc. Natl. Acad. Sci.* **69**: 1555-1559.

Hsieh, J.-C., S.K. Yoo, and J. Ito. 1990. An essential arginine residue for initiation of protein-primed DNA replication. *Proc. Natl. Acad. Sci.* **87**: 8665-8669.

Hsieh, J.-C., G. Jung, M.C. Leavitt, and J. Ito. 1987. Primary structure of the DNA terminal protein of bacteriophage PRD1. *Nucleic Acids Res.* **15**: 8999-9009.

Hu, J.C., and J.E. Dahlberg. 1983. Structural features required for the binding of tRNA^{Trp} to avian myeloblastosis virus reverse transcriptase. *Nucleic Acids Res.* **11**: 4823-4833.

Huang, Y., J. Mak, Q. Cao, Z. Li, M.A. Wainberg, and L. Kleiman. 1994. Incorporation of excess wild-type and mutant tRNA$_3^{Lys}$ into human immunodeficiency virus type 1. *J. Virol.* **68**: 7676-7683.

Hunter, T. and B. Francke. 1974. In vitro polyoma DNA synthesis: Involvement of RNA in discontinuous chain growth. *J. Mol. Biol.* **83**: 123-130.

Isel, C., R. Marquet, G. Keith, C. Ehresmann, and B. Ehresmann. 1993. Modified nucleotides of tRNA$_3^{Lys}$ modulate primer/template loop-loop interaction in the initiation complex of HIV-1 reverse transcription. *J. Biol. Chem.* **268**: 25269-25272.

Ishimi, Y., A. Claude, P. Bullock, and J. Hurwitz. 1988. Complete enzymatic synthesis of DNA containing the SV40 origin of replication. *J. Biol. Chem.* **263**: 19723-19733.

Janetzky, B. and Lehle, L. 1992. Ty4, a new retrotransposon from *Saccharomyces cerevisiae*, flanked by *tau*-elements. *J. Biol. Chem.* **267**: 19798-19805.

Jung, G., M.C. Leavitt, J.-C. Hsieh, and J. Ito. 1987. Bacteriophage PRD1 DNA

polymerase: Evolution of DNA polymerases. *Proc. Natl. Acad. Sci.* **84:** 8287–8291.
Katz, R.A., G. Merkel, J. Kulkosky, J. Leis, and A.M. Skalka. 1990. The avian retroviral IN protein is both necessary and sufficient for integrative recombination *in vitro*. *Cell* **63:** 87–95.
Kaufmann, G. 1981. Characterization of initiator RNA from replicating SV40 DNA synthesized in isolated nuclei. *J. Mol. Biol.* **147:** 25–39.
Kaufmann, G., S. Anderson, and M.L. DePamphilis. 1977. RNA primers in simian virus 40 DNA replication: Distribution of 5′-terminal oligoribonucleotides in nascent DNA. *J. Mol. Biol.* **116:** 549–567.
Kaufmann, G., R. Bar-Shavit, and M.L. DePamphilis. 1978. Okazaki pieces grow opposite to the replication fork direction during simian virus 40 replication. *Nucleic Acids Res.* **5:** 2535–2545.
Kawasaki K., T. Enomoto, M. Suzuki, M. Seki, F. Hanaoka, and M. Yamada. 1986. Detection and characterization of a novel factor that stimulates DNA polymerase-α. *Biochemistry* **25:** 3044–3050.
Kelly, R.E., M.L. DeRose, B.W. Draper, and G.M. Wahl. 1995. Identification of an origin of bidirectional replication within the coding region of the ubiquitously expressed *CAD* gene. *Mol. Cell. Biol.* **15:** 4136–4148.
Kempken, F., F. Meinhardt, and K. Esser. 1989. In *organello* replication and viral affinity of linear, extrachromosonal DNA of the ascomycete *Ascobolus immersus*. *Mol. Gen. Genet.* **218:** 523–530.
Kennell, J.C., H. Wang, and A. Lambowitz, 1994. The Mauriceville plasmid of *Neurospora* spp. uses novel mechanisms for initiating reverse transcription in vivo. *Mol. Cell. Biol.* **14:** 3094–3107.
Khudyakov, Y.E. and A.M. Makhov. 1989. Prediction of terminal protein and ribonuclease H domains in the gene P product of hepadnaviruses. *FEBS Lett.* **243:** 115–118.
Kikuchi, Y. and N. Sasaki. 1992. Hyperprocessing of tRNA by the catalytic RNA of RNase P. *J. Biol. Chem.* **267:** 11972–11976.
Kikuchi, Y., Y. Ando, and T. Shiba. 1986. Unusual priming mechanism of RNA-directed DNA synthesis in copia retrovirus-like particles of *Drosophila*. *Nature* **323:** 824–826.
King, A.J. and P.C. van der Vliet. 1994. A precursor terminal protein-trinucleotide intermediate during initiation of adenovirus DNA replication: Regeneration of molecular ends in vitro by a jumping-back mechanism. *EMBO J.* **13:** 5786–5792.
Kobayashi, H., K. Kitabayaski, K. Matsumoto, and H. Hirokawa. 1991a. Primer protein of bacteriophage M2 exposes the RGD receptor site upon linking the first deoxynucleotide. *Mol. Gen. Genet.* **226:** 65–69.
———. 1991b. Receptor sequence of the terminal protein of bacteriophage M2 that interacts with an RGD (Arg-Gly-Asp) sequence of the primer protein. *Virology* **185:** 901–903.
Kohlstaedt, L.A. and T.A. Steitz. 1992. Reverse transcriptase of human immunodeficiency virus can use either tRNA$_3^{Lys}$ or *Escherichia coli* tRNA$_2^{Gln}$ as a primer in an *in vitro* primer-utilization assay. *Proc. Natl. Acad. Sci.* **89:** 9652–9656.
Kornberg, A. and T. Baker. 1992. *DNA replication*. W.H. Freeman, New York.
Kuchta, R.D., B. Reid, and L.M. Chang. 1990. DNA primase processivity and the primase to polymerase-α activity switch. *J. Biol. Chem.* **265:** 16158–16165.
Kumble, K.D., P.L. Iversen, and J.K. Vishwanatha. 1992. The role of primer recognition proteins in DNA replication: Inhibition of cellular proliferation by antisense

oligodeoxyribonucleotides. *J. Cell Sci.* **101**: 35–41.
Lauermann, V. and J.D. Boeke. 1994. The primer tRNA sequence is not inherited during Ty1 retrotransposition. *Proc. Natl. Acad. Sci.* **91**: 9847–9851.
Lehman, I.R. and L.S. Kaguni. 1989. DNA polymerase α. *J. Biol. Chem.* **264**: 4265–4268.
Leis, J., A. Aiyar, and D. Cobrinik. 1993. Regulation of initiation of reverse transcription of retroviruses. In *Reverse transcriptase* (ed. A.M. Skalka and S.P. Goff), pp. 33–47. Cold Spring Harbor Laboratory Press, Cold Spring Harbor, New York.
Levin, J.G. and J.G. Seidman. 1979. Selective packaging of host tRNA's by murine leukemia virus particles does not require genomic RNA. *J. Virol.* **29**: 328–335.
———. 1981. Effect of polymerase mutations on packaging of primer tRNA^{Pro} during murine leukemia virus assembly. *J. Virol.* **38**: 403–408.
Li, X., J. Mak, E.J. Arts, Z. Gu, L. Kleiman, M.A. Wainberg, and M.A. Parniak. 1994. Effects of alterations of primer-binding site sequences on human immunodeficiency virus type 1 replication. *J. Virol.* **68**: 6198–6206.
Lichy, J.H., M.S. Horwitz, and J. Hurwitz. 1981. Formation of a covalent complex between the 80,000-dalton adenovirus terminal protein and 5′-dCMP *in vitro*. *Proc. Natl. Acad. Sci.* **78**: 2678–2682.
Lin, Y.S., H.M. Kieser, D.A. Hopwood, and C.W. Chen. 1993. The chromosomal DNA of *Streptomyces lividans* 66 is linear. *Mol. Microbiol.* **10**: 923–933.
Lindauer, A., D. Fraser, M. Bruderlein, and R. Schmitt. 1993. Reverse transcriptase families and a copia-like retrotransposon, *osser*, in the green alga *Volvox carteri*. *FEBS Lett.* **319**: 261–266.
Lindenbaum, J.D., J. Field, and J. Hurwitz. 1986. The adenovirus DNA-binding protein and adenovirus DNA polymerase interact to catalyze elongation of primed DNA templates. *J. Biol. Chem.* **261**: 10218–10227.
Liu, M.-H., Y. Yuan, and R. Reddy. 1994. Human RNase P RNA and nucleolar 7-2 RNA share conserved 'T$_o$' antigen-binding domains. *Mol. Cell. Biochem.* **130**: 75–82.
Mak, J., M. Jiang, M.A. Wainberg, M.-L. Hammarskjold, D. Rekosh, and L. Kleiman. 1994. Role of Pr160$^{gag-pol}$ in mediating the selective incorporation of tRNA^{Lys} into human immunodeficiency virus type 1 particles. *J. Virol.* **68**: 2065–2072.
Martín, A.C., R. López, and P. García. 1995. Nucleotide sequence and transcription of the left early region of *Streptococcus pneumoniae* bacteriophage Cp-1 coding for the terminal protein and the DNA polymerase. *Virology* **211**: 21–32.
Martín, G., J.M. Lázaro, E. Méndez, and M. Salas. 1989. Characterization of the phage φ29 protein p5 as a single-stranded DNA binding protein. Function in φ29 DNA-protein p3 replication. *Nucleic Acids Res.* **17**: 3663–3672.
Matsumoto, K., T. Saito, and H. Kirokawa. 1983. *In vitro* initiation of bacteriophage φ29 and M2 DNA replication: Genes required for formation of a complex between the terminal protein and 5′ dAMP. *Mol. Gen. Genet.* **191**: 26–30.
Matsumoto, K., T. Saito, C.I. Kim, T. Ando, and H. Hirokawa. 1984. Bacteriophage φ29 DNA replication in vitro: Participation of the terminal protein and the gene 2 product on elongation. *Mol. Gen. Genet.* **196**: 381–386.
McNeel, D.G. and F. Tamanoi. 1991. Terminal region recognition factor 1, a DNA-binding protein recognizing the inverted terminal repeats of the pGKL1 linear DNA plasmids. *Proc. Natl. Acad. Sci.* **88**: 11398–11402.
Meinhardt, F., F. Kempken, J. Kämper, and K. Esser. 1990. Linear plasmids among eukaryotes: Fundamentals and application. *Curr. Genet.* **17**: 89–95.

Mendelman, L.V. and C.C. Richardson. 1991. Requirements for primer synthesis by bacteriophage T7 63-kDa gene 4 protein. *J. Biol. Chem.* **266:** 23240-23250.

Méndez, E., L. Blanco, J.A. Esteban, A. Bernad, and M. Salas. 1992. Initiation of φ29 DNA replication occurs at the second 3′ nucleotide of the linear template: A slidingback mechanism for protein-primed DNA replication. *Proc. Natl. Acad Sci.* **89:** 9579-9583.

Mindich, L. and D.H. Bamford. 1988. Lipid-containing bacteriophages. In *The bacteriophages* (ed. R. Calendar), vol. 2, pp. 475-520. Plenum Press, New York.

Molnar-Kimber, K.L., J.W. Summers, J.M. Taylor, and W.S. Mason. 1983. Protein covalently bound to minus strand DNA intermediates of duck hepatitis B virus. *J. Virol.* **45:** 165-172.

Mougel, M., N. Tounekti, J.-L. Darlix, J. Paoletti, B. Ehresmann, and C. Ehresmann. 1993. Conformational analysis of the 5′ leader and the *gag* initiation site of Mo-MuLV RNA and allosteric transitions induced by dimerization. *Nucleic Acids Res.* **21:** 4677-4684.

Mul, Y.M. and P.C. van der Vliet. 1993. The adenovirus DNA binding protein effects the kinetics of DNA replication by a mechanism distinct from NFI or Oct-1. *Nucleic Acids Res.* **21:** 641-647.

Murphy, J.E. and S.P. Goff. 1989. Construction and analysis of deletion mutations in the U5 region of Moloney murine leukemia virus: Effects on RNA packaging and reverse transcription. *J. Virol.* **63:** 319-327.

Nasheuer, H.P. and F. Grosse. 1988. DNA polymerase α-primase from calf thymus. Determination of the polypeptide responsible for primase activity. *J. Biol. Chem.* **263:** 8981-8988.

Nethanel, T. and G. Kaufmann. 1990. Two DNA polymerases may be required for synthesis of the lagging strand of SV40. *J. Virol.* **64:** 5912-5918.

Nethanel, T., T. Zlotkin, and G. Kaufmann. 1992. Assembly of SV40 Okazaki pieces from DNA primers is reversibly arrested by ATP depletion. *J. Virol.* **66:** 6634-6640.

Nethanel, T., S. Reisfeld, G. Dinter-Gottlieb, and G. Kaufmann. 1988. An Okazaki piece of SV40 may be synthesized by ligation of shorter precursor chains. *J. Virol.* **62:** 2867-2873.

Oeser, B. and P. Tudzynski. 1989. The linear mitochondrial plasmid pClK1 of the phytopathogenic fungus *Claviceps purpurea* may code for a DNA polymerase and an RNA polymerase. *Mol. Gen. Genet.* **217:** 132-140.

Ollis, D.L., R. Brick, R. Hamlin, N.G. Xuong, and T.A. Steitz. 1985. Structure of the large fragment of *Escherichia coli* DNA polymerase I complexed with dTMP. *Nature* **313:** 762-766.

Olson, P., H.M. Temin, and R. Dornburg. 1992. Unusually high frequency of reconstitution of long terminal repeats in U3-minus retrovirus vectors by DNA recombination or gene conversion. *J. Virol.* **66:** 1336-1343.

Ortín, J., E. Viñuela, M. Salas, and C. Vásquez. 1971. DNA-protein complex in circular DNA from phage φ29. *Nat. New Biol.* **234:** 275-277.

Paillard, M., R.A. Sederoff, and C.S. Levings III. 1985. Nucleotide sequence of the S-1 mitochondrial DNA from the S cytoplasm of maize. *EMBO J.* **4:** 1125-1128.

Panet, A., W.A. Haseltine, D. Baltimore, G. Peters, F. Harada, and J.E. Dahlberg. 1975. Specific binding of tryptophan transfer RNA to avian myeloblastosis virus RNA-dependent DNA polymerase (reverse transcriptase). *Proc. Natl. Acad. Sci.* **72:** 2535-2539.

Peñalva, M.A. and M. Salas. 1982. Initiation of phage φ29 DNA replication *in vitro*: Formation of a covalent complex between the terminal protein, p3, and 5'-dAMP. *Proc. Natl. Acad. Sci.* **79**: 5522-5526.

Perlman, D. and J.A. Huberman. 1977. Asymmetric Okazaki piece synthesis during replication of SV40 DNA in vivo. *Cell* **12**: 1029-1043.

Peters, G.G. and J. Hu. 1980. Reverse transcriptase as the major determinant for selective packaging of tRNA's into avian sarcoma virus particles. *J. Virol.* **36**: 692-700.

Pettit, S.C., M.S. Horwitz, and J.A. Engler. 1989. Mutations of the precursor to the terminal protein of adenovirus serotype 2 and 5. *J. Virol.* **63**: 5244-5250.

Pochart, P., B. Agoutin, C. Fix, G. Keith, and T. Heyman. 1993. A very poorly expressed tRNA^{Ser} is highly concentrated together with replication primer initiator tRNA^{Met} in the yeast Ty1 virus-like particles. *Nucleic Acids Res.* **21**: 1517-1521.

Prieto, I., J.M. Lázaro, J.A. García, J.M. Hermoso, and M. Salas. 1984. Purification in a functional form of the terminal protein of *Bacillus subtilis* phage φ29. *Proc. Natl. Acad. Sci.* **81**: 1639-1643.

Prieto, I., M. Serrano, J.M. Lázaro, M. Salas, and J.M. Hermoso. 1988. Interaction of the bacteriophage φ29 protein p6 with double-stranded DNA. *Proc. Natl. Acad. Sci.* **85**: 314-318.

Pritchard, C.G., D.T. Weaver, E.F. Baril, and M.L. DePamphilis. 1983. DNA polymerase-α cofactors C_1C_2 function as primer recognition proteins. *J. Biol. Chem.* **258**: 9810-9819.

Radziwill, G., W. Tuckek, and H. Schaller. 1990. Mutational analysis of the hepatitis B virus P gene product: Domain structure and RNAse H activity. *J. Virol.* **64**: 613-620.

Rattray, A.J. and J.J. Champoux. 1987. The role of Moloney murine leukemia virus RNase H activity in the formation of plus-strand primers. *J. Virol.* **61**: 2843-2851.

Reichard, P., R. Eliasson, and G. Soderman. 1974. Initiator RNA in discontinuous polyoma DNA synthesis. *Proc. Natl. Acad. Sci.* **71**: 4901-4905.

Rekosh, D.M.K., W.C. Russell, A.J.D. Bellet, and A.J. Robinson. 1977. Identification of a protein linked to the ends of adenovirus DNA. *Cell* **11**: 283-295.

Rohe, M., J. Schründer, P. Tudzynski, and F. Meinhardt. 1992. Phylogenetic relationship of linear, protein-primed replicating genomes. *Curr. Genet.* **21**: 173-176.

Romero, A., R. López, R. Lurz, and P. García. 1990. Temperate bacteriophages of *Streptococus pneumoniae* that contain protein covalently linked to the 5' ends of their DNA. *J. Virol.* **64**: 5149-5155.

Rothnie, H.M., K.J. McCurrah, L.A. Glover, and N. Hardman. 1991. Retrotransposon-like nature of Tp1 elements: Implications for the organization of highly repetitive, hypermethylated DNA in the genome of *Physarum polycephalum*. *Nucleic Acids Res.* **19**: 279-286.

Sakaguchi, K. 1990. Invertrons, a class of structurally and functionally related genetic elements that includes linear DNA plasmids, transposable elements, and genomes of adeno-type viruses. *Microbiol. Rev.* **54**: 66-74.

Salas, M. 1991. Protein-priming of DNA replication. *Annu. Rev. Biochem.* **60**: 39-71.

Salas, M., R.P. Mellado, E. Viñuela, and J.M. Sogo. 1978. Characterization of a protein covalently linked to the 5' termini of the DNA of *Bacillus subtilis* phage φ29. *J. Mol. Biol.* **119**: 269-291.

Salas, M., J. Méndez, J.A. Esteban, M. Serrano, C. Gutierrez, J.M. Hermoso, A. Bravo, M.S. Soengas, J.M. Lázaro, M.A. Blasco, R. Freire, A. Bernad, J.M. Sogo, and L. Blanco. 1993. Terminal protein priming of DNA replication: Bacteriophage φ29 as a model system. In *Virus strategies, molecular biology and pathogenesis* (ed. W. Doerfler and P. Böhm), pp. 3–19. Verlag Chemie. Weinheim, Germany.

Santocanale, C., M. Foiani, G. Lucchini, P. Plevani. 1993. The isolated 48,000-dalton subunit of yeast DNA primase is sufficient for RNA primer synthesis. *J. Biol. Chem.* **268:** 1343–1348.

Savilahti, H. and D. Bamford. 1987. The complete nucleotide sequence of the left very early region of *Escherichia coli* bacteriophage PRD1 coding for the terminal protein and the DNA polymerase. *Gene* **57:** 121–130.

Savilahti, H., J. Caldentey, K. Lundström, J.E. Syväoja, and D.H. Bamford. 1991. Overexpression, purification and characterization of *Escherichia coli* bacteriophage PRD1 DNA polymerase. *J. Biol. Chem.* **266:** 18737–18744.

Schmitt, M.E., J.L. Bennett, D.J. Dairaghi, and D.A. Clayton. 1993. Secondary structure of RNase MRP RNA predicted by phylogenetic comparison. *FASEB J.* **7:** 208–213.

Seeger, C., J. Summers, and W.S. Mason. 1991. Viral DNA synthesis. *Curr. Top. Microbiol. Immunol.* **168:** 41–59.

Selsing, E. and R.D. Wells. 1979. Polynucleotide block polymers consisting of a DNA•RNA hybrid joined to a DNA•DNA duplex. *J. Biol. Chem.* **254:** 5410–5416.

Serrano, M., M. Salas, and J.M. Hermoso. 1990. A novel nucleoprotein complex at a replication origin. *Science* **248:** 1012–1016.

Serrano, M., C. Gutiérrez, M. Salas, and J.M. Hermoso. 1993. Superhelical path of the DNA in the nucleoprotein complex that activates the initiation of phage φ29 DNA replication. *J. Mol. Biol.* **230:** 248–259.

Serrano, M., J. Gutiérrez, I. Prieto, J.M. Hermoso, and M. Salas. 1989. Signals at the bacteriophage φ29 DNA replication origins required for protein p6. binding and activity. *EMBO J.* **8:** 1879–1885.

Sheaff, R.J. and R.D. Kuchta. 1993. Mechanism of calf thymus DNA primase: Slow initiation, rapid polymerization, and intelligent termination. *Biochemistry* **32:** 3027–3037.

———. 1994. Misincorporation of nucleotides by calf thymus DNA primase and elongation of primers containing multiple noncognate nucleotides by DNA polymerase-α. *J. Biol. Chem.* **269:** 19225–19231.

Shih, M.F., K. Watabe, and J. Ito. 1982. *In vitro* complex formation between bacteriophage φ29 terminal protein and deoxynucleotide. *Biochem. Biophys. Res. Commun.* **105:** 1031–1036.

Shimada, M., H. Hosaka, H. Takaku, J.S. Smith, M.J. Roth, S. Inouye, and M. Inouye. 1994. Specificity of priming reaction of HIV-1 reverse transcriptase, 2-OH or 3′ OH. *J. Biol. Chem.* **269:** 3925–3927.

Shimamoto, T., M.-Y. Hsu, S. Inouye, and M. Inouye. 1993. Reverse transcriptases from bacterial retrons require specific secondary structures at the 5′ end of the template for the cDNA priming reaction. *J. Biol. Chem.* **268:** 2684–2692.

Shiue, S.Y., J.C. Hsieh, and J. Ito. 1991. Mapping of the linking tyrosine residue of the PRD1 terminal protein. *Nucleic Acids Res.* **19:** 3805–3810.

Singh, H., R.G. Brooke, M.H. Pausch, G.T. Williams, C. Trainor, and L.B. Dumas. 1986. Yeast DNA primase and DNA polymerase activities. An analysis of RNA priming and its coupling to DNA synthesis. *J. Biol. Chem.* **261:** 8564–8569.

Smart, J.E. and B.W. Stillman. 1982. Adenovirus terminal protein precursor. *J. Biol. Chem.* **257:** 13499–13506.

Smith, J.K., A. Cywinski, and J.M. Taylor. 1984. Specificity of initiation of plus-strand DNA by Rous sarcoma virus. *J. Virol.* **52:** 314–319.

Soengas, M.S., C. Gutiérrez, and M. Salas. 1995. Helix-destabilizing activity of φ29 single-stranded DNA binding protein: Effect on the elongation rate during strand displacement DNA replication. *J. Mol. Biol.* **253:** 517-529.

Soengas, M.S., J.A. Esteban, M. Salas, and C. Gutiérrez. 1994. Complex formation between phage φ29 single-stranded DNA binding protein and DNA. *J. Mol. Biol.* **239:** 213-226.

Soengas, M.S., J.A. Esteban, J.M. Lázaro, A. Bernad, M.A. Blasco, M. Salas, and L. Blanco. 1992. Site-directed mutagenesis at the Exo III motif of φ29 DNA polymerase. Overlapping structural domains for the 3' → 5' exonuclease and strand-displacement activities. *EMBO J.* **11:** 4227-4237.

Steenbergh, P.H., J. Maat, H. van Ormondt, and J.S. Sussenbach. 1977. The nucleotide sequence at the termini of adenovirus type 5 DNA. *Nucleic Acids Res.* **4:** 4371-4389.

Stewart. L., G. Schatz, and V.M. Vogt. 1990. Properties of avian retrovirus particles defective in viral protease. *J. Virol.* **64:** 5076-5092.

Stucka, R., C. Schwarzlose, H. Lochmuller, U. Hacker, and H. Feldmann. 1992. Molecular analysis of the yeast Ty4 element: Homology with Ty1, copia, and plant retrotransposons. *Gene* **122:** 119-128.

Summers, J. and W.S. Mason. 1982. Replication of the genome of a hepatitis B-like virus by reverse transcription of an RNA intermediate. *Cell* **29:** 403-415.

Suzuki, M., E. Savoysky, S. Izuta, M. Tatebe, T. Okajima, and S. Yoshida. 1993. RNA priming coupled with DNA synthesis on natural template by calf thymus DNA polymerase-α-primase. *Biochemistry* **32:** 12782-12792.

Swart, J.R. and M.A. Griep. 1993. Primase from *E. coli* primes single stranded templates in the absence of single stranded DNA binding protein or other auxiliary proteins. *J. Biol. Chem.* **268:** 12970-12976.

Tasheva, E.S. and D.J. Roufa. 1994. A mammalian origin of bidirectional DNA replication within the Chinese hamster RPS14 locus. *Mol. Cell. Biol.* **14:** 5628-5635.

Temperley, S.M. and R.T. Hay. 1992. Recognition of the adenovirus type 2 origin of DNA replication by the virally encoded DNA polymerase and preterminal proteins. *EMBO J.* **11:** 761-768.

Tenney, D.J., A.K. Sheaffer, W.W. Hurlburt, M. Bifano, and R.K. Hamatake. 1995. Sequence-dependent primer synthesis by the herpes simplex virus helicase primase complex. *J. Biol. Chem.* **270:** 9129-9136.

Tseng, B.Y. and C.N. Ahlem. 1983. A DNA primase from mouse cells: Purification and partial characterization. *J. Biol. Chem* **258:** 9845-9849.

Tseng, B.Y. and M. Goulian. 1975. Evidence for covalent association of RNA with nascent DNA in human lymphocytes. *J. Mol. Biol.* **99:** 339-346.

———. 1980. Initiator RNA synthesis upon ribonucleotide depletion: Evidence for base substitutions. *J. Biol. Chem.* **255:** 2062-2066.

Tseng, B.Y. and C.E. Prussak. 1989. Sequence and structural requirements for primase initiation in the SV40 origin of replication. *Nucleic Acids Res.* **17:** 1953-1963.

Tseng, B., J.M. Erickson, and M. Goulian. 1979. Initiator RNA of nascent DNA from animal cells. *J. Mol. Biol.* **129:** 531.

Vishwanatha, J.K., M. Yamaguchi, M.L. DePamphilis, and E.F. Baril. 1986. Selection of template initiation sites and lengths of RNA primers synthesized by DNA primase are strongly affected by its organization in a multiprotein DNA polymerase-α complex. *Nucleic Acids Res.* **14:** 7305-7323.

Von Pawel-Rammingen, U., S. Astrom, and A.S. Bystrom. 1992. Mutational analysis of

conserved positions potentially important for initiator tRNA function in *Saccharomyces cerevisiae. Mol. Cell. Biol.* **12:** 1432-1442.
Voytas, D.F. and J.D. Boeke. 1992. Yeast retrotransposon revealed (letter). *Nature* **358:** 717.
———. 1993. Yeast retrotransposons and tRNAs. *Trends Genet.* **9:** 421-427.
Wakefield, J.K., H. Rhim, and C.D. Morrow. 1994. Minimal sequence requirements of a functional human immunodeficiency virus 1 primer binding site. *J. Virol.* **68:** 1605-1614.
Wang, G.H. and C. Seeger. 1993. Novel mechanism for reverse transcription in hepatitis B viruses. *J. Virol.* **67:** 6507-6512.
Wang, H. and A. Lambowitz. 1993. The Mauriceville plasmid reverse transcriptase can initiate cDNA synthesis de novo and may be related to reverse transcriptase and DNA polymerase progenitor. *Cell* **75:** 1071-1081.
Waqar, M.A., and J.A. Huberman. 1975. Covalent attachment of RNA to nascent DNA in mammalian cells. *Cell* **6:** 551-557.
Warmington, J.R., R.B. Waring, C.S. Newlon, K.J. Indge, and S.G. Oliver. 1985. Nucleotide sequence characterization of Ty1-17, a class II transposon from yeast. *Nucleic Acids Res.* **13:** 6679-6693.
Watabe, K., M. Leusch, and J. Ito. 1984a. Replication of bacteriophage φ29 DNA in vitro: The roles of terminal protein and DNA polymerase. *Proc. Natl. Acad. Sci.* **81:** 5374-5378.
———. 1984b. A 3′ to 5′ exonuclease activity is associated with phage φ29 DNA polymerase. *Biochem. Biophys. Res. Commun.* **123:** 1019-1026.
Watabe, K., M.F. Shih, and J. Ito. 1983. Protein-primed initiation of phage φ29 DNA replication. *Proc. Natl. Acad. Sci.* **80:** 4248-4252.
Weiser, B., D. Ganem, C. Seeger, and H.E. Varmus. 1983. Closed circular viral DNA and asymmetrical heterogeneous forms in livers from animals infected with ground squirrel hepatitis virus. *J. Virol.* **48:** 1-19.
Weiss, R., N. Teich, H. Varmus, and J. Coffin, eds. 1985. *RNA tumor viruses,* 2nd edition, parts 1 and 2. Cold Spring Harbor Laboratory, Cold Spring Harbor, New York.
Wohrl, B.M., B. Ehresmann, G. Keith, S.F.J. LeGrice. 1993. Nuclease footprinting of human immunodeficiency virus reverse transcriptase/tRNA^{Lys-3} complexes. *J. Biol. Chem.* **268:** 13617-13624.
Yamaguchi, M., E.A. Hendrickson, and M.L. DePamphilis. 1985a. DNA primase-DNA polymerase-α from simian cells: Sequence specificity of initiation sites on simian virus 40 DNA. *Mol. Cell. Biol.* **5:** 1170-1183.
———. 1985b. DNA primase-DNA polymerase-α from simian cells: Modulation of RNA primer synthesis by ribonucleoside triphosphates. *J. Biol. Chem.* **260:** 6254-6263.
Yanofsky, S., F. Kawamura, and J. Ito. 1976. Thermolabile transfecting DNA from temperature-sensitive mutant of phage φ29. *Nature* **259:** 60-63.
Yoda, K. and T. Okazaki. 1991. Specificity of the recognition sequence for *E. coli* primase. *Mol. Gen. Genet.* **227:** 1-8.
Yoo, S.K. and J. Ito. 1991. Sequence requirements for protein-primed DNA replication of bacteriophage PRD1. *J. Mol. Biol.* **218:** 779-789.
Zakrzewska-Czekwinska, J. and H. Schrempf. 1992. Characterization of an autonomously replicating region from the *Streptomyces lividans* chromosome. *J. Bacteriol.* **147:** 2688-2693.

Zhu, W. and J. Ito. 1994. Purification and characterization of PRD1 DNA polymerase. *Biochim. Biophys. Acta* **1219**: 267–276.

Zhu, W., M.C. Leavitt, G. Jung, and J. Ito. 1994. Mutagenesis of a highly conserved lysine 340 of the PRD1 DNA polymerase. *Biochim. Biophys. Acta* **1219**: 260–266.

Zoulim, F. and C. Seeger. 1994. Reverse transcription in hepatitis B viruses is primed by a tyrosine residue of the polymerase. *J. Virol.* **68**: 6–13.

6
Mechanisms for Completing DNA Replication

Deepak Bastia and Bidyut K. Mohanty
Department of Microbiology
Duke University Medical Center
Durham, North Carolina 27710

DNA replication consists of three steps: initiation, ongoing replication, and termination. The termination of replication is important because the synchrony of the termination process with subsequent cell division should be a significant factor in the equal and orderly distribution of genetic material to the daughter cells of both prokaryotes and eukaryotes.

Termination in a circular chromosome in prokaryotes involves arrest of replication forks at specific sequences called replication arrest sites (for reviews of this process, see Hill 1992; Baker 1995). There are some exceptions to the phenomenon of sequence-specific replication arrest in both prokaryotic and eukaryotic chromosomes; e.g., the early stage of bacteriophage λ DNA replication (Valenzuela et al. 1976) and SV40 DNA replication (Lai and Nathans 1975). Replication fork arrest is the first step in the termination process. Termination also involves, in *Escherichia coli*, decatenation of the arrested, catenated daughter molecules of DNA by a special topoisomerase called topo IV (Kato et al. 1988, 1990; Schmid 1990; Adams et al. 1992; Hiasa and Marians 1994b). Mutants that are defective in topo IV, a heterodimeric enzyme, are also defective in nucleoid segregation (Kato et al. 1988; Schmid 1990; Adams et al. 1992). More information on topoisomerases can be found in the chapter by Hangaard Andersen et al. (this volume). A third step in termination and proper segregation of the newly completed daughter molecules involves a site-specific recombination system. The system resolves multimers generated by an odd number of recombinations which may occur between the two separating daughter, circular DNA molecules (Blakely et al. 1991; Kuempel et al. 1991).

Although a great deal is known about the termination of replication in prokaryotes, very little is known about this process in eukaryotes. In eukaryotic chromosomes, the linear DNA molecule faces two major problems for completing a round of replication. First, the newly synthesized strand at its 5' end would have a gap created by the removal of

Concepts in Eukaryotic DNA Replication
© 1999 Cold Spring Harbor Laboratory Press 0-87969-557-9/99

the last primer RNA. The filling-in of the gap to complete a round of replication requires special structures at the ends of the DNA molecules, such as terminal redundancy, inverted repeats, or telomeres (for a discussion of the problem, see Kornberg and Baker 1992). Telomeres have been reviewed in a separate chapter in this volume (see Greider et al.). Second, since initiation of replication occurs at many origins in linear eukaryotic DNA, each replication unit must terminate or merge with the preceding and the succeeding units at internal termini. Whether these internal termini are sequence-specific remains unknown.

Since the movement of the replication fork is driven by two key enzymes, namely DNA helicase and DNA polymerase, interference with the activities of either of the two enzymes would cause pausing or arrest of replication forks. Interference with the movement of the replication fork may be due to specific DNA sequences, lesions on DNA, or specific DNA/protein complexes; interference may be transient (pausing) or may lead to arrest of the replisome complex for a longer period.

We first discuss pausing or arrest of replication forks at certain DNA sequences and the nature of these sequences. These pause or arrest sites do not necessarily define authentic replication termination sites but have interesting biological implications and, therefore, are discussed. A major part of this chapter focuses on our present knowledge of sequence-specific replication arrest systems in the prokaryotic organisms *E. coli* and *Bacillus subtilis*. Replication arrest at these sequences to which specific proteins bind has been referred to as termination process in the past. In fact, the fork arrest constitutes only the first step of replication termination. The few examples of known sequence-specific internal arrest sites in linear eukaryotic chromosomes are discussed at the end of the chapter.

PAUSING OF REPLICATION FORKS CAUSED BY DNA SEQUENCE/DNA POLYMERASE INTERACTION

Replication Pause Sites

Both prokaryotic and eukaryotic DNA polymerases pause or are arrested at certain DNA sequences (LaDuca et al. 1983; for review, see Bierne and Michel 1994). The arresting sequences are hairpin loops, polypurine stretches, or sequences that could adopt triple helix structures when subjected to negative supercoiling (Wells et al. 1988; Lindsey and Leach 1989; Dayn et al. 1992).

Weaver and DePamphilis (1984) have systematically analyzed the role of palindromic and nonpalindromic sequences in arresting DNA synthesis in vivo and in vitro. They observed that DNA polymerase-α was arrested before entering a palindromic sequence (class I sites). The arrest was nonpolar. There were also nonpalindromic (class II) sites that arrested DNA polymerase-α, and the sequences in front of and behind the point of arrest were necessary for the arrest to occur. Interestingly, the complementary sequence did not arrest polymerase-α-catalyzed DNA synthesis. The physiological role of these pause sites is mostly unknown at this time. An interesting sequence that causes fork arrest was discovered by the investigation of mitomycin-C-induced "onion skin" replication of mammalian chromosomal DNA initiated from the origin of replication of an integrated polyomavirus DNA. The pause-inducing site present in the adjacent host DNA contained the sequence $(dG-dA)_{27}$-$(dT-dC)_{27}$ (Baran et al. 1983, 1987). This sequence, when cloned into SV40 DNA, caused altered plaque morphology and arrested SV40 replication in vivo for approximately a minute. It is conceivable that such sequences are positioned to limit onion skin replication to certain regions of the chromosome (Rao et al. 1988).

Tapper and DePamphilis (1980) examined the pausing of the two replication forks approaching the replication terminus of SV40 and discovered that replication pause sites were spread over several hundred base pairs. In the absence of a specific termination site in SV40 (Lai and Nathans 1975), it is conceivable that the pause sites promote confluence of the two forks within a given region of SV40 DNA.

G-rich polypurine tracts have been shown to arrest DNA synthesis in vitro, and the arrest of DNA polymerase at the homopurine tracts was not relieved by the addition of single-stranded DNA-binding proteins or other accessory proteins (d'Ambrosio and Furano 1987). Near the left end of the linear chromosome of bacteriophage ϕ29, a strong pause site has been observed in vitro (M. Salas, pers. comm.) and, interestingly, this pause site arrests the fork moving left to right to prevent it from moving in a direction opposite to the direction of transcription of strongly transcribed early genes. Thus, the pause site could serve to prevent DNA replication from entering actively transcribed regions.

The mechanistic aspects of DNA polymerase pausing caused by DNA sequences are not entirely clear and thus remain as a potentially interesting topic for future investigation.

Possible Biological Roles of Pause Sites

Analysis of the available data provides very few clues to the possible physiological roles of replication pause sites. The pausing of replication forks in the regions near the replication termini could be a mechanism to terminate replication preferentially in that segment of the chromosome, in the absence of a sequence-specific replication terminus. The early and late transcription of SV40 DNA is controlled by promoters located near the origin, and the early and late transcription proceeds counterclockwise and clockwise, respectively, toward the terminus. It is conceivable that the pause sites fine-tune the bidirectional fork movement to proceed in the same directions as that of early and late transcripts and, thus, prevent the replication forks from entering actively transcribed regions from the opposite direction and from colliding with the transcriptional apparatus.

In summary, several types of DNA sequences can cause pausing of the movement of DNA polymerases in vivo and in vitro. In some cases, the pausing could serve physiological functions such as limiting DNA amplification by onion skin replication to a certain region of the chromosome or preventing replication forks from entering actively transcribed regions from a direction opposite to that of the transcribing RNA polymerase (Brewer 1988). The polypurine pause site that was discovered at the termini of adjacent chromosomal DNA that was replicated by polyomavirus-initiated onion skin replication of transformed rat cells could be acting to limit gene amplification to a limited region of the chromosome. However, since polyoma DNA can integrate into many chromosomal sites, it is likely that the replication is arrested wherever the forks find a polypurine stretch by chance. It is not known what function the polypurine stretches perform in host replication or recombination. Since pause sites are potentially recombinogenic, these sites could have evolved to make certain DNA regions available for the initiation of the recombinogenic process.

REPLICATION FORK ARREST AT THE TERMINI OF PROKARYOTES

Methods for Detection of Replication Termini

Most prokaryotic chromosomes initiate replication from one origin at a time, and the replication forks usually move bidirectionally until they meet each other at the terminus to generate two daughter molecules. The terminus region in most of the bacterial chromosomes and plasmids contains sequence-specific replication arrest sites that limit the end of the replication cycle to this region by blocking the fork progression (for review, see Hill 1992). To locate the origin and the terminus and to determine the direction of the fork movement, the frequencies of genetic markers located near the origin, termini, and the region in between are measured, during either exponential or synchronized growth. More than three decades ago, the idea of marker frequency analysis was elegantly used by Yoshikawa and Sueoka (1963) to localize the replication origin and the terminus and to determine the direction of fork progression in *B. subtilis*. The principle behind the work is shown in Figure 1A. In a linear chromosome of unit length, when replication is proceeding unidirectionally from an origin located at 0 toward the terminus located at 1, the frequency $g(x)$ of a marker x is obtained by solving the integral shown in Figure 1A. The solution of the integral yields 2^{1-x}. When $x = 0$, $g(x) = 2$; when $x = 1$, $g(x) = 1$. Thus, the frequency of markers located close to the origin will be two times that of the markers located at or near the terminus. The frequencies of the markers located between the *ori* and the terminus follow a gradient of values between 1 and 2 shown in the curve of $g(x)$ plotted as a function of the map location. In practice, Yoshikawa and Sueoka (1963) determined the marker frequencies by extracting the sheared DNA from an exponentially growing prototrophic strain of *B. subtilis*, transforming a multiply marked auxotroph and measuring the number of transformants for each marker. The marker frequencies were normalized using a marker located in the middle of the chromosome, such as leucine (Fig. 1A). The same idea can be extended to investigate the movement of a bidirectionally replicating chromosome.

If the replication terminus is located at a position that is 180° from the origin, the symmetry of the replication fork movement will normally make it difficult to determine if the replication fork arrest is sequence-specific. If the movement of the forks is synchronous, the two forks meet each other at a location diametrically opposite to the origin even in the absence of an arresting sequence at the terminus. Thus, an asymmetry has to be generated in the fork progression to detect a sequence-specific terminus. The asymmetry can be created by moving the origin with respect to the terminus, either by introducing a replication origin at a new location or by introducing deletions or duplications into one arm of the chromosome between the *ori* and the terminus. A replication terminus that arrests forks can be detected by the arrest of one fork before the arrival of the second fork at an asymmetric location with respect to *ori*.

In *B. subtilis*, asymmetry was generated by making one arm from the *ori* to the terminus longer (than the other arm) by 25% by a chromosomal duplication (see Schnieder et al. 1983). Using this strain, O'Sullivan and Anagnostopoulus (1982) performed marker frequency analysis to localize an arresting terminus. Similarly, asymmetry in the *E. coli* chromosome was created by initiating replication from an integrated copy of P2 phage origin and by introducing deletions (Kuempel et al. 1977; Louarn et al. 1977; Kuempel and Duerr 1979; Francois et al. 1989, 1990). His-

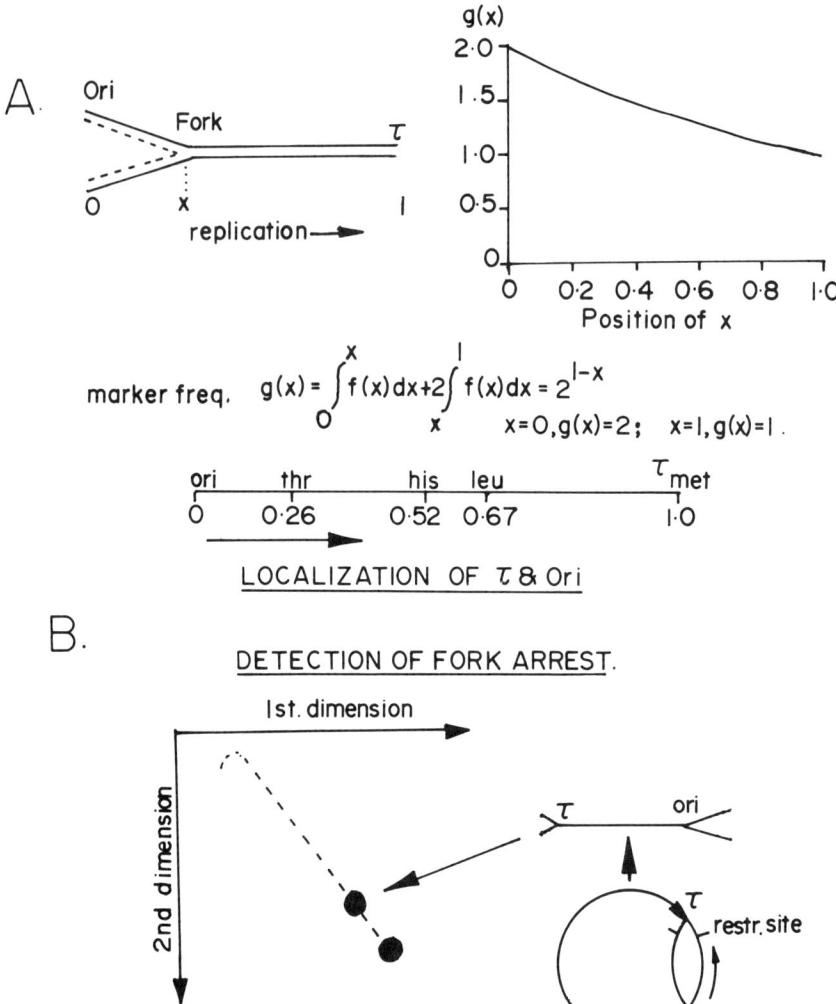

Figure 1 Methods for detection of replication origins and termini. (A) Marker frequency analysis developed by Yoshikawa and Sueoka (1963). The chromosome is linear with a replication origin at 0 and the terminus at 1.0. The fork movement is unidirectional. The equation for marker frequency $g(x)$, the plot of $g(x)$ as a function of map location x, and the chromosomal map of a segment of B. subtilis chromosome are shown. The equation predicts that the frequency of the markers at ori, in a log-phase cell population, would be twice that of the markers at the terminus. (B) Two-dimensional Brewer-Fangman gel electrophoresis of a unidirectional replicon. The ori and terminus (τ) are shown. The circular replicon is linearized at the unique restriction site, and the DNA is fractionated in the first dimension by molecular mass. The second dimension includes ethidium bromide, resulting in shape discrimination.

torically, the discovery of the specific terminus in *E. coli* was made by Kuempel et al. (1977) and Louarn et al. (1977). The replication terminus was identified in plasmid R6K by electron microscopy of the forks impeded by an asymmetrically located terminus (Lovett et al. 1975; Crosa et al. 1976). The bacterial termini were localized initially by measuring marker frequency by DNA-DNA hybridizations and comparing the frequency of markers located immediately before and after the terminus.

The termini of R6K were cloned into a unidirectionally replicating vector, and the replication was visualized by electron microscopy of replication intermediates (Kolter and Helinski 1978). The results showed that the replication termini transiently arrested the replication fork, and then the forks were released to finish replication near the *ori*. Using smaller and smaller pieces of the terminal DNA, the termini were localized to a 216-bp piece of DNA (Bastia et al. 1981a). Subsequent work revealed that the DNA fragment contained two sites of opposite polarity (Horiuchi and Hidaka 1988), each of which could block the replication fork approaching from one direction.

Another technique developed to map an arrested fork in larger bacterial chromosomes is one-dimensional gel electrophoresis designed to detect a Y-shaped stalled fork (Weiss and Wake 1984; Horiuchi and Hidaka 1988; Pelletier et al. 1988).

A technique that has greatly aided the investigation of replication origins, fork movement analysis, localization of replication arrest sites, and the termini is the two-dimensional neutral agarose gel electrophoresis developed by Brewer and Fangman (1987). A variant procedure using electrophoresis in the first dimension in a neutral agarose gel followed by a second dimension in alkaline condition has been developed by Huberman and colleagues (Linskens and Huberman 1988). For detailed discussion of the procedures, the reader is referred to the original papers. The Brewer-Fangman technique, as it applies to detecting replication termini in a unidirectionally replicating plasmid, is shown in Figure 1B. A replication terminus (τ) was cloned into a pUC19 vector approximately 450 bp away from the unidirectional replication origin. The replication intermediates were isolated from cells growing in the exponential phase, and the plasmid DNA was restricted at a unique restriction site between the *ori* and the terminus. Note that the arrested intermediate should have a double Y structure. The DNA was resolved by electrophoresis in the first dimension in an agarose gel at neutral pH and was run on the second dimension at the same pH in a gel containing ethidium bromide. The dye intercalates between DNA base pairs to different extents, depending on the shape of the DNA, and helps discriminate between linear, Y-shaped, or X-shaped DNA molecules. The expected distribution is a hook-shaped pattern of DNA that is visualized by Southern blotting and hybridization by a labeled DNA probe. The monomeric linear DNA forms the prominent lower spot, whereas the termination intermediate is marked by the spot just above and to the left of the monomer spot. By cutting the DNA at other unique restriction sites, one can localize the terminator site (in this case already predetermined) on the plasmid DNA (Fig. 1B). The neutral-alkaline two-dimensional gel procedure is equally effective in detecting and localizing termination sites (Linskens and Huberman 1988).

Structure and Chromosomal Location of Replication Arrest Sequences of Prokaryotes

Definitive information on replication fork arrest sequences has come mostly from the studies of prokaryotic chromosomes. The relatively small amount of available information from eukaryotes such as ribosomal DNA of yeast, human, and plants, and from mammalian mitochondrial DNA, is discussed in a later section.

The replication arrest sites of chromosomes of *E. coli* and *B. subtilis* and of plasmid R6K are shown in Figure 2,A–C. The information on the replication fork arrest in *E. coli* comes from the work of Hill, Kuempel, and their associates (Hill et al. 1988) and from the work of Horiuchi and his associates (Hidaka et al. 1988, 1991). Most of the information on the structure and in vivo analysis of the replication arrest system of *B. subtilis* has come from the work done in the laboratory of Wake and coworkers (for review, see Lewis and Wake 1991 and many other papers by Wake and colleagues cited in this chapter). The replication arrest sites of plasmids of *E. coli* are almost indentical to those of the *E. coli* host chromosomes (Bastia et al. 1981b; Hidaka et al. 1988; Sista et al. 1989).

There are six known replication arrest sites in both *E. coli* (Hidaka et al. 1991; Hill 1992) and *B. subtilis* (Franks et al. 1995). In both *E. coli* and *B. subtilis* the sites are arranged in two sets of three, and in each set all three sites have the same polarity (Fig. 2). The polarity is defined as the orientation of the site for fork arrest with respect to the origin of replication (Hidaka et al. 1988; Hill et al. 1988; Sista et al. 1989; Smith and Wake 1992; Sahoo et al. 1995a). The location and orientation of the arrest sites are such that a replication fork approaching from left to right passes through the first set of three sites in both *E. coli* and *B. subtilis* and through one of the two sites of R6K and is arrested at the first site of

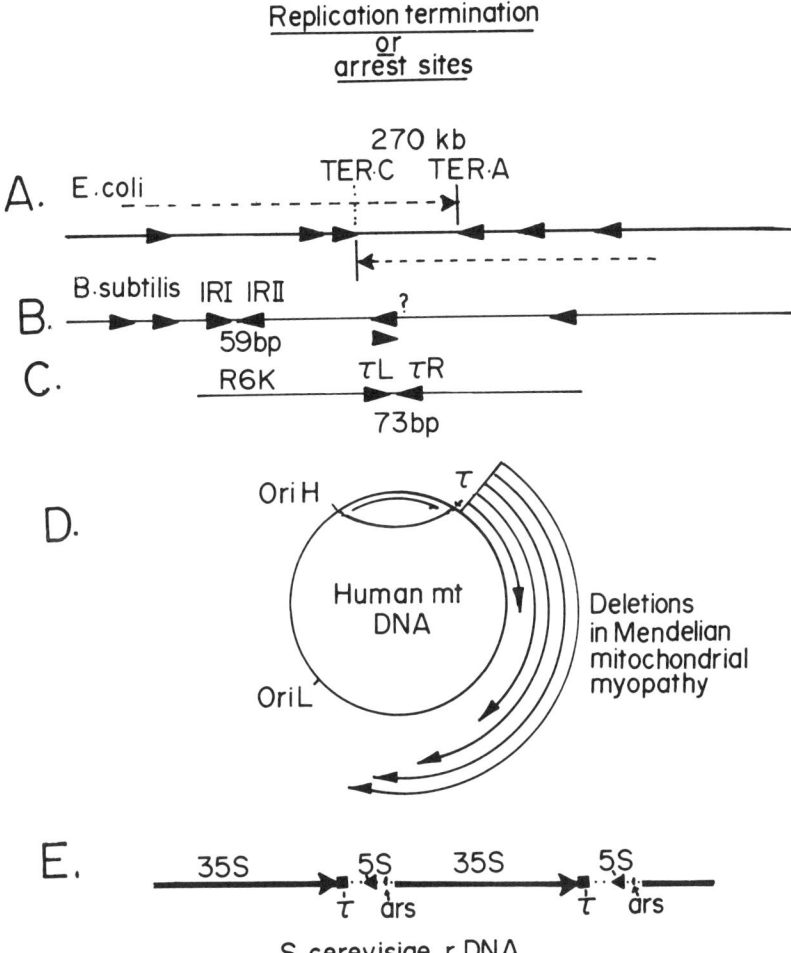

Figure 2 Replication termini in various systems. (*A*) Replication termini of *E. coli*. The dotted lines with arrows show the points of arrest of the two replication forks of the bidirectional replicon. (*B, C*) Replication termini of *B. subtilis* and the plasmid R6K. (*D*) D-loop of the human mitochondrial DNA. τ indicates the point of arrest of the ~600-nt-long newly synthesized H strand. In Mendelian mitochondrial myopathy, deletions extend from near τ, clockwise toward the origin of light-strand synthesis *oriL*. (*E*) Replication fork arrest sites (called replication fork barrier or RFB) τ in yeast rDNA. Replication forks moving in a direction opposite to the transcription of 35S rRNA are arrested at τ (RFB).

the second set that the fork encounters. The converse is true for the fork approaching from right to left. Once the first fork is arrested at a given site, the second fork approaching from the opposite direction stops at the first site, thus preventing the reduplication of the region between two sets of arrest sites. The sites TerA, TerB, and TerC (also called τ1, τ2, and τ3) have been reported to be the most frequently used sites in *E. coli* under laboratory conditions (Pelletier et al. 1988; Louarn et al. 1991). TerI (also called IRI) of *B. subtilis* is reported to be the site that arrests the replication fork moving right to left, which usually arrives first at the terminus (Carrigan et al. 1991; Smith and Wake 1992; Franks et al. 1995).

The replication arrest sites of *B. subtilis* are related to one another by a consensus sequence, as are the sites of *E. coli* and plasmid R6K (Table 1). There is no sequence homology between the sites of gram-positive *B. subtilis* and gram-negative *E. coli*. Unlike the sites of *E. coli* and R6K, the sites of *B. subtilis* have two overlapping sequences called the core (or B site) and the auxiliary sequence (or A site) (Lewis et al. 1990; Sahoo et al. 1995a). The sequences of the fork arrest sites are binding sites for the replication terminator proteins, Ter (Tus) and RTP, encoded by *E. coli* and *B. subtilis*, respectively.

Replication Terminator Protein of *E. coli*

The existence of a terminator protein that causes replication fork arrest was first suggested by in vitro replication experiments carried out by Germino and Bastia (1981). A hybrid replicon containing the plasmid ColE1 origin of replication and the replication terminus of R6K was replicated in cell extracts of an *E. coli* strain that did not harbor a resident R6K plasmid. Replication fork initiated from the ColE1 origin in vitro was arrested at the R6K terminus. Considering that the plasmid did not encode a terminator protein and that the terminus sequence did not have features (e.g., hairpins, polypurine stretches) that are known to impede polymerase movement, the work suggested that the host cell extract probably contained a terminator protein that recognized the R6K terminus.

The terminator protein was first purified from cell extracts of *E. coli*, and the protein bound to two sites present in a 216-bp DNA fragment of R6K (Hidaka et al. 1989; Sista et al. 1989). The sites corresponded to the left and the right arrest sites of R6K (Horiuchi and Hidaka 1988). The *tus* gene encoding the Ter (Tus) protein was discovered by Hill et al. (1989) and Hidaka et al. (1989) and the protein, by direct analysis and as deduced from the DNA sequence, was found to be 36 kD in molecular mass. Cross-linking studies (Sista et al. 1991) and sedimentation equilibrium studies (Gottlieb et al. 1992) both showed a monomeric protein in solution.

Table 1 Sequences of replication fork arrest sites of different organisms

E. coli chromosome
TerA	5'-AATTAGTATGTTGTAACTAAAGT-3'
TerB	5'-AATAAGTATGTTGTAACTAAAGT-3'
TerC	5'-ATATAGGATGTTGTAACTAATAT-3'
TerD	5'-CATTAGTATGTTGTAACTAAATG-3'
TerE	5'-TTAAAGTATGTTGTAACTAAGNN-3'
TerF	5'-CCTTCGTATGTTGTAACGACGAT-3'

R6K plasmid
TerR1	5'-CTCTTGTGTGTTGTAACTAAATC-3'
TerR2	5'-CTATTGAGTGTTGTAACTACTAG-3'

```
Consensus       AAA  AA         A
           5'-NN  G   TGTTGTAACTA NNN-3'
              TTT TG              C
```

B. subtilis chromosome (strain 168)
TerI (IRI)	5'-ACTAAGAAAACTATGTACCAAATGTTCAGT-3'
TerII (IRII)	5'-ACTGACAACACTAGTTACTAAATATTCAAT-3'
TerIII	5'-ACTAATTGATCTATGTACTAAATATTCATA-3'
TerIV	5'-ACTAACTAAACTATGTACTAAATATTCACT-3'
TerV	5'-ACTAAATAAATAATGTACTAAATATTCAAC-3'
TerVI	5'-ACTAAATAATCTATGTACCAAATGTTCAAT-3'

B. subtilis chromosome (strain W23)
IRI	5'-ACTAAGTGAACTGTGAACCAAATGTTCACT-3'
IRII	5'-ACTGAGAACACTATGTACTAAATATTCAAT-3'

```
Consensus      A    TAAACTA   T  T    A
          5'-ACT AN         TG AC AAAT TTCANN-3'
             G   AGCTTAG    A  C      G
```

Saccharomyces cerevisiae rRNA
5'-TTGCCCGGACAGTTTGCTTCATGGAGCAGTTTTTTCCGCACCATC
AGAGCGGCAAACATGAGTGCTTGTATAAGTTTAGAGAATTGAGA-3'

Mitochondrial D-loop sequence that may be a termination sequence
5'-TTGACTGTACATAGTACATTATGTCAAATTC-3'

The contact points of Ter protein with the replication arrest sites of R6K (Sista et al. 1991) and of *E. coli* (Gottlieb et al. 1992) have been analyzed showing a pattern of asymmetric contacts involving both strands of the cognate sites. The equilibrium dissociation constant varied between the Ter sites. Sista et al. (1991) have found the K_d for the R6K Ter site to be 5 × 10^{-9} moles/liter, whereas Gottlieb et al. (1992) found the K_d for the *E. coli* Ter site to be 10^{-11} moles/liter. For the more efficient TerB sites of *E. coli*, Gottlieb et al. (1992) found the K_d value to be 3.4 × 10^{-13} moles/liter. The differences between two different Ter sites may be due to differences in their contact points, and the differences in the values for the same site in different observations may be due to buffer conditions and methods used in the different experiments. Table 2 shows characteristics of different Ter sites. Gottlieb et al. (1992) have compared K_d, dissociation rate constant, and half-life values of Ter-Tus complex with that of the *lac* repressor-operator complex. Although the K_d values for both the complexes were found to be similar, the half-life of the Ter-Tus complex was very high in comparison to that of *lac* repressor-operator complex.

The Ter protein, added either to cell extracts of *tus*$^-$ *E. coli* (Khatri et al. 1989; MacAllister et al. 1990) or to a defined replication system for ColE1 type origins (Hill and Marians 1990) or for the *oriC* system of *E. coli* (Lee et al. 1989; Lee and Kornberg 1992), caused orientation-specific arrest of replication forks. Although both the leading and the lagging strands were arrested at the terminus, a transient intermediate, consistent with a D-loop structure, could be detected in the in vitro reaction (MacAllister et al. 1990). Both Hill and Marians (1990) and Lee and Kornberg (1992) mapped the point of arrest in vitro of the newly synthesized DNA at the terminator sites. The site was located just within the region protected in hydroxyradical and DNase I footprinting experiments.

A significant develoment in understanding the biochemistry of fork arrest emerged from the discovery by Khatri et al. (1989) and Lee et al. (1989) that the Ter (Tus) protein of *E. coli* was able to impede, in one orientation of the terminus, the activity of the main replicative helicase, DnaB. The DnaB helicase translocates in the 5'→3' direction (LeBowitz and McMacken 1986), whereas the helicase PriA (factor Y) of *E. coli* translocates in the 3'→5' direction (Lee and Marians 1987). The Ter (Tus) protein of *E. coli* was also able to impede PriA helicase in an orientation-specific manner (Hiasa and Marians 1992; Lee and Kornberg 1992).

Interestingly, the Ter protein was able to impede the helicase activity of SV40 large T antigen in an orientation-dependent mode in vitro (Bedrosion and Bastia 1991; Amin and Hurwitz 1992; Hidaka et al. 1992). A hybrid replicon with SV40 origin and the R6K terminus, when

Table 2 Characteristics of replication fork arrest sites and the terminator proteins

Sequence	Name	Source	Protein	Mol. mass (kD)	Gene	K_d (moles/l)
5'-TGAGTGTTGTAACTACTA-3' 3'-ACTCACAACATTGATGAT-5'	TerR	R6K	Ter (Tus)	36	*Tus*	1×10^{-11} 5×10^{-9}
5'-AATAAGTATGTTGTAACTAAAGT-3' 3'-TTATTCATACAACATTGATTTCA-5'	TerB	*E. coli*	Ter (Tus)	36	*Tus*	3.4×10^{-13}
5'-ACTAAGAAAACTATGTACCAAATGTTCAGT-3' 3'-TGATTCTTTTGATACATGGTTTACAAGTCA-5'	TerI (IRI)	*B. subtilis*	RTP	14	*rtp*	1.2×10^{-11} 6×10^{-11}

Closed circles show methylation protection of G residues. Arrows and vertical lines indicate the site and polarity of replication blockage.

replicated in HeLa cell extract containing the SV40 T antigen and Ter protein, showed polar fork arrest at the R6K terminus (Bedrosian and Bastia 1991; Amin and Hurwitz 1992).

What is the mechanism of orientation-specific fork arrest? Does it involve strictly Ter protein/terminus DNA interaction that imposes a polar roadblock to most helicases? Alternatively, are both terminator protein/DNA interaction and helicase/terminator protein interaction involved in polar fork arrest?

The issue of helicase specificity of the Ter protein has been debated (Khatri et al. 1989; Lee et al. 1989; Hiasa and Marians 1992; Lee and Kornberg 1992). Some laboratories have reported that Ter protein blocks all helicases, including DnaB and SV40 Tag, that promote Cairns-type replication to generate θ-shaped replication intermediates and also the helicases involved in rolling-circle replication, e.g., Rep helicase; in DNA repair, e.g., helicase II; and in conjugative DNA transfer, e.g., helicase I (Lee et al. 1989; Hidaka et al. 1992; Lee and Kornberg 1992).

Khatri et al. (1989) and Hiasa and Marians (1992) have reported that Ter protein does not block helicase II or Rep helicase. The inhibition of helicase II reported by Bedrosian and Bastia (1991) was due to high concentrations of Ter protein used in the work. In a detailed study of the helicase specificity of the contrahelicase (antihelicase) activity of Ter protein, Sahoo et al. (1995b) discovered that the Ter protein did not impede helicase I or Rep helicase over a wide range of enzyme (helicase)-to-substrate and terminator protein-to-substrate ratios. Under the same conditions, the Ter protein readily impeded activities of both DnaB and PriA helicases in a polar fashion.

Lee and Kornberg (1992) have reported that Ter protein impedes the activities of the Klenow fragment of DNA polymerase I and the DNA polymerases of T7 and T5 phages. The blocks showed minimal orientation dependence of the terminus sequence (a twofold difference). It is likely that the polymerase block is mainly due to a nonspecific roadblock created on DNA by Ter protein and is not considered to be of any major physiological significance.

Recently, Mohanty et al. (1996) have discovered that the terminator proteins of both *E. coli* and *B. subtilis* block the elongation of RNA chains by several prokaryotic RNA polymerases in a completely polar fashion. The antihelicase activity and the RNA polymerase antielongation activities were isopolar. Mohanty et al. (1996) have investigated the possible biological significance of the RNA polymerase impedance by the terminator proteins. Using model substrates, they discovered that passage of an RNA transcript through the non-blocking end of the terminus abrogates the contrahelicase activity of both Ter and RTP. Transcriptional passage also releases the replication forks arrested at the τ2 (TerB) terminus of *E. coli*. The in vivo significance of the results can be understood by considering the finding along with the observations of Roecklein and Kuempel (1992) that transcription elongation initiated from an upstream promoter (P1) and directed toward the TerB (τ2) sequence in vivo in *tus*⁺ cells was arrested at the τ2 site but invaded the τ2 terminus in *tus*⁻ cells. Thus, the anti-transcriptase activity of the Ter protein and RTP most likely has evolved to protect the functional integrity of the replication arrest sites from transcriptional invasion.

Consistent with this notion, Hidaka et al. (1988) had observed that the non-blocking ends of each of the three frequently used replication termini τ1, τ2, and τ3 (Fig. 2A) were flanked by sequences with a GC-rich hairpin loop and an AT-rich tail characteristic of ρ-independent terminator sites. Considered together, these results support the idea that the ability of the terminator protein to block transcriptional elongation is a mechanism to protect the replication termini from functional inactivation by invading transcripts.

Replication Terminator Protein of *B. subtilis*

Unlike the Ter protein of *E. coli*, RTP of *B. subtilis* is a dimer with subunit molecular mass of 14.5 kD (Lewis et al. 1989, 1990). Two dimers of RTP bind to each arrest site of *B. subtilis* in a stepwise manner. A single dimer first binds to the core site (B site) and then, by apparent cooperative protein-protein interaction, promotes the binding of a second dimer to the auxiliary (A) site (Fig. 3). The core site by itself, binding to a single monomer, is incapable of arresting replication forks (Smith and Wake 1992; Sahoo et al. 1995a).

The mode of interaction of RTP with the cognate binding site has been studied both by missing nucleoside hydroxyradical footprinting (Langley et al. 1993) and by methylation protection and interference studies (Sahoo et al. 1995a). The results show that the protein dimer contacts the core site (B site) more frequently than the auxiliary site (A site), and also that the contacts at the core site are on both strands of DNA whereas the contacts at the auxiliary site were mostly one-stranded. Sahoo et al. (1995a) reported that RTP showed purine contacts on both strands of the core but with only one strand of the auxiliary site (Table 2).

There have been two major breakthroughs in the biochemical and structure-function analysis of RTP. Although there is very little primary

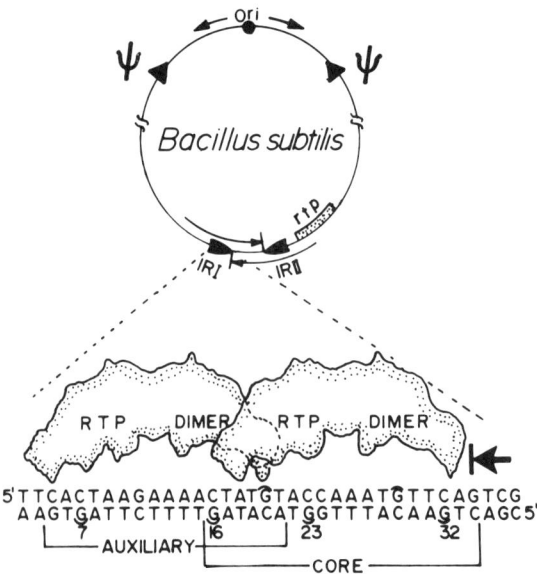

Figure 3 Replication of a *B. subtilis* chromosome showing the termini ψ that are active during stringent response. The IRI and IRII termini contain overlapping core and auxiliary sequences. A single dimer of RTP binds to the core and, by cooperative protein-protein interaction, promotes the binding of a second dimer to the auxiliary site. RTP is encoded by the *rtp* gene. (Reprinted, with permission, from Manna et al. 1996.)

structural homology between Ter of *E. coli* and RTP of *B. subtilis*, Kaul et al. (1994) discovered that RTP functions both in vivo and in vitro in *E. coli*. It blocks the activity of both DnaB and PriA helicases of *E. coli* (Kaul et al. 1994; Sahoo et al. 1995b). Independently, Wake and colleagues have also reported that RTP functions in vivo in *E. coli* (Young and Wake 1994). Since there is no currently available in vitro replication system for *B. subtilis*, it is now possible to analyze the biochemical functions of RTP using the well-defined surrogate *E. coli* in vitro replication system. A second major breakthrough has been the solving of the crystal structure of RTP apoprotein at 2.6 Å (Bussiere et al. 1995).

Crystal Structure of RTP

The crystal structure of the RTP apoprotein has the following features: (1) an amino-terminal disordered region; (2) four α helices with the carboxy-terminal, longest α helix (α4) forming an antiparallel coiled-coil dimerization domain; (3) one short and two long β sheets, the β2 and β3 being connected by an extended loop (Fig. 4).

On the basis of the known activities of RTP, one would expect the protein to have the following domains: (1) a DNA-binding domain, (2) a dimerization domain, (3) a dimer-dimer interaction surface, (4) a helicase-blocking surface, and (5) a region that blocks elongation of RNA chains by prokaryotic RNA polymerases unless, of course, the postulated helicase-blocking domain is also involved in blocking chain extension by RNA polymerase. An interesting question to consider is how a symmetric dimer of RTP is able to act in an asymmetric fashion by blocking helicase activity in a polar mode.

The "Winged Helix" DNA-binding Domain of RTP

The DNA-binding domain of RTP is of interest for at least two reasons: First, the DNA/protein interaction at the terminus presents the terminator protein to the approaching helicase; second, the DNA/protein interaction is likely to generate functional asymmetry from two interacting, symmetric dimers, since DNA/protein contacts are different at the core (B site) and the auxiliary site (A site). Figure 3 shows diagrammatically the interaction between two dimers of RTP on the IRI (BS3) terminus. IRI stands for inverted repeat I and is synonymous with binding site 3, BS3 (see Sahoo et al. 1995a).

Pai et al. (1996) have localized the DNA-binding domain by systematic saturation mutagenesis of DNA encoding RTP and then by screening of the mutants for any defect in DNA binding by employing the genetic selection scheme of Elledge et al. (1990). The mutants that showed defects in DNA binding by the in vivo genetic assay were further characterized by biochemical analysis. By site-directed mutagenesis, mutants of RTP were isolated with cysteine residues substituted at selected sites, derivatized with azido-phenacyl bromide, and cross-linked to the DNA. The cross-linking experiments provided direct evidence for either the contact of α3, β2, and the amino-terminal arm with DNA or the very close proximity of these regions to the DNA. One mutant protein showing a severe DNA-binding defect was crystallized, and the crystal structure was solved and compared with the wild-type RTP structure. The structures were almost identical, thereby showing no misfolding caused by the mutation. The combination of these approaches showed that the amino-terminal unstructured arm, the β2 sheet, and the α3 helix make contacts with the terminus DNA (see Fig. 5). The structure has been referred to as a "winged helix." The β2-β3 sheets and the extended loops

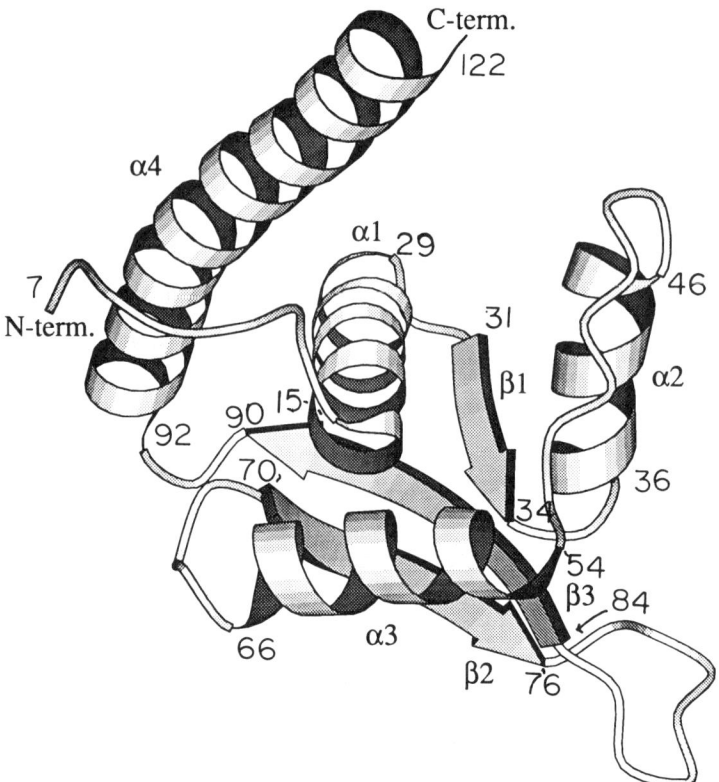

Figure 4 Ribbon diagram of an RTP monomer. The protein contains an unstructured amino-terminal region, four α helices, three β strands, and an extended loop connecting β2 with β3.

connecting the two form the wings (Figs. 4 and 5). Swindels (1995) has compared the structure of RTP with that of the eukaryotic fork head transcription factor, histone H5, and LexA, and has noted the remarkable similarity in the tertiary structures of these, otherwise unrelated, proteins.

Dimerization Domain

The carboxy-terminal α helices form the antiparallel, coiled-coil dimerization domain of RTP (Bussiere et al. 1995).

Dimer-dimer Interaction Domain

Wake and his colleagues (Lewis et al. 1990; Carrigan et al. 1991; Langley et al. 1993) have observed that in vivo, a single dimer of RTP

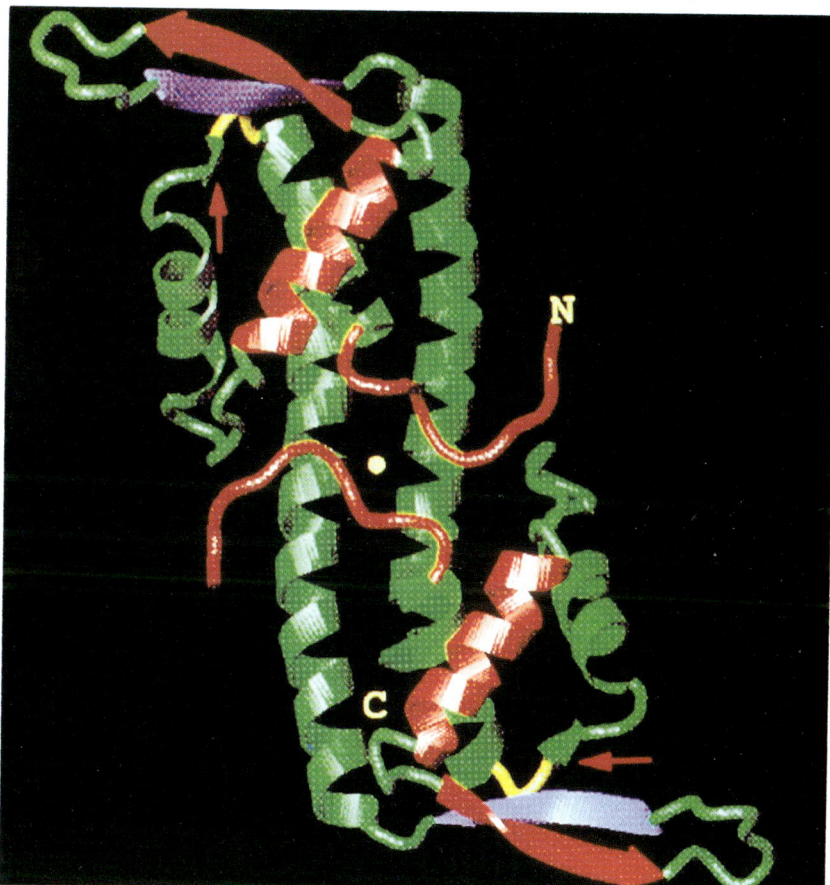

Figure 5 Ribbon diagram of an RTP dimer. (*Red*) DNA-binding regions; (*blue*) dimer-dimer interaction region; (*yellow*) helicase-blocking region. Mutations at the regions marked by red arrows result in loss of RNA polymerase block with no change in DNA binding. The carboxy-terminal α4 helices form an antiparallel, coiled-coil dimerization domain.

binding to a core site fails to impede replication forks. Sahoo et al. (1995a) have extended these results in an in vitro system and have shown that a single dimer of RTP, binding to a core site of a terminator DNA, cannot impede DnaB helicase or replication forks. Thus, two dimers of RTP binding to the core and the auxiliary sites (see Fig. 3) are needed to form a functional replication terminus.

The dimer-dimer interaction domain is therefore an important region of RTP structure. Since RTP crystallizes as a dimer, crystallography did not provide any clue as to the location of the dimer-dimer interaction domain. We resorted to mutagenesis and biochemical analysis of the mutant forms of the protein to localize the dimer-dimer interaction site. The site seems to be located in the β3 strand of RTP (blue region, Fig. 5). A tyrosine residue, located at the coordinate 88 on the β3 strand, upon mutation to a phenylalanine, yielded a protein that binds to the core site but fails to bind to the auxiliary site. The tyrosine at the position 88 is within hydrogen-bond-forming distance of a glycine residue at position 34. Mutation of the glycine at 34 to an arginine also abolishes dimer-dimer interaction. It appears that β3-β3 stacking between two dimers may be the basis of dimer-dimer interaction (Manna et al. 1996). The mutant is completely defective in impeding helicases, replication forks, and RNA polymerases. In addition, a valine to glycine substitution at position 85 and leucine to serine substitution at 82, at the loop connecting β2 to β3, also abolishes dimer-dimer interaction.

Helicase-blocking Surface

Mechanistically, the helicase-blocking surface is of considerable importance, because the existence of such a surface would strongly support a mechanism of fork arrest that involves specific blocking of the helicase activity by protein-protein interaction with RTP. The following criteria were used to look for such a mutant. The mutations affecting the putative helicase-blocking surface of RTP should not affect DNA binding. If the impedance of RNA polymerase-catalyzed chain elongation is controlled by a separate region of RTP, the mutants impairing helicase block should not affect the ability to block RNA polymerase. Using the above criteria, we examined mutants of the region of an exposed hydrophobic patch with a few charged residues located between the α1 helix and the β1 sheet (Figs. 4 and 5). We discovered that a mutation in the region colored yellow in Figure 5 met the criteria stated above. The mutant bound to DNA almost normally, blocked RNA chain extension normally, but failed to block DnaB helicase, PriA helicase, and replication forks in vitro (A.C. Manna et al., in prep.). This result is significant in two ways. First, the mutant RTP marks the helicase-blocking domain of RTP and supports a mechanism of contrahelicase activity that involves RTP/helicase interaction. Second, the results indicate that different regions of the protein are probably involved in the antihelicase and RNA chain anti-elongation activities. A third mutation at the site marked by red arrows in Figure 5 greatly reduced RNA polymerase block without detectably affecting DNA binding. Thus, this region marked by the mutation identifies the RNA polymerase-blocking surface of RTP (Fig. 5). The mutant defective in RNA polymerase block was also defective in helicase block.

The structure-function analysis of RTP, guided by the crystal structure, has been very informative with regard to the mechanism of replication fork impedance. Future work will be directed toward mapping of the sites of DnaB helicase, of SV40 T antigen, and of T7 RNA polymerase that interact with RTP.

It should be noted that the replication terminator proteins of *E. coli* and *B. subtilis*, without the aid of any other protein, block not only the translocation of DnaB and PriA helicases, but also the authentic unwinding of DNA, regardless of the length of the duplex region in helicase substrates (Sahoo et al. 1995a,b). These findings disagree with an earlier report that the replication terminator protein of *E. coli* by itself can inhibit helicase translocation on DNA, but for blockage of DNA unwinding, needed the participation of other replisomal proteins (Hiasa and Marians 1992). It would be surprising, however, if the fork arrest is not more efficient in the presence of a full complement of proteins that drive the replication fork.

Replication Arrest Sites of *B. subtilis* Active under Stringent Conditions

Simone Seror and coworkers (Henckes et al. 1989) have shown that conditions that generate high levels of the alarmone ppGpp in the cell, either by treating *B. subtilis* with hydroxamate or by shifting a temperature-sensitive mutant of the DnaB gene to the nonpermissive temperature (note the DnaB of *B. subtilis*, unlike the protein with the same designation in *E. coli*, is not a helicase but is an initiator protein), cause the replication forks initiated at the *ori* to be arrested approximately 200 kb away on either side at the so-called stringent termini (marked ψ in Fig. 3). Seror's group has shown further that the arrest of replication forks at the stringent termini ψ requires RTP (Levine et al. 1995). Although the marker frequency analysis by DNA-DNA hybridization used to map the arrest site did not allow a precise localization of the ψ sites, the requirement for RTP might indicate the presence of arrest sites such as those present at the normal replication terminus (Fig. 2) that is used under relaxed conditions. How does RTP arrest forks at ψ sites under high ppGpp but not under relaxed conditions? It is known that many promoters such as that for rRNA are shut off during stringent conditions (Cashel and Rudd 1987). Combining this observation with the finding of Mohanty et al. (1996) that replication arrest sites of *B. subtilis* can be rendered ineffective in blocking replication forks by passage of an RNA transcript, a possible mechanism of conditional usage of a replication arrest site can be hypothesized. It is tempting to suggest that the ψ sites are normally kept in a nonfunctional state by transcripts directed through the sequences by promoters that are sensitive to high ppGpp. Under stringent conditions, we suggest that these promoters are turned off, thus allowing RTP bound to the ψ sites to block replication forks. Under low ppGpp, we suggest that the promoters are turned on, and the passage of RNA transcript through the ψ sequences keeps them inactive. In this scheme, ψ sites may be simply regular terminator sequences that bind RTP. Alternative models would involve modification of RTP under stringent conditions. The scheme proposed can be experimentally tested once the ψ sites have been more precisely localized. Levine et al. (1995) have pointed out that the ψ sites serve as checkpoints of replication under stringent conditions. Interestingly, in *E. coli*, ppGpp causes replication arrest at the origin of the *E. coli* chromosome and not at other sites (Levine et al. 1991).

Regulation of Synthesis of Ter Protein

Both in vitro studies by Natarajan et al. (1991) and in vivo studies by Roecklein et al. (1991) showed that Ter protein is a transcriptional repressor of its own synthesis and prevents RNA polymerase from binding to a promoter that is present immediately upstream of the sequence encoding the Ter (Tus) protein. Thus, Ter (Tus) autoregulates its synthesis by binding to the τ2 site that acts as an operator of the autoregulated promoter. Interestingly, Roecklein and Kuempel (1992) also discovered that transcription from an upstream promoter is impeded at the elongation stage by the τ2 sequence in a *tus*$^+$ but not in a *tus*$^-$ host. Natarajan et al. (1993) have reported that a partially purified protein fraction of *E. coli* was able to abrogate both the contrahelicase activity and the fork arrest by the Ter protein in vitro. This opens the possibility of regulation of replication arrest function at a different level by protein-protein interaction.

Possible Physiological Roles of Sequence-specific Replication Arrest System

Since the *tus* gene of *E. coli* and the *rtp* gene of *B. subtilis* can be deleted without causing lethality, specific fork arrest and termination of replication do not appear to be essential for cell viability. Yet the fact that multiple replication arrest sites have been maintained against mutational drift suggests that there is a physiological need for the sites. At this time, there are no definitive clues as to why specific arrest sites have been maintained in the chromosomes. Several suggestions have been made regarding the possible physiological role of these sites. Brewer (1988) has suggested that the sites prevent replication forks from entering the chromosome from a direction opposite to that of transcription, in order to avoid collision between RNA polymerase and the replisomal machinery. Liu and Alberts (1995) have shown that such collision can cause pausing of replication forks. The pausing may invite recombination involving the free DNA ends exposed by falling off of the replication proteins, thus exposing free DNA ends that may lead to genomic instability. Lee et al. (1989) have suggested that the replication arrest sites prevent the θ-type replication from turning into a σ-type or rolling-circle replication. Lewis and Wake (1991) have speculated that the arrest sites might provide the proper structure for more efficient decatenation of daughter molecules. Dasgupta et al. (1991) integrated a copy of the R1 plasmid into the *E. coli* chromosome and caused chromosomal replication to be initiated from the R1 origin in the *tus*+ cell of *E. coli*. The location of the R1 *ori* was such that the almost unidirectional initiation caused the replication fork to travel over a much greater length of DNA to reach the terminus. Under this condition, cell division was perturbed, causing a lack of proper septation and filament formation. Deletion of the *tus* gene greatly reduced filament formation by eliminating fork arrest at the terminus, allowing the forks to pass through and finish replication. Dasgupta et al. (1991) have speculated that the Ter sites restrict each fork to travel one-half the length of the chromosome before meeting each other at a site located diametrically opposite to the *ori* and thus maintain symmetry of replication. Perhaps this also maintains proper gene dosage of markers on both arms of the chromosome. Thus, the Ter system ensures the symmetry of replication. Do termination events signal subsequent cell division at the end of a replication cycle? Although the existence of such a mechanism has been proposed, recent work using synchronized initiation of a single cycle of replication and a DNA synthesis inhibitor shows that septa can form on partially replicated chromosomes that have not reached the terminus. Thus, under these experimental conditions, no linkage of cell septation to completion of the termination of replication was observed. Nordstrom et al. (1991) have also argued for separate mechanisms controlling termination from that of cell division. Recently Hiasa and Marians (1994a), using minichromosome templates containing *oriC* and Ter-binding sites, have shown that the Tus-Ter (Ter-τ) complex prevents overreplication of bidirectionally replicating template. In summary, information on the physiological role of the replication arrest system remains largely obscure.

Decatenation of Circular Daughter Chromosomes

The separation of the daughter molecules after replication requires decatenation; i.e., separation of the two intertwined DNA molecules. The separation at the termini could occur in two steps: In the first, melting of the hydrogen bonds between the remaining parental duplex by a helix-destabilizing protein or a helicase is followed by repair of the gaps of the two catenated rings; in the second step, a topoisomerase decatenates the two daughter molecules (Adams et al. 1992). Topo IV has been identified as the enzyme responsible for decatenation (Kato et al. 1988, 1990; Peng and Marians 1993). Topo IV mutants are defective in partition and accumulate nucleoids in the middle of the cell (Kato et al. 1988; Schmid 1990). Inhibition of topo IV in temperature-sensitive mutants of the two structural genes *parC* and *parE* of *Salmonella* causes the accumulation of catenated dimers. The catenanes were right-handed and parallel, consistent with a melting step followed by a decatenation step of chromosome separation at the terminus (see Fig. 6) (Adams et al. 1992; Zechiedrich and Cozzarelli 1995). A more detailed discussion of topoisomerases can be found in the chapter by Hangaard Andersen et al. (this volume).

The movement of the fork generates positive superhelical stress that can be removed by gyrase, which is uniquely capable of putting in negative superhelicity. However, gyrase can not substitute for topo IV in decatenation (Zechiedrich and Cozzarelli 1995). It is conceivable that topo IV is compartmentalized in a termination complex that gyrase cannot penetrate (Adams et al. 1992). Under certain conditions, topo IV can carry out *oriC* replication in vitro, replacing the need for gyrase, but only under substoichiometric ratios of enzymes to substrate (Hiasa and Marians 1994b). At stoichiometric ratios, the enzyme has been found to relax the substrate.

Pulse-chase experiments with bacterial plasmids under conditions where topo IV is inhibited yielded only about 10% accumulation of catenated dimers, raising questions as to whether topo IV was the true decatenating enzyme at the terminal stages of replication. Recent work in which such pulse-chase experiments were carried out using topo IV ts-gyrA^R (quinoline resistant) and topo IV ts-gyrA$^+$ double mutants showed that substantial amounts of catenated dimers accumulated when gyrase activity was blocked by quinolines under conditions that also blocked topo IV. Thus, catenated dimers are authentic and major kinetic intermediates of termination of replication. These intermediates are subject to decay by gyrase at a rate that is one one-hundredth of that of the decatenating activity of topo IV (Zechiedrich and Cozzarelli 1995).

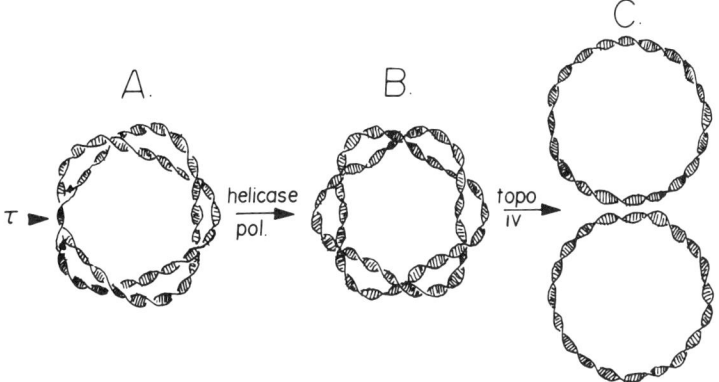

Decatenation of terminated daughter molecules

Figure 6 Schematic representation of the decatenation and segregation of two terminated daughter molecules. The scheme is redrawn from Adams et al. (1992) and Zechiedrich and Cozzarelli (1995). (*A*) Two daughter molecules with the replication forks arrested at the terminus τ. The molecules are still held together by base-pairing of a single turn of the unreplicated parental DNA. (*B*) Unreplicated parental DNA is separated by melting of the base pairs by a helicase, followed by synthesis to fill in the gap, thus generating two fully replicated catenated daughter molecules. (*C*) Topo IV acts on the catenated daughter molecules, thus separating the two. The two-step process of decatenation is supported by topological evidence (Adams et al. 1992).

In plasmid DNA replication in eukaryotic cells (driven by the SV40 origin of replication), the sequence at the termination site strongly affects the fraction of catenated dimers that form (Weaver et al. 1985; Fields-Berry and DePamphilis 1989). For example, the normal termination site for SV40 DNA replication and the yeast CEN3 sequence both promote formation of catenated intertwines when replication terminates in these sequences, but not when these sequences were placed elsewhere on the plasmid and termination occurred outside of them. Therefore, topo II must act behind the replication forks as they enter the termination region, and these two sequences must impede the ability of topo II to resolve catenated intertwines. If the action of topo II did not occur until replication was completed and catenated dimers were formed, then the ability of topo II to resolve catenation should not be affected by the sequence at the termination region, because the catenated intertwines would be distributed throughout the two interlocked sibling molecules.

Multimer Resolution

Because of an odd number of random crossovers between the catenated daughter molecules, multimers are generated. The terminal region of the *E. coli* chromosome has a 33-bp-long site called *dif* (deletion induced filamentation) that interacts with the host-encoded recombinases *XerC* and *XerD* to resolve the multimers into monomers. Deletions of the *dif* site or mutations in *XerC* or *XerD* cause filamentation of cells (Blakely et al. 1991; Kuempel et al. 1991; Leslie and Sherratt 1995). The 33-bp *dif* site can substitute for the natural *dif* region, but only in a relatively location-specific manner (Tecklenberg et al. 1995).

The Replication Terminus Is a Recombinogenic Hot Spot

Several groups of workers have reported that the replication terminus of *E. coli* is a hot spot for recombination (Bierne et al. 1991; Louarn et al. 1991; Horiuchi et al. 1994; Horiuchi and Fujimura 1995). Using a λ cI 857 temperature-sensitive repressor-encoding prophage integrated at several sites on the *E. coli* chromosome, Louarn et al. (1991) discovered that the prophage inserted at the terminus was excised at a significantly higher frequency than phage insertions elsewhere on the chromosome. The excision was compatible with recombination events occurring at each replication cycle. The authors have proposed a model which postulates that RecA-mediated recombination plays a critical role in the resolution of catenated dimers. The model is an interesting one but does not seem to take into account the known requirement for topo IV in catenane resolution (Kato et al. 1990; Adams et al. 1992). A hybrid replicon of pBR322 and M13 origins containing the replication arrest site with blocking end toward both origins yielded, in a tus^+ cell, deletions of which the majority mapped to within 5–6 nucleotides of the replication arrest site of leading-strand synthesis. The authors pointed out the recombinogenic potential of the replication arrest sites and the consequent propensity to cause genome instability (Bierne et al. 1991).

Horiuchi and Fujimura (1995) constructed *E. coli* strains with two replication arrest sites that blocked both forks of the bidirectionally expanding replication bubble. These strains were hyperrecombinogenic and were dependent on $recA^+$ and $recB^+(C^+)$ genes for growth and induced SOS response constitutively. The authors have postulated a recombinogenic event caused by a double strand break at the stalled fork and reconstruction of the fork by recBCD-dependent recombination events. In summary, stalled recombination forks at the replication arrest sites generate a hyperrecombinogenic state that would cause genome instability and genome rearrangement.

REPLICATION ARREST SITES IN EUKARYOTES

Arrest of DNA Replication Forks at Specific Sites

Replication forks do not travel at a continuous rate along the genome, but pause at various sequences (Tapper and DePamphilis 1980). Some sequences have been identified at specific sites in the genomes of animal viruses, eukaryotic cells, and mitochondria that can arrest progress of DNA replication forks. Brewer (1988) has suggested that the primary function of arrest sites for DNA replication forks in the rRNA gene repeats is to prevent collision between replication forks and the RNA polymerase I transcription apparatus. Whether or not this hypothesis can be generally applied to other genes awaits further investigation.

The Epstein-Barr virus (EBV) genome contains a tandem repeat of 20 copies of a 30-bp sequence that binds tightly the virally encoded EBNA1 protein and functions both as an enhancer for viral promoters and as a component of the EBV origin of DNA replication (*oriP*). Replication forks originating at *oriP* do not traverse this EBV enhancer region (Gahn and Schildkraut 1989). Termination of DNA replication in circular plasmids driven by the EBV *oriP* and EBNA1 terminate replication within the enhancer. The EBV enhancer, in the presence of EBNA1 protein, also can arrest replication forks originating from the SV40 replication origin (Dhar and Schildkraut 1991). Arrest of DNA replication forks required at least two, but not more than six, tandem repeats. More recent studies show that this EBNA1-binding sequence can prevent SV40 T-antigen helicase from unwinding DNA (C. Schildkraut, pers. comm.). Whether or not the ability of EBNA1 DNA-binding sites to arrest replication forks exhibits polarity remains to be determined.

The best-studied replication fork barriers in eukaryotes are found in the nontranscribed spacer regions of yeast rRNA genes (Brewer and Fangman 1988; Linskens and Huberman 1988; Brewer et al. 1992). Actively transcribed rRNA gene tandem repeats exhibit a barrier to replication forks traveling upstream (i.e., opposite to the direction of transcription), as diagrammed in Figure 2E. Replication forks seldom originate from origins of replication located immediately downstream from inactive genes (Lucchini and Sogo 1994). A site in the intergenic spacer sequence exhibits polarity by blocking forks coming from one direction but not from the other (Brewer et al. 1992; Kobayashi et al. 1992). Transcription by yeast RNA polymerase I of a DNA fragment containing a transcription terminator near the putative replication arrest site revealed three sites that could arrest transcription (Lang and Reeder 1993). The first site bound the yeast Reb1 protein and arrested RNA polymerase I arriving from one direction, but not the other. The second site did not exhibit polarity. It may bind as-yet-unidentified proteins. Interestingly, the third site, which also lacked polarity, overlapped the putative replication arrest site. Recently, Horiuchi and Kobayashi (cited in Horiuchi and Fujimura 1995) have confirmed the earlier observations of Voelkel-Meimon et al. (1987) that the putative replication arrest site in yeast rRNA gene repeats is associated with a recombination hot spot (HOT 1 sequence). One of the mutants with reduced recombination activity at this HOT spot also showed reduced arrest of DNA replication forks. These results suggest that DNA replication forks in yeast may be subject to the same type of orientation-specific replication arrest sites that have been described in bacteria. DNA replication arrest sites analogous to those described in budding yeast have been mapped in the tandem repeats of rRNA genes in human cells (Little et al. 1993), plant cells (Hernandez et al. 1988, 1993), mouse cells and fission yeast (Lopez-Estrano et al. 1996), and the mouse rDNA replication fork arrest sites exhibit polarity. Thus, DNA replication fork arrest sites appear to be a common feature of rRNA gene repeats in nature.

Another example of a site-specific DNA replication arrest site is found in mitochondrial DNA. Mammalian (and other animal) mitochondrial DNA carry out replication in two stages. The parental light strand serves as the template for initiation from Ori_H, and the fork, after copying about 600 nucleotides, is arrested at a specific site forming a D-loop (Fig. 2E) (for reviews, see Clayton 1991a,b). Eventually, the daughter H strand elongates to expose Ori_L, which initiates the daughter light strand. The arrest of the heavy strand occurs near the terminus-associated sequences (TAS), which are evolutionarily conserved in vertebrates. A protein isolated from bovine cells binds to the mitochondrial TAS (Madsen et al. 1993). No further information is presently available on the physiological role of the putative protein in D-loop biogenesis.

The region of the D-loop, as in cases of replication termination sites, appears to be recombinogenic. In a Mendelian, autosomal dominant disorder called mitochondrial myopathy, the carriers develop an accumulation of deletions in the mitochondrial DNA, starting from a unique location to within a few nucleotides of the 3′ end of the arrested nascent H-strand DNA (Zeviani et al. 1989). In a more recent report, Zeviani (1992) has shown that some of the deletions in the mitochondrial DNA of myopathic patients are found downstream from the D-loop.

The origins of deletions in mitochondrial DNA triggered by a mutant nuclear gene are interesting to contemplate. First, the deletions could be caused by a mutant replication terminator protein, encoded by a nuclear gene, but acting to cause H-strand arrest. If the mutant protein has a propensity to allow formation of more free ends or gaps or a hyperrecombinogenic substrate, deletions could be generated by nonhomologous recombination. The lesions found downstream from the 3' end of the arrested H strand (Zeviani 1992) could also be potentiated by the extension and downstream pausing of the daughter strand. The isolation and analysis of the probable human mitochondrial replication arrest protein and the nuclear gene encoding the protein are interesting problems to tackle in the future.

Sequence-specific arrest sites for DNA replication forks may also exist in regions where DNA amplification events occur. For example, cells in culture and cancer cells in animals undergo constant genomic rearrangements and gene amplification events that can be demonstrated by selecting for cells that continue to proliferate in the presence of various metabolic inhibitors (for review, see Schimke 1984; Stark et al. 1989; Di Leonardo et al. 1993; Stark 1993; Wintersberger 1994). Developmentally orchestrated gene amplification occurs in the chorion genes of *Drosophila* (Heck and Spradling 1990) and the polytene chromosomes of *Chironomous* and *Sciara* (Gerbi et al. 1993). A mechanism that can account for selective gene amplification is the "onion skin model" first proposed to explain the excision of integrated SV40 genomes from the chromosomes of virus-transformed cell lines (Botchan et al. 1979). Direct evidence for this model comes from analysis of chorion gene amplification in *Drosophila* embryos (Kafatos et al. 1985).

The onion skin mechanism for gene amplification requires that multiple initiation events occur at the same origin of replication before the completion of replication of the replication unit, and that replication forks emanating from this origin are arrested at sites on either side of the sequence that is amplified. Thus, the second set of forks resulting from reinitiation at the origin eventually catches up with the first set of forks, creating multiple, concentric replication bubbles (like the layers of an onion). At this time, there is no direct evidence for specific sequences that arrest replication forks flanking known amplification loci, but one could anticipate finding them. In addition, there may be specific terminator proteins produced by the specialized cells in which programmed gene amplification occurs. Such systems should provide fertile ground for future experiments.

Two Pathways for Separation of Sibling Molecules at Termination Sites

Termination of DNA replication occurs whenever two oncoming replication forks collide. Since DNA replication begins at thousands of sites distributed throughout the chromosomes of eukaryotic cells, termination of DNA replication must also occur at thousands of sites during each S phase. However, in contrast to initiation of DNA replication where specific sequences determine where replication begins in the genomes of eukaryotic cells, animal viruses, and mitochondria (see DePamphilis, this volume), termination of DNA replication can occur at virtually any sequence. However, it is possible, and perhaps likely, that some replication units (e.g., rDNA) terminate replication at specific sequences. In SV40, although termination of replication does not require specific sequences (Lai and Nathans 1975), the sequence at the termination site does strongly influence the pathway by which the two sibling molecules are separated (Weaver et al. 1985; Fields-Berry and DePamphilis 1989).

As replication forks advance, DNA unwinding introduces positive superhelical turns in front of the fork that are relaxed by topoisomerase I (Minden and Marians 1986). As two oncoming replication forks approach each other, the length of unreplicated DNA between them grows shorter, and therefore the target for topo I grows smaller. Eventually, topo I can no longer act in front of the forks. When this happens, DNA unwinding at the replication fork produces one catenated intertwine in the two sibling molecules behind the replication fork for each 10 bp of DNA unwound in front of the replication fork. In eukaryotes, topo II is required to remove these catenated intertwines. Mutations in yeast topo II that inactivate its decatenation activity result in interlocked DNA cellular chromosomes during S phase and in catenated dimers in plasmids that replicate in yeast cells (DiNardo et al. 1984; Uemura et al. 1987). Inhibition of topo II in mammalian cells either by indirect means (Sundin and Varshavsky 1981; Weaver et al. 1985; Fields-Berry and DePamphilis 1989) or by the use of specific inhibitors (Richter et al. 1987; Snapka et al. 1988; Ishimi et al. 1992) results in accumulation of late-replicating intermediates (θ structures that have completed >90% of their replication) of circular DNA molecules as well as catenated dimers containing multiple intertwines. Depending on experimental conditions, the primary product can be almost all late-replicating intermediates (Ishimi et al. 1992) or almost entirely catenated dimers (Sundin and Varshavsky 1981). The fact that late-replicating intermediates accumulate when topo II is inhibited demonstrates that catenated intertwines are normally removed prior to separation of sibling DNA molecules (Weaver et al. 1985).

Neither the SV40 termination region nor the yeast CEN3 sequence promotes formation of catenated intertwines under normal physiological conditions (Koshland and Hartwell 1987; Fields-Berry and DePamphilis

1989). Therefore, catenated intertwines are normally removed prior to or concomitant with termination of DNA replication when two replication forks collide. The presence of sequence-specific termination sites in bacterial chromosomes may ensure that this occurs in bacterial chromosomes prior to cell division.

CONCLUSION

The two known replication terminator proteins, namely Ter (Tus) of *E. coli* and RTP of *B. subtilis*, arrest replication forks by their polar contrahelicase activity. These proteins also block RNA chain elongation in a polar fashion. It is interesting to note that programmed cell death of bacterial cells that lose a resident plasmid also involves a plasmid-coded contrahelicase and an antidote protein. The contrahelicase is long-lived, but the antidote (contra-contrahelicase) protein is short-lived. Thus, in the plasmid-free segregants, the antidote decays and the longer-lived contrahelicase neutralizes the helicase, thereby causing replication arrest and cell death (Ruiz-Echevarria et al. 1995).

A polar contrahelicase from a eukaryote has not yet been discovered. We strongly suspect that such proteins exist in eukaryotes and are likely to be involved in fork arrest at specific sequences. The search for such protein(s) should prove to be interesting and rewarding.

It has been noted that fork arrest sites are very recombinogenic and therefore are regions of potential chromosome instability. That certain chromosome instabilities lead to the induction of cancer has been recognized (Hartwell 1992). This fact should provide additional incentive to search for eukaryotic replication terminator proteins.

ACKNOWLEDGMENTS

We thank Drs. T. Sahoo, A. Manna, and S. Pai for permission to quote unpublished data. We thank Mrs. Waltraud Bastia for editorial work and Drs. S. White and D.E. Bussiere for many useful discussions. Preparation of this chapter became a much easier task because of the interest, encouragement, and advice given by Dr. Mel DePamphilis. Work in our laboratory was supported by a grant (GM-49264) and a Merit Award (R37 AI-19881) from the National Institutes of Health.

REFERENCES

Adams, D.E., E.M. Schekhtman, E.L. Zechiedrich, M.B. Schmid, and N. Cozzarelli. 1992. The role of topoisomerase IV in partitioning bacterial replicons and the structure

of catenated intermediates in DNA replication. *Cell* **71:** 277-288.
Amin, A.A. and J. Hurwitz. 1992. Polar arrest of the simian virus 40 tumor antigen-mediated replication fork movement *in vitro* by the tus protein-terB complex of *E. coli. J. Biol. Chem.* **267:** 18612-18622.
Baker, T. 1995. Replication arrest. *Cell* **80:** 521-524.
Baran, N., A. Lapidot, and H. Manor. 1983. Onion skin replication of integrated polyoma-transformed rat cells: Termination within a specific cellular DNA segment. *Proc. Natl. Acad. Sci.* **80:** 105-109.
———. 1987. Unusual sequence element found at the end of an amplicon. *Mol. Cell. Biol.* **7:** 2636-2640.
Bastia, D., J. Germino, J.H. Crossa, and P. Hale. 1981a. Molecular cloning of the replication terminus of the plasmid R6K. *Gene* **14:** 81-89.
Bastia, D., J. Germino, J.H. Crossa, and J. Ram. 1981b. The nucleotide sequence surrounding the replication terminus of R6K. *Proc. Natl. Acad. Sci.* **78:** 2095-2099.
Bedrosian, C.L. and D. Bastia. 1991. *E coli* replication terminator protein impedes simian virus 40 SV40 replication fork movement and SV40 large tumor antigen helicase activity *in vitro* at a prokaryotic terminus sequence. *Proc. Natl. Acad. Sci.* **88:** 2618-2622.
Bierne, H. and B. Michel. 1994. When replication forks stop. *Mol. Microbiol.* **13:** 17-23.
Bierne, H., S.D. Ehrlich, and B. Michel. 1991. The replication termination signal terB of the *Escherichia coli* chromosome is a deletion hot spot. *EMBO J.* **10:** 2699-2705.
Blakely, G., S.D. Colloms, G. May, M. Burk, and D.J.C. Sherratt. 1991. *E. coli* XerC recombinase is required for chromosomal segregation at cell division. *New Biol.* **3:** 789-798.
Botchan, M., W. Topp, and J. Sambrook. 1979. Studies on simian virus 40 excision from cellular chromosomes. *Cold Spring Harbor Symp. Quant. Biol.* **43:** 709-719.
Brewer, B. 1988. When polymerases collide: Replication and transcriptional organization of the *Escherichia coli* chromosome. *Cell* **53:** 679-686.
Brewer, B. and W.L. Fangman. 1987. The localization of replication origins on ARS plasmids in *Saccharomyces cerevisiae*. *Cell* **51:** 463-471.
———. 1988. A replication fork barrier at the 3' end of yeast ribosomal RNA genes. *Cell* **55:** 637-743.
Brewer, B.J., D. Lockshon, and W.L. Fangman. 1992. The arrest of replication forks in the rDNA of yeast occurs independently of transcription. *Cell* **71:** 267-276.
Bussiere, D.E., D. Bastia, and S. White. 1995. Crystal structure of the replication terminator protein from *Bacillus subtilis* at 2.6 Å. *Cell* **80:** 651-660.
Carrigan, C.M., R.A. Peck, M.T. Smith, and R.G. Wake. 1991. Normal *ter*C region of the *Bacillus subtilis* chromosome acts in a polar manner to arrest the clockwise replication fork. *J. Mol. Biol.* **222:** 197-207.
Cashel, M. and K. Rudd. 1987. The stringent response. In Escherichia coli *and* Salmonella typhimurium: *Cellular and molecular biology* (ed. F.C. Neidhardt et al.), pp. 1410-1438. American Society for Microbiology, Washington, D.C.
Clayton, D. 1991a. Nuclear gadgets in mitochondrial DNA replication and transcription. *Trends Biochem. Sci.* **16:** 107-111.
———. 1991b. Replication and transcription of vertebrate mitochondrial DNA. *Annu. Rev. Cell Biol.* **7:** 453-478.
Crosa, J.H., L. Luttropp, and S. Falkow. 1976. Mode of replication of the conjugative R-plasmid RSF1040 in *Escherichia coli*. *J. Bacteriol.* **26:** 454-466.

d'Ambrosio, E. and A.V. Furano. 1987. DNA synthesis arrest sites at the right terminus of rat long interspersed repeated [LINE or L1Rn] DNA family members. *Nucleic Acids Res.* **15:** 3155-3175.

Dasgupta, S., R. Bernander, and K. Nordstrom. 1991. In vivo effect of the *tus* mutation on cell division in an *Escherichia coli* strain where chromosome replication is under control of plasmid R1. *Res. Microbiol.* **142:** 177-180.

Dayn, A., G.M. Samadashwily, and S. Mirkin. 1992. Intramolecular DNA triplexes: Unusual sequence requirements and influence on DNA polymerization. *Proc. Natl. Acad. Sci.* **89:** 11406-11410.

Dhar, V. and C.L. Schildkraut. 1991. Role of EBNA-1 in arresting replication forks at the Epstein-Barr virus oriP family of tandem repeats. *Mol. Cell. Biol.* **11:** 6268-6278.

Di Leonardo, A., S.P. Linke, Y. Yin, and G.M. Wahl. 1993. Cell cycle regulation of gene amplification. *Cold Spring Harbor Symp. Quant. Biol.* **58:** 655-667.

DiNardo, S., K. Voelkel, and R. Sternglanz. 1984. DNA topoisomerase mutant of *Saccharomyces cerevisiae:* Topoisomerase II is required for segregation of daughter molecules at the termination of DNA replication. *Proc. Natl. Acad. Sci.* **81:** 2616-2620.

Elledge, S.J., P. Sugino, L. Guarente, and R.W. Davis. 1990. Genetic selection for genes encoding sequence-specific DNA-binding proteins. *Proc. Natl. Acad. Sci.* **86:** 3689-3693.

Fields-Berry, S.C. and M.L. DePamphilis. 1989. Sequences that promote formation of catenated intertwines during termination of DNA replication. *Nucleic Acids Res.* **17:** 3261-3273.

Francois, V., J. Louarn, and J.-M. Louarn. 1989. The terminus region of *E. coli* chromosome is flanked by several polar replication pause sites. *Mol. Microbiol.* **3:** 995-1002.

Francois, V., J. Louarn, J.-E. Rebollo, and J.-M. Louarn. 1990. Replication termination, nondivisible zones and structure of the *Escherichia coli* chromosome. In *The bacterial chromosome* (ed. K. Drilca and M. Riley), pp. 351-359. American Society for Microbiology, Washington, D.C.

Franks, A.H., A.A. Griffith, and R.G. Wake. 1995. Identification and characterization of new DNA replication terminators in *Bacillus subtilis. Mol. Microbiol.* **17:** 13-23.

Gahn, T.A. and C.L. Schildkraut. 1989. The Epstein-Barr virus origin of replication, *ori*P, contains both the initiation and termination sites of DNA replication. *Cell* **58:** 527-535.

Gerbi, S.A., C. Liang, N. Wu, S.M. DiBartolomeis, B. Bienz-Tadmor, H.S. Smith, and F.D. Urnov. 1993. DNA amplification in DNA puff II/9A of *Sciaria coprophila. Cold Spring Harbor Symp. Quant. Biol.* **58:** 487-494.

Germino, J. and D. Bastia. 1981. Termination of DNA replication *in vitro* at a sequence-specific replication terminus. *Cell* **23:** 681-687.

Gottlieb, P.A., S. Wu, X. Zhang, M. Tecklenberg, P.L. Kuempel, and T.M. Hill. 1992. Equilibrium, kinetic, and footprinting studies of the Tus-Ter protein DNA interaction. *J. Biol. Chem.* **267:** 7434-7443.

Hartwell, L. 1992. Defects in a cell cycle checkpoint may be responsible for the genomic instability of cancer cells. *Cell* **71:** 543-546.

Heck, M.M.S. and A.C. Spradling. 1990. Multiple replication origins are used during *Drosophila* chorion gene amplification. *J. Cell Biol.* **110:** 903-914.

Henckes, G., F. Harper, A. Levine, F. Vannier, and S.J. Seror. 1989. Overreplication of the origin region in the dnaB37 mutant of *Bacillus subtilis*: Postinitiation control of chromosomal replication. *Proc. Natl. Acad. Sci.* **86:** 8660-8664.

Hernandez, P., S.S. Lamm, C.A. Bjerknes, and J. Van't Hof. 1988. Replication termini in the rDNA of the synchronized pea root cells (*Pisum sativum*). *EMBO J.* **7:** 303-308.

Hernandez, P., L. Martin-Parras, M.L. Martinez-Robles, and J.B. Schvartzman. 1993. Conserved features in the mode of replication of eukaryotic ribosomal RNA genes. *EMBO J.* **12:** 1475-1485.

Hiasa, H. and K.J. Marians. 1992. Differential inhibition of the DNA translocation and DNA unwinding activities of DNA helicases by the *E. coli* tus protein. *J. Biol. Chem.* **267:** 11379-11385.

———. 1994a. Tus prevents overreplication of *oriC* plasmid DNA. *J. Biol. Chem.* **269:** 26959-26968.

———. 1994b. Topoisomerase IV can support *oriC* DNA replication *in vitro*. *J. Biol. Chem.* **269:** 16371-16375.

Hidaka, M., M. Akiyama, and T. Horiuchi. 1988. A consensus sequence of three DNA replication terminus sites on the *E. coli* chromosome is highly homologous to the terR sites of the R6K plasmid. *Cell* **55:** 467-475.

Hidaka, M., T. Kobayashi, and T. Horiuchi. 1991. A newly identified DNA replication terminus site, TerE, on the *Escherichia coli* chromosome. *J. Bacteriol.* **173:** 391-393.

Hidaka, M., T. Kobayashi, S. Takenaka, H. Takeya, and T. Horiuchi. 1989. Purification of a DNA replication terminus (ter) site-binding protein in *Escherichia coli* and identification of the structural gene. *J. Biol. Chem.* **264:** 21031-21037.

Hidaka, M., T. Kobayashi, Y. Ishimi, M. Seki, T. Enomoto, M. Abdel-Monem, and T. Horiuchi. 1992. Termination complex in *Escherichia coli* inhibits SV40 DNA replication *in vitro* by impeding the action of T antigen helicase. *J. Biol. Chem.* **267:** 5361-5365.

Hill, T.M. 1992. Arrest of bacterial DNA replication. *Annu. Rev. Microbiol.* **46:** 603-633.

Hill, T.M. and K.J. Marians. 1990. *Escherichia coli* tus protein acts to arrest the progression of DNA replication forks *in vitro*. *Proc. Natl. Acad. Sci.* **87:** 2481-2485.

Hill, T.M., A.J. Pelletier, M.L. Tecklenberg, and P.L. Kuempel. 1988. Identification of DNA sequence from the *E. coli* terminus region that halts replication forks. *Cell* **55:** 459-466.

Hill, T.M., M. Tecklenberg, A.J. Pelletier, and P.L. Kuempel. 1989. *tus*, the *trans*-acting gene required for termination of DNA replication in *Escherichia coli*, encodes a DNA-binding protein. *Proc. Natl. Acad. Sci.* **86:** 1593-1597.

Horiuchi, T. and Y. Fujimura. 1995. Recombinational rescue of the stalled fork: A model based on analysis of an *Escherichia coli* strain with a chromosome region difficult to replicate. *J. Bacteriol.* **177:** 783-791.

Horiuchi, T. and M. Hidaka. 1988. Core sequence of two separable terminus sites of the R6K plasmid that exhibit polar inhibition of replication is a 20 bp inverted repeat. *Cell* **54:** 515-523.

Horiuchi, T., Y. Fujimura, H. Nishitani, T. Kobayashi, and M. Hidaka. 1994. The DNA replication fork blocked at the Ter site may be an entrance for the RecBCD enzyme into duplex DNA. *J. Bacteriol.* **176:** 4656-4663.

Ishimi, Y., R. Ishida, and T. Andoh. 1992. Effect of ICRF-193, a novel DNA topoisomerase II inhibitor, on simian virus 40 DNA and chromosome replication *in vitro*. *Mol. Cell. Biol.* **12:** 4007-4012.

Kafatos, F.C., W. Orr, and C. Delidakis. 1985. Developmentally regulated gene

amplification. *Trends Genet.* **1:** 301-306.
Kato, J., Y. Nishimura, M. Yamada, and H. Suzuki. 1990. A new topoisomerase essential for chromosome segregation in *E. coli. Cell* **63:** 393-404.
Kato, J., Y. Nishimura, R. Imamura, H. Niki, S. Hiraga, and H. Suzuki. 1988. Gene organization in a region containing a new gene involved in chromosome partitioning in *Escherichia coli. J. Bacteriol.* **170:** 3967-3977.
Kaul, S., B.K. Mohanty, T. Sahoo, I. Patel, S. Khan, and D. Bastia. 1994. The replication terminator protein of the gram-positive bacterium *B. subtilis* functions as a polar contrahelicase in gram-negative *E. coli. Proc. Natl. Acad. Sci.* **91:** 11143-11147.
Khatri, G.S., T. MacAllister, P.R. Sista, and D. Bastia. 1989. The replication termination protein of *E. coli* is a DNA sequence specific contra-helicase. *Cell* **59:** 667-674.
Kobayashi, T., M. Hidaka, M. Nishizawa, and T. Horiuchi. 1992. Identification of a site required for DNA replication fork blocking activity in the rRNA gene cluster in *Saccharomyces cerevisiae. Mol. Gen. Genet.* **233:** 355-362.
Kolter, R. and D.R. Helinski. 1978. Activity of the replication terminus of plasmid in hybrid replicons in *Escherichia coli. J. Mol. Biol.* **124:** 425-441.
Kornberg, A. and T.A. Baker. 1992. *DNA replication*, 2nd edition. W.H. Freeman, New York.
Koshland, D. and L.H. Hartwell. 1987. The structure of sister minichromosome DNA before anaphase in *Saccharomyces cerevisiae. Science* **238:** 1713-1716.
Kuempel, P.L. and S.A. Duerr. 1979. Chromosome replication in *Escherichia coli* is inhibited in the terminus region near the *rac* locus. *Cold Spring Harbor Symp. Quant. Biol.* **43:** 563-567.
Kuempel, P.L., S.A. Duerr, and N.R. Seely. 1977. Terminus region of the chromosome of *Escherichia coli* inhibits replication forks. *Proc. Natl. Acad. Sci.* **74:** 3927-3931.
Kuempel, P.L., J.M. Henson, L. Dircks, M. Tecklenberg, and D.F. Lim. 1991. *dif*, a *rec*A-independent recombination site in the terminus region of the chromosome of *Escherichia coli. New Biol.* **3:** 799-811.
Lai, C.S. and D. Nathans. 1975. Non-specific termination of SV40 DNA replication. *J. Mol. Biol.* **97:** 113-118.
LaDuca, R.J., P.J. Fay, C. Chuang, C.S. McHenry, and R.A. Bambara. 1983. Site-specific pausing of deoxyribonucleic acid synthesis catalyzed by four forms of *Escherichia coli* DNA polymerase III. *Biochemistry* **22:** 5177-5188.
Lang, W.H. and R.H. Reeder. 1993. The Reb1 site is an essential component of a terminator for RNA polymerase I in *Saccharomyces cerevisiae. Mol. Cell. Biol.* **13:** 649-658.
Langley, D.B., M.T. Smith, P.J. Lewis, and R.G. Wake. 1993. Protein-nucleoside contacts in the interaction between the replication terminator protein of *Bacillus subtilis* and the DNA terminator. *Mol. Microbiol.* **10:** 771-779.
LeBowitz, J.O. and R. McMacken. 1986. The *Escherichia coli* DnaB protein is a DNA helicase. *J. Biol. Chem.* **261:** 4738-4748.
Lee, E.H. and A. Kornberg. 1992. Features of replication fork blockage by the *Escherichia coli* terminus binding protein. *J. Biol. Chem.* **267:** 8778-8784.
Lee, E.H., A. Kornberg, M. Hidaka, T. Kobayashi, and T. Horiuchi. 1989. *Escherichia coli* replication termination protein impedes the action of helicases. *Proc. Natl. Acad. Sci.* **86:** 9104-9108.
Lee, M.S. and K.J. Marians. 1987. *Escherichia coli* replication factor Y, a component of the primosome, can act as a DNA helicase. *Proc. Natl. Acad. Sci.* **84:** 8345-8349.

Leslie, N.R. and D.J. Sherratt. 1995. Site-specific recombination in the replication terminus region of *Escherichia coli*: Functional replacement of *dif*. *EMBO J.* **14:** 1561–1570.
Levine, A., S. Autret, and S.J. Seror. 1995. A checkpoint involving RTP, the replication terminator protein, arrests replication downstream of the origin during the stringent response in *Bacillus subtilis*. *Mol. Microbiol.* **15:** 287–295.
Levine, A., F. Vannier, M. Dehbi, G. Henckes, and S.J. Seror. 1991. The stringent reponse blocks DNA replication outside the *ori* region in *Bacillus subtilis* and at the origin in *Escherichia coli*. *Mol. Microbiol.* **219:** 605–613.
Lewis, P.J. and R.G. Wake. 1991. Termination of chromosome replication in *Bacillus subtilis*. *Res. Microbiol.* **142:** 893–900.
Lewis, P.J., M.T. Smith, and R.G. Wake. 1989. A protein involved in termination of chromosome replication in *Bacillus subtilis* binds specifically to the *ter*C site. *J. Bacteriol.* **171:** 3564–3567.
Lewis, P.J., G.B. Ralston, R.I. Christopherson, and R.G. Wake. 1990. Identification of the replication terminator protein binding sites in the terminus region of the *Bacillus subtilis* chromosome and stoichiometry of the binding. *J. Mol. Biol.* **214:** 73–84.
Lindsey, J.C. and J.R.F.C. Leach. 1989. Slow replication of palindrome-containing DNA. *J. Mol. Biol.* **206:** 779–782.
Linskens, M.H.K. and J.A. Huberman. 1988. Organization of origin of replication of rDNA in *Saccharomyces cerevisiae*. *Mol. Cell. Biol.* **8:** 4927–4935.
Little, R.D., T.H.K. Platt, and C.L. Schildkraut. 1993. Initiation and termination of DNA replication in human rRNA genes. *Mol. Cell. Biol.* **13:** 6600–6613.
Liu, B. and B. Alberts. 1995. Head-on collision between a DNA replication apparatus and RNA polymerase transcription complex. *Science* **267:** 1131–1137.
Lopez-Estrano, C., J. Schvartzman, and P. Hernandez. 1996. The replication of ribosomal RNA genes in eukaryotes. In *Chromosomes today* (ed. M. Puertas et al.). Chapman and Hall, London. (In press.)
Louarn, J., J. Patte, and J.-M. Louarn. 1977. Evidence for a fixed termination site of chromosome replication in *Escherichia coli* K12. *J. Mol. Biol.* **115:** 295–314.
Louarn, J.-M., J. Louarn, V. Francois, and J. Patte. 1991. Analysis and possible role of hyperrecombination in the termination region of the *Escherichia coli* chromosome. *J. Bacteriol.* **173:** 5097–5104.
Lovett, M.A., R.A. Sparks, and D.R. Helinski. 1975. Bidirectional replication of plasmid R6K DNA in *Escherichia coli*: Correspondence between origin of replication and a single strand break in relaxed complex. *Proc. Natl. Acad. Sci.* **72:** 2905–2909.
Lucchini, R. and J.M. Sogo. 1994. Chromatin structure and transcriptional activity around the replication forks at the 3' end of the yeast rRNA genes. *Mol. Cell. Biol.* **14:** 318–326.
MacAllister, T., G.S. Khatri, and D. Bastia. 1990. Sequence-specific and polarized replication termination *in vitro*: Complementation of extracts of tus$^-$ *Escherichia coli* by purified Ter protein and analysis of termination intermediates. *Proc. Natl. Acad. Sci.* **87:** 2828–2832.
Madsen, C.S., S.C. Ghivizzani, and W.W. Hauswirth. 1993. Protein binding to a single termination-associated sequence in the mitochondrial DNA D-loop region. *Mol. Cell. Biol.* **13:** 2162–2171.
Manna, A.C., K.S. Pai, D.E. Bussiere, S.W. White, and D. Bastia. 1996. The dimer-dimer interaction surface of the replication terminator protein of *Bacillus subtilis* and termina-

tion of DNA replication. *Proc. Natl. Acad. Sci.* (in press).
Minden, J.S. and K.J. Marians. 1986. *Escherichia coli* topoisomerase I can segregate replicating pBR322 daughter DNA molecules *in vitro*. *J. Biol. Chem.* **261:** 11906–11917.
Mohanty, B.K., T. Sahoo, and D. Bastia. 1996. Relationship between sequence-specific termination of DNA replication and transcription. *EMBO J.* (in press).
Natarajan, S., W.I. Kelly, and D. Bastia. 1991. Replication terminator protein of *Escherichia coli* is a transcriptional repressor of its own synthesis. *Proc. Natl. Acad. Sci.* **88:** 3867–3871.
Natarajan, S., S. Kaul, A. Miron, and D. Bastia. 1993. A 27 kd protein of *E. coli* promotes antitermination of replication *in vitro* at a sequence-specific replication terminus. *Cell* **72:** 113–120.
Nordstrom, K., R. Bernander, and S. Dasgupta. 1991. Analysis of the bacterial cell cycle using strains in which chromosome replication is controlled by plasmid R1. *Res. Microbiol.* **142:** 181–188.
O'Sullivan, M.A. and C. Anagnostopoulos. 1982. Replication terminus of the *Bacillus subtilis* chromosome. *J. Bacteriol.* **151:** 135–143.
Pai, K.S., D.E. Bussiere, F. Wang, C. Hutchison, S. White, and D. Bastia. 1996. The structure and function of the replication terminator protein of *Bacillus subtilis:* Identification of the "winged-helix" DNA binding domain. *EMBO J.* (in press).
Pelletier, A.J., T.M. Hill, and P.L. Kuempel. 1988. Location of sites that inhibit progression of replication forks in the terminus region of *Escherichia coli*. *J. Bacteriol.* **170:** 4295–4298.
Peng, H. and K.J. Marians. 1993. Decatenation activity of topoisomerase IV during *oriC* and pBR322 DNA replication in vitro. *Proc. Natl. Acad. Sci.* **90:** 8571–8575.
Rao, P.S., H. Manor, and R.G. Martin. 1988. Pausing in SV40 DNA replication by a sequence containing $(dG-dA)_{27}(dT-dc)_{27}$. *Nucleic Acids Res.* **16:** 8077–8094.
Richter, A., U. Strausfeld, and R. Knippers. 1987. Effects of VM26 (teniposide), a specific inhibitor of type II DNA topoisomerase, on SV40 DNA replication *in vivo*. *Nucleic Acids Res.* **15:** 3455–3468.
Roecklein, B.A. and P.L. Kuempel. 1992. *In vivo* characterization of *tus* gene expression in *Escherichia coli*. *Mol. Microbiol.* **6:** 1655–1661.
Roecklein, B., A. Pelletier, and P.L. Kuempel. 1991. The *tus* gene of *Escherichia coli*: Autoregulation, analysis of flanking sequences and identification of a complementary system in *Salmonella typhimurium*. *Res. Microbiol.* **142:** 169–175.
Ruiz-Echevarria, M.J., G. Gimenez-Gallego, R. Sabariegos-Jareno, and R. Diaz-Orejas. 1995. Kid, a small protein of the parD stability system of plasmid R1, is an inhibitor of DNA replication acting at the initiation of DNA synthesis. *J. Mol. Biol.* **247:** 568–577.
Sahoo, T., B.K. Mohanty, I. Patel, and D. Bastia. 1995a. Termination of DNA replication *in vitro*: Requirement for stereospecific interaction between two dimers of the replication termination protein of *Bacillus subtilis* and with the terminator site to elicit polar contrahelicase and fork impedance. *EMBO J.* **14:** 619–628.
Sahoo, T., B.K. Mohanty, A. Manna, M. Lobert, and D. Bastia. 1995b. The contrahelicase activities of the replication terminator proteins of *Escherichia coli* and *Bacillus subtilis* are helicase-specific and impede both helicase translocation and authentic DNA unwinding. *J. Biol. Chem.* **270:** 29138–29144.
Schimke, R.T. 1984. Gene amplification in cultured cells. *Cell* **37:** 705–713.
Schmid, M.B. 1990. A locus affecting nucleoid segregation in *Salmonella typhimurium*.

J. Bacteriol. **172:** 5416–5424.

Schneider, A.M., M. Gaisne, and C. Anagnostopoulos. 1983. Genetic structure and internal rearrangements of stable merodiploids from *Bacillus subtilis* strains carrying the trpE26 mutation. *Genetics* **101:** 189–210.

Sista, P.R., C.A. Hutchison, and D. Bastia. 1991. DNA-protein interaction at the replication termini of plasmid R6K. *Genes Dev.* **5:** 74–82.

Sista, P.R., S. Mukherjee, P. Patel, G.S. Khatri, and D. Bastia. 1989. A host-encoded DNA-binding protein promotes termination of plasmid replication at a sequence-specific replication terminus. *Proc. Natl. Acad. Sci.* **86:** 3026–3030.

Smith, M.T. and R.G. Wake. 1992. Definition and polarity of action of DNA replication terminators in *Bacillus subtilis*. *J. Mol. Biol.* **227:** 648–657.

Snapka, R.M., M.A. Powelson, and J.M. Strayer. 1988. Swivelling and decatenation of replicating simian virus 40 genomes *in vivo*. *Mol. Cell. Biol.* **8:** 515–521.

Stark, G.R. 1993. Regulation and mechanisms of mammalian gene amplification. *Adv. Cancer Res.* **61:** 87–113.

Stark, G.R., M. Debatisse, E. Giulotto, and G.M. Wahl. 1989. Recent progress in understanding mechanisms of mammalian DNA amplification. *Cell* **57:** 901–908.

Sundin, O. and A. Varshavsky. 1981. Terminal stages of SV40 DNA replication proceed via multiply intertwined catenated dimers. *Cell* **21:** 103–114.

Swindels, M.B. 1995. Identification of a common fold in the replication terminator protein suggests a possible mode for DNA binding. *Trends Biochem. Sci.* **20:** 300–302.

Tapper, D.P. and M.L. DePamphilis. 1980. Preferred sites are involved in the arrest and initiation of DNA synthesis during the replication of SV40 DNA. *Cell* **22:** 97–108.

Tecklenberg, M., A. Naumer, O. Nagappan, and P. Kuempel. 1995. The *dif* resolvase locus of the *Escherichia coli* chromosome can be replaced by a 33-bp sequence, but function depends on location. *Proc. Natl. Acad. Sci.* **92:** 1352–1356.

Uemura, T., H. Ohkura, Y. Adachi, K. Morino, K. Shiozaki, and M. Yanagida. 1987. DNA topoisomerase II is required for condensation and separation of mitotic chromosomes in *S. pombe*. *Cell* **50:** 917–925.

Valenzuela, M.S., D. Freifelder, and R. Inman. 1976. Lack of a unique terminator site for the first round of bacteriophage DNA replication. *J. Mol. Biol.* **102:** 569–589.

Voelkel-Meimon, K., R.L. Keil, and G.S. Roeder. 1987. Recombination-stimulating sequences in yeast ribosomal DNA correspond to sequences regulating transcription by RNA polymerase I. *Cell* **48:** 1071–1079.

Weaver, D.T. and M.L. DePamphilis. 1984. The role of palindromic and non-palindromic sequences in arresting DNA synthesis *in vitro* and *in vivo*. *J. Mol. Biol.* **180:** 961–986.

Weaver, D.T., S.C. Fields-Berry, and M.L. DePamphilis. 1985. The termination region for SV40 DNA replication directs the mode of separation for the two sibling molecules. *Cell* **41:** 565–575.

Weiss, A.S. and R.G. Wake. 1984. A unique DNA intermediate associated with termination of chromosome replication in *Bacillus subtilis*. *Cell* **39:** 683–689.

Wells, R.D., D.A. Collier, J.C. Hanvey, M. Shimizu, and F. Wohlrab. 1988. The chemistry and biology of unusual DNA structures adopted by oligopurine.oligopyrimidine sequences. *FASEB J.* **2:** 2939–2949.

Wintersberger, E. 1994. DNA amplification: New insights into its mechanism. *Chromosoma* **103:** 73–81.

Yoshikawa, H. and N. Sueoka. 1963. Sequential replication of *Bacillus subtilis* chromosome I: Comparison of marker frequencies in exponential and stationary growth

phases. *Proc. Natl. Acad. Sci.* **49:** 559-566.
Young, P.A. and R.G. Wake. 1994. The *Bacillus subtilis* replication terminator system functions in *Escherichia coli*. *J. Mol. Biol.* **240:** 275-280.
Zechiedrich, E.L. and N.R. Cozzarelli. 1995. Roles of topoisomerase IV and DNA gyrase in DNA unlinking during replication in *Escherichia coli*. *Genes Dev.* **9:** 2859-2869.
Zeviani, M. 1992. Nucleus-driven mutations of human mitochondrial DNA. *J. Inherited Metab. Dis.* **15:** 456-471.
Zeviani, M., S. Servidei, C. Gellera, E. Bertini, S. DiMauro, and S. DiDonato. 1989. An autosomal dominant disorder with multiple deletions of mitochondrial DNA starting at the D-loop region. *Nature* **339:** 309-311.

7
Fidelity of DNA Replication

John D. Roberts[1] and Thomas A. Kunkel[2]
[1]Laboratory of Molecular Carcinogenesis
[2]Laboratory of Molecular Genetics
National Institute of Environmental Health Sciences
Research Triangle Park, North Carolina 27709

The six billion nucleotides of the diploid human genome are replicated in only a few hours while generating so few errors that the spontaneous mutation rate may be less than 1 mutation per genome per cell division (Loeb 1991). This incredible accuracy results from three major error-avoidance processes: the high selectivity of DNA polymerases, exonucleolytic proofreading, and postreplication mismatch repair. In this chapter, we review our current understanding of the first two of these processes. Readers interested in eukaryotic mismatch repair are referred to a recent review (Modrich 1994).

We begin by describing the steps in the polymerization reaction cycle that discriminate against base substitution errors, then review studies of the substitution fidelity of the five classes of eukaryotic DNA polymerases. We then consider several ways to make errors by template-primer slippage and review what is known about eukaryotic DNA polymerase frameshift error rates. Finally, we present information on the fidelity with which the multiprotein replication machinery replicates undamaged DNA and DNA containing adducts of known carcinogens.

BASE SUBSTITUTION FIDELITY

Discrimination Steps in a Polymerization Cycle

The error discrimination steps that operate during incorporation of a single nucleotide have been worked out primarily with prokaryotic and viral DNA polymerases. The steps (Fig. 1) include binding of the polymerase to the DNA, formation of a ternary complex with the incoming deoxyribonucleoside triphosphate (dNTP), a conformational change in this complex to position the substrates for phosphodiester bond formation, the chemical reaction step to form the bond, a second conformational change following the chemical reaction, and release of pyrophos-

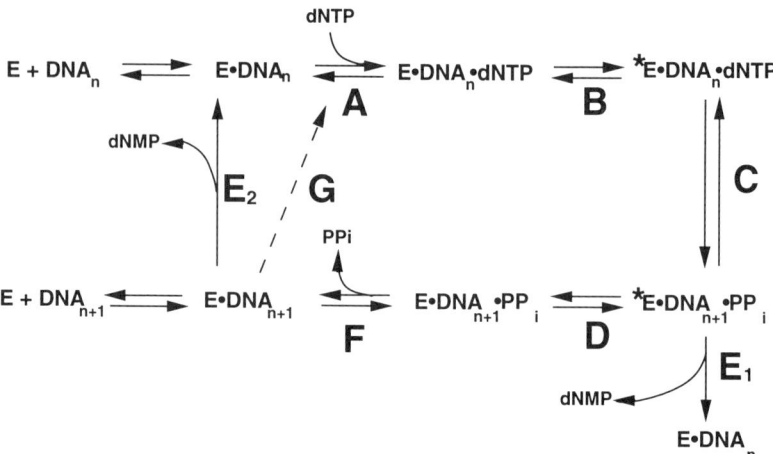

Figure 1 Reaction pathway for exonuclease-proficient DNA polymerases. Asterisks represent enzyme (E) in a different conformation. The enzyme conformations in the ternary E·DNA$_n$·dNTP and E·DNA$_{n+1}$·PPi complexes are unknown and not necessarily the same. Entry into the next cycle of polymerization is indicated by G. Translocation has not been assigned to a particular step in the reaction. Steps A–F are discussed in the text. (Reprinted, with permission, from Kunkel 1992.)

phate (PPi). After the chemical step, the polymerase translocates to commence the next cycle of polymerization, extending the nascent strand in the 5' to 3' direction.

DNA Polymerase Selectivity

There are several points in this cycle where discrimination against errors by the DNA polymerase occurs. The first is during binding of the dNTP to the complex formed between the polymerase and a correctly paired template primer (Fig. 1, step A). Incorrect dNTPs bind much less avidly than do correct dNTPs. At this step, discrimination against several mispairs, quantitated by kinetic analyses includes: Klenow polymerase, the large fragment of *Escherichia coli* DNA polymerase I, 0 to 23-fold (Eger and Benkovic 1992); T7 DNA polymerase, 200- to 400-fold (Wong et al. 1991; for review, see Johnson 1993); T4 DNA polymerase, ≥250-fold (Capson et al. 1992); and human immunodeficiency virus type 1 reverse transcriptase (HIV-1 RT), 25- to 400-fold (Johnson 1993; Zinnen et al. 1994). The amount of discrimination varies over a considerable range, depending on the composition of the mispair and the surrounding sequence context. This reflects differences in hydrogen-bonding potentials for various combinations of dNTPs and template bases, sequence-dependent differences in base stacking, and a demand for equivalent base-pair geometry (for review, see Echols and Goodman 1991).

For the three viral DNA polymerases mentioned above, the amount of binding discrimination exceeds that predicted from the free-energy differences between correct and incorrect base pairs in aqueous solution. Within a DNA polymerase active site, hydrogen-bonded water molecules are displaced from the transition state for complementary base pairs (Fersht 1985), and this exclusion of water from the active site may amplify base-pair free-energy differences, thus enhancing fidelity (Petruska et al. 1986; Abbotts et al. 1991; Johnson 1993). A differential ability to exclude water from the active site might partly explain DNA polymerase-dependent differences in dNTP-binding discrimination. Klenow polymerase utilizes dNTP-binding discrimination to a lesser extent than do the other three polymerases examined. This has led to the suggestion that in the "ground state" (i.e., prior to the inferred conformational change at step B), initial binding of the dNTP to Klenow polymerase may not involve base-pairing (Johnson 1993). It has also been suggested (Capson et al. 1992; Johnson 1993) that any reduction in enzymatic efficiency of *E. coli* pol I resulting from competitive inhibition by the three incorrect dNTPs might be tolerated by the cell because the primary role for this polymerase in vivo is in DNA repair or replacement of RNA primers during replication. In contrast, reduced efficiency might be less acceptable for polymerases that replicate entire genomes. All three viral polymerases mentioned above, whose role is genomic replication, discriminate more strongly against incorrect dNTP binding than does Klenow polymerase.

The next selectivity step in the polymerization cycle, inferred from several lines of evidence (Johnson 1993; Zinnen et al. 1994), involves a conformational change in the ternary complex to position the dNTP for subsequent phosphodiester bond formation (Fig. 1, step B). For T7 DNA polymerase, this change to a "closed" structure in which the polymerase is thought to lock onto the template-primer·dNTP complex (Johnson 1993) is much more rapid for correct base pairs that can adopt Watson-Crick geometry than for incorrect base pairs that cannot. This leads to a 2000- to 4000-fold faster rate of incorporation of correct nucleotides (Johnson 1993). Similarly, a difference in the rate of change in protein conformation with correctly versus incorrectly bound dNTPs has been suggested to enhance the selectivity of HIV-1 RT by 7- to 17,000-fold (Johnson 1993; Zinnen et al. 1994), depending on the mispair considered.

The situation is somewhat different with Klenow polymerase. Although a change in polymerase conformation has been inferred to limit the catalytic rate for correct incorporation by this enzyme, this step is not rate-limiting for incorrect incorporation. Rather, discrimination against incorrect incorporation is mostly due to a strong reduction in the rate of phosphodiester bond formation (Fig. 1, step C; Kuchta et al. 1988).

Steps A, B, and C in Figure 1 all contribute to polymerase selectivity against insertion of incorrect nucleotides, with the relative importance of each step dependent on the polymerase examined. The contributions of these individual steps to the fidelity of eukaryotic DNA polymerase remain to be established. However, steady-state kinetic analyses have established misinsertion and mispair extension rates of some eukaryotic DNA polymerases (see below).

Exonucleolytic Proofreading

With the Klenow polymerase, a slow step has been detected after chemistry and prior to pyrophosphate release (Carroll and Benkovic 1990; Polesky et al. 1992). This delay, suggested to result from a second conformational change in the enzyme-template·primer complex, provides an opportunity for its intrinsic $3' \rightarrow 5'$ exonuclease to remove misinserted nucleotides (step E_1). Once a nucleotide is incorporated, the E·DNA_{n+1} complex can enter the next cycle of polymerization (Fig. 1, step G). However, the rate of correct incorporation onto a terminal mispair is much slower than onto a correctly paired terminus. This slow step provides another opportunity for excision (Fig. 1, step E_2), after transferring the misinserted nucleotide from the polymerase active site to the exonuclease active site. These sites are on separate domains of Klenow polymerase and are spatially separated in other DNA polymerases. Movement into the exonuclease active site can occur with or without enzyme dissociation from the template primer, depending on the polymerase (Joyce 1989; Donlin et al. 1991; Capson et al. 1992; Reddy et al. 1992). If the enzyme does dissociate, it may rebind to the exonuclease active site and edit the misinsertion. It is also formally possible that a terminal misinsertion may be bound and proofread by another exonuclease, one associated with a different polymerase or one not associated with a polymerase at all.

Prokaryotic polymerases containing associated $3' \rightarrow 5'$ exonuclease activities have average base substitution error rates of about 10^{-6} to 10^{-7}. The proofreading contribution to these rates has been assessed by selectively stimulating the polymerase activity relative to the exonuclease activity by increasing the concentration of the next correct nucleotide to be incorporated after a misinsertion. Alternatively, exonuclease activity can be inhibited by adding to the reaction a nucleotide monophosphate (dNMP), the end product of exonuclease action. Proofreading can also be eliminated by changing amino acid residues essential for exonuclease activity. These residues are found in three conserved DNA sequence motifs common to the coding sequences of DNA polymerases containing intrinsic 3'→5' exonucleases (Ito and Braithwaite 1990; Blanco et al. 1991; Chung et al. 1991; Morrison et al. 1991; Simon et al. 1991; Zhang et al. 1991). The results obtained from these three approaches suggest that proofreading contributes on average about 100-fold to fidelity. This value is consistent with estimates from in vivo studies using *E. coli* strains deficient in proofreading activity (Schaaper 1993) and with calculations suggesting that the energetic cost of improving fidelity by more than this amount using exonucleolytic activity could be unacceptably high for an organism due to too much excision of correctly paired bases (Fersht et al. 1982).

The contribution of exonucleolytic proofreading to base substitution fidelity can vary over a wide range, from only a few-fold (Bebenek et al. 1990) to almost 1000-fold (West Frey et al. 1993). This results partly from the different rate constants for polymerization from the 12 possible mispairs and also reflects enzyme- and sequence-specific influences. These are expected on the basis of the idea originally proposed by Brutlag and Kornberg (1972) and subsequently supported by extensive data (see, e.g., Bloom et al. 1994; Carver et al. 1994 and references therein) that a terminus containing a terminal mispair has a higher probability of being single-stranded ("frayed") than does a correctly paired terminus. Such a frayed end will preferentially bind to the exonuclease active site, which prefers single-stranded DNA. Similarly, a matched and, therefore, double-stranded terminus will preferentially bind to the polymerase active site, which prefers double-stranded DNA (for review, see Joyce and Steitz 1994). Because the stability of the duplex region of the template-primer will depend on its DNA sequence, proofreading efficiency is expected to differ in different sequence contexts having differing stabilities. Moreover, the degree of fraying needed to allow single-stranded DNA to bind to the exonuclease active site may vary, depending on the distance between the polymerase and exonuclease active sites. This distance, estimated to be 25–30 Å for the Klenow polymerase, could be greater for some enzymes than others (e.g., compare data in Cowart et al. [1989] to data in Capson et al. [1992]; for review, see Joyce and Steitz 1994), leading to enzyme-mediated differences in proofreading efficiency. It has also been proposed that the Klenow polymerase active site promotes movement of DNA into the exonuclease active site by rejecting aberrant primer termini (Carver et al. 1994).

Base Substitution Error Rates of Eukaryotic DNA Polymerases

The above concepts provide a framework for considering what is known about the substitution fidelity of eukaryotic DNA polymerases. Error rates during catalysis in vitro by the five template-dependent eukaryotic DNA polymerases (see Wang, this volume) are shown in Table 1. These were obtained using fidelity assays that require both misinsertion and mispair extension to score an error. The "average" values are from copying a 250-base single-stranded template sequence of the *lacZ* α-complementation gene in bacteriophage M13mp2 DNA (Kunkel 1985a). This assay scores all stable misincorporations that yield an M13 plaque

Table 1 Error rates of eukaryotic DNA polymerases

DNA polymerase	Average error rate ($\times 10^{-6}$)	Refs.	Range of error rates ($\times 10^{-6}$)	Refs.
Substitution errors				
pol-α	160[a]	1–3	1.2–380	1–13
pol-β	670	2	45–1000	2, 9
pol-δ	~10[b]	4	≤2.3–29	4
pol-ε	≤6.7	4	≤1–19[c]	4, 14
pol-γ	1.8[d]	15, 16	1.8–1200[e]	2, 9, 15, 16
One-base frameshift errors				
pol-α	50[a]	1, 4, 17		
pol-β	900	17		
pol-δ	18[b]	4		
pol-ε	5[c]	4		
pol-γ	2.4	24		

References: (1) Kunkel et al. 1989; (2) Kunkel and Alexander 1986; (3) Roberts and Kunkel 1988; (4) Thomas et al. 1991b; (5) Copeland and Wang 1991; (6) Reyland and Loeb 1987; (7) Grosse et al. 1983; (8) Kaguni et al. 1984; (9) Kunkel and Loeb 1981; (10) Brooke et al. 1991; (11) Copeland et al. 1993; (12) Dong et al. 1993a; (13) Perrino and Loeb 1989b; (14) Kunkel et al. 1987; (15) Wernette et al. 1988; (16) Kunkel and Mosbaugh 1989; (17) Kunkel 1986.
[a]Measurements have been made with pol-α preparations from several sources; the values given are averages.
[b]Reactions contained PCNA to stimulate gap-filling synthesis by pol-δ.
[c]Higher values obtained with reactions in which proofreading was compromised; thus, they probably represent a minimal estimate of the accuracy of the enzyme.
[d]Data are from reversion assays.
[e]See comment in Reference 16 regarding the higher error rate values.

with reduced blue-color intensity. The type of polymerase error is defined by sequencing the identified M13 mutants. The substitution error rates are thus average values per detectable nucleotide polymerized, for all 12 possible single-base mispairs in a variety of sequence contexts (Bebenek and Kunkel 1995). The wide range of error rates reported in the literature (Table 1) illustrates that substitution fidelity depends on the DNA polymerase, the composition of the mispair, and the local template-primer sequence. The influence of these parameters can be examined at specific template positions using steady-state kinetic approaches (Mendelman et al. 1989, 1990). This allows the amount of discrimination at both the misinsertion and mispair extension steps to be estimated separately. The fidelity of the eukaryotic DNA polymerases is considered in more detail below.

DNA Polymerase-α

This is the most extensively studied eukaryotic polymerase. DNA polymerase-α (pol-α) isolated from several sources has an average base substitution error rate of 160×10^{-6} when a variety of errors in numerous sequence contexts are considered collectively (Table 1). Similar values have been obtained when the fidelity of the yeast or human p180 catalytic subunit alone is compared to that of the four-subunit DNA polymerase-α:RNA primase complex (Kunkel et al. 1989; Copeland et al. 1993). Similar values for misinsertion fidelity, representing the product of steps A, B, and C in Figure 1, have also been observed by kinetic analyses of the *Drosophila melanogaster* DNA polymerase-α:DNA primase (Mendelman et al. 1989). Thus, highly purified DNA pol-α is not particularly accurate relative to the high fidelity required to replicate eukaryotic genomes (see below). However, a 10^{-4} error rate may more than suffice if pol-α is only responsible for synthesis of a small number of nucleotides from an RNA primer. Mistakes made here could also be removed during the RNA primer excision-replacement synthesis reaction. Moreover, pol-α fidelity estimates are thus far limited to synthesis initiated from exogenously supplied DNA primers. Since several recent observations suggest that pol-α synthesis coupled to primase activity may differ in some respects (Sheaff et al. 1994), it is possible that the fidelity of RNA-primed DNA synthesis could be higher (or lower) than present data suggest.

An average error rate of about 10^{-4} is consistent with the fact that many preparations of purified pol-α lack $3' \rightarrow 5'$ exonuclease activity and that pol-α genes lack the three conserved sequence motifs characteristic of such exonucleases. Nonetheless, evidence exists for pol-α-associated proofreading. A 3'→5' exonuclease activity is present in preparations of *D. melanogaster* pol-α from which the associated 70-kD subunit has been removed (Cotterill et al. 1987), and the resulting polymerase is more accurate than the pol-α:primase complex (Cotterill et al. 1987; Reyland et al. 1988). Proofreading activity may be removed during some purification schemes but retained during others. For example, mouse and human pol-α preparations have been reported that contain 3'→5' exonuclease activity (Chen et al. 1979; Bialek et al. 1989), and the latter catalyzes high-fidelity synthesis. Precedent for proofreading by exonuclease activity of a separate gene product comes from studies in *E. coli*, where the polymerase (α subunit) and exonuclease (ε subunit) activities of the replicative Pol III holoenzyme are encoded by two different genes, designated *dnaE* and *dnaQ*, respectively (for review, see Echols and Goodman 1991). Thus, the difficulty that pol-α has in extending certain mispairs (Perrino and Loeb 1989a; Mendelman et al. 1990) could provide the opportunity for a separate exonuclease activity to proofread misinserted nucleotides, especially if pol-α dissociates from the template primer. This possibility is supported by the observation of high-fidelity DNA synthesis by pol-α in the presence of the *E. coli* ε subunit (Perrino and Loeb 1989b) or pol-δ (Perrino and Loeb 1990), which contains an intrinsic 3'→5' exonuclease activity. The extraordinary range of error rates reported in the literature for synthesis by pol-α (Table 1) thus reflects the large number of studies performed and the use of different assay methods and pol-α preparations. It also clearly reflects differences in discrimination for mispairs of varying composition and in different local sequence contexts.

There has been much recent progress in defining the structure of DNA polymerases (for review, see Joyce and Steitz 1994). Moreover, several conserved amino acid motifs have been identified by sequence alignments of DNA polymerase genes (Delarue et al. 1990; Braithwaite and Ito 1993 and references therein). These data are the starting points to identify amino acid residues that, when changed, alter replication fidelity. Among the eukaryotic polymerases, this approach has been applied first to human pol-α. Several mutant enzymes have been constructed containing single amino acid differences in the most conserved motifs, designated I and II (for review, see Wang 1991). The proteins were overproduced, and their biochemical properties were examined. Two of the motif I mutants, having single-amino-acid differences in residues important for binding the divalent metal ion required for catalysis, have higher discrimination than the wild-type pol-α for nucleotide misinsertion and for mispair extension in Mn^{++}-activated kinetic assays (Copeland et al. 1993). Likewise, two of the motif II mutants, having single-amino-acid differences in residues suggested to be important for binding the incoming dNTP and for interacting with the primer, have improved insertion fidelity (Dong et al. 1993a), whereas another has reduced discrimination against mispair extension (Dong et al. 1993b). A continuation of this approach with pol-α should increase our understanding of the fidelity of this polymerase. Polymerases with reduced or enhanced fidelity should also be useful "biomarkers" for defining their roles in replication and repair.

DNA Polymerase-β

This smallest of the eukaryotic DNA polymerases is also the least accurate. The single subunit polymerases purified from rat hepatoma cells or chicken embryos have average substitution error rates of 670×10^{-6} (Kunkel and Alexander 1986). For individual mispairs, error rate values range from 45×10^{-6} to 1000×10^{-6}. Similar, and in a few instances even higher, error rates have been obtained for direct misinsertion by the rat enzyme, using steady-state kinetic analyses of rat pol-β (Boosalis et al. 1989). These rates are consistent with the fact that purified pol-β lacks associated $3' \rightarrow 5'$ exonuclease activity (and the three exonuclease motifs conserved in proofreading polymerases). However, although pol-β shares this property with pol-α, it is even less accurate, demonstrating that, independent of proofreading, selectivity against substitution errors depends on the DNA polymerase.

Low-fidelity synthesis by pol-β may be consistent with a modest catalytic role in vivo, i.e., filling gaps of one or a few nucleotides during base excision repair. Alternatively, pol-β may have higher accuracy than current estimates suggest. Thus far, pol-β fidelity has been measured using template primers containing long single-stranded template regions, where synthesis is distributive rather than processive. Recently, pol-β has been shown to catalyze processive synthesis on a template adjacent to a $5'$-phosphoryl end (Singhal and Wilson 1993). As this type of synthesis may more closely resemble that occurring during base excision repair, it will be interesting to determine pol-β fidelity using substrates containing single-stranded gaps of one or a few nucleotides. It is also possible that pol-β fidelity is influenced by accessory proteins. For example, mammalian DNase V, a 12-kD protein having both $3' \rightarrow 5'$ and $5' \rightarrow 3'$ exonuclease activity (Mosbaugh and Meyer 1980), associates with pol-β in vitro.

X-ray crystallographic structural information is now available for the ternary complex of pol-β-template·primer-dNTP (Pelletier et al. 1994). This structure provides an excellent opportunity for polymerase engineering to understand fidelity.

DNA Polymerase-δ

The fidelity of DNA synthesis in vitro by this polymerase is remarkably understudied, given its central role in eukaryotic replication. In one study (Thomas et al. 1990), the average substitution error rate in vitro of pol-δ plus proliferating cell nuclear antigen (PCNA) is about 10×10^{-6} (Table 1). The error rate of pol-δ alone was not determined because pol-δ would not complete gap-filling synthesis in the absence of PCNA. The fact that pol-δ (plus PCNA) is more accurate than pol-α or pol-β is consistent with proofreading of misinsertions by its associated $3' \rightarrow 5'$ exonuclease activity, which has properties expected of an editing exonuclease (for review, see Bambara and Jessee 1991). The coding sequences of the yeast and human genes have the three conserved exonuclease motifs characteristic of proofreading enzymes (Chung et al. 1991; Zhang et al. 1991).

DNA Polymerase-ε

This polymerase also has an associated exonuclease activity, and the coding sequence of the gene has the three conserved exonuclease motifs characteristic of proofreading enzymes (Morrison et al. 1991). Pol-ε is highly accurate, having a very low rate for substitutions that revert a termination codon in the *lacZ* gene (Kunkel et al. 1987)[1] and an average error rate of $\leq 6.7 \times 10^{-6}$ in the forward mutation assay. In fact, substitution errors by pol-ε are only detected when dGMP, a putative inhibitor of exonucleolytic proofreading, is included in the synthesis reaction. This is consistent with the idea that the $3' \rightarrow 5'$ exonuclease of pol-ε removes misinsertions during synthesis in vitro.

DNA Polymerase-γ

The mitochondrial replicative polymerase from several sources has high base-substitution fidelity (Table 1). The polymerase also has an associated $3' \rightarrow 5'$ exonuclease activity (Kunkel and Soni 1988a; Insdorf and Bogenhagen 1989; Kaguni and Olsen 1989; Kunkel and Mosbaugh 1989;

---

[1] At the time this study was performed, what is now known to be DNA polymerase-ε was then designated DNA polymerase-δII, hence the title of the article.

Foury and Vanderstraeten 1992; Gray and Wong 1992), and the coding sequence of the yeast gene has the three conserved exonuclease motifs characteristic of proofreading enzymes (Foury and Vanderstraeten 1992). The catalytic properties of the exonuclease and the fact that fidelity is reduced in reactions containing a high dNTP concentration or dGMP (Kunkel and Soni 1988a; Kunkel and Mosbaugh 1989) are consistent with a proofreading role in vitro.

ERRORS INVOLVING TEMPLATE-PRIMER MISALIGNMENT

In addition to direct misincorporation of noncomplementary nucleotides, base-addition, -deletion, and even -substitution errors can be generated by processes involving template-primer misalignments during replication.

Frameshifts[2] Initiated by Template-primer Slippage

Strand slippage during replication of iterated[3] sequences results in misaligned intermediates stabilized by correct base pairs (Fig. 2A). Subsequent polymerization from the misaligned intermediate leads to deletion if the unpaired nucleotide(s) is in the template strand (Fig. 2A) or to addition if the unpaired nucleotide(s) is in the primer strand (not shown). This mechanism predicts (Streisinger et al. 1967) that the error rate should increase as the length of the run increases, because the potential number of correct base pairs that could stabilize the misaligned intermediates increases, as does the number of potential misaligned intermediates that can form (Fig. 2). Furthermore, the longer the run, the greater the distance between the extra nucleotide and the 3'-OH primer terminus, potentially reducing interference by the extra base during phosphodiester bond formation within the enzyme active site.

In support of this logic, frameshift error rates, expressed per nucleotide polymerized to correct for differences in the number of nucleotides in runs of different lengths, do indeed increase as the length of a homopolymeric run increases, for DNA pol-α (Kunkel 1990) and T7 DNA polymerase (Kunkel et al. 1994). In addition, error rates for one-base deletions in homopolymeric runs by pol-β (Kunkel 1986) and HIV-

---

[2]Although the term frameshift mutation usually refers to changes in the number of base pairs in a protein-coding sequence that are not multiples of three, for convenience it refers here to mutations resulting from any difference in the number of base pairs, regardless of location.
[3]A non-iterated nucleotide is one having 5' and 3' neighbors that are not identical to the nucleotide considered. An iterated nucleotide has at least one identical neighbor. Iterated means repeated; reiterated means re-repeated.

1 RT (Bebenek et al. 1993) decrease when the template sequence is altered to either shorten or eliminate the repetitive sequence. That the slippage mechanism operates in vivo is suggested by the instability in repetitive sequences associated with cancer (for review, see Loeb 1994) and neurodegenerative diseases (see, e.g., Willems 1994).

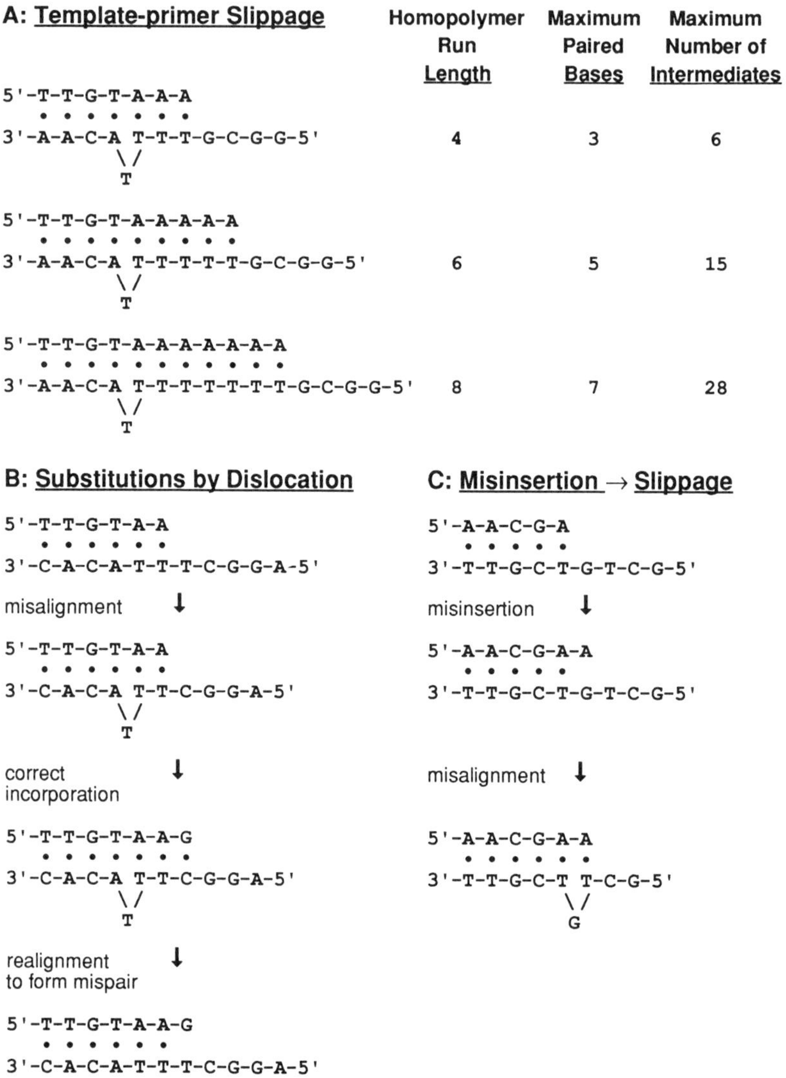

A: Template-primer Slippage

```
5'-T-T-G-T-A-A-A
   • • • • • • •
3'-A-A-C-A T-T-T-G-C-G-G-5'
         \ /
          T

5'-T-T-G-T-A-A-A-A-A
   • • • • • • • • •
3'-A-A-C-A T-T-T-T-T-G-C-G-G-5'
         \ /
          T

5'-T-T-G-T-A-A-A-A-A-A-A
   • • • • • • • • • • •
3'-A-A-C-A T-T-T-T-T-T-T-G-C-G-G-5'
         \ /
          T
```

Homopolymer Run Length	Maximum Paired Bases	Maximum Number of Intermediates
4	3	6
6	5	15
8	7	28

B: Substitutions by Dislocation

```
5'-T-T-G-T-A-A
   • • • • • •
3'-C-A-C-A-T-T-T-C-G-G-A-5'
```
misalignment ↓
```
5'-T-T-G-T-A-A
   • • • • • •
3'-C-A-C-A T-T-C-G-G-A-5'
         \ /
          T
```
correct ↓
incorporation
```
5'-T-T-G-T-A-A-G
   • • • • • • •
3'-C-A-C-A T-T-C-G-G-A-5'
         \ /
          T
```
realignment ↓
to form mispair
```
5'-T-T-G-T-A-A-G
   • • • • • •
3'-C-A-C-A-T-T-T-C-G-G-A-5'
```

C: Misinsertion → Slippage

```
5'-A-A-C-G-A
   • • • • •
3'-T-T-G-C-T-G-T-C-G-5'
```
misinsertion ↓
```
5'-A-A-C-G-A-A
   • • • • •
3'-T-T-G-C-T-G-T-C-G-5'
```
misalignment ↓
```
5'-A-A-C-G-A-A
   • • • • •
3'-T-T-G-C-T T-C-G-5'
              \ /
               G
```

Figure 2 Pathways for errors involving misaligned template primers. See text for description.

Base Substitutions Initiated by Strand Slippage

Following slippage, correct incorporation of another nucleotide followed by realignment before continued incorporation generates a terminal mispair (Kunkel 1985a). This can yield a base substitution, but in this case, initiated by slippage rather than by misinsertion. This process has been termed dislocation mutagenesis (Kunkel and Alexander 1986) by analogy with a dislocated shoulder joint that pops out of alignment but ultimately resumes a normal position. Strong support for the model comes from fidelity studies with pol-β (Kunkel and Soni 1988b; Boosalis et al. 1989) and HIV-1 RT (Bebenek et al. 1993) in vitro. With both polymerases, strong base substitution hot spots are observed at the ends of several different homopolymeric runs, and the substitution specificity depends on the immediate template neighbor. The dislocation concept is not limited to the situation shown in Figure 2, but can involve an extra nucleotide in the primer strand (Fig. 5A in Bebenek et al. 1993), two unpaired template nucleotides (Fig. 5C in Bebenek et al. 1993), or many nucleotides (Fig. 3C in Kunkel and Soni 1988a; for review, see also Ripley 1990).

Frameshifts Initiated by Misinsertion

A distinctly different way to generate a misaligned substrate is misincorporation followed by template-primer rearrangement to provide a correct terminal base pair for continued polymerization (Fig. 2C). The resulting misalignment ultimately leads to a frameshift error, but in this case it is initiated by misinsertion rather than strand slippage. This model was suggested by several observations (for review, see Kunkel 1990) indicating that "difficult-to-extend" mispairs may realign such that synthesis proceeds from a substrate containing an extra nucleotide in the template strand but a correct base pair at the terminus.

In principle, this mechanism is possible at any template position and is not limited to the production of minus-one-base errors. Thus, frameshift errors at template runs, as well as plus and minus errors of varying numbers of nucleotides, may initiate by misincorporation. The model has been supported by fidelity studies with yeast pol-α (Kunkel et al. 1989), Klenow DNA polymerase (Bebenek et al. 1990), and HIV-1 RT (Bebenek et al. 1992) in vitro. The concept of difficult-to-extend termini led to the suggestion (Kunkel and Soni 1988b) that incorporation opposite damaged templates might also yield frameshifts by this mechanism. Studies involving replication of DNA containing several different lesions (Wang and Taylor 1992; Shibutani and Grollman 1993; Lindsley and Fuchs 1994; Napolitano et al. 1994) strongly support this suggestion.

Frameshift Error Rates of Eukaryotic DNA Polymerases

The average frameshift error rates of the five template-dependent eukaryotic DNA polymerases are shown in Table 1. These rates are errors per detectable nucleotide polymerized with the *lacZ* α-complementation gene target and are for single-base frameshift errors. Similar to the situation with substitution errors, DNA pol-α and pol-β, which lack associated proofreading activity, are less accurate than are pol-δ, pol-ε, and pol-γ, which have intrinsic exonucleases. Direct comparison of the frameshift fidelity of wild-type versus exonuclease-deficient derivatives of Klenow polymerase (Bebenek et al. 1990) and T7 DNA polymerase (Kunkel et al. 1994) strongly suggests that frameshift intermediates at both non-run sequences and in homopolymeric runs of up to 5 bp are subject to exonucleolytic proofreading. Moreover, the frameshift fidelity of pol-ε is reduced in reactions containing a high concentration of dNTPs and dGMP (Thomas et al. 1991b), suggesting that frameshift errors by this enzyme are also proofread.

Polymerase rates for minus-one-base errors are higher than for plus-one-base errors. Just as for substitution errors, frameshift error rates are sequence-dependent and polymerase-dependent, with error rates for the same mistake varying over 1000-fold (for detailed discussion, see Kunkel 1990). Polymerases can also delete or add more than a single nucleotide. Some of these errors can be explained by strand slippage between perfectly repeated DNA sequences separated by a variable number of intervening nucleotides (for review, see Kunkel 1990; Ripley 1990). More complicated models involving strand-switching, primer loop-back, and palindromic DNA sequences have also been invoked to explain complex frameshift mutations generated by DNA polymerases.

Processivity and Frameshift Fidelity

One property of polymerization relevant to polymerase frameshift fidelity at homopolymeric runs is processivity, the number of nucleotides incorporated per polymerase association/dissociation with the template primer. This was suggested by the observation that pol-α is both more accurate and more processive than pol-β (Kunkel 1985b). This correlation has been examined in greater detail with HIV-1 RT. This polymerase is inaccurate for one-base frameshifts within some but not all template runs. These hot spots for frameshift errors are also template positions where the probability of termination of processive synthesis is high (Bebenek et al. 1989). Moreover, when changes were introduced into the sequences flanking the runs, increases or decreases in frameshift error rates were observed, and these often correlated with concomitant increases or decreases in termination of processive synthesis within the run (Bebenek et al. 1993). The data thus reveal a consistent pattern wherein low processivity correlates with low frameshift fidelity, consistent with the idea that the formation and/or utilization of misaligned template primers is increased during the dissociation-reinitiation phase of a polymerization reaction.

Effect of Accessory Proteins on DNA Polymerase Fidelity

Further support for a relationship between processivity and fidelity comes from a study of the frameshift fidelity of T7 DNA polymerase with and without its processivity protein, thioredoxin (Kunkel et al. 1994). T7 DNA polymerase alone has low processivity, adding only 1–50 nucleotides before dissociating. However, when it is complexed with its accessory subunit thioredoxin, polymerization proceeds for thousands of nucleotides without dissociation. Fidelity measurements with an exonuclease-deficient mutant of T7 pol showed that the rate for one-base addition frameshifts at homopolymeric runs was 46-fold higher in the absence of thioredoxin than in its presence. This may have general significance, given that accessory proteins that enhance processivity are a general feature of multiprotein replication complexes (see Stillman, this volume). Frameshift fidelity conferred by the polymerase clamp protein is particularly interesting inasmuch as replication infidelity is one possible explanation for the instability of repetitive genomic sequences reported for several diseases (Loeb 1994; Willems 1994).

In the absence of thioredoxin, the exonuclease-deficient T7 DNA polymerase was found to be more accurate during DNA synthesis in vitro for substitutions and 1- and 2-nucleotide deletions (Kunkel et al. 1994). One possible explanation is that the premutational intermediates formed during polymerization are not successfully extended unless the polymerase is complexed with thioredoxin. The biological implication is that, under some circumstances, an accessory protein-mediated alteration in the extension rate from an unusual template primer, e.g., following incorporation opposite a damaged base or slippage at a damaged site, could actually serve a mutator function by enhancing extension synthesis to seal the error before transfer to the exonuclease active site for removal.

Another accessory protein that logically could influence fidelity is single-stranded DNA-binding protein (SSB). The substitution fidelity of synthesis by several polymerases is increased a few fold in reactions containing *E. coli* SSB (Kunkel et al. 1979, 1983). Similarly, the rate of deletions between direct repeats generated by yeast pol-α is reduced in reactions containing a yeast SSB (Roberts et al. 1990). Human replication protein A (RP-A) is a 3-subunit SSB required for replication of SV40-origin-containing DNA in HeLa cell extracts. This protein increased the fidelity of pol-α-catalyzed gap-filling synthesis by 4-fold in one study (Carty et al. 1992) but had little effect on the frameshift fidelity of yeast pol-α:primase complex in a different study (Roberts et al. 1990). Results to date thus suggest that SSB does affect fidelity, but the effects are small relative to the high degree of discrimination imposed by the polymerase and exonuclease.

FIDELITY OF MULTIPROTEIN REPLICATION COMPLEXES

A full appreciation of how genomes are stably replicated, and how instability may arise to generate disease, requires a better understanding of the fidelity of the multiprotein replication machinery. An important step toward achieving this understanding has been the development of systems that replicate double-stranded DNA in vitro (numerous chapters in this volume). One system for studying human genomic replication depends on the SV40 origin of replication (Hassell and Brinton, this volume). Circular, double-stranded DNA substrates containing the SV40 origin can be fully replicated by the proteins present in primate cells, with only the addition of SV40 T antigen needed to initiate replication at the origin. For studies in vitro, replication can be performed in extracts of cells grown in culture or by reconstitution with purified proteins (see, e.g., Waga and Stillman 1994). The latter approach has defined two polymerases (pol-α and pol-δ) and several additional proteins required for complete replication.

High Replication Fidelity with Undamaged DNA

Using forward mutation assays with either the *lacZ* (Roberts and Kunkel 1988, 1993) or *supF* (Hauser et al. 1988) reporter gene, DNA replicated in human HeLa and simian CV-1 cell extracts was found to have a mutant frequency that was not increased significantly above the background mutant frequency of unreplicated DNA. Sequence analysis of *lacZ* mutants recovered from the unreplicated as well as replicated DNA showed no significant differences, yielding error rates varying from ≤6.2 × 10^{-6} to ≤0.1 × 10^{-6}, depending on the substitution or frameshift error considered (Thomas et al. 1991b). Inasmuch as these HeLa cell extracts have mismatch repair activity (Thomas et al. 1991a), these rates represent the sum of both replication fidelity and any heteroduplex repair occurring in the extract. Two studies (see below) suggest that mismatch repair in the extract only affects error rate determinations by 2- to 3-fold, suggesting that replication fidelity itself is high. Because no replication errors are detected with undamaged DNA and equimolar dNTP concentrations, further understanding of how high replication fidelity is achieved and how it can be perturbed requires manipulation of reaction components in order to obtain replication errors. Several approaches have been used to address specific questions.

Proofreading and the Fidelity of Leading- and Lagging-strand Replication

One obvious question is whether proofreading is partly responsible for high replication fidelity. To answer this question, reactions were performed with unequal dNTP concentrations to force specific misinsertions. For example, misincorporation of dGTP to revert a TGA opal codon can be obtained by increasing the concentration of dGTP relative to dATP (Fig. 3). The contribution of proofreading to replication fidelity can then be examined either by increasing absolute dNTP concentrations to stimulate polymerization at the expense of proofreading, or by inhibiting proofreading by adding deoxynucleoside monophosphate to the replication reaction. Results from both approaches suggest that proofreading

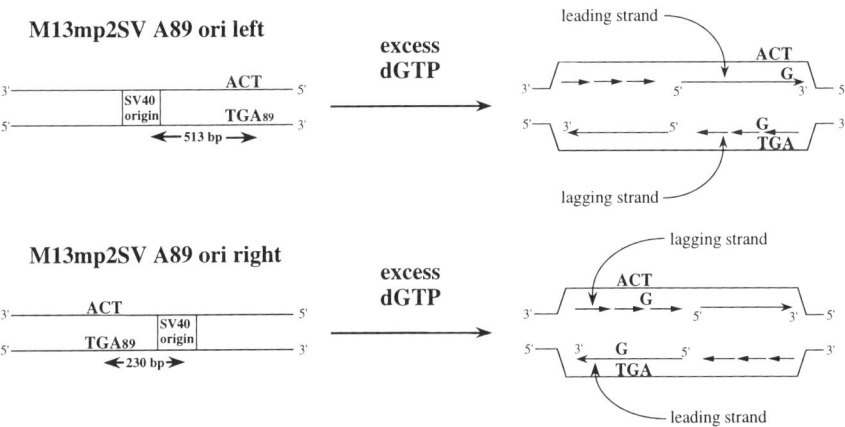

Figure 3 Representation of leading- and lagging-strand synthesis across an opal codon. See text for description.

contributes substantially to replication fidelity for base substitution (Roberts et al. 1991) and frameshift errors (Roberts et al. 1993).

The approach has been extended to examine the fidelity of the leading- and lagging-strand replication machinery. This requires comparing results with two vectors. The first (ori left, Fig. 3) contains the origin a few hundred nucleotides to the left of the reporter gene. This distance is small relative to the size of the vector (7398 bp). Previous studies (Edenberg and Huberman 1975; Li and Kelly 1985) have indicated that the rate of replication fork movement is similar in both directions from the origin. Thus, with this vector the (+) viral strand within the *lacZ* target is likely to be replicated by the lagging-strand replication apparatus. The second vector contains the origin to the right of the target, again only a few hundred nucleotides from the reporter gene (Fig. 3). In this vector, the (+) viral strand in the *lacZ* target is assumed to be replicated by leading-strand replication proteins. Fidelity measurements with these two vectors allow determination of the fidelity of replication of the same sequence by either the leading- or lagging-strand apparatus, provided that the strand on which the error was made can be assigned. This is done with dNTP substrate pool imbalances (Fig. 3), where the dNTP provided in excess is assumed to be responsible for the substitution error observed, or with template DNA damage (see below).

When this approach was combined with the reaction conditions that diminish proofreading, the data suggested that proofreading contributes to substitution fidelity during both leading- and lagging-strand replication (Roberts et al. 1991). The strategy has also been used to describe average leading- and lagging-strand replication rates for errors resulting from misinsertion of either TTP (Roberts et al. 1994) or dGTP (Izuta et al. 1995). Average error rates for replication of the *lacZ* template in a HeLa cell extract suggest that leading- and lagging-strand replication fidelity is similar for several errors but different for others (Table 2) (for additional data, see Izuta et al. 1995). There are several possible explanations for leading- and lagging-strand error rate differences. One is that mismatch repair in the extract is responsible for the asymmetry. A second is that replication of the two strands is highly asymmetric, providing unequal opportunities to make mistakes. Replication of the two strands may be performed by different DNA polymerases or perhaps the same polymerase but with a different complement of accessory proteins (for models, see Waga and Stillman 1994; Stillman; Hassell and Brinton; both this volume). This could yield differences in misinsertion rates and/or ability to extend rather than proofread mispaired or misaligned template primers. Replication on the leading strand is highly processive

Table 2 Comparative leading- and lagging-strand replication error rates

Error considered	Strand	Mispair frequency $\times 10^{-6}$ leading	Mispair frequency $\times 10^{-6}$ lagging	Ratio lagging:leading
Average from extract reaction				
G·dTTP	−	18	39	2:1
C·dTTP	−	28	32	1:1
G·dTTP	+	2.5	82	33:1
C·dTTP	+	2.5	21	8:1
Average from reconstituted reaction				
G·dTTP	−	43	60	1:1
C·dTTP	−	≤7.1	43	≥6:1
G·dTTP	+	≤5.4	200	≥37:1
C·dTTP	+	11	43	4:1
Error rates in extract at specific sites				
T·dGTP	+ (121)	1.6	19	12:1
G·dTTP	+ (145)	≤4.8	71	≥15:1

as compared to discontinuous synthesis of Okazaki fragments on the lagging strand, which involves more than one DNA polymerase and/or one or more switches between enzymes as well as the synthesis and eventual replacement of RNA primers.

The possible influence of mismatch repair on replication fidelity in extracts has been examined by separating a HeLa cell extract into two fractions, neither of which has replication activity. When combined, these fractions reconstitute replication activity that is devoid of mismatch repair activity (Roberts et al. 1994). When the fidelity of this reaction was examined using excess dTTP (Roberts et al. 1994) or excess dGTP (Izuta et al. 1995), error rates were increased by only 2- to 3-fold over those observed in extracts having mismatch repair activity, confirming that replication is indeed highly accurate. Moreover, the error specificity was similar to that in extracts (Table 2) (Roberts et al. 1994; Izuta et al. 1995), including unequal leading- and lagging-strand rates for G·dTTP and C·dTTP errors on the plus strand (Table 2). Thus, at least these error rate asymmetries are not due to differential mismatch repair. The availability of human cell extracts having high replication activity but defective in mismatch repair due to mutant mismatch repair genes (Umar et al. 1994) will facilitate future studies of the fidelity of the human replication machinery in the absence of mismatch repair.

In another study of replication reconstituted from individual components, reactions were performed with pol-α as the only DNA polymerase (Carty et al. 1990). Fidelity was found to be intermediate between that of the unfractionated replication system and gap-filling synthesis by purified HeLa cell pol-α. This suggests that pol-α can replicate DNA with high fidelity when carrying out semiconservative DNA replication, but that additional cellular factors not present in the reconstituted system may be contributing to the higher replication fidelity of the unfractionated system. Even unfractionated extracts may be missing fidelity components that are active in vivo or simply not functioning in the extract.

Leading- and lagging-strand error rates differ at template positions 121 and 145 on the plus strand of the *lacZ* gene in M13mp2SV (Table 2). Two observations suggest that these differences may be due to differential proofreading. First, at both positions, the next correct dNTP to be incorporated following the misinsertion is the nucleotide present in excess (and, therefore, at high concentration) during replication. As explained above (Fig. 1), this situation would favor polymerization at the expense of excision of the terminal mispair. Thus, both hot spots for lagging-strand replication errors are sites where proofreading might be partially suppressed. Second, when replication reactions were repeated with added dGMP to inhibit proofreading, the error rates at both positions increased with both ori left and ori right substrates (Izuta et al. 1995). If one assumes that the addition of dGMP does not affect the inherent base selectivity of the insertion step, then the dGMP-dependent increase in error rate suggests that misinsertions are indeed occurring that, in the absence of monophosphate, are removed by the exonuclease. Inasmuch as fidelity in the absence of dGMP is higher during leading-strand replication, this suggests that leading-strand misinsertions are more effectively excised than are the lagging-strand errors that are readily detected even when dGMP is absent. Differential proofreading thus provides one mechanism to explain differences in leading- and lagging-strand replication fidelity. Since the assignment of the leading- and lagging-strand DNA polymerases during eukaryotic replication is not yet definitive, proofreading on the two strands could be carried out by any of several exonucleases.

Replication Fidelity with Damaged Substrates

A large number of genes in eukaryotic cells either control or catalyze the repair of a wide variety of physical and chemical insults, some resulting from normal cellular processes (e.g., deamination, depurination, oxidative stress, alkylation) and others from exposure to the external environment (for recent reviews, see Hanawalt 1994; Hartwell and Kastan 1994; Sancar 1994 and numerous references therein). When these repair systems fail, lesions may persist in DNA or in dNTP precursor pools. A plethora of lesions have been described over the years that have different structures and thus potentially affect replication fidelity by altering different discrimination steps. For example, an alkylated base may have altered hydrogen-bonding potential and lead to direct misinsertion errors, whereas an abasic site has lost base hydrogen-bonding potential altogether. Bulky lesions may perturb base-pair geometry or base-stacking interactions leading to misinsertion, or their structures may be inconsistent with continued replication, leading to termination or template-primer rearrangement and frameshift errors. The presence of lesions in the template strand could also affect communication between the polymerase and exonuclease active sites that is critical for proofreading. Lesion-induced replication infidelity has been examined in a number of studies with purified DNA polymerases (for review, see Echols and Goodman 1991). Discussing all this information is beyond the scope of this chapter; we briefly review only a few recent studies of the fidelity of SV40 origin-dependent replication using damaged substrates.

Replication Infidelity with 8-O-dGTP

Oxidative metabolism is known to generate mutagenic compounds within cells, one of which is 8-oxo-deoxyguanosine. The presence of several lines of defense against mutations resulting from this base analog (for review, see Michaels and Miller 1992; Grollman and Moriya 1993) suggests that it is biologically important. A variety of DNA polymerases, as well as the replication complex in HeLa cell extracts, misincorporate the triphosphate form of this base analog, 8-O-dGTP, opposite template adenines, yielding A→C transversions (Cheng et al. 1992; Pavlov et al. 1994; Minnick et al. 1995). The data suggest that 8-O-dGTP could be highly mutagenic during genomic replication in eukaryotes. This may be the case for other modified dNTPs as well (see, e.g., Feig et al. 1994). The amount of 8-O-dGTP in human cells may be modulated by hydrolysis by the human homolog (Sakumi et al. 1993) of the *E. coli mutT* protein. If so, inactivating mutations in this gene (or other enzymes that sanitize dNTP pools) might result in a mutator phenotype in human cells.

Mutagenic Translesion Replication of DNA Containing Cyclobutane Pyrimidine Dimers

Among the insults that generate lesions in DNA, ultraviolet radiation has received perhaps the greatest attention, partly due to its established role as a skin carcinogen (Brash et al. 1991). The mutagenic potential of UV photoproducts during replication by eukaryotic proteins has been examined in several studies. Pol-α is unable to synthesize past UV lesions (Moore et al. 1981), whereas DNA pol-δ in the presence of PCNA is able to replicate past *cis*-syn and *trans*-syn TT dimers (O'Day et al. 1992). Three studies have demonstrated translesion replication of cyclobutane pyrimidine dimers in cell extracts using SV40 origin-containing vectors (Carty et al. 1993; Thomas and Kunkel 1993; Thomas et al. 1993). This replication is mutagenic, with an error specificity similar to that observed in vivo (Keyse et al. 1988; Armstrong and Kunz 1990).

Mutagenic Translesion Replication of DNA Containing AAF Adducts

The ability of the replication complex in a HeLa cell extract to bypass site-specific N-2-acetylaminofluorene (AAF) adducts has also been examined (Thomas et al. 1994). The major effect was inhibition of replication, with termination occurring immediately before incorporation opposite the adduct. Among the replicated products was a higher proportion of those representing replication of the undamaged strand, leaving open the possibility that the first fork to encounter the lesion became uncoupled, i.e., replication of the damaged strand ceased while replication of the undamaged strand continued. Product analysis suggested that translesion bypass had occurred and that frameshift errors had been generated by the mechanism involving correct incorporation opposite the lesion followed by slippage (Fig. 2C).

FIDELITY OF DNA REPLICATION IN VIVO

How accurate is replication in vivo? This question has been elegantly addressed in *E. coli* by measuring mutation rates for a variety of sequence changes, using strains selectively disabled in key fidelity processes. The wild-type spontaneous mutation rate in the *lacI* reporter gene is 10^{-10} mutations per base pair replicated per generation (Table 3) (see, e.g., Schaaper 1993). A *mutL* mutant lacking methyl-directed postreplication mismatch repair has a 20- to 400-fold higher spontaneous mutation rate, depending on the type of mutation considered. With some simplifying assumptions (Schaaper 1993), the resulting mutation rate of about 10^{-7} can be considered as the fidelity of chromosomal replication. Analysis of a double mutant lacking mismatch repair and defective in proofreading by the ε subunit of the replicative DNA polymerase III holoenzyme (which replicates both the leading and lagging strands) suggests that

Table 3 Estimated contributions of the three major discrimination steps to replication fidelity

Discrimination step	Contribution		
	E. coli in vivo	eukaryotic in vitro	eukaryotic in vivo
Nucleotide selectivity	$2 \times 10^{-5} - 2 \times 10^{-6}$	$10^{-3} - 10^{-6}$	?
Exonucleolytic proofreading	40–200	0–200	10–≥200
Mismatch repair	20–400	–	10–≥700
Mutation rate	10^{-10}		$\leq 10^{-10}$

The values listed depend on simplifying assumptions and often involve caveats that are discussed in the text and in the references cited therein.

proofreading contributes between 40- and 200-fold to this rate, with the balance (factors of 200,000 to 2,000,000) representing the base selectivity of the replication machinery (Table 3).

Estimating replication fidelity in eukaryotes is complicated by the possibility of several types of mismatch repair, the influence of spontaneous damage, and the likelihood of multiple damage-repair pathways. Moreover, estimates are based on a few reporter genes, providing a limited view of replication fidelity for large eukaryotic genomes. Despite these qualifications, existing mutation rate data in mutant cells reveal a similar picture to that seen in *E. coli*. Mutation rates in eukaryotic cells are generally $\leq 10^{-10}$ mutations per base pair replicated per generation (see Loeb 1991 and references therein). The mutation rate in mismatch-repair-defective yeast (see, e.g., Strand et al. 1993; Prolla et al. 1994 and references therein) and human tumor cells (Kat et al. 1993; Bhattacharyya et al. 1994; Farber et al. 1994; Eshleman et al. 1995) are elevated up to several hundred-fold. Moreover, yeast mutants of pol-δ, pol-ε, and pol-γ containing substitutions for conserved exonuclease residues also have spontaneous mutation rates that are increased by up to several hundred-fold (Morrison et al. 1991, 1993; Foury and Vanderstraeten 1992), emphasizing the importance of proofreading in vivo during both nuclear and mitochondrial DNA replication.

CONCLUDING REMARKS

In many ways, our current view of eukaryotic replication fidelity is quite limited. Eukaryotic genomes are huge compared to the few hundred nucleotides scanned by current reporter genes. For example, the human genome contains a wide variety of repetitive sequence elements (see, e.g., Beckman and Weber 1992) whose instability is associated with cancer (for review, see Loeb 1994) and neurodegenerative diseases (see, e.g., Willems 1994). Despite this association, we know very little about replication error rates in repetitive sequences except for short homopolymeric runs. The fact that eukaryotic cells devote an enormous amount of energy to checkpoints in the cell cycle and a multiplicity of DNA repair processes clearly indicates the biological risk associated with unrepaired lesions. Appropriately, a great deal of attention has been paid to studies of these processes, as exemplified by the fact that DNA repair was the "molecule of the year" in 1994 (Culotta and Koshland 1994). To fully appreciate the effects of unrepaired lesions, more studies are needed to define the consequences of an encounter between an unrepaired lesion and a eukaryotic replication fork, and, possibly, the fidelity of DNA synthesis associated with the repair processes themselves. Our understanding should increase as studies are performed using a variety of systems, eventually including those that mimic replication of highly organized nuclear and mitochondrial DNA.

ACKNOWLEDGMENTS

We thank William C. Copeland and Rémi Palmantier for helpful comments on the manuscript.

REFERENCES

Abbotts, J., M. Jaju, and S.H. Wilson. 1991. Thermodynamics of A:G mismatch poly(dG) synthesis by human immunodeficiency virus 1 reverse transcriptase. *J. Biol. Chem.* **266**: 3937–3943.

Armstrong, J.D. and B.A. Kunz. 1990. Site and strand specificity of UVB mutagenesis in the *SUP-o* gene of yeast. *Proc. Natl. Acad. Sci.* **87**: 9005–9009.

Bambara, R.A. and C.B. Jessee. 1991. Properties of DNA polymerases δ and ε, and their roles in eukaryotic DNA replication. *Biochim. Biophys. Acta* **1088**: 11–24.

Bebenek, K. and T.A. Kunkel. 1995. Analyzing the fidelity of DNA polymerases. *Methods Enzymol.* (in press).

Bebenek, K., J.D. Roberts, and T.A. Kunkel. 1992. The effects of dNTP pool imbalances on frameshift fidelity during DNA replication. *J. Biol. Chem.* **267**: 3589–3596.

Bebenek, K., J. Abbotts, S.H. Wilson, and T.A. Kunkel. 1993. Error-prone polymerization by HIV-1 reverse transcriptase. Contribution of template-primer misalignment, miscoding, and termination probability to mutational hot spots. *J. Biol. Chem.* **268**: 10324–10334.

Bebenek, K., C.M. Joyce, M.P. Fitzgerald, and T.A. Kunkel. 1990. The fidelity of DNA synthesis catalyzed by derivatives of *Escherichia coli* DNA polymerase I. *J. Biol. Chem.* **265**: 13878–13887.

Bebenek, K., J. Abbotts, J.D. Roberts, S.H. Wilson, and T.A. Kunkel. 1989. Specificity

and mechanism of error-prone replication by human immunodeficiency virus 1 reverse transcriptase. *J. Biol. Chem.* **264:** 16948-16956.
Beckman, J.S. and J.L. Weber. 1992. Survey of human and rat microsatellites. *Genomics* **12:** 627-631.
Bhattacharyya, N.P., A. Skandalis, A. Ganesh, J. Groden, and M. Meuth. 1994. Mutator phenotypes in human colorectal carcinoma cell lines. *Proc. Natl. Acad. Sci.* **91:** 6319-6323.
Bialek, G., H.-P. Nasheuer, H. Goetz, and F. Grosse. 1989. Exonucleolytic proofreading increases the accuracy of DNA synthesis by human lymphocyte DNA polymerase α-DNA primase. *EMBO J.* **8:** 1833-1839.
Blanco, L., A. Bernad, and M. Salas. 1991. MIP1 DNA polymerase of *S. cerevisiae*: Structural similarity with the *E. coli* DNA polymerase I-type enzymes. *Nucleic Acids Res.* **19:** 955.
Bloom, L.B., M.R. Otto, R. Eritja, L.J. Reha-Krantz, M.F. Goodman, and J.M. Beechem. 1994. Pre-steady-state kinetic analysis of sequence-dependent nucleotide excision by the 3'-exonuclease activity of bacteriophage T4 DNA polymerase. *Biochemistry* **33:** 7576-7586.
Boosalis, M.S., D.W. Mosbaugh, R. Hamatake, A. Sugino, T.A. Kunkel, and M.F. Goodman. 1989. Kinetic analysis of base substitution mutagenesis by transient misalignment of DNA and by miscoding. *J. Biol. Chem.* **264:** 11360-11366.
Braithwaite, D.K. and J. Ito. 1993. Compilation, alignment, and phylogenetic relationships of DNA polymerases. *Nucleic Acids Res.* **21:** 787-802.
Brash, D.E., J.A. Rudolph, J.A. Simon, A. Lin, G.J. McKenna, H.P. Baden, A.J. Halperin, and J. Pontén. 1991. A role for sunlight in skin cancer: UV-induced p53 mutations in squamous cell carcinoma. *Proc. Natl. Acad. Sci.* **88:** 10124-10128.
Brooke, R.G., R. Singhal, D.C. Hinkle, and L.B. Dumas. 1991. Purification and characterization of the 180 kDa and 86 kDa subunits of the *Saccharomyces cerevisiae* DNA primase-DNA polymerase protein complex. The 180 kDa subunit has both DNA polymerase and 3'→5'-exonuclease activities. *J. Biol. Chem.* **266:** 3005-3015.
Brutlag, D. and A. Kornberg. 1972. Enzymatic synthesis of deoxyribonucleic acid. XXXVI. A proofreading function for the 3'→5' exonuclease activity in deoxyribonucleic acid polymerases. *J. Biol. Chem.* **247:** 241-248.
Capson, T.L., J.A. Peliska, B. Fenn Kaboord, M. West Frey, C. Lively, M. Dahlberg, and S.J. Benkovic. 1992. Kinetic characterization of the polymerase and exonuclease activities of the gene 43 protein of bacteriophage T4. *Biochemistry* **31:** 10984-10994.
Carroll, S.S. and S.J. Benkovic. 1990. Mechanistic aspects of DNA polymerases: *Escherichia coli* DNA polymerase I (Klenow fragment) as a paradigm. *Chem. Rev.* **90:** 1291-1307.
Carty, M.P., A.S. Levine, and K. Dixon. 1992. HeLa cell single-stranded DNA-binding protein increases the accuracy of DNA synthesis by DNA polymerase α. *Mutat. Res.* **274:** 29-43.
Carty, M.P., J. Hauser, A.S. Levine, and K. Dixon. 1993. Replication and mutagenesis of UV-damaged DNA templates in human and monkey cell extracts. *Mol. Cell. Biol.* **13:** 533-542.
Carty, M.P., Y. Ishimi, A.S. Levine, and K. Dixon. 1990. DNA polymerase α from HeLa cells synthesizes DNA with high fidelity in a reconstituted replication system. *Mutat. Res.* **232:** 141-153.
Carver, T.E., Jr., R.A. Hochstrasser, and D.P. Millar. 1994. Proofreading DNA: Recogni-

tion of aberrant DNA termini by the Klenow fragment of DNA polymerase I. *Proc. Natl. Acad. Sci.* **91:** 10670-10674.
Chen, Y.-C., E.W. Bohn, S.R. Planck, and S.H. Wilson. 1979. Mouse DNA polymerase α. Subunit structure and identification of a species with associated exonuclease. *J. Biol. Chem.* **254:** 11678-11687.
Cheng, K.C., D.S. Cahill, H. Kasai, S. Nishimura, and L.A. Loeb. 1992. 8-Hydroxyguanine, an abundant form of oxidative DNA damage, causes G→T and A→C substitutions. *J. Biol. Chem.* **267:** 166-172.
Chung, D.W., J. Zhang, C.K. Tan, E.W. Davie, A.G. So, and K.M. Downey. 1991. Primary structure of the catalytic subunit of human DNA polymerase δ and chromosomal location of the gene. *Proc. Natl. Acad. Sci.* **88:** 11197-11201.
Copeland, W.C. and T.S.-F. Wang. 1991. Catalytic subunit of human DNA polymerase α overproduced from Baculovirus-infected insect cells. *J. Biol. Chem.* **266:** 22739-22748.
Copeland, W.C., N.K. Lam, and T.S.-F. Wang. 1993. Fidelity studies of the human DNA polymerase α. The most conserved region among α-like DNA polymerases is responsible for metal-induced infidelity in DNA synthesis. *J. Biol. Chem.* **268:** 11041-11049.
Cotterill, S.M., M.E. Reyland, L.A. Loeb, and I.R. Lehman. 1987. A cryptic proofreading 3ʹ → 5ʹ exonuclease associated with the polymerase subunit of the DNA polymerase-primase from *Drosophila melanogaster*. *Proc. Natl. Acad. Sci.* **84:** 5635-5639.
Cowart, M., K.J. Gibson, D.J. Allen, and S.J. Benkovic. 1989. DNA substrate structural requirements for the exonuclease and polymerase activities of procaryotic and phage DNA polymerases. *Biochemistry* **28:** 1975-1983.
Culotta, E. and D.E. Koshland Jr. 1994. DNA repair works its way to the top. *Science* **266:** 1926-1929.
Delarue, M., O. Poch, N. Tordo, D. Moras, and P. Argos. 1990. An attempt to unify the structure of polymerases. *Protein Eng.* **3:** 461-467.
Dong, Q., W.C. Copeland, and T.S.-F. Wang. 1993a. Mutational studies of human DNA polymerase α. Identification of residues critical for deoxynucleotide binding and misinsertion fidelity of DNA synthesis. *J. Biol. Chem.* **268:** 24163-24174.
―――. 1993b. Mutational studies of human DNA polymerase α. Serine 867 in the second most conserved region among α-like DNA polymerases is involved in primer binding and mispair primer extension. *J. Biol. Chem.* **268:** 24175-24182.
Donlin, M.J., S.S. Patel, and K.A. Johnson. 1991. Kinetic partitioning between the exonuclease and polymerase sites in DNA error correction. *Biochemistry* **30:** 538-546.
Echols, H. and M.F. Goodman. 1991. Fidelity mechanisms in DNA replication. *Annu. Rev. Biochem.* **60:** 477-511.
Edenberg, H.J. and J.A. Huberman. 1975. Eukaryotic chromosome replication. *Annu. Rev. Genet.* **9:** 245-284.
Eger, B.T. and S.J. Benkovic. 1992. Minimal kinetic mechanism for misincorporation by DNA polymerase I (Klenow fragment). *Biochemistry* **31:** 9227-9236.
Eshleman, J.R., E.Z. Lang, G.K. Bowerfind, R. Parsons, B. Vogelstein, J.K.V. Wilson, M.L. Veigl, W.D. Sedwick, and S.D. Markowitz. 1995. Increased mutation rate at the *hprt* locus accompanies microsatellite instability in colon cancer. *Oncogene* **10:** 33-37.
Farber, R.A., T.D. Petes, M. Dominska, S.S. Hudgens, and R.M. Liskay. 1994. Instability of simple sequence repeats in a mammalian cell line. *Hum. Mol. Genet.* **3:** 253-256.
Feig, D.I., L.C. Sowers, and L.A. Loeb. 1994. Reverse chemical mutagenesis: Identifica-

tion of the mutagenic lesions resulting from reactive oxygen species-mediated damage to DNA. *Proc. Natl. Acad. Sci.* **91:** 6609-6613.

Fersht, A.R. 1985. *Enzyme structure and mechanism.* W.J. Freeman, New York.

Fersht, A.R., J.W. Knill-Jones, and W.-C. Tsui. 1982. Kinetic basis of spontaneous mutation. Misinsertion frequencies, proofreading specificities and cost of proofreading by DNA polymerases of *Escherichia coli. J. Mol. Biol.* **156:** 37-51.

Foury, F. and S. Vanderstraeten. 1992. Yeast mitochondrial DNA mutators with deficient proofreading exonucleolytic activity. *EMBO J.* **11:** 2717-2726.

Gray, H. and T.W. Wong. 1992. Purification and identification of subunit structure of the human mitochondrial DNA polymerase. *J. Biol. Chem.* **267:** 5835-5841.

Grollman, A.P. and M. Moriya. 1993. Mutagenesis by 8-oxoguanine: An enemy within. *Trends Genet.* **9:** 246-249.

Grosse, F., G. Krauss, J.W. Knill-Jones, and A.R. Fersht. 1983. Accuracy of DNA polymerase-α in copying natural DNA. *EMBO J.* **2:** 1515-1519.

Hanawalt, P.C. 1994. Transcription-coupled repair and human disease. *Science* **266:** 1957-1958.

Hartwell, L.H. and M.B. Kastan. 1994. Cell cycle control and cancer. *Science* **266:** 1821-1828.

Hauser, J., A.S. Levine, and K. Dixon. 1988. Fidelity of DNA synthesis in a mammalian *in vitro* replication system. *Mol. Cell. Biol.* **8:** 3267-3271.

Insdorf, N.F. and D.F. Bogenhagen. 1989. DNA polymerase ψ from *Xenopus laevis.* 2. A $3' \rightarrow 5'$ exonuclease is tightly associated with the DNA polymerase activity. *J. Biol. Chem.* **264:** 21498-21503.

Ito, J. and D.K. Braithwaite. 1990. Yeast mitochondrial DNA polymerase is related to the family A DNA polymerases. *Nucleic Acids Res.* **18:** 6716.

Izuta, S., J.D. Roberts, and T.A. Kunkel. 1995. Replication error rates for G•dGTP, T•dGTP and A•dGTP mispairs: Evidence for differential proofreading by leading and lagging strand DNA replication complexes. *J. Biol. Chem.* **270:** 2595-2600.

Johnson, K.A. 1993. Conformational coupling in DNA polymerase fidelity. *Annu. Rev. Biochem.* **62:** 685-713.

Joyce, C.M. 1989. How DNA travels between the separate polymerase and $3'$-$5'$-exonuclease sites of DNA polymerase I (Klenow fragment). *J. Biol. Chem.* **264:** 10858-10866.

Joyce, C.M. and T.A. Steitz. 1994. Function and structure relationships in DNA polymerases. *Annu. Rev. Biochem.* **63:** 777-822.

Kaguni, L.S. and M.W. Olsen. 1989. Mismatch-specific $3' \rightarrow 5'$ exonuclease associated with the mitochondrial DNA polymerase from *Drosophila* embryos. *Proc. Natl. Acad. Sci.* **86:** 6469-6473.

Kaguni, L.S., R.A. DiFrancesco, and I.R. Lehman. 1984. The DNA polymerase-primase from *Drosophila melanogaster.* Rate and fidelity of polymerization on single-stranded DNA templates. *J. Biol. Chem.* **259:** 9314-9319.

Kat, A., W.G. Thilly, W.-H. Fang, M.J. Longley, G.-M. Li, and P. Modrich. 1993. An alkylation-tolerant, mutator human cell line is deficient in strand-specific mismatch repair. *Proc. Natl. Acad. Sci.* **90:** 6424-6428.

Keyse, S.M., F. Amaudruz, and R.M. Tyrrell. 1988. Determination of the spectrum of mutations induced by defined-wavelength solar UVB (313-nm) radiation in mammalian cells by use of a shuttle vector. *Mol. Cell. Biol.* **8:** 5425-5431.

Kuchta, R., P. Benkovic, and S.J. Benkovic. 1988. Kinetic mechanism whereby DNA polymerase I (Klenow) replicates DNA with high fidelity. *Biochemistry* **27:** 2716-2725.

Kunkel, T.A. 1985a. The mutational specificity of DNA polymerase-β during *in vitro* DNA synthesis. Production of frameshift, base substitution, and deletion mutants. *J. Biol. Chem.* **260:** 5787-5796.

———. 1985b. The mutational specificity of DNA polymerases-α and -γ during *in vitro* DNA synthesis. *J. Biol. Chem.* **260:** 12866-12874.

———. 1986. Frameshift mutagenesis by eukaryotic DNA polymerases *in vitro*. *J. Biol. Chem.* **261:** 13581-13587.

———. 1990. Misalignment-mediated DNA synthesis errors. *Biochemistry* **29:** 8003-8011.

———. 1992. DNA replication fidelity. *J. Biol. Chem* **267:** 18251-18254.

Kunkel, T.A. and P.S. Alexander. 1986. The base substitution fidelity of eucaryotic DNA polymerases. Mispairing frequencies, site preferences, insertion preferences, and base substitution by dislocation. *J. Biol. Chem.* **261:** 160-166.

Kunkel, T.A. and L.A. Loeb. 1981. Fidelity of mammalian DNA polymerases. *Science* **213:** 765-767.

Kunkel, T.A. and D.W. Mosbaugh. 1989. Exonucleolytic proofreading by a mammalian DNA polymerase γ. *Biochemistry* **28:** 988-995.

Kunkel, T.A. and A. Soni. 1988a. Exonucleolytic proofreading enhances the fidelity of DNA synthesis by chick embryo DNA polymerase-γ. *J. Biol. Chem.* **263:** 4450-4459.

———. 1988b. Mutagenesis by transient misalignment. *J. Biol. Chem.* **263:** 14784-14789.

Kunkel, T.A., R.R. Meyer, and L.A. Loeb. 1979. Single-strand binding protein enhances fidelity of DNA synthesis *in vitro*. *Proc. Natl. Acad. Sci.* **76:** 6331-6335.

Kunkel, T.A., S.S. Patel, and K.A. Johnson. 1994. Error-prone replication of repeated DNA sequences by T7 DNA polymerase in the absence of its processivity subunit. *Proc. Natl. Acad. Sci.* **91:** 6830-6834.

Kunkel, T.A., R.D. Sabatino, and R.A. Bambara. 1987. Exonucleolytic proofreading by calf thymus DNA polymerase δ. *Proc. Natl. Acad. Sci.* **84:** 4865-4869.

Kunkel, T.A., R.M. Schaaper, and L.A. Loeb. 1983. Depurination-induced infidelity of deoxyribonucleic acid synthesis with purified deoxyribonucleic acid replication proteins *in vitro*. *Biochemistry* **22:** 2378-2384.

Kunkel, T.A., R.K. Hamatake, J. Motto-Fox, M.P. Fitzgerald, and A. Sugino. 1989. Fidelity of DNA polymerase I and the DNA polymerase I-DNA primase complex from *Saccharomyces cerevisiae*. *Mol. Cell. Biol.* **9:** 4447-4458.

Li, J.J. and T.J. Kelly. 1985. Simian virus 40 DNA replication *in vitro*: Specificity of initiation and evidence for bidirectional replication. *Mol. Cell. Biol.* **5:** 1238-1246.

Lindsley, J.E. and R.P.P. Fuchs. 1994. Use of single-turnover kinetics to study bulky adduct bypass by T7 DNA polymerase. *Biochemistry* **33:** 764-772.

Loeb, L.A. 1991. Mutator phenotype may be required for multistage carcinogenesis. *Cancer Res.* **51:** 1-5.

———. 1994. Microsatellite instability: Marker of a mutator phenotype in cancer. *Cancer Res.* **54:** 5059-5063.

Mendelman, L.V., J. Petruska, and M.F. Goodman. 1990. Base mispair extension kinetics: Comparison of DNA polymerase α and reverse transcriptase. *J. Biol. Chem.* **265:** 2338-2346.

Mendelman, L.V., M.S. Boosalis, J. Petruska, and M.F. Goodman. 1989. Nearest neigh-

bor influences on DNA polymerase insertion fidelity. *J. Biol. Chem.* **264**: 14415–14423.
Michaels, M.L. and J.H. Miller. 1992. The GO system protects organisms from the mutagenic effect of the spontaneous lesion 8-hydroxyguanine (7,8-dihydro-8-oxoguanine). *J. Bacteriol.* **174**: 6321–6325.
Minnick, D.T., Y.I. Pavlov, and T.A. Kunkel. 1995. The fidelity of the human leading and lagging strand DNA replication apparatus with 8-oxodeoxyguanosine triphosphate. *Nucleic Acids Res.* **22**: 5658–5664.
Modrich, P. 1994. Mismatch repair, genetic stability, and cancer. *Science* **266**: 1959–1960.
Moore, P.D., K.K. Bose, S.D. Rabkin, and B.S. Strauss. 1981. Sites of termination of *in vitro* DNA synthesis on ultraviolet- and N-acetylaminofluorene-treated ϕX174 templates by prokaryotic and eukaryotic DNA polymerases. *Proc. Natl. Acad. Sci.* **78**: 110–114.
Morrison, A., J.B. Bell, T.A. Kunkel, and A. Sugino. 1991. Eukaryotic DNA polymerase amino acid sequence required for 3'→5' exonuclease activity. *Proc. Natl. Acad. Sci.* **88**: 9473–9477.
Morrison, A., A.L. Johnson, L.H. Johnston, and A. Sugino. 1993. Pathway correcting DNA replication errors in *Saccharomyces cerevisiae*. *EMBO J.* **12**: 1467–1473.
Mosbaugh, D.W. and R.R. Meyer. 1980. Interaction of mammalian deoxyribonuclease V, a double-strand 3' to 5' and 5' to 3' exonuclease with deoxyribonucleic acid polymerase-β from Novikoff hepatoma. *J. Biol. Chem.* **255**: 10239–10247.
Napolitano, R.L., I.B. Lambert, and R.P.P. Fuchs. 1994. DNA sequence determinants of carcinogen-induced frameshift mutagenesis. *Biochemistry* **33**: 1311–1315.
O'Day, C.L., P.M.J. Burgers, and J.-S. Taylor. 1992. PCNA-induced synthesis past *cis*-syn and *trans*-syn-1 thymine dimers by calf thymus DNA polymerase δ *in vitro*. *Nucleic Acids Res.* **20**: 5403–5406.
Pavlov, Y.I., D.T. Minnick, S. Izuta, and T.A. Kunkel. 1994. DNA replication fidelity with 8-oxodeoxyguanosine triphosphate. *Biochemistry* **33**: 4695–4701.
Pelletier, H., M.R. Sawaya, A. Kumar, S.H. Wilson, and J. Kraut. 1994. Structures of ternary complexes of rat DNA polymerase β, a DNA template-primer, and ddCTP. *Science* **264**: 1891–1903.
Perrino, F.W. and L.A. Loeb. 1989a. Differential extension of 3' mispairs is a major contribution to the high fidelity of calf thymus DNA polymerase-α. *J. Biol. Chem.* **264**: 2898–2905.
———. 1989b. Proofreading by the ϵ subunit of *Escherichia coli* DNA polymerase III increases the fidelity of calf thymus DNA polymerase α. *Proc. Natl. Acad. Sci.* **86**: 3085–3088.
———. 1990. Hydrolysis of 3'-terminal mispairs *in vitro* by the 3'→5' exonuclease of DNA polymerase δ permits subsequent extension by DNA polymerase α. *Biochemistry* **29**: 5226–5231.
Petruska, J., L.C. Sowers, and M.F. Goodman. 1986. Comparison of nucleotide interactions in water, proteins, and vacuum: Model for DNA polymerase fidelity. *Proc. Natl. Acad. Sci.* **83**: 1559–1562.
Polesky, A.H., M.D. Dahlberg, S.J. Benkovic, N.D.F. Grindley, and C.M. Joyce. 1992. Side chains involved in catalysis of the polymerase reaction of DNA polymerase I from *Escherichia coli*. *J. Biol. Chem.* **267**: 8417–8428.
Prolla, T.A., D.-M. Christie, and R.M. Liskay. 1994. Dual requirement in yeast DNA

mismatch repair for *MLH1* and *PMS1*, two homologs of the bacterial *mutL* gene. *Mol. Cell. Biol.* **14:** 407–415.

Reddy, M.K., S.E. Weitzel, and P.H. von Hippel. 1992. Processive proofreading is intrinsic to T4 DNA polymerase. *J. Biol. Chem.* **267:** 14157–14166.

Reyland, M.E. and L.A. Loeb. 1987. On the fidelity of DNA replication. Isolation of high fidelity DNA polymerase-primase complexes by immunoaffinity chromatography. *J. Biol. Chem.* **262:** 10824–10830.

Reyland, M.E., I.R. Lehman, and L.A. Loeb. 1988. Specificity of proofreading by the 3'→5' exonuclease of the DNA polymerase-primase of *Drosophila melanogaster*. *J. Biol. Chem.* **263:** 6518–6524.

Ripley, L.S. 1990. Frameshift mutation: Determinants of specificity. *Annu. Rev. Genet.* **24:** 189–213.

Roberts, J.D. and T.A. Kunkel. 1988. Fidelity of a human cell DNA replication complex. *Proc. Natl. Acad. Sci.* **85:** 7064–7068.

———. 1993. Fidelity of DNA replication in human cells. In *Gene and chromosome analysis*, Part B, (ed. K.W. Adolph), pp. 295–313. Academic Press, San Diego.

Roberts, J.D., D. Nguyen, and T.A. Kunkel. 1993. Frameshift fidelity during replication of double-stranded DNA in HeLa cell extracts. *Biochemistry* **32:** 4083–4089.

Roberts, J.D., D.C. Thomas, and T.A. Kunkel. 1991. Exonucleolytic proofreading of leading and lagging strand DNA replication errors. *Proc. Natl. Acad. Sci.* **88:** 3465–3469.

Roberts, J.D., S. Izuta, D.C. Thomas, and T.A. Kunkel. 1994. Mispair-, site-, and strand-specific error rates during simian virus 40 origin-dependent replication *in vitro* with excess deoxythymidine triphosphate. *J. Biol. Chem.* **269:** 1711–1717.

Roberts, J.D., R.K. Hamatake, M.S. Fitzgerald, A. Sugino, and T.A. Kunkel. 1990. Effect of accessory proteins on the fidelity of DNA synthesis by eukaryotic replicative polymerases. In *Mutation and the environment,* part A: *Basic mechanisms* (ed. M.L. Mendelsohn and R.J. Albertini), pp. 91–100. Wiley-Liss, New York.

Sakumi, K., M. Furuichi, T. Tsuzuki, T. Kakuma, S. Kawabata, H. Maki, and M. Sekiguchi. 1993. Cloning and expression of cDNA for a human enzyme that hydrolyzes 8-oxo-dGTP, a mutagenic substrate for DNA synthesis. *J. Biol. Chem.* **268:** 23524–23530.

Sancar, A. 1994. Mechanisms of DNA excision repair. *Science* **266:** 1954–1956.

Schaaper, R.M. 1993. Base selection, proofreading, and mismatch repair during DNA replication in *Escherichia coli*. *J. Biol. Chem.* **268:** 23762–23765.

Sheaff, R.J., R.D. Kuchta, and D. Ilsley. 1994. Calf thymus DNA polymerase α-primase: "Communication" and primer•template movement between the two active sites. *Biochemistry* **33:** 2247–2254.

Shibutani, S. and A.P. Grollman. 1993. On the mechanism of frameshift (deletion) mutagenesis *in vitro*. *J. Biol. Chem.* **268:** 11703–11710.

Simon, M., L. Giot, and G. Faye. 1991. The 3' to 5' exonuclease activity located in the DNA polymerase δ subunit of *Saccharomyces cerevisiae* is required for accurate replication. *EMBO J.* **10:** 2165–2170.

Singhal, R.K. and S.H. Wilson. 1993. Short gap-filling synthesis by DNA polymerase β is processive. *J. Biol. Chem.* **268:** 15906–15911.

Strand, M., T.A. Prolla, R.M. Liskay, and T.D. Petes. 1993. Destabilization of tracts of simple repetitive DNA in yeast by mutations affecting DNA mismatch repair. *Nature* **365:** 274–276.

Streisinger, G., Y. Okada, J. Emrich, J. Newton, A. Tsugita, E. Terzaghi, and M. Inouye. 1967. Frameshift mutations and the genetic code. *Cold Spring Harbor Symp. Quant. Biol.* **31:** 77–84.

Thomas, D.C. and T.A. Kunkel. 1993. Replication of UV-irradiated DNA in human cell extracts: Evidence for mutagenic bypass of pyrimidine dimers. *Proc. Natl. Acad. Sci.* **90:** 7744–7748.

Thomas, D.C., J.D. Roberts, and T.A. Kunkel. 1991a. Heteroduplex repair in extracts of human HeLa cells. *J. Biol. Chem.* **266:** 3744–3751.

Thomas, D.C., D.C. Nguyen, W.W. Piegorsch, and T.A. Kunkel. 1993. Relative probability of mutagenic translesion synthesis on the leading and lagging strands during replication of UV-irradiated DNA in a human cell extract. *Biochemistry* **32:** 11476–11482.

Thomas, D.C., J.D. Roberts, M.P. Fitzgerald, and T.A. Kunkel. 1990. Fidelity of animal cell DNA polymerases α and δ and of a human DNA replication complex. In *Antimutagenesis and anticarcinogenesis mechanisms*, vol. 2 (ed. K. Kuroda et al.), pp. 289–297. Plenum Press, New York.

Thomas, D.C., X. Veaute, T.A. Kunkel, and R.P.P. Fuchs. 1994. Mutagenic replication in human cell extracts of DNA containing site specific N-2-acetylaminofluorene adducts. *Proc. Natl. Acad. Sci.* **91:** 7752–7756.

Thomas, D.C., J.D. Roberts, R.D. Sabatino, T.W. Myers, C.K. Tan, K.M. Downey, A.G. So, R.A. Bambara, and T.A. Kunkel. 1991b. Fidelity of mammalian DNA replication and replicative DNA polymerases. *Biochemistry* **30:** 11751–11759.

Umar, A., J.C. Boyer, D.C. Thomas, D.C. Nguyen, J.I. Risinger, J. Boyd, Y. Ionov, M. Perucho, and T.A. Kunkel. 1994. Defective mismatch repair in extracts of colorectal and endometrial cancer cell lines exhibiting microsatellite instability. *J. Biol. Chem* **269:** 14367–14370.

Waga, S. and B. Stillman. 1994. Anatomy of a DNA replication fork revealed by reconstitution of SV40 DNA replication in vitro. *Nature* **369:** 207–212.

Wang, C.-I. and J.-S. Taylor. 1992. In vitro evidence that UV-induced frameshift and substitution mutations at T tracts are the result of misalignment-mediated replication past a specific thymine dimer. *Biochemistry* **31:** 3671–3681.

Wang, T.S.-F. 1991. Eukaryotic DNA polymerases. *Annu. Rev. Biochem.* **60:** 513–552.

Wernette, C.M., M.C. Conway, and L.S. Kaguni. 1988. Mitochondrial DNA polymerase from *Drosophila melanogaster* embryos: Kinetics, processivity, and fidelity of DNA polymerization. *Biochemistry* **27:** 6046–6054.

West Frey, M., N.G. Nossal, T.L. Capson, and S.J. Benkovic. 1993. Construction and characterization of a bacteriophage T4 DNA polymerase deficient in 3ʹ→5ʹ exonuclease activity. *Proc. Natl. Acad. Sci.* **90:** 2579–2583.

Willems, P.J. 1994. Dynamic mutations hit double figures. *Nat. Genet.* **8:** 213–215.

Wong, I., S.S. Patel, and K.A. Johnson. 1991. An induced-fit kinetic mechanism for DNA replication fidelity: Direct measurement by single-turnover kinetics. *Biochemistry* **30:** 526–537.

Zhang, J., D.W. Chung, C.-K. Tan, K.M. Downey, and A.G. So. 1991. Primary structure of the catalytic subunit of calf thymus DNA polymerase δ: Sequence similarities with other DNA polymerases. *Biochemistry* **30:** 11742–11750.

Zinnen, S., J.-C. Hsieh, and P. Modrich. 1994. Misincorporation and mispaired primer extension by human immunodeficiency virus reverse transcriptase. *J. Biol. Chem* **269:** 24195–24202.

8
DNA Excision Repair Pathways

Errol C. Friedberg
Laboratory of Molecular Pathology
Department of Pathology
University of Texas
Southwestern Medical Center
Dallas, Texas 75235

Richard D. Wood
Imperial Cancer Research Fund
Clare Hall Laboratories
South Mimms, Herts, EN6 3LD
United Kingdom

Most life forms have the ability to respond to alterations in genomic DNA that occur spontaneously or are caused by environmental agents. Generally, these responses take one of two forms. Cells can either repair the damage and restore the genome to its normal physical and functional state, or they can tolerate lesions in a way that reduces their lethal effects (Friedberg et al. 1995). This brief overview exclusively considers the former cellular response to DNA damage, which represents true DNA repair. However, the tolerance of base damage, typically by replicative bypass, sets the stage for permanent mutations in DNA. In fact, a major function of DNA repair is the prevention of mutations, which can have significant phenotypic consequences, including neoplastic transformation in mammalian cells (Friedberg et al. 1995).

The repair of altered bases in DNA is frequently classified into two major categories that have important mechanistic distinctions. A relatively limited group of lesions in DNA can be repaired in single-step reactions that directly reverse the damage. The light-dependent monomerization of cyclobutane pyrimidine dimers by DNA photolyase is a well-characterized example (Kim and Sancar 1993). DNA photolyases have been extensively characterized from many prokaryotes and fungi, and from vertebrates, including fish (Yasuhira and Yasui 1992) and marsupials (Yasui et al. 1994). However, this mode for the repair of the quantitatively major form of DNA damage induced by ultraviolet (UV) radiation seems to have been lost in placental mammals (Li et al. 1993).

Concepts in Eukaryotic DNA Replication
© 1999 Cold Spring Harbor Laboratory Press 0-87969-557-9/99

The direct removal of small alkyl groups (such as methyl groups) specifically from the O^6 position of guanine and the O^4 position of thymine in DNA is another notable example of repair by the reversal of base damage. The enzyme that removes alkyl groups from these specific sites in guanine and thymine is designated O^6-methylguanine-DNA methyltransferase and is ubiquitous in nature.

A more general mode of DNA repair, which is the primary topic of this review, is the physical excision of damaged or inappropriate bases from the genome by multistep biochemical reactions. At present, three specific modes of such repair have been identified and characterized in eukaryotes. These are referred to as base excision repair, nucleotide excision repair, and mismatch repair.

BASE EXCISION REPAIR OF DNA IN EUKARYOTES

The term base excision repair (BER) was coined to emphasize that this DNA repair mechanism is characterized by the excision of nucleic acid base residues in the free form (Friedberg et al. 1995). In contrast, nucleotide excision repair (NER) removes damaged nucleotides as part of fragments which are about 30 nucleotides long. The primary and initiating event of BER is the hydrolysis of the N-glycosyl bond linking a nitrogenous base to the deoxyribose-phosphate chain, thereby releasing the free base (Fig. 1). The hydrolysis of N-glycosyl bonds in DNA is catalyzed by a class of enzymes called DNA glycosylases. Multiple DNA glycosylases have been identified in eukaryotic cells (Table 1). Each enzyme removes a limited spectrum of base alterations. Uracil-DNA glycosylase is particularly specific, as it catalyzes the excision exclusively of the base uracil (and 5-fluorouracil) from DNA. Uracil in DNA usually results from the spontaneous or chemically induced deamination of cytosine, although it can occasionally arise by incorporation from small intracellular pools of dUTP or from the deamination of dCTP (Friedberg et al. 1995). Since deamination of cytosine in DNA generates a U·G mispair, excision of uracil is important to avoid G·C→A·T transition mutations during subsequent semiconservative DNA replication. The enzyme from human cells is encoded by the *UNG* gene, which maps to chromosome 12 (Aasland et al. 1990). The amino acid sequence of human uracil-DNA glycosylase shares extensive amino acid sequence identity with the *ung* gene product from a variety of other organisms.

Another highly specific DNA glycosylase is the thymine mismatch-DNA glycosylase, which to date has been identified only in extracts of human cells (Neddermann and Jiricny 1993). This enzyme specifically

Excision Repair in Eukaryotes 251

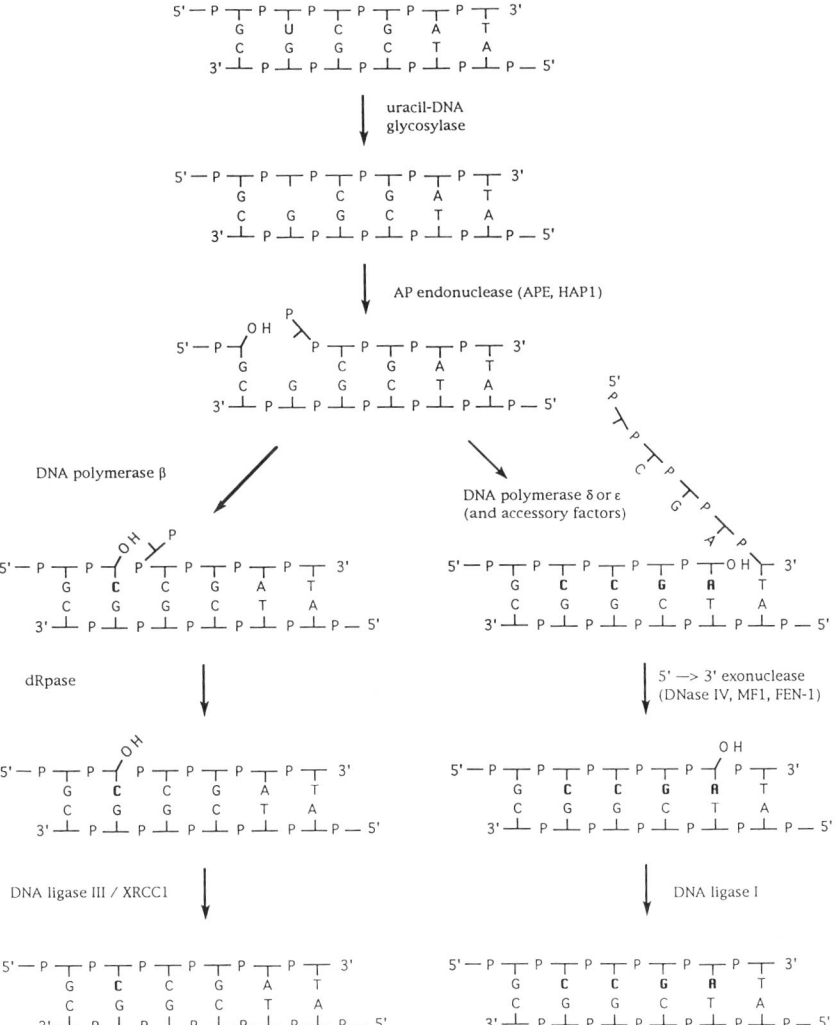

Figure 1 Base excision repair (BER) of DNA in eukaryotes. Repair of a uracil residue initiated by uracil-DNA glycosylase is shown as an example. A branched pathway of repair synthesis is depicted, resulting in either single nucleotide repair patches (*left*) or longer patches of variable size (*right*), depending on which enzymes are utilized. Enzymes implicated in the longer patch repair mode are tentatively assigned. (Adapted, with permission, from Lindahl et al. 1995.)

catalyzes the excision of T when it is mispaired with G, a mispairing that results from the deamination of 5-methylcytosine in DNA. Hence, this form of BER represents one of the several biochemical strategies that hu-

Table 1 DNA glycosylases for base excision repair in eukaryotes

DNA glycosylase	*E. coli* homolog	Substrates
Uracil-DNA glycosylase	*ung*	uracil or fluorouracil in DNA
3-MeA-DNA glycosylase	*alkA*	3-methyladenine hypoxanthine O^2-methylthymine (minor groove alterations)
FaPy-DNA glycosylase	*fpg (mutM)*	formamidopyrimidines 8-hydroxyguanine (oxidized and ring-opened purines)
Pyrimidine hydrate-DNA glycosylase	*nth (endoIII)*	thymine glycol cytosine hydrate urea (oxidized and ring-opened pyrimidines)
G:T mismatch glycosylase	–	thymine when paired to guanine, O^6-methylguanine, or 6-thioguanine

man cells possess for the repair of mismatched bases (see below).

The excision of bases from duplex DNA generates apurinic or apyrimidinic (AP) sites (Fig. 1). The repair of these sites of base loss utilizes a specific class of endonucleases designated AP endonucleases (Friedberg et al. 1995). Prokaryotes such as *Escherichia coli* have at least two such enzymes. However, studies of yeast and mammalian cells have thus far resulted in the purification and characterization of a single major endonuclease that catalyzes the incision of phosphodiester linkages exclusively 5' to AP sites, generating 3'-OH and 5'-deoxyribose-phosphate residues (Fig. 1). This enzyme also has 3'-phosphatase activity, and, in some organisms, a weak 3'→5' exonuclease activity. In mammalian cells, the gene that encodes this AP endonuclease is variously called *BAP1* (bovine AP endonuclease), *APEX* (AP endonuclease/ exonuclease), *HAP1* (human AP endonuclease), or *APE* (AP endonuclease) (Demple et al. 1991; Robson et al. 1991, 1992; Seki et al. 1991). Intriguingly, the HAP1/APE protein was also independently isolated as REF-1, a redox factor that can regulate the Fos-Jun transcriptional activation proteins (Walker et al. 1993; Xanthoudakis et al. 1994). Thus, in addition to its role in BER, this protein may play a role in transducing signals associated with oxidative stress to a regulatory network which involves genes associated with the metabolism of reactive oxygen species. Like human uracil-DNA glycosylase, the HAP1 protein shows extensive evolutionary conservation at the amino acid sequence level, and the protein can correct some of the mutant phenotypes of *E. coli* cells defective in Xth protein, the major AP endonuclease in that organism.

The completion of BER requires the removal of the 5'-terminal deoxyribose-phosphate residue generated by the AP endonuclease, followed by repair synthesis and DNA ligation. A 47-kD enzyme activity designated DNA deoxyribophosphodiesterase (dRpase) has been identified in human cells and can remove such sugar-phosphate residues from duplex DNA (Price and Lindahl 1991). If this enzyme is indeed utilized in vivo, a single nucleotide gap would be generated that can be filled in by a DNA polymerase (Fig. 1). In vitro, such repair synthesis is efficiently catalyzed by DNA polymerase-β (Dianov et al. 1992). It has been suggested that because of its limited fidelity, polymerase-β may be utilized in this particular very short patch mode of repair synthesis during BER and not in repair synthesis associated with longer repair patches.

BER is sometimes associated with the generation of longer repair patches. There are experimental indications that DNA polymerases δ and ε are primarily involved in this second BER synthesis mode in both yeast and human cells. These longer repair tracts are thought to result from a nick translation reaction accompanied by strand displacement in the 5'→3' direction, thereby generating a flap type of structure (Fig. 1). Removal of the overhanging 5'-terminal single-stranded region of DNA is believed to be effected by a 5' single-strand/duplex junction-specific nuclease originally designated DNase IV (Lindahl et al. 1969; Robins et al. 1994) and more recently as FEN-1 (Harrington and Lieber 1994). This nuclease, which is conserved in the yeasts *Saccharomyces cerevisiae* as Rad27 protein and in *Schizosaccharomyces pombe* as rad2 protein (Murray et al. 1994), has also been implicated in lagging-strand DNA synthesis during semiconservative replication (Ishimi et al. 1988; Turchi et al. 1994; Waga et al. 1994). Inspection of the predicted amino acid sequences of DNase IV, Rad27, *S. pombe* rad2 protein, and various prokaryotic DNA polymerases endowed with 5'→3' exonuclease activity (such as *E. coli* DNA polymerase I), shows limited regions of amino acid sequence homology (Table 2). Interestingly, this homology is shared with junction-specific nucleases that are involved in NER in *S. cerevisiae* and in mammalian cells (Table 2; see later discussion).

The final biochemical event during BER is DNA ligation. Mammalian cells (and possibly the yeast *S. cerevisiae*) contain several DNA ligases. These are fully discussed in chapter 20. A recently recognized

Table 2 Amino acid sequence homology between various nuclease families

E. coli pol I	(101)	-MGLPLL--SGV**EADD**-IG-LAR-A---G
Bacillus caldotenax pol I	(97)	-Y-IP-Y----Y**EADD**-IG-LA--A---G
Streptococcus pneumoniae pol I	(102)	-MGI--Y--A-Y**EADD**-IG-L-K-A---G
Thermus flavus pol I	(104)	LLGL--L--PGF**EADD**-LA-LAK-A---G
T4 rnh (orfA)	(119)	YMPY-VM----Y**EADD**-IAVL-K---L-G
T5 D15	(116)	---FP-F---GV**EADD**--AYI-K----L-
T7 6	(121)	---F-CI--P-L**EGDD**-MGVIA--P--FG
S. cerevisiae Rad27	(145)	LMGIPYI--AP-**EAEA**QCA-LAK-GKVYA
S. pombe Rad2	(147)	LMGIPFV--APC**EAEA**QCA-LARSGKVYA
Mus musculus FEN-1	(145)	LMGIPYL--AP-**EAEA**-CA-LAKAGKVYA
Homo sapiens DNase IV	(147)	LMGIPYL--AP-**EAEA**-CA-L-KAGKVYA
S. cerevisiae Rad2	(781)	-FGIPYI--APM**EAEA**QCA-L-----V-G
S. pombe Rad13	(767)	LFGLPYI--AP-**EAEA**QCS-L-----V-G
Xenopus laevis XPG	(811)	LFGIPYI--APM**EAEA**QCAIL--T----G
H. sapiens XPG	(778)	LFGIPYI--APM**EAEA**QCAIL--T----G

The various nucleases shown are grouped according to known functions. The top group are DNA polymerases with associated nuclease activities. The next group are members of the eukaryotic DNase IV family, followed by the S. cerevisiae Rad2 nuclease family. The numbers in parentheses refer to amino acid position. Only identical or related amino acids are shown. The dashes indicate nonconserved positions. The EA(G)E(D) motif that is conserved in all 15 sequences is shown in bold. (Adapted, with thanks, from Stuart G. Clarkson.)

aspect of DNA metabolism that can affect the kinetics of BER is the potential competition for strand breaks in DNA between the enzymatic machinery required for the completion of this repair process and the enzyme poly(ADP-ribose) polymerase. This enzyme normally has a high affinity for strand breaks in DNA. However, in the presence of NAD^+, the bound enzyme undergoes extensive autoribosylation, which results in decreased binding affinity for DNA and its eventual dissociation (Molinete et al. 1993; Satoh et al. 1993). It has been suggested that the binding of non-ribosylated poly(ADP ribose) polymerase to strand breaks introduced by the sequential action of a DNA glycosylase and AP endonuclease during BER may signal a slowing or cessation of DNA replication while BER takes place. Alternatively, or additionally, the binding of poly(ADP-ribose) polymerase to DNA may serve to reduce the potential for the initiation of recombination at sites of strand breakage (Lindahl et al. 1995).

NUCLEOTIDE EXCISION REPAIR OF DNA IN EUKARYOTES

NER is characterized by the excision of damaged bases in oligonucleotide fragments. In contrast to the limited substrate specificity of most DNA glycosylases, NER operates on a large spectrum of base damage, particularly that produced by environmental mutagenic and carcinogenic agents which produce bulky, helix-distorting perturbations in DNA structure. In human cells, NER is the principal mechanism by which base damage produced by UV radiation is removed from DNA. Individuals with the rare inherited disorder xeroderma pigmentosum (XP) have defects in NER genes (Table 3), generally leading to a greatly increased risk of sunlight-induced skin cancer (Cleaver and Kraemer 1989).

There is now compelling evidence that in both prokaryotes and eukaryotes, the oligonucleotides excised during NER are generated by dual incisions which flank sites of base damage. Unlike BER, which is believed to require no more than 5 proteins to complete the entire process, there is good evidence that in eukaryotes the events that precede repair synthesis and DNA ligation during NER require the participation of between 15 and 20 gene products (Table 3). This degree of biochemical complexity is reminiscent of that associated with the initiation of basal transcription by RNA polymerase II. Indeed, it has been suggested that the generation of a transcription "bubble" as part of the process of promoter clearance during RNA polymerase II transcription, and the generation of a bubble that defines the single-strand/duplex DNA junctions required for bimodal incision during NER (Fig. 2), may be mechanistically related (Hoeijmakers and Bootsma 1994). This suggestion is to a large extent prompted by the recent discovery that both in the yeast *S. cerevisiae* and in human cells the processes of NER and RNA polymerase II-mediated transcription share multiple proteins in common (Table 3) (Bootsma and Hoeijmakers 1993; Chalut et al. 1994; Drapkin and Reinberg 1994; Friedberg et al. 1994).

Among the many proteins required for the initiation of RNA polymerase II transcription in *S. cerevisiae* (all of which are encoded by essential genes) are a complex of six polypeptides designated core TFIIH (Table 3) (Svejstrup et al. 1995). This core complex is believed to assemble with three other proteins (TFIIK) endowed with kinase activity for the carboxy-terminal domain of the largest subunit of RNA polymerase II, to yield a holoTFIIH supercomplex, the form of TFIIH that is functional in transcription initiation (Svejstrup et al. 1995). Four of the six subunits of core TFIIH have been directly shown to be indispensable for NER in this yeast (Table 3) (Svejstrup et al. 1995). In crude extracts of *S. cerevisiae* that are competent for RNA polymerase II basal transcription,

256 E.C. Friedberg and R.D. Wood

Table 3 Eukaryotic nucleotide excision repair genes and proteins

Human gene	Human map position	S. cerevisiae homolog	S. pombe homolog	M_r (human gene product unless indicated otherwise)	Comments
XPA	9q34.1	RAD14		31 kD (40/42 kD on gels)	binds damaged DNA
XPB/ERCC3	2q21	SSL2 (RAD25)	ercc3^{sp+}	89 kD (89 kD on gels)	3'→5' DNA helicase; in TFIIH
XPC	3p25	RAD4		106 kD (125 kD on gels)	binds ssDNA
HHR23B	3p25	RAD23		43 kD (58 kD on gels)	associated with XPC
XPD/ERCC2	19q13.2	RAD3	rad15$^+$	87 kD (80 kD on gels)	5'→3' DNA helicase; in TFIIH
XPG/ERCC5	13q33	RAD2	rad13$^+$	133 kD (180–200 kD on gels)	CS in some affected individuals; DNA nuclease
XPF/ERCC4? (ERCC4)	16p13.13	RAD1	rad16$^+$	126 (S. cerevisiae)	component of DNA nuclease
ERCC1	19q13.2	RAD10	swi10$^+$	31 (human) (39 kD on gels)	component of DNA nuclease
p44	5q13	SSL1		44 kD (yeast 50 kD)	in TFIIH
p62	11p14-15.1	TFB1		70–73 kD in yeast	in TFIIH
p52	6p21.3-22.2	TFB2		55 kD in yeast	in TFIIH
CSA/ERCC8	5	RAD28		44 kD	WD-repeat protein
CSB/ERCC6	10q11.2	RAD26		168 kD	DNA helicase? transcription coupling?
RPAp70	17p13	RFA1		68 kD (70 kD on gels)	binding to single-stranded DNA
RPAp32	1p35-36.1	RFA2		29 kD (34 kD on gels)	
RPAp14	7p21-22	RFA3		13.6 kD	
LIG1	19q13.2-3	CDC9	cdc17$^+$	102 kD (120 kD on gels)	DNA ligase I
RFC1		CDC44		140 kD	RF-C large subunit
PCNA	20	POL30	pcn1$^+$	29 kD (36 kD on gels)	toroidal sliding clamp

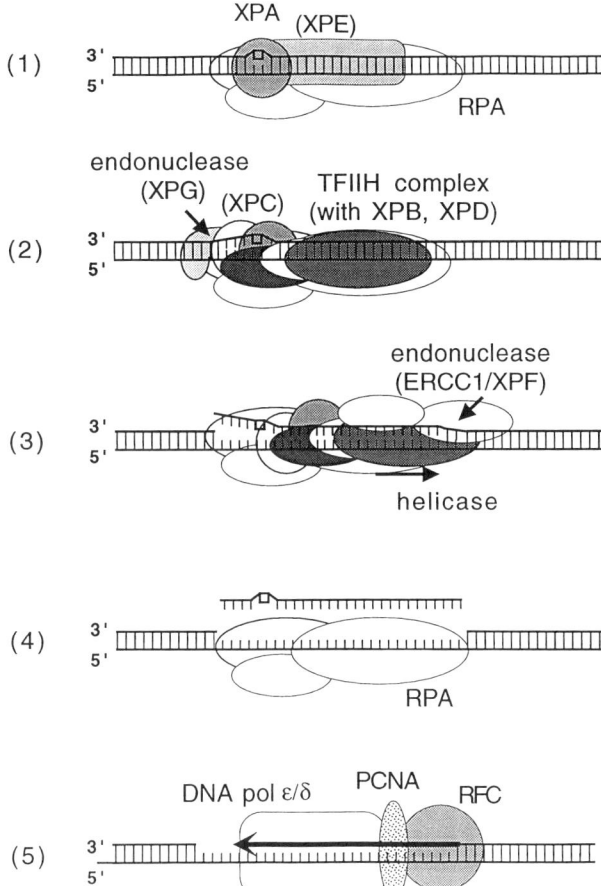

Figure 2 Nucleotide excision repair (NER) in mammalian cells. The steps shown are (1) DNA damage recognition; (2) incision on the 3' side of the damage; (3) the generation of a further opened structure by the helicase function(s) of TFIIH and incision on the 5' side of the damage; (4) release of a damage-containing oligonucleotide; (5) repair synthesis.

a different supercomplex can be identified that includes all six subunits of core TFIIH plus at least five other proteins which are indispensable for NER but are not known to be required for transcription, namely, Rad1, Rad2, Rad4, Rad10, and Rad14. This supercomplex comprising at least these 11 proteins is referred to as the nucleotide excision repairosome (Svejstrup et al. 1995). There are indications of comparable protein complexes for transcription and NER in human cells (Roy et al. 1994).

The observation that core TFIIH is common to both repair and transcription may help explain the fact that in both yeast and mammalian cells NER takes place significantly faster in the template strand of transcriptionally active genes than in the nontranscribed strand (Bohr 1992; Hanawalt 1992). However, it remains to be determined whether the particular form of TFIIH that participates in transcriptionally coupled NER is the same as that loaded onto the DNA during transcription initiation as part of the TFIIH holocomplex. It is intuitively compelling to consider that when the TFIIH holocomplex is loaded onto promoter sites, the NER proteins are retained in the transcription elongation complex. Thus, if transcription is arrested at sites of base damage in the template strand, the core TFIIH complex would constitute a strategically positioned nucleation site for the assembly of a functionally active repairosome. However, there is as yet no direct experimental evidence for this mechanism. An alternative possibility is that TFIIH does not participate in the process of transcription elongation. There is indeed some evidence for this view (Drapkin and Reinberg 1994; Goodrich and Tjian 1994). In this event, repair proteins would be recruited to sites of arrested transcription at base damage by a mechanism(s) that is yet to be specifically determined. Additionally, in mammalian cells, most of the genome is transcriptionally silent, so NER frequently occurs in the absence of a coupling to transcription. Yet components of TFIIH are still required, and it remains to be discovered precisely how these are delivered to damaged sites in transcriptionally silent regions of the genome.

Biochemical functions have been identified for several NER proteins (Table 3). The bimodal incision mechanism involves single-strand/duplex junction-specific nucleases. In *S. cerevisiae* the nuclease that is believed to incise DNA 5' to sites of base damage is carried in the Rad1/Rad10 protein complex (Tomkinson et al. 1993). The mammalian homologs of Rad1/Rad10 are ERCC4 and ERCC1, respectively (Table 3). In vitro, the yeast Rad1/Rad10 endonuclease (and presumably the ERCC1/XPF complex in mammalian cells) cuts splayed-arm substrates specifically at 3' single-strand/duplex junctions where the single strand has a 3' end (Bardwell et al. 1994). Reciprocally, purified human XPG protein cuts splayed-arm substrates specifically at single-strand/duplex junctions where the single strand has a 5' end, and additionally cuts bubble substrates with a consistent polarity (O'Donovan et al. 1994). Thus, XPG protein is believed to incise DNA on the 3' side of damage during NER. Presumably the homologous Rad2 protein acts similarly (Harrington and Lieber 1994). Rad2 protein (Habraken et al. 1993), Rad1/Rad10 complex (Tomkinson et al. 1993; Sung et al. 1993), and XPG protein (O'Donovan et al. 1994; Habraken et al. 1994) also can cut bacteriophage M13 DNA, presumably at single-strand/duplex junctions at hairpin loops in such DNA. Thus, it seems reasonable to conclude that bimodal damage-specific incision during NER in eukaryotes is achieved by the concomitant or sequential actions of the Rad1/Rad10 and Rad2 nucleases in yeast, and by the ERCC1/XPF and XPG nucleases in mammalian cells (Fig. 2). There are indications that the XPC protein may also participate in the incision process (Shivji et al. 1994).

Studies on NER suggest that the excised oligonucleotide fragments have a precise size of approximately 30 ± 2 nucleotides in human (and presumably in yeast) cells. Based on the established specificity of the endonucleases just discussed for single-strand/duplex junctions, we are led to the model that an open structure of about 30 nucleotides is somehow generated in damaged DNA during NER in eukaryotes. Regardless of precisely how the TFIIH core complex is loaded onto DNA, the known biochemical properties of two components of this complex may help explain how such an open structure might arise. Both the yeast Rad3 (human XPD) and Ssl2 (human XPB) proteins are DNA helicases with opposite directionality. It is therefore possible that one or both of these helicases unwind limited regions of duplex DNA on either side of a damaged site during NER (Fig. 2). There is indeed extensive evidence that the $5'\rightarrow3'$ helicase function of Rad3 protein is specifically required for NER but not for transcription. The $3'\rightarrow5'$ helicase function of yeast Ssl2 (XPB) protein is essential for the viability of yeast cells and is therefore presumably indispensable for transcription as well.

It is not yet established how endonucleolytic cleavage is directed specifically to the DNA strand containing a lesion. The yeast Rad14 and homologous human XPA proteins are DNA-binding proteins with preferential affinity for certain types of DNA damage (Robins et al. 1991; Guzder et al. 1993; Jones and Wood 1993; Asahina et al. 1994). Presumably, these proteins play some role in the recognition of base damage. The XPE protein may also participate in this process (Chu and Chang 1988; Hirschfeld et al. 1990; Keeney et al. 1993; Takao et al. 1993; Payne and Chu 1994). It has also been shown that the helicase function of purified Rad3 (human XPD) protein is arrested by the presence of many types of base damage, specifically in the strand on which the protein translocates (Naegeli et al. 1992; Sung et al. 1994). Hence, Rad3 (XPD) protein may also be an important player in damage recognition, both in the non-transcribed bulk of the genome and in transcribed regions of DNA. Specific protein-protein interactions, such as found between XPA and ERCC1, may help direct endonucleolytic cleavage to the correct strand (Li et al. 1994; Park and Sancar 1994).

In addition to the proteins discussed above, there are indications that other polypeptides are involved in NER. In yeast, these include the *RAD7*, *RAD16*, and *RAD23* gene products, but the precise functional role(s) of these proteins remains to be determined. Recent studies using a cell-free system for NER in yeast indicate an absolute requirement for Rad7 protein (Z. Wang and E.C. Friedberg, unpubl.). The single-stranded DNA-binding replication protein A (RP-A) is necessary for NER supported by mammalian cell extracts in vitro, where it participates during DNA repair synthesis. Additionally, RP-A is required for damage-specific incision (Coverley et al. 1992; Shivji et al. 1992). The homolog of this heterotrimeric protein in *S. cerevisiae* is encoded by the *RFA* genes, and yeast strains carrying mutations in the *RFA1* gene are abnormally sensitive to UV radiation, suggesting a role for RP-A in NER in vivo (Longhese et al. 1994).

The repair synthesis step of NER in mammalian cells requires the DNA polymerase accessory factor proliferating cell nuclear antigen (PCNA) in vitro (Shivji et al. 1992), and there is also evidence that PCNA is involved in vivo (Celis and Madsen 1986; Toschi and Bravo 1988; Miura et al. 1992; Hall et al. 1993; Jackson et al. 1994). These observations, together with experiments with chemical inhibitors, implicate the PCNA-dependent DNA polymerases δ or ϵ in the repair synthesis step. Perhaps either enzyme works in this function, but DNA polymerase-ϵ appears to be the most suitable candidate in vivo (Nishida et al. 1988; Syväoja et al. 1990). Recent studies have shown that DNA polymerase-ϵ is also functionally well suited for NER in vitro, and in the presence of RP-A, replication protein C (RP-C) is also required (R.D. Wood et al., unpubl.). RP-C functions to load PCNA onto a DNA template in order to initiate DNA synthesis (Podust et al. 1994). The repair synthesis patch in vitro (Hansson et al. 1989; Shivji et al. 1992) and in vivo (Cleaver et al. 1991) is about 30 nucleotides long, reflecting precise filling of the gap created by the excision of oligonucleotides.

MISMATCH REPAIR IN EUKARYOTES

Mismatches can arise in DNA by two primary mechanisms. Replicative DNA polymerases do not copy templates with complete accuracy, so mismatches can arise because of replicative errors. Additionally, heteroduplexes formed as recombination intermediates between two homologous pieces of DNA (such as two alleles of a gene) can contain mismatches arising from polymorphisms. The former mechanism is the most important source of mismatches in somatic cells.

General (so-called long-patch) mismatch repair is best understood in *E. coli*, where the core enzymes of the system are the products of the *mutH*, *mutL*, and *mutS* genes (Modrich 1991; Friedberg et al. 1995). Mismatch repair can only protect cells from permanent mutations if the parental strand (containing the correct information) can be accurately distinguished from the daughter strand. In *E. coli*, the strand discrimination signal is provided by adenine methylation in GATC sequences; newly replicated DNA is not yet methylated on the daughter strand (Modrich 1991; Friedberg et al. 1995). The MutH protein binds to DNA at hemimethylated GATC sequences and effects incision on the unmethylated strand. MutS protein recognizes and binds to the mismatch, and the intervening region (often hundreds of nucleotides long) is excised and recopied by a DNA polymerase (Modrich 1991; Friedberg et al. 1995). The MutL protein mediates communication between the distantly bound MutH and MutS products, bringing them together by looping out the intervening region of DNA (Fig. 3) (Modrich 1991).

Long-patch mismatch excision repair is a highly conserved process that appears to work in a similar way in eukaryotes as in *E. coli*, except that strand discrimination does not appear to be methyl-directed in eukaryotes (Friedberg et al. 1995). Thus, there may be no eukaryotic MutH homolog. Instead, the strand-discrimination signal is thought to be provided by single-stranded nicks or gaps in newly replicated DNA, which have not yet been joined by a DNA ligase. A protein that recognizes these nicks or gaps would replace the function of MutH. However, several structural homologs of MutL and MutS have been isolated from yeast and from mammalian cells. Of the two known MutS homologs, one (hMSH2) is a nuclear protein and the other (hMSH1) is mitochondrial. Interestingly, there are at least three MutL homologs in humans (hMLH1, hPMS1, and hPMS2). hMSH2 has been demonstrated to be a DNA mismatch-binding protein in vitro (Palombo et al. 1994). It seems probable that the MutS and MutL proteins form complexes with one another in different combinations to facilitate recognition of a wide range of different types of mismatches, ranging from common single base-pair mismatches such as G·T, through loop-outs of one, two, or more nucleotides.

Inactivation of mismatch repair genes in humans is clearly implicated in the pathogenesis of hereditary non-polyposis colon cancer (HNPCC). Individuals with this condition inherit an inactivated allele of a mismatch repair gene, and colorectal carcinomas in these individuals (as well as sporadic tumors from non-HNPCC patients) have two inactivated alleles. The mismatch repair defect increases the spontaneous mutation rate in

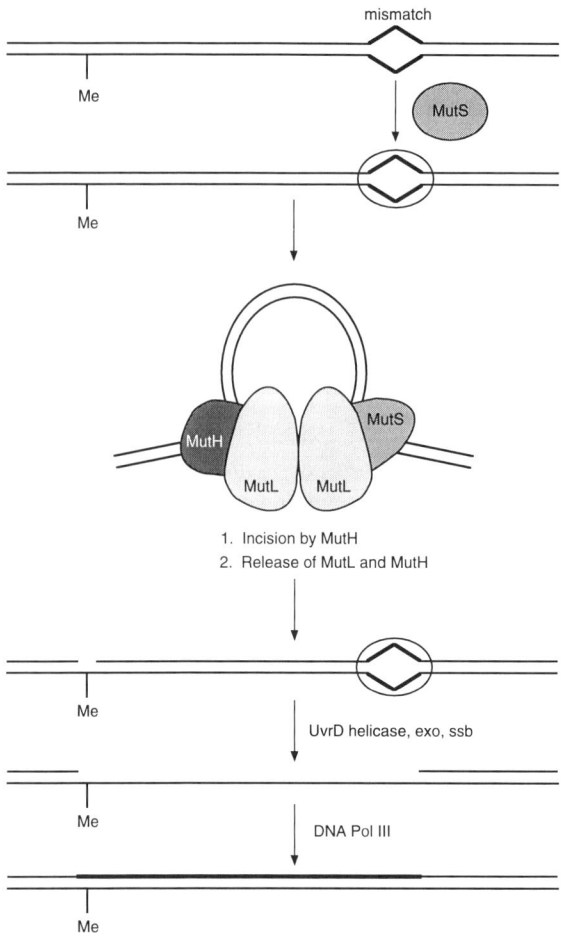

Figure 3 General (long-patch) mismatch repair of DNA in *E. coli*. The eukaryotic process is believed to share many of the features shown here. See text for details.

the cells. This hypermutable state is thought to be an early event in the progression of tumors to malignancy, as it greatly accelerates the acquisition of mutations in other tumor suppressor genes and oncogenes. Thus far, HNPCC families have been found with mutations in the hMSH2, hMLH1, hPMS1, and hPMS2 genes (Fishel et al. 1993; Leach et al. 1993; Bronner et al. 1994; Nicolaides et al. 1994; Papadopoulos et al. 1994). A notable characteristic of colorectal carcinoma cell lines is their high rate of polymorphism in microsatellite repeat sequences. During DNA replication, di- or trinucleotide repeat units in such sequences can be accidentally lost or gained by replication fork slippage. In the absence of mismatch repair, the lengths of the microsatellite sequences change more rapidly than in normal cells, reflecting the hypermutable state (Aaltonen et al. 1993).

An interesting feature of mismatch repair-defective cells is an association with tolerance to simple DNA N-nitroso-methylating agents such as MNNG and MNU (Karran and Bignami 1994). This tolerance results because the most toxic DNA adduct produced by such agents, O^6-methylguanine, can pair with thymine and the resulting $G^{O6Me} \cdot T$ base pair is recognized as a mismatch. However, G·T mismatches are nearly always repaired by removal of the T residue, so futile cycles of mismatch repair are initiated. The major alternative pathway for the removal of O^6-methylguanine is by the O^6-methylguanine-DNA methyltransferase referred to earlier, but many cells, particularly those that are transformed, spontaneously lose expression of this enzyme. In such cases, the repeated excision cycles of futile mismatch repair eventually lead to lethal strand breaks (Karran and Bignami 1994). Thus, there may be selective pressure for loss of mismatch repair in cells that are frequently exposed to methylating agents. Colon cells are exposed to bile acids that can be converted into methylating compounds (Karran and Bignami 1994). Cells with defective O^6-methylguanine-DNA methyltransferase and defective mismatch binding proteins become more resistant to DNA-methylating agents, but are hypermutable and show microsatellite instability (Branch et al. 1993; Kat et al. 1993; Aquilina et al. 1994).

Mismatch repair can be studied in vitro and the entire repair process can be carried out in mammalian cell extracts, so details of the biochemistry of this excision repair mode are emerging rapidly.

CONCLUDING REMARKS

The general topic of excision repair of DNA in eukaryotes has undergone many exciting developments in recent years. The biochemistry of BER is essentially fully defined in vitro and is consonant with the genetics of this process. It has additionally been firmly established that NER is a complex biochemical process involving a large number of gene products. The findings in a number of laboratories that some components of the NER machinery are also components of the RNA polymerase II basal transcription apparatus have added new and exciting dimensions to this DNA repair mode, including new insights into the possible molecular pathogenesis of human hereditary diseases associated with defective NER (Friedberg et al. 1994; Vermeulen et al. 1994). Equally dramatic strides have been made in deciphering the mechanism of strand-directed mismatch repair in eukaryotes. The association of defective NER and mismatch repair with a variety of human cancers provides convincing support for the somatic mutation hypothesis of neoplastic transformation and the crucial role of DNA repair in protecting against this consequence of DNA damage.

ACKNOWLEDGMENTS

We apologize to many of our colleagues for the restrictive reference list necessitated by space limitations. We thank our laboratory colleagues for critical review of the manuscript. E.C.F. and R.D.W. acknowledge grant support from the U.S. Public Health Service and the Imperial Cancer Research Fund, respectively.

REFERENCES

Aaltonen, L.A., P. Peltomäki, F.S. Leach, P. Sistonen, L. Pylkkänen, J.-P. Mecklin, H. Järvinen, S.M. Powell, J. Jen, S.R. Hamilton, G.M. Petersen, K.W. Kinzler, B. Vogelstein, and A. de la Chapelle. 1993. Clues to the pathogenesis of familial colorectal cancer. *Science* **260:** 812–816.

Aasland, R., L.C. Olsen, N.K. Spurr, H.E. Krokan, and D.E. Helland. 1990. Chromosomal assignment of human uracil-DNA glycosylase to chromosome 12. *Genomics* **7:** 139–141.

Aquilina, G., P. Hess, P. Branch, C. MacGeoch, I. Casciano, P. Karran, and M. Bignami. 1994. A mismatch recognition defect in colon-carcinoma confers DNA microsatellite instability and a mutator phenotype. *Proc. Natl. Acad. Sci.* **91:** 8905–8909.

Asahina, H., I. Kuraoka, M. Shirakawa, E.H. Morita, N. Miura, I. Miyamoto, E. Ohtsuka, Y. Okada, and K. Tanaka. 1994. The XPA protein is a zinc metalloprotein with an ability to recognize various kinds of DNA damage. *Mutat. Res.* **315:** 229–237.

Bardwell, A.J., L. Bardwell, A.E. Tomkinson, and E.C. Friedberg. 1994. Specific cleavage of model recombination and repair intermediates by the yeast Rad1-Rad10 DNA endonuclease. *Science* **265:** 2082–2085.

Bohr, V.A. 1992. Gene-specific DNA repair: Characteristics and relations to genomic instability. In *DNA repair mechanisms* (ed. V.A. Bohr et al.), pp. 217–227. Munksgaard, Copenhagen.

Bootsma, D. and J.H.J. Hoeijmakers. 1993. Engagement with transcription. *Nature* **363:** 114–115.

Branch, P., G. Aquilina, M. Bignami, and P. Karran. 1993. Defective mismatch binding and a mutator phenotype in cells tolerant to DNA damage. *Nature* **362:** 652–654.

Bronner, C.E., S.M. Baker, P.T. Morrison, G. Warren, L.G. Smith, M.K. Lescoe, M. Kane, C. Earabine, J. Lipford, A. Lindblom, P. Tannergård, R.J. Bollag, A.R. Godwin, D.C. Ward, M. Nordenskjöld, R. Fishel, R. Kolodner, and R.M. Liskay. 1994. Mutation in the DNA mismatch repair gene homologue *hMLH1* is associated with hereditary non-polyposis colon cancer. *Nature* **368:** 258–261.

Celis, J.E., and P. Madsen. 1986. Increased nuclear cyclin/PCNA antigen staining of non S-phase transformed human amnion cells engaged in nucleotide excision DNA repair. *FEBS Lett.* **209:** 277–283.

Chalut, C., V. Moncollin, and J.M. Egly. 1994. Transcription by RNA polymerase II: A process linked to DNA repair. *BioEssays* **16:** 651–655.

Chu, G. and E. Chang. 1988. Xeroderma pigmentosum group E cells lack a nuclear factor that binds to damaged DNA. *Science* **242:** 564–567.

Cleaver, J.E. and K.H. Kraemer. 1989. Xeroderma pigmentosum. In *The metabolic basis of inherited disease*, 6th edition (ed. C.R. Scriver et al.), pp. 2949–2971. McGraw-Hill, New York.

Cleaver, J.E., J. Jen, W.C. Charles, and D.L. Mitchell. 1991. Cyclobutane dimers and (6-4)photoproducts in human-cells are mended with the same patch sizes. *Photochem. Photobiol.* **54:** 393–402.

Coverley, D., M.K. Kenny, D.P. Lane, and R.D. Wood. 1992. A role for the human single-stranded DNA binding protein HSSB/RPA in an early stage of nucleotide excision repair. *Nucleic Acids Res.* **20:** 3873–3880.

Demple, B., T. Herman, and D.S. Chen. 1991. Cloning and expression of ape, the cDNA-encoding the major human apurinic endonuclease-definition of a family of DNA-repair enzymes. *Proc. Natl. Acad. Sci.* **88:** 11450–11454.

Dianov, G., A. Price, and T. Lindahl. 1992. Generation of single-nucleotide repair patches following excision of uracil residues from DNA. *Mol. Cell. Biol.* **12:** 1605–1612.

Drapkin, R. and D. Reinberg. 1994. The multifunctional TFIIH complex and transcriptional control. *Trends Biochem. Sci.* **19:** 504–508.

Fishel, R., M.K. Lescoe, M.R.S. Rao, N.G. Copeland, N.A. Jenkins, J. Garber, M. Kane, and R. Kolodner. 1993. The human mutator gene homolog *MSH2* and its association with hereditary nonpolyposis cancer. *Cell* **75:** 1027–1038.

Friedberg, E.C., G.C. Walker, and W. Siede. 1995. *DNA repair and mutagenesis*. ASM Press, Washington, D.C.

Friedberg, E.C., A.J. Bardwell, L. Bardwell, Z. Wang, and G. Dianov. 1994. Transcription and nucleotide excision repair-reflections, considerations and recent biochemical insights. *Mutat. Res.* **307:** 5–14.

Goodrich, J.A. and R. Tjian. 1994. Transcription factors IIE and IIH and ATP hydrolysis direct promoter clearance by RNA polymerase II. *Cell* **77:** 145–156.

Guzder, S.N., P. Sung, L. Prakash, and S. Prakash. 1993. Yeast DNA-repair gene *RAD14* encodes a zinc metalloprotein with affinity for ultraviolet-damaged DNA. *Proc. Natl. Acad. Sci.* **90:** 5433–5437.

Habraken, Y., P. Sung, L. Prakash, and S. Prakash. 1993. Yeast excision repair gene *RAD2* encodes a single-stranded DNA endonuclease. *Nature* **366:** 365–368.

———. 1994. Human xeroderma-pigmentosum group-G gene encodes a DNA endonuclease. *Nucleic Acids Res.* **22:** 3312–3316.

Hall, P.A., P.H. McKee, H. Menage, R. Dover, and D.P. Lane. 1993. High levels of p53 protein in UV-irradiated normal human skin. *Oncogene* **8:** 203–207.

Hanawalt, P.C. 1992. Transcription-dependent and transcription-coupled DNA repair responses. In *DNA repair mechanisms* (ed. V.A. Bohre et al.), pp. 231–242. Munksgaard, Copenhagen.

Hansson, J., M. Munn, W.D. Rupp, R. Kahn, and R.D. Wood. 1989. Localization of DNA repair synthesis by human cell extracts to a short region at the site of a lesion. *J.*

Biol. Chem. **264:** 21788-21792.

Harrington, J.J. and M.R. Lieber. 1994. Functional domains within FEN-1 and Rad2 define a family of structure-specific endonucleases—Implications for nucleotide excision-repair. *Genes Dev.* **8:** 1344-1355.

Hirschfeld, S., A.S. Levine, K. Ozato, and M. Protic. 1990. A constitutive damage-specific DNA-binding protein is synthesized at higher levels in UV-irradiated primate cells. *Mol. Cell. Biol.* **10:** 2041-2048.

Hoeijmakers, J.H.J. and D. Bootsma. 1994. Incisions for excision. *Nature* **371:** 654-655.

Ishimi, Y., A. Claude, P. Bullock, and J. Hurwitz. 1988. Complete enzymatic synthesis of DNA containing the SV40 origin of replication. *J. Biol. Chem.* **263:** 19723-19733.

Jackson, D.A., A.B. Hassan, R.J. Errington, and P.R. Cook. 1994. Sites in human nuclei where damage induced by ultraviolet-light is repaired—Localization relative to transcription sites and concentrations of proliferating cell nuclear antigen and the tumor-suppressor protein, p53. *J. Cell Sci.* **107:** 1753-1760.

Jones, C.J. and R.D. Wood. 1993. Preferential binding of the xeroderma pigmentosum group A complementing protein to damaged DNA. *Biochemistry* **32:** 12096-12104.

Karran, P. and M. Bignami. 1994. DNA damage tolerance, mismatch repair and genome instability. *BioEssays* **16:** 833-839.

Kat, A., W.G. Thilly, W.H. Fang, M.J. Longley, G.M. Li, and P. Modrich. 1993. An alkylation-tolerant, mutator human cell line is deficient in strand-specific mismatch repair. *Proc. Natl. Acad. Sci.* **90:** 6424-6428.

Keeney, S., G.J. Chang, and S. Linn. 1993. Characterization of a human DNA-damage binding protein implicated in xeroderma pigmentosum E. *J. Biol. Chem.* **268:** 21293-21300.

Kim, S.T. and A. Sancar. 1993. Photochemistry, photophysics, and mechanism of pyrimidine dimer repair by DNA photolyase. *Photochem. Photobiol.* **57:** 895-904.

Leach, F.S., N.C. Nicolaides, N. Papadopoulos, B. Liu, J. Jen, R. Parsons, P. Peltomäki, P. Sistonen, L.A. Aaltonen, M. Nyström-Lahti, X.-Y. Guan, J. Zhang, P.S. Meltzer, J.-W. Yu, F.-T. Kao, D.J. Chen, K.M. Cerosaletti, R.E.K. Fournier, S. Todd, T. Lewis, R.J. Leach, S.L. Naylor, J. Weissenbach, J.-P. Mecklin, H. Järvinen, G.M. Petersen, S.R. Hamilton, J. Green, J. Jass, P. Watson, H.T. Lynch, J.M. Trent, A. de la Chappelle, K.W. Kinzler, and B. Vogelstein. 1993. Mutations of a *mutS* homolog in hereditary nonpolyposis colorectal cancer. *Cell* **75:** 1215-1225.

Li, L., S.J. Elledge, C.A. Peterson, E.S. Bales, and R.J. Legerski. 1994. Specific association between the human DNA repair proteins XPA and ERCC1. *Proc. Natl. Acad. Sci.* **91:** 5012-5016.

Li, Y.F., S.T. Kim, and A. Sancar. 1993. Evidence for lack of DNA photoreactivating enzyme in humans. *Proc. Natl. Acad. Sci.* **90:** 4389-4393.

Lindahl, T., J.A. Gally, and G.M. Edelman. 1969. Deoxyribonuclease IV: A new exonuclease from mammalian tissues. *Proc. Natl. Acad. Sci.* **62:** 597-603.

Lindahl, T., M.S. Satoh, and G. Dianov. 1995. Enzymes acting at strand interruptions in DNA. *Philos. Trans. R. Soc. Lond. Biol. Sci.* **347:** 57-62.

Longhese, M.P., P. Plevani, and G. Lucchini. 1994. Replication factor A is required in vivo for DNA replication, repair, and recombination. *Mol. Cell. Biol.* **14:** 7884-7890.

Miura, M., M. Domon, T. Sasaki, and Y. Takasaki. 1992. Induction of proliferating cell nuclear antigen (PCNA) complex-formation in quiescent fibroblasts from a xeroderma pigmentosum patient. *J. Cell. Physiol.* **150:** 370-376.

Modrich, P. 1991. Mechanisms and biological effects of mismatch repair. *Annu. Rev.*

Genet. 25: 229-253.

Molinete, M., W. Vermeulen, A. Burkle, J. Menissier-De Murcia, J.H. Kupper, J.H.J. Hoeijmakers, and G. de Murcia. 1993. Overproduction of the poly(ADP-ribose) polymerase DNA-binding domain blocks alkylation-induced DNA-repair synthesis in mammalian cells. *EMBO J.* **12:** 2109-2117.

Murray, J.M., M. Tavassoli, R. Al-Harithy, K.S. Sheldrick, A.R. Lehmann, A.M. Carr, and F.Z. Watts. 1994. Structural and functional conservation of the human homologue of the *Schizosaccharomyces pombe rad2* gene, which is required for DNA repair and chromosome segregation. *Mol. Cell. Biol.* **14:** 4878-4888.

Naegeli, H., L. Bardwell, and E.C. Friedberg. 1992. The DNA helicase and adenosine-triphosphatase activities of yeast rad3 protein are inhibited by DNA damage—A potential mechanism for damage-specific recognition. *J. Biol. Chem.* **267:** 392-398.

Neddermann, P. and J. Jiricny. 1993. The purification of a mismatch-specific thymine-DNA glycosylase from HeLa cells. *J. Biol. Chem.* **268:** 21218-21224.

Nicolaides, N.C., N. Papadopolous, B. Liu, Y.F. Wei, K.C. Carter, S.M. Ruben, C.A. Rosen, W.A. Haseltine, R.D. Fleischmann, and C.M. Fraser. 1994. Mutations of 2 pms homologs in hereditary nonpolyposis colon cancer. *Nature* **371:** 75-80.

Nishida, C., P. Reinhard, and S. Linn. 1988. DNA repair synthesis in human fibroblasts requires DNA polymerase δ. *J. Biol. Chem.* **263:** 501-510.

O'Donovan, A., A.A. Davies, J.G. Moggs, S.C. West, and R.D. Wood. 1994. XPG endonuclease makes the 3' incision in human DNA nucleotide excision repair. *Nature* **371:** 432-435.

Palombo, F., M. Hughes, J. Jiricny, O. Truong, and J. Hsuan. 1994. Mismatch repair and cancer. *Nature* **367:** 417.

Papadopoulos, N., N.C. Nicolaides, Y.-F. Wei, S.M. Ruben, K.C. Carter, C.A. Rosen, W.A. Haseltine, R.D. Fleischmann, C.M. Fraser, M.D. Adams, J.C. Venter, S.R. Hamilton, G.M. Petersen, P. Watson, H.T. Lynch, P. Peltomaki, J.-P. Mecklin, A. de la Chapelle, K.W. Kinzier, and B. Vogelstein. 1994. Mutation of a *mutL* homolog in hereditary colon cancer. *Science* **263:** 1625-1629.

Park, C.H. and A. Sancar. 1994. Formation of a ternary complex by human XPA, ERCC1, and ERCC4(XPF) excision-repair proteins. *Proc. Natl. Acad. Sci.* **91:** 5017-5021.

Payne, A. and G. Chu. 1994. Xeroderma-pigmentosum group-e binding factor recognizes a broad spectrum of DNA damage. *Mutat. Res.* **310:** 89-102.

Podust, L.M., V.N. Podust, C. Floth, and U. Hübscher. 1994. Assembly of DNA polymerase δ and ε holoenzymes depends on the geometry of the DNA template. *Nucleic Acids Res.* **22:** 2970-2975.

Price, A. and T. Lindahl. 1991. Enzymatic release of 5'-terminal deoxyribose phosphate residues from damaged DNA in human cells. *Biochemistry* **30:** 8631-8637.

Robins, P., D.J.C. Pappin, R.D. Wood, and T. Lindahl. 1994. Structural and functional homology between mammalian DNase IV and the 5' nuclease domain of *Escherichia coli* DNA polymerase I. *J. Biol. Chem.* **269:** 28535-28538.

Robins, P., C.J. Jones, M. Biggerstaff, T. Lindahl, and R.D. Wood. 1991. Complementation of DNA repair in xeroderma pigmentosum group A cell extracts by a protein with affinity for damaged DNA. *EMBO J.* **10:** 3913-3921.

Robson, C.N., A.M. Milne, D.J.C. Pappin, and I.D. Hickson. 1991. Isolation of cDNA clones encoding an enzyme from bovine cells that repairs oxidative DNA damage *in vitro*—Homology with bacterial repair enzymes. *Nucleic Acids Res.* **19:** 1087-1092.

Robson, C.N., D. Hochhauser, R. Craig, K. Rack, V.J. Buckle, and I.D. Hickson. 1992. Structure of the human DNA-repair gene *HAP1* and its localization to chromosome 14q 11 2-12. *Nucleic Acids Res.* **20:** 4417–4421.

Roy, R., J.P. Adamczewski, T. Seroz, W. Vermeulen, J.-P. Tassan, L. Schaeffer, E.A. Nigg, J.H.J. Hoeijmakers, and J.-M. Egly. 1994. The MO15 cell cycle kinase is associated with the TFIIH transcription-DNA repair factor. *Cell* **79:** 1093–1101.

Satoh, M.S., G.G. Poirier, and T. Lindahl. 1993. NAD+-dependent repair of damaged DNA by human cell extracts. *J. Biol. Chem.* **268:** 5480–5487.

Seki, S., S. Ikeda, S. Watanabe, M. Hatsushika, K. Tsutsui, K. Akiyama, and B. Zhang. 1991. A mouse DNA-repair enzyme (APEX nuclease) having exonuclease and apurinic apyrimidinic endonuclease activities—Purification and characterization. *Biochim. Biophys. Acta* **1079:** 57–64.

Shivji, M.K.K., A.P.M. Eker, and R.D. Wood. 1994. DNA repair defect in xeroderma pigmentosum group C and complementing factor from HeLa cells. *J. Biol. Chem.* **269:** 22749–22757.

Shivji, M.K.K., M.K. Kenny, and R.D. Wood. 1992. Proliferating cell nuclear antigen is required for DNA excision repair. *Cell* **69:** 367–374.

Sung, P., P. Reynolds, L. Prakash, and S. Prakash. 1993. Purification and characterization of the *Saccharomyces cerevisiae RAD1-RAD10* endonuclease. *J. Biol. Chem.* **268:** 26391–26399.

Sung, P., J.F. Watkins, L. Prakash, and S. Prakash. 1994. Negative superhelicity promotes ATP-dependent binding of yeast Rad3 protein to ultraviolet-damaged DNA. *J. Biol. Chem.* **269:** 8303–8308.

Svejstrup, J.Q., Z. Wang, W.J. Feaver, X. Wu, T.F. Donahue, E.C. Friedberg, and R.D. Kornberg. 1995. Different forms of TFIIH for transcription and DNA repair: Holo-TFIIH and a nucleotide excision repairosome. *Cell* **80:** 21–28.

Syväoja, J., S. Suomensaari, C. Nishida, J.S. Goldsmith, G.S.J. Chui, S. Jain, and S. Linn. 1990. DNA polymerases α, δ, and ε: Three distinct enzymes from HeLa cells. *Proc. Natl. Acad. Sci.* **87:** 6664–6668.

Takao, M., M. Abramic, M. Moos, V.R. Otrin, J.C. Wootton, M. McLenigan, A.S. Levine, and M. Protic. 1993. A 127 kDa component of a UV-damaged DNA-binding complex, which is defective in some xeroderma pigmentosum group E patients, is homologous to a slime-mold protein. *Nucleic Acids Res.* **21:** 4111–4118.

Tomkinson, A.E., A.J. Bardwell, L. Bardwell, N.J. Tappe, and E.C. Friedberg. 1993. Yeast DNA repair and recombination proteins Rad1 and Rad10 constitute a single-stranded-DNA endonuclease. *Nature* **362:** 860–862.

Toschi, L. and R. Bravo. 1988. Changes in cyclin/proliferating cell nuclear antigen distribution during DNA repair synthesis. *J. Cell Biol.* **107:** 1623–1628.

Turchi, J.J., L. Huang, R.S. Murante, Y. Kim, and R.A. Bambara. 1994. Enzymatic completion of mammalian lagging-strand DNA replication. *Proc. Natl. Acad. Sci.* **91:** 9803–9807.

Vermeulen, W., A.J. van Vuuren, M. Chipoulet, L. Schaeffer, E. Appeldoorn, G. Weeda, N.G.J. Jaspers, A. Priestly, C.F. Arlett, A.R. Lehmann, M. Stefanini, M. Mezzina, A. Sarasin, D. Bootsma, J.-M. Egly, and J.H.J. Hoeijmakers. 1994. Three unusual repair deficiencies associated with transcription factor BTF2(TFIIH). Evidence for the existence of a transcription syndrome. *Cold Spring Harbor Symp. Quant. Biol.* **59:** 317–329.

Waga, S., G. Bauer, and B. Stillman. 1994. Reconstitution of complete SV40 DNA-replication with purified replication factors. *J. Biol. Chem.* **269**: 10923-10934.

Walker, L.J., C.N. Robson, E. Black, D. Gillespie, and I.D. Hickson. 1993. Identification of residues in the human DNA-repair enzyme HAP1 (REF-1) that are essential for redox regulation of jun DNA-binding. *Mol. Cell. Biol.* **13**: 5370-5376.

Xanthoudakis, S., G.G. Miao, and T. Curran. 1994. The redox and DNA-repair activities of REF-1 are encoded by nonoverlapping domains. *Proc. Natl. Acad. Sci.* **91**: 23-27.

Yasuhira, S. and A. Yasui. 1992. Visible light-inducible photolyase gene from the goldfish *Carassius auratus. J. Biol. Chem.* **267**: 25644-25647.

Yasui, A., A.P.M. Eker, S. Yasuhira, H. Yajima, T. Kobayashi, M. Takao, and A. Oikawa. 1994. A new class of DNA photolyases present in various organisms including aplacental mammals. *EMBO J.* **13**: 6143-6151.

9
Chromatin Structure and DNA Replication: Implications for Transcriptional Activity

Alan P. Wolffe
Laboratory of Molecular Embryology
National Institute of Child Health and Human Development
National Institutes of Health
Bethesda, Maryland 20892-2710

The period in the cell cycle when the genome is replicated (S phase) is crucially important for the establishment and maintenance of programs of differential gene activity. Not only must DNA be replicated, but the chromosome itself must be duplicated. The majority of genes in the proliferating cell of a defined type retain the same states of transcriptional activity through cell division. This requires the duplication of the precise nucleoprotein complexes directing gene transcription or repression on the nascent DNA templates. The maintenance of these specific regulatory complexes through replication reflects the commitment of a defined cell type or line to a particular state of determination. Preexisting chromosomal structures are transiently disrupted by transit through the replication elongation complex. Most of these structures are faithfully reassembled following replication through mechanisms discussed in this chapter. However, the transient disruption of these structures also offers a window of opportunity for modifying regulatory nucleoprotein complexes. These alterations can either activate genes through the disruption of repressed states, or direct the repression of previously active genes. Thus, cell division offers a molecular mechanism to redirect the commitment of a cell toward a particular determined state. A consideration of the processes occurring at the eukaryotic replication fork suggests how this important development process might be accomplished.

Concepts in Eukaryotic DNA Replication
© 1999 Cold Spring Harbor Laboratory Press 0-87969-557-9/99

IMPLICATIONS OF DNA REPLICATION FOR STABLE STATES OF TRANSCRIPTIONAL ACTIVITY

Active and Repressed States of Eukaryotic Genes

The local nucleoprotein complexes required to maintain a eukaryotic gene in an active or repressed state have been defined in some detail (Tjian and Maniatis 1994; Wolffe 1994a). Transcriptional activity for a given gene depends on a number of sequence-specific transcription factors (e.g., SP1), structural proteins (e.g., HMGI/Y), and non-DNA-binding proteins associated with the promoter interacting to recruit general transcription factors (e.g., TFIIA, TFIIB) together with the TFIID complex (containing TBP [*T*ATA *b*inding *p*rotein] and the TAFs [*T*ATA *a*ssociated *f*actors]). The assembly of this large nucleoprotein complex is initiated through the association of DNA-binding proteins and requires many intermediate steps leading to the recruitment of RNA polymerase II and, eventually, to transcription itself. Conversely, several features may determine a gene to be transcriptionally inactive. A common mechanism appears to be a deficiency in an essential component required for the assembly of the active complex. If this component is a DNA-binding protein, the cognate DNA sequence might become associated with the histone proteins. Specific nucleosomal structures assembled by the core histones (H2A, H2B, H3, and H4) might restrict the subsequent association of either sequence-specific DNA-binding proteins or the basal transcriptional machinery (Simpson 1991; Wolffe 1994b). Other proteins may stabilize repressive higher-order chromatin structures dependent on prior association of the core histones (Hansen and Wolffe 1994); these include linker histone variants (Khochbin and Wolffe 1994) or the chromodomain (*chro*matin *mo*dification *o*rganizer) proteins such as HP1 and Polycomb (James and Elgin 1986; Paro and Hogness 1991).

Generally, the assemblies of nucleosomes or transcription complexes on the promoter of a eukaryotic gene are mutually exclusive. The prior assembly of nucleosomes can prevent transcription factors from binding to DNA and, conversely, the prior assembly of a transcription complex prevents nucleosome formation from repressing transcription (Fig. 1). Although these results provide an excellent molecular basis for the maintenance of stable states of gene expression in a terminally differentiated nondividing cell, they do not explain why either transcriptionally active or inactive states are assembled onto DNA in the first place, nor do they explain how such states can be propagated through cell division. Clearly, because both nucleoprotein structures can incorporate the same DNA molecule, the possibility exists of a competition occurring between the assembly of the two structures. This competition, in fact, occurs dur-

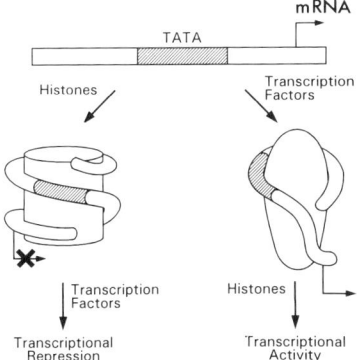

Figure 1 Nucleosome assembly and transcription complex assembly are often mutually exclusive. Two alternate pathways are shown for the association of DNA-binding proteins with a promoter containing a TATA homology. The start site of transcription of mRNA is indicated by the bent arrow.

ing the staged assembly of either active or repressed states following replication (see Almouzni et al. 1990a; Aparicio and Gottschling 1994). Molecular mechanisms that influence the outcome of this competition direct the commitment of a cell to a particular state of determination or facilitate developmentally regulated switches in cell fate. However, to appreciate how this competition occurs, we must first discuss the consequences of DNA replication for preexisting chromatin structures.

Impact of DNA Replication on Preexisting Chromatin Structures

Chromatin consists of long arrays of nucleosomal DNA interspersed with specific regulatory nucleoprotein complexes. The replication fork moves through chromatin without apparent impediment. Replication fork progression disrupts preexisting nucleosomes; however, the fate of regulatory nucleoprotein complexes depends on the particular structure examined.

Nucleosomes

Major considerations for preexisting nucleosomes during the replication process are whether the histones present in the nucleosome stay together on nascent DNA, and whether nucleosomes are randomly or conservatively segregated to daughter DNA strands. DNA replication requires the transient unwinding of duplex parental DNA into two single-stranded regions. Although histones associate with single-stranded DNA (Almouzni et al. 1990b), they do not assemble nucleosomes. This property, coupled with the competing protein-DNA interactions involved in DNA synthesis at the replication fork, probably accounts for nucleosome disruption. Histones released from the parental chromatin during replication in vitro can be easily sequestered onto competitor DNA (Gruss et al. 1993). However, in vivo these histones are sequestered onto daughter DNA molecules close to the replication fork (see Fig. 4) (Sogo et al. 1986; Perry et al. 1993).

A nucleosome contains an octamer consisting of two molecules each of the four core histones (H2A, H2B, H3, and H4) and a single molecule of a fifth linker histone (H1). The four core histones and the linker histone have very selective interactions with each other (Fig. 2A) (Arents et al. 1991; Arents and Moudrianakis 1993). Our most detailed understanding of nucleosomal architecture and construction has relied on in vitro experiments that have attempted to reconstruct nucleosomes with purified histones. These experiments have been informative, although they involve dialysis from high salt to low salt concentrations and do not employ the molecular chaperones used in vivo (see below). The central "kernel" of the nucleosome is made up of two heterodimers of histones H3 and H4. Only when this "tetramer" is bound to DNA can two heterodimers of H2A and H2B bind to complete assembly of the histone octamer (Hayes et al. 1991). One heterodimer of H2A and H2B binds to either side of the histone tetramer in an interaction dependent on both protein-protein and protein-DNA contacts. Only when the complete octamer of core histones has assembled on DNA can a single molecule of linker histone be stably bound (Fig. 2B) (Hayes and Wolffe 1993). The exact position of the linker histone within the nucleosome is currently the subject of controversy (Pruss et al. 1995); however, in the one case in which it has been mapped within a specific nucleosome, it occupies an asymmetric position within the nucleosome (Fig. 2C) (Hayes et al. 1994).

In vivo a comparable assembly of the nucleosome occurs. A tetramer (H3, H4)$_2$ and a dimer (H2A, H2B) are stable at physiological ionic strengths (Eickbusch and Mondrianakis 1978). However, they will not associate together in the absence of DNA under physiological conditions. The tetramer (H3, H4)$_2$ must again associate with DNA on newly replicated DNA before H2A and H2B can complete the nucleosome core (Almouzni et al. 1990a). Histone H1 is the last protein to be stably sequestered, completing the nucleosome (Fig. 3) (Worcel et al. 1978). I expand on the details of replication-coupled chromatin assembly later in this chapter.

Chromatin Structure, Replication, and Transcription 275

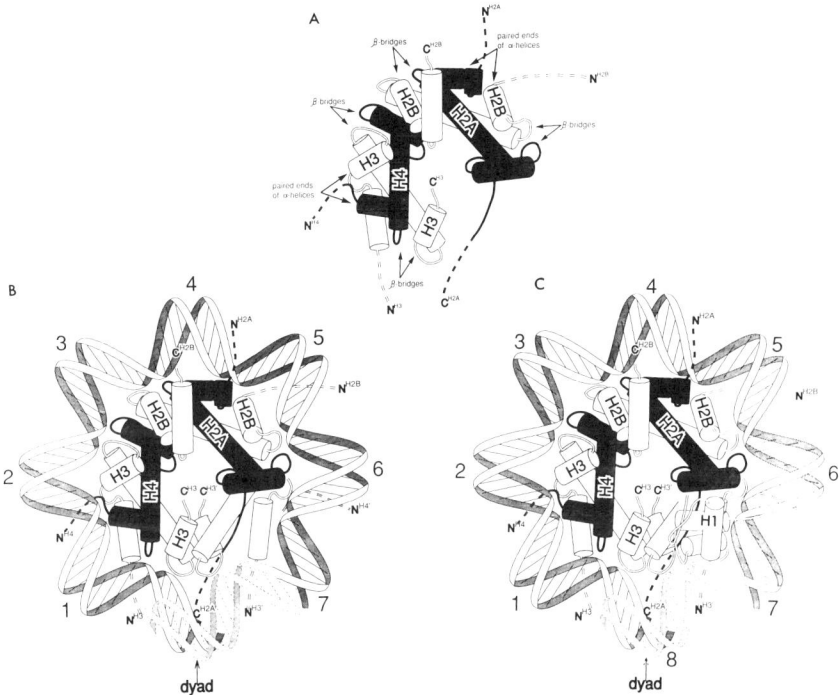

Figure 2 Nucleosomal architecture. (*A*) The histone fold and DNA-binding motifs. The relative juxtaposition of the two histone heterodimers as viewed "from the top" (i.e., along the superhelical axis of the DNA) is shown. The approximate positions of the flexible histone tails are shown by broken lines. Note the six regularly spaced domains (*double arrows*) predicted to be involved in DNA binding. (*B*) A view down the superhelical axis of the nucleosome core. The helical turns of DNA are numbered relative to the dyad axis (0). (*C*) One potential position for the linker histone H1 globular domain within the nucleosome (Pruss et al. 1995).

Jackson (1987, 1990) determined that a substantial fraction of the preexisting octamers associated with DNA within the chromosome in vivo fell apart following replication into dimers (H2A, H2B) and tetramers (H3, H4)$_2$. Tetramers from these preexisting nucleosomes rapidly reassociate with daughter DNA duplexes (Gruss et al. 1993). Newly synthesized dimers (H2A, H2B) can then be sequestered to complete the octamer, mixing old and new histones into a single structure (see Fig. 4). The disruption of preexisting nucleosomal structure at the replication fork, coupled to dissociation of the histones from DNA,

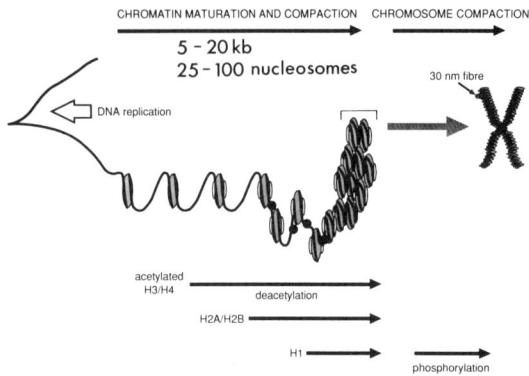

Figure 3 An in vivo pathway for de novo chromatin assembly coupled to replication. Chromatin assembly on nascent DNA. Acetylated histones H3 and H4 (*stippled ellipsoids*) are sequestered by the DNA first, histones H2A and H2B (*open ellipsoids*) follow, and, finally, histone H1 (*dark circle*) binds, stabilizing chromatin folding within the irregular 30-nm fiber. During this progressive assembly of chromatin, DNA is compacted, nucleosome formation leads to a sevenfold compaction of DNA, and the subsequent formation of the 30-nm fiber contributes a further sevenfold compaction. These compactions represent the major topological constraints of DNA in a eukaryotic nucleus. Mature chromatin is predominantly deacetylated. During mitosis, histone H1 is phosphorylated and 30-nm fibers pack together to assemble the mitotic chromosome. As much as 5–20 kb of nascent DNA, including 25–100 nucleosomes, may be present as "immature" chromatin associated with the replication fork at various levels of compaction (see text for details).

strongly suggests that the dispersive segregation of these histones to both daughter DNA duplexes occurs during replication (Cusick et al. 1984; Sogo et al. 1986; Burhans et al. 1991; Krude and Knippers 1991; Randall and Kelly 1992; Sugasawa et al. 1992). Importantly, the incorporation of preexisting histone tetramers (H3, H4)$_2$ into nascent chromatin provides a means of maintaining and propagating a stable state of gene activity. The old H3 and H4 present in the nascent chromatin retain their preexisting posttranslational modification state (Perry et al. 1993). This differs from that of newly synthesized H3 and H4 and can potentially influence subsequent transcription of the associated DNA (see below). The dispersive segregation of "old" histones coupled to maintenance of their preexisting states of modification provides a molecular mechanism whereby an epigenetic imprint might be propagated through replication (Fig. 4, see below).

Chromatin Structure, Replication, and Transcription 277

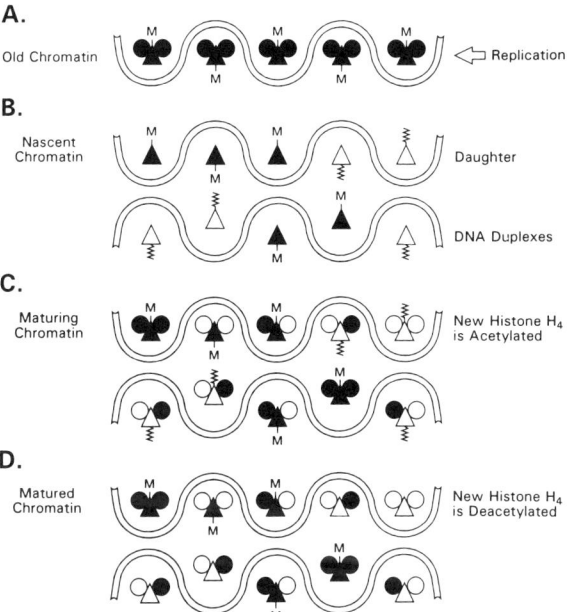

Figure 4 Nucleosome disruption during replication and reassembly following replication. (*A*) Old chromatin consisting of preexisting nucleosomes (histone octamer plus DNA) containing a tetramer (H3, H4)$_2$ (*filled triangle*) and two dimers (H2A, H2B) (*filled circles*). The histones in the tetramer are modified (M). Replication displaces these histones from DNA; the octamer can fall apart into tetramers and dimers. (*B*) Nascent chromatin. Old tetramers associate with both daughter DNA duplexes. Newly synthesized tetramers (*open triangles*) containing diacetylated histone H4 (*zigzag line*) also associate with daughter DNA in a process facilitated by CAF-1. (*C*) Maturing chromatin. Old and new dimers (*open circles*) bind to the tetramers. (*D*) Matured chromatin. New tetramers are deacetylated.

Regulatory Complexes

A special case for a regulatory nucleoprotein complex maintaining association with DNA throughout the cell cycle is the protein assembly that regulates use of an origin of replication itself. Stable association of proteins with an origin through the cell cycle has been established through in vivo footprinting methodologies on the replication origin of Epstein-Barr virus (Hsieh et al. 1993) and on a yeast chromosomal ARS element (Diffley and Cocker 1992). Implicit in the maintenance of these regulatory complexes through S phase is the concept that they duplicate themselves. An attractive mechanism for the maintenance of regulatory complexes through replication requires multiple copies of a given *trans-*

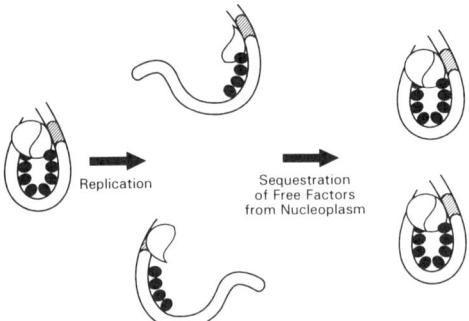

Figure 5 A regulatory nucleoprotein complex could make use of multiple protein-DNA interactions to maintain integrity through replication. Following replication, proteins partition to daughter DNA duplexes. Free factors are then sequestered from the nucleoplasm to reassemble two daughter complexes.

acting factor to bind to the regulatory DNA sequences (Fig. 5) (Brown 1984). This could be determined by sequence or structural selectivity. If the preexisting multimeric protein complex is split during replication, copies of the *trans*-acting factors could be segregated to both daughter DNA duplexes. These *trans*-acting factors could then either directly sequester other factors from the nucleoplasm making use of protein-protein interactions, or they could maintain the regulatory DNA sequences accessible in the face of ongoing chromatin assembly, such that when other factors became available they could bind to DNA. Structurally driven protein association is consistent with the maintenance of DNA distortion throughout the cell cycle at the Epstein-Barr viral origin (Hsieh et al. 1993).

In contrast to origin complexes, the basal transcriptional machinery appears to be removed from promoter elements by the passage of a replication fork (Wolffe and Brown 1986). Replication is found to be dominant to the transcription process, and a direct consequence of replication fork progression through an active 5S rRNA gene is the displacement of transcription factors. Several correlations from in vivo work support the generality of this observation. There is a clear antagonism between transcription and replication on efficiently replicating SV40 DNA molecules (Lebkowski et al. 1985; Lewis and Manley 1985). Replication forks invade the transcriptionally active ribosomal RNA genes in yeast (Saffer and Miller 1986; Lucchini and Sogo 1994). Thus, replication apparently resets the transcriptional status of a chromosome to "ground zero." The component protein molecules that determine transcriptional activity have to reassemble regulatory complexes de novo on the daughter DNA duplexes. This reassembly occurs not on naked DNA, but on a nascent chromatin template.

Chromatin Assembly Has Replication-dependent and -independent Pathways

Replication-independent Pathways

Early work on physiological chromatin assembly pathways made use of cell-free preparations from *Xenopus* oocytes and eggs (Laskey et al. 1978; Glikin et al. 1984). More recently, extracts of *Drosophila* embryos have been used with similar results (Becker and Wu 1992). For both systems, chromatin assembly on nonreplicating DNA is relatively slow, taking several hours to assemble nucleosomes to a physiological density (one nucleosome per 180–200 bp). This contrasts with the rapid assembly of chromatin in vivo during early embryogenesis in *Xenopus* and *Drosophila*, where entire cell cycles take only 30 minutes and 10 minutes, respectively. Thus, the molecular mechanisms that mediate chromatin assembly in the absence of DNA replication have questionable physiological relevance. Nevertheless, these systems have provided useful information on the biochemistry of the assembly process.

In *Xenopus* oocytes, histones are synthesized under the control of distinct regulatory mechanisms that operate outside of S phase. Tetramers (H3, H4)$_2$ are stored in a complex with the molecular chaperone N1/N2 (Kleinschmidt et al. 1986). Dimers (H2A, H2B) are stored in a complex with the chaperone nucleoplasmin (Dilworth et al. 1987). Both chaperones exchange histones onto DNA at physiological ionic strength. N1/N2 must function before nucleoplasmin to assemble a nucleosome (Kleinschmidt et al. 1990). During normal development, nucleoplasmin has a specialized role in the remodeling of *Xenopus* sperm chromatin, where it facilitates the exchange of sperm-specific basic proteins for histones H2A and H2B (Philpott and Leno 1992). Nucleoplasmin and N1/N2 allow large amounts of histones to be stably sequestered in the *Xenopus* oocyte and egg; however, a role for these proteins in directly mediating chromatin assembly during early embryogenesis remains to be established.

Replication-dependent Pathways

In vivo in normal somatic cells, the vast bulk of the histone proteins are synthesized during S phase. These histones are immediately assembled onto nascent DNA at the replication fork (Ruiz-Carrillo et al. 1975; Jackson et al. 1976). Stillman (1986) discovered that the chromatin assembly process is coupled to replication. The molecular chaperone mediating the process is chromatin assembly factor 1 (CAF-1), which requires ongoing DNA replication to function (Smith and Stillman 1989). CAF-1 directs the association of the tetramer $(H3, H4)_2$ with replicating DNA. Dimers (H2A/H2B) then bind in a CAF-1-independent process to complete the histone octamer (Smith and Stillman 1991; see also Fotedar and Roberts 1989). CAF-1 requires a modified tetramer $(H3, H4)_2$ from the cytosol of human cells in order to function (Kaufman and Botchan 1994). This is potentially a key regulatory event in distinguishing the biochemistry of replication-dependent and -independent chromatin assembly pathways. It is possible that the phosphorylation and diacetylation of histone H4 coupled to its synthesis (Ruiz-Carrillo et al. 1975; Jackson et al. 1976; Dimitrov et al. 1994) may be necessary for chromatin assembly. Whether CAF-1 has specific interactions either with highly modified H4 and/or with the replication machinery itself are important questions yet to be resolved.

Almouzni and colleagues (Almouzni and Méchali 1988a,b; Almouzni et al. 1990b, 1991) established that replication-coupled pathways of chromatin assembly also exist in *Xenopus*. However, the molecular chaperones that couple replication to chromatin assembly, such as the CAF-1 found in somatic cells, remain to be defined. These replication-dependent pathways direct the efficient assembly of nucleosomes both in vitro and in vivo with kinetics that could easily accommodate a cell cycle duration of 30 minutes (Almouzni and Méchali 1988a; Almouzni et al. 1990b; Almouzni and Wolffe 1993). The mechanism of enhanced assembly involves both the rapid deposition of the histone tetramer (H3, $H4)_2$ and facilitation of the subsequent deposition of dimers (H2A, H2B) (Almouzni et al. 1990b). Similar results consistent with a facilitated two-step assembly of chromatin have been obtained in mammalian systems (Gruss et al. 1990).

The de novo assembly of chromatin on replicating templates in vitro provides a useful independent confirmation of earlier work on the staged assembly of chromatin during S phase in vivo. As discussed earlier, DNA replication disrupts preexisting nucleoprotein structures within the chromosome. Histones that are displaced during replication reassociate with newly synthesized DNA, but do so randomly on both daughter DNA duplexes. A consequence of this segregation is that nascent chromatin has a 50% enrichment of preexisting histones. The remainder of the histones incorporated into chromatin are newly synthesized. Radiolabeling of these newly synthesized histones has allowed the kinetics of their incorporation into chromatin and subsequent modification to be determined.

Newly synthesized and preexisting histone tetramers (H3, H4)$_2$ associate with nascent DNA (Worcel et al. 1978; Jackson 1987, 1990); this is followed over the space of several minutes by the sequestration of both preexisting and newly synthesized histone dimers (H2A, H2B). Thus, the majority of nucleosomes behind a replication fork are hybrids of both old and new core histones. Finally, a mixture of newly synthesized and preexisting histone H1 stably associates with the nascent chromatin.

The overall process of chromatin maturation as assayed by nuclease sensitivity requires as long as 10–20 minutes in a rapidly proliferating mammalian cell (Cusick et al. 1983). Assuming a rate of replication fork movement of 0.5–1 kb of DNA per minute, this implies that 25–100 nucleosomes are present on both of the nascent DNA duplexes as "immature" chromatin during S phase (see Fig. 3). The initial rapid deposition of old and new histones H3 and H4 on newly synthesized DNA reflects the nuclease-sensitive stage, whereas the subsequent deposition of histone dimers (H2A, H2B) and histone H1 correlates with the appearance of regular nucleosomal arrays and nuclease resistance (Smith et al. 1984). The sequential sequestration of histones is clearly once again related to the structure of the nucleosome, since the tetramer (H3, H4)$_2$ forms the core of the structure, whereas histones H2A and H2B bind at the periphery of the nucleosome, and histone H1 can only associate in its proper place after two turns of DNA are wrapped around the core histones (Fig. 2) (Hayes et al. 1991; Hayes and Wolffe 1993).

Newly synthesized histone H4 is phosphorylated and acetylated in the amino-terminal tail domain (Ruiz-Carillo et al. 1975; Jackson et al. 1976). Approximately 30 minutes after deposition during chromatin assembly, the diacetylated H4 is deacetylated to its mature form. If H4 deacetylation is inhibited, chromatin never achieves the nuclease resistance of bulk chromatin, indicative of the formation of stable higher-order structures. Histone H1 may be less efficiently incorporated into chromatin containing acetylated H4 (Perry and Annunziato 1989; but see Ura et al. 1994). Thus, histone diacetylation is likely to maintain nascent chromatin in a structure that is more accessible to other DNA-binding proteins.

In summary, chromatin assembly in vivo is coupled to replication, most probably through the activity of specific molecular chaperones such as CAF-1. Nucleosome assembly occurs in stages and involves transient posttranslational modifications of core histones synthesized during S phase (Figs. 3, 4).

Epigenetic Mechanisms: The Assembly of Active and Repressed Transcriptional States

In vivo experiments using *Saccharomyces cerevisiae* suggest that replication disassembles repressed chromatin states and facilitates the access of *trans*-acting factors to DNA (Aparicio and Gottschling 1994). Other experiments using yeast suggest that replication has an essential role in facilitating the repression of specific genes (Miller and Nasmyth 1984). We have discussed how biochemical experiments indicate that replication introduces a dynamic aspect to chromosomal structure, both directing the disassembly of preexisting structures and facilitating the assembly of nucleosomes. A central issue in gene regulation is how the assembly of nucleoprotein structures following replication can maintain or alter states of potential transcriptional activity.

Repression

Replication and transcription are most clearly seen to be linked in yeast. Components of the yeast origin recognition complex (ORC) regulate both the initiation of replication within the chromosome and the repression of transcription within the same chromosomal domain (Bell et al. 1993). The molecular mechanisms responsible for the repression of transcription directed by ORC are unknown. Two possible explanations are (1) that the ORC compartmentalizes adjacent chromatin into a transcriptionally incompetent environment within the nucleus, or (2) that the ORC influences the type of chromatin assembled adjacent to it. The ORC complex may be a greatly streamlined version of the replication factories of larger eukaryotes (Cook 1991). These replication factories represent special nuclear compartments at which proteins involved in the replication process are sequestered. It is possible that a gene adjacent to the origin is directed by the ORC to reside in a replication-competent but transcriptionally incompetent environment. Alternatively, if replication itself is essential for transcriptional repression (Miller and Nasmyth 1984; Laurensen and Rine 1992), then the coupling of chromatin assembly to the replication process could contribute to repression. Preexisting transcriptionally active complexes would be displaced by the replication fork. *trans*-Acting factors would then have to compete for assembly against the deposition of histones. In vivo and in vitro experiments in *Xenopus* demonstrate that the coupling of nucleosome assembly to replication can very effectively repress basal transcription (Almouzni et al. 1990a; Almouzni and Wolffe 1993). As discussed earlier, the ORC complex provides one biological example of the maintenance of sequence-specific or structure-dependent protein-DNA interactions through the replication process. However, since the ORC also serves to initiate the replication process, maintenance of the ORC may occur under circumstances distinct from the transcription complexes or chromatin structures that are exposed to the fully assembled replication-elongation complex.

We have discussed how the histones already on the template during replication are segregated randomly to the daughter DNA duplexes, but within the context of small groups of nucleosomes. This maintenance of histone modification states potentially influences *trans*-acting factor access to DNA. Moreover, if proteins that modify the subsequent folding of nucleosomal arrays or that modify histones themselves, for example, by acetylation or deacetylation, are also partitioned in this way, the properties of a chromatin domain might be stably propagated. For example, histone H4 acetylation may interfere with the association of histone H1 with chromatin (Perry and Annunziato 1989). Histone H1 is known to repress specific genes in vivo (Bouvet et al. 1994). Other proteins that might recognize properties of the "old" histones within nascent chromatin include the chromodomain proteins that initiate the formation of heterochromatin. These are also good candidates for propagating preexisting states of chromatin-mediated transcriptional repression (Fig. 6).

Activation

In vitro experiments using cell-free preparations of *Xenopus* eggs indicate that stable states of gene activity can be propagated in a nuclear environment (Wolffe 1993; Barton and Emerson 1994). How might this occur?

The simplest situation leading to continued gene activity would be the case in which a superabundance of transcription factors specific for a given gene was available within the nucleus throughout the cell cycle, including S phase. The factors would always be able to bind to their regulatory elements should they become accessible, recruiting the basal transcriptional machinery to the nascent promoter DNA, and thereby preventing histones or other proteins from binding to the TATA homology. Several features of nascent chromatin facilitate the association of transcription factors (Wolffe 1991). For example, the complex of the 5S rRNA gene with the tetramer $(H3, H4)_2$ is not repressive to transcription (Wolffe 1989; Tremethick et al. 1990; Almouzni et al. 1991), whereas the complete octamer of core histones $(H2A, H2B, H3, H4)_2$ is repressive at high densities of octamers bound to DNA (Clark and Wolffe 1991; Hayes and Wolffe 1992). Moreover, acetylation of the core

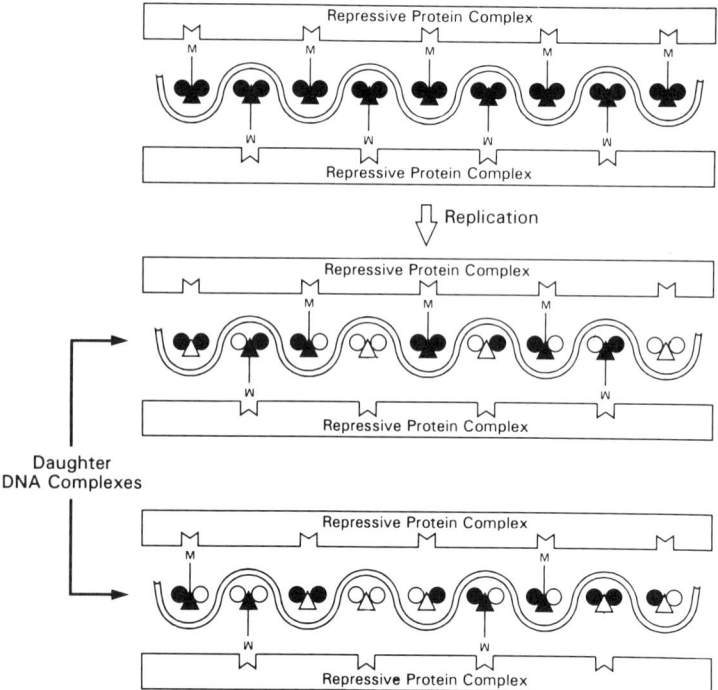

Figure 6 Preexisting histone modifications could provide an epigenetic imprint. A repressive protein complex (e.g., containing chromodomain proteins) recognizes a histone modification (M). Following replication, modified histones are segregated to both daughter DNA duplexes sufficient to sustain interaction with the repressive protein complex.

histones facilitates transcription factor access to DNA even when the complete octamer is bound (Lee et al. 1993). The histone tetramer (H3, H4)$_2$ recognizes the DNA sequences that position the nucleosome containing the 5S rRNA gene (Hayes et al. 1991), hence it is probable that the formation of a specific chromatin structure also has a role in allowing transcription factors access to the template. Thus, following replication, it is probable that sequence-specific DNA-binding proteins have an opportunity to reassociate with daughter DNA molecules (Fig. 7). Replication might under certain circumstances facilitate gene activation (Enver et al. 1988; Wilson and Patient 1993; Aparicio and Gottschling 1994).

The regulated activity of a transcription factor, such that it becomes able to bind to DNA or to function during S phase, could lead to transcription activation in a way that is replication-dependent. In a developmental context, this event might be coupled to a particular embryonic

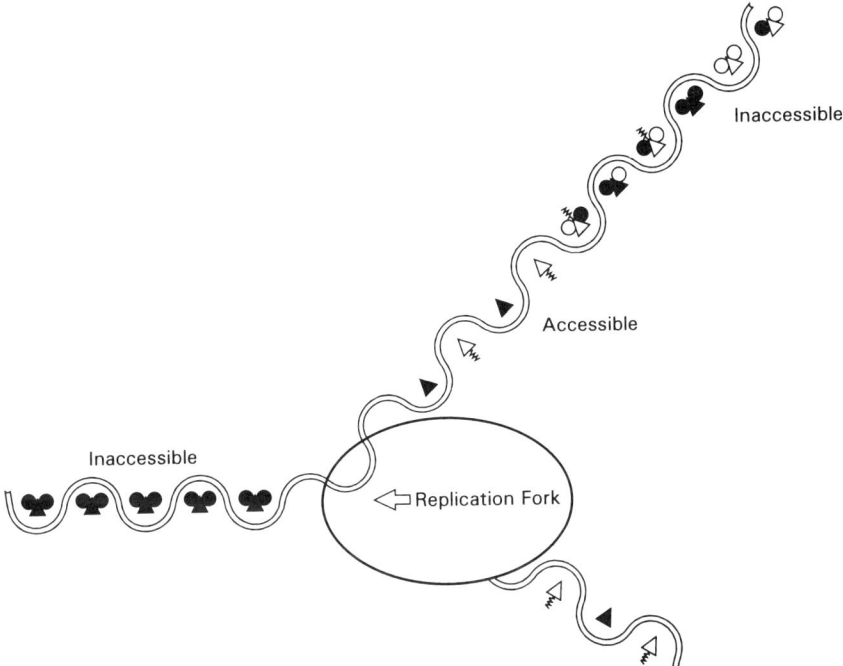

Figure 7 Replicative disruption of preexisting chromatin structures provides a window of opportunity for transcription factors to program genes. Nucleosome assembly is represented as in Fig. 4. The accessibility of nascent, maturing, and mature chromatin to *trans*-acting factors is indicated.

cleavage cycle or to a regulated period of cell division. For example, in *Caenorhabditis elegans* and the sea urchin, replication events are correlated with changes in the commitment of cells to a particular developmental fate (Mita-Miyazawa et al. 1985; Edgar and McGhee 1988). Similar changes can occur in differentiated cells that express one set of specialized genes and that can switch to another program of gene expression only after one or more cell divisions (e.g., Wolffian regeneration of the lens; Takata et al. 1964). However, replication events are not necessarily essential for changing gene expression within a particular cell (Chiu and Blau 1984; Blau et al. 1985). This is not surprising, since chromatin structure is not completely inert in vivo. Histones H2A, H2B, and H1 are known to exchange with a pool of free histones in a cell (Louters and Chalkley 1985). Complexes of DNA with only histones H3 and H4 therefore exist for a limited amount of time. However, a comparison of the rate and efficiency of gene activation in the presence or absence of cell proliferation has not yet been made.

The maintenance of specific transcription factor-DNA interactions through replication, as discussed earlier for the ORC, might be facilitated by considering the promoter, the enhancer, and locus control regions not as separate entities, but as contributory components to a single structure (Wolffe 1990). This could be achieved through protein-protein interactions between the distinct nucleoprotein complexes assembled at each regulatory element. One reason for the separation of these regulatory elements over extensive distances may be that any single structure might be independently disrupted by DNA replication, while the other would remain intact. If protein binding to one sequence element influences the binding of proteins to the other, then the intact nucleoprotein complex might facilitate the re-formation of the disrupted one.

Replication Timing

Chromatin organization outside the ORC may also have significance for the initiation of replication and the timing of this initiation in S phase. If replication disrupts both active and repressed chromatin structures, then the entire nucleus has to be remodeled after each replication event. I have suggested a means of accomplishing this remodeling; however, the re-formation of nuclear structures has other implications. If there are limiting transcription factors available in a cell, then a gene that is replicated early in S phase has more opportunity for the assembly of an active transcription complex than a gene that replicates late. This is simply because the gene that replicates early is available for transcription factors to bind before all of the early-replicating portion of the genome has sequestered these factors. A late-replicating gene therefore experiences a relative deficiency in factor availability (Gottesfeld and Bloomer 1982; Wormington et al. 1982). Conversely, it is also possible that the type of chromatin assembled early in S phase is more accessible to transcription factors than chromatin assembled late in S phase. For example, early-replicating chromatin may sequester histones that are more highly acetylated and, consequently, more accessible to the transcription factors that maintain continued transcription activity. Transcriptionally active genes replicate early in S phase (Goldman et al. 1984; Gilbert 1986; Wolffe 1993). The reason for this early replication is unknown, but possibilities include the local disruption of chromatin structure by transcription complexes, such that the DNA within those chromatin domains becomes more accessible to the replication machinery (Wolffe and Brown 1988). Many transcription factors may also be replication factors (DePamphilis 1988, 1993); consequently, local concentrations of transcription factors may favor the assembly of replication initiation complexes.

The issue very much is one of which came first: the chicken or the egg, or both? It is possible to argue that active transcription complexes open chromatin to admit replication factors, or, alternatively, that these sites are replicated first and are thus more accessible to transcription factors. Whether either or both of the much discussed mechanisms operate in vivo remains to be established.

CURRENT PROBLEMS AND FUTURE PROSPECTS

Chromatin structure is now realized to reflect a dynamic interaction between the many protein complexes that both organize DNA and fulfill regulatory roles. A much simplified picture suggests that replication disrupts local chromatin structures that preexist on the chromosome before replication. The subsequent reassembly of the nucleosome necessitates a staged process using modified histones that might be more accessible to transcription factors. This would provide a window of opportunity for reestablishing particular states of transcriptional activity. On a more global scale (>1–2 kb), chromatin proteins that retain a particular modification (e.g., acetylation) or that cooperatively influence chromosome structure toward activation or repression could provide an imprint on chromatin activity through DNA replication and chromosome duplication. Replication is established as having a major impact on preexisting nucleoprotein structures and a major role in their reassembly. Although significant attention has been given to the enzymology of the duplication of DNA, relatively little progress has been made concerning the enzymology of chromosomal duplication. The molecular mechanisms of chromatin assembly are not defined in any detail. The definition of molecular chaperones such as CAF-1 is a major advance; however, how CAF-1 functions is unknown. Does CAF-1 have a catalytic or structural role? Does it interact with the elongation complex? What are the special features of the histones that allow CAF-1 to utilize them for nucleosome assembly? On a more mundane level, we do not know the precise sequences or structures of the histone proteins necessary for chromatin assembly. The enzymes that transform nascent chromatin into a mature structure are yet to be defined at the molecular level. How mature chromatin is recognized by other proteins that influence states of gene repression, such as the chromodomain proteins, is unknown.

At this time, only the simple 5S rRNA genes of *Xenopus* have been extensively analyzed with respect to the significance of intermediates in chromatin assembly for the capacity to program genes for future transcription. This work is greatly in need of extension to a broader spectrum of eukaryotic genes. Preliminary work in *Drosophila* is consistent with a transition from programmable to stable repressed states as chromatin matures (Kamakaka et al. 1993). Caution must be used with many of the interpretations concerning the general impact of chromatin structure on transcription, since in vitro chromatin assembly systems currently make use of oocyte or embryonic extracts. These systems contain histone variants or modifications not found in normal somatic cells (Dimitrov et al. 1994). It is to be hoped that chromatin assembly systems coupled to replication that employ biochemically defined histones from normal somatic cells will be developed.

The impact of DNA replication on gene expression is readily analyzed through yeast genetics (Miller and Nasmyth 1984; Aparicio and Gottschling 1994; Laurenson and Rine 1992); however, homologous biochemical systems are currently lacking to test the many hypotheses proposed to explain the phenomena observed. Much progress in the biochemistry of yeast replication can be anticipated. The further reconstruction of determinative events in development will require continued consideration of the fate of regulatory nucleoprotein complexes during replication (Diffley and Cocker 1992). This is an important focus for future research. At a biochemical level, in vitro systems capable of maintaining states of gene expression through replication offer considerable promise (Wolffe 1993; Barton and Emerson 1994). The future clearly has the exciting prospect of understanding and thus reconstructing chromosomal duplication in all its complexity at a molecular level.

REFERENCES

Almouzni, G. and M. Méchali. 1988a. Assembly of spaced chromatin promoted by DNA synthesis in extracts from *Xenopus* eggs. *EMBO J.* **7:** 664–672.

———. 1988b. Assembly of spaced chromatin: Involvement of ATP and DNA topoisomerase activity. *EMBO J.* **7:** 4355–4365.

Almouzni, G. and A.P. Wolffe. 1993. Replication coupled chromatin assembly is required for the repression of basal transcription *in vivo. Genes Dev.* **7:** 2033–2047.

Almouzni, G., M. Méchali, and A.P. Wolffe. 1990a. Competition between transcription complex assembly and chromatin assembly on replicating DNA. *EMBO J.* **9:** 573–582.

———. 1991. Transcription complex disruption caused by a transition in chromatin structure. *Mol. Cell. Biol.* **11:** 655–665.

Almouzni, G., D.J. Clark, M. Méchali, and A.P. Wolffe. 1990b. Chromatin assembly on replicating DNA *in vitro. Nucleic Acids Res.* **18:** 5767–5774.

Aparicio, O.M. and D.E. Gottschling. 1994. Overcoming telomeric silencing: A *trans*-activator competes to establish gene expression in a cell cycle dependent way. *Genes Dev.* **8:** 1133–1146.

Arents, G. and E.N. Moudrianakis. 1993. Topography of the histone octamer surface:

Repeating structural motifs utilized in the docking of nucleosomal DNA. *Proc. Natl. Acad. Sci.* **90:** 10489-10493.

Arents, G., R.W. Burlingame, B.W. Wang, W.E. Love, and E.N. Moudrianakis. 1991. The nucleosomal core histone octamer at 3.1Å resolution: A tripartite protein assembly and a left-handed superhelix. *Proc. Natl. Acad. Sci.* **88:** 10148-10152.

Barton, M.C. and B.M. Emerson. 1994. Regulated expression of the β-globin gene locus in synthetic nuclei. *Genes Dev.* **8:** 2453-2465.

Becker, P.B. and C. Wu. 1992. Cell-free system for assembly of transcriptionally repressed chromatin from *Drosophila* embryos. *Mol. Cell. Biol.* **12:** 2241-2249.

Bell, S.P., R. Kobayashi, and B. Stillman. 1993. Yeast origin recognition complex functions in transcription silencing and DNA replication. *Science* **262:** 1844-1849.

Blau, H.M., G.K. Pavlath, E.C. Hardeman, C.P. Chiu, L. Silberstein, S.G. Webster, S.C. Miller, and C. Webster. 1985. Plasticity of the differentiated state. *Science* **230:** 758-766.

Bouvet, P., S. Dimitrov, and A.P. Wolffe. 1994. Specific regulation of chromosomal 5S rRNA gene transcription *in vivo* by histone H1. *Genes Dev.* **8:** 1147-1159.

Brown, D.D. 1984. The role of stable complexes that repress and activate eukaryotic genes. *Cell* **37:** 359-365.

Burhans, W.C., L.T. Vassilev, J. Wu, J.M. Sogo, F. Nallaseth, and M.L. DePamphilis. 1991. Emetine allows identification of origins of mammalian DNA replication by imbalanced DNA synthesis, not through conservative nucleosome segregation. *EMBO J.* **10:** 4351-4360.

Chiu, C.P. and H.M. Blau. 1984. Reprogramming cell differentiation in the absence of DNA synthesis. *Cell* **37:** 879-887.

Clark, D.J. and A.P. Wolffe. 1991. Superhelical stress and nucleosome mediated repression of 5S RNA gene transcription *in vitro*. *EMBO J.* **10:** 3419-3428.

Cook, P.R. 1991. The nucleoskeleton and the topology of replication. *Cell* **66:** 627-635.

Cusick, M.E., M.L. DePamphilis, and P.M. Wasserman. 1984. Dispersive segregation of nucleosomes during replication of simian virus 40 chromosomes. *J. Mol. Biol.* **178:** 249-271.

Cusick, M.E., K.S. Lee, M.L. DePamphilis, and P.M. Wasserman. 1983. Structure of chromatin at deoxyribonucleic acid replication forks: Nuclease hypersensitivity results from both prenucleosomal deoxyribonucleic acid and an immature chromatin structure. *Biochemistry* **22:** 3873-3884.

DePamphilis, M.L. 1988. Transcriptional elements as components of eukaryotic origins of DNA replication. *Cell* **52:** 635-638.

———. 1993. How transcription factors regulate origins of DNA replication in eukaryotic cells. *Trends Cell Biol.* **3:** 161-167.

Diffley, J.F.X. and J.H. Cocker. 1992. Protein DNA interactions at a yeast replication origin. *Nature* **357:** 169-172.

Dilworth, S.M., S.J. Black, and R.A. Laskey. 1987. Two complexes that contain histones are required for nucleosome assembly *in vitro*: Role of nucleoplasmin and N1 in *Xenopus* egg extracts. *Cell* **51:** 1009-1018.

Dimitrov, S., M.C. Dasso, and A.P. Wolffe. 1994. Remodeling sperm chromatin in *Xenopus laevis* egg extracts: The role of core histone phosphorylation and linker histone B4 in chromatin assembly. *J. Cell Biol.* **126:** 591-601.

Edgar, L.G. and J.D. McGhee. 1988. DNA synthesis and the control of embryonic gene expression in *C. elegans*. *Cell* **53:** 589-599.

Eickbusch, T.H. and E.N. Moudrianakis. 1978. The histone core complex: An octamer assembled by two sets of protein-protein interactions. *Biochemistry* **17:** 4955–4965.

Enver, T., A.C. Brewer, and R.K. Patient. 1988. Role of DNA replication in β-globin gene activation. *Mol. Cell. Biol.* **8:** 1301–1308.

Fotedar, R. and J.M. Roberts. 1989. Multistep pathway for replication dependent nucleosome assembly. *Proc. Natl. Acad. Sci.* **86:** 6459–6463.

Gilbert, D.M. 1986. Temporal order of replication of *Xenopus laevis* 5S ribosomal RNA genes in somatic cells. *Proc. Natl. Acad. Sci.* **83:** 2924–2928.

Glikin, G.C., I. Ruberti, and A. Worcel. 1984. Chromatin assembly in *Xenopus* oocytes: *In vitro* studies. *Cell* **37:** 33–41.

Goldman, M.A., G.P. Holmquist, M.C. Gray, L.A. Caston, and A. Nag. 1984. Replication timing of genes and middle repetitive sequences. *Science* **224:** 686–692.

Gottesfeld, J.M. and L.S. Bloomer. 1982. Assembly of transcriptionally active 5S RNA gene chromatin *in vitro*. *Cell* **28:** 781–791.

Gruss, C., J. Wu, T. Koller, and J.M. Sogo. 1993. Disruption of nucleosomes at replication forks. *EMBO J.* **12:** 4533–4545.

Gruss, C., C. Gutierrez, W.C. Burhans, M.L. De Pamphilis, T. Koller, and J.M. Sogo. 1990. Nucleosome assembly in mammalian cell extracts before and after DNA replication. *EMBO J.* **9:** 2911–2922.

Hansen, J.C. and A.P. Wolffe. 1994. A role for histones H2A/H2B in chromatin folding and transcriptional repression. *Proc. Natl. Acad. Sci.* **91:** 2339–2343.

Hayes, J.J. and A.P. Wolffe. 1992. Histones H2A/H2B inhibit the interactions of transcription factor IIIA with the *Xenopus borealis* somatic 5S RNA gene in a nucleosome. *Proc. Natl. Acad. Sci.* **89:** 1229–1233.

———. 1993. Preferential and asymmetric interaction of linker histones with 5S DNA in the nucleosome. *Proc. Natl. Acad. Sci.* **90:** 6415–6419.

Hayes, J.J., D.J. Clark, and A.P. Wolffe. 1991. Histone contributions to the structure of DNA in the nucleosome. *Proc. Natl. Acad. Sci.* **88:** 6829–6833.

Hayes, J.J., D. Pruss, and A.P. Wolffe. 1994. Contacts of the globular domain of histone H5 and core histones with DNA in a chromatosome. *Proc. Natl. Acad. Sci.* **91:** 7817–7821.

Hsieh, D.-J., S.M. Camiolo, and Y.L. Yates. 1993. Constitutive binding of EBNA1 protein to the Epstein-Barr virus replication origin, oriP, with distortion of DNA structure during latent infection. *EMBO J.* **12:** 4933–4944.

Jackson, V., A. Shires, N. Tanphaichitr, and R. Chalkley. 1976. Modification of histones immediately after synthesis. *J. Mol. Biol.* **104:** 471–483.

Jackson, V. 1987. Deposition of newly synthesized histones: New histones H2A and H2B do not deposit in the same nucleosome with new histones H3 and H4. *Biochemistry* **26:** 2315–2325.

———. 1990. *In vivo* studies on the dynamics of histone-DNA interaction: Evidence for nucleosome dissolution during replication and transcription and a low level of dissolution independent of both. *Biochemistry* **29:** 719–731.

James, T.C. and S.C. Elgin. 1986. Identification of a non histone chromosomal protein associated with heterochromatin in *Drosophila melanogaster* in its gene. *Mol. Cell. Biol.* **6:** 3862–3872.

Kamakaka, R.T., M. Bulger, and J.T. Kadonaga. 1993. Potentiation of RNA polymerase II transcription by Gal4-VP16 during but not after DNA replication and chromatin assembly. *Genes Dev.* **7:** 1779–1795.

Kaufman, P.D. and M.R. Botchan. 1994. Assembly of nucleosomes: Do multiple assembly factors mean multiple mechanisms? *Curr. Opin. Genet. Dev.* **4:** 229–235.

Khochbin, S. and A.P. Wolffe. 1994. Developmentally regulated expression of linker-histone variants in vertebrates. *Eur. J. Biochem.* **225:** 501–510.

Kleinschmidt, J.A., A. Seiter, and H. Zentgraf. 1990. Nucleosome assembly *in vitro*: Separate histone transfer and synergistic interaction of native histone complexes purified from nuclei of *Xenopus laevis* oocytes. *EMBO J.* **9:** 1309–1318.

Kleinschmidt, J.A., C. Dingwall, G. Maier, and W.W. Franke. 1986. Molecular characterization of a karyophilic histone-binding protein: cDNA cloning amino acid sequence and expression of nuclear protein N1/N2 of *Xenopus laevis*. *EMBO J.* **5:** 3547–3552.

Krude, T. and R. Knippers. 1991. Transfer of nucleosomes from parental to replicated chromatin. *Mol. Cell. Biol.* **11:** 6257–6267.

Laskey, R.A., B.M. Honda, A.D. Mills, and J.T. Finch. 1978. Nucleosomes are assembled by an acidic protein which binds histones and transfers them to DNA. *Nature* **275:** 416–420.

Laurensen, P. and J. Rine. 1992. Silencers, silencing and heritable transcriptional states. *Microbiol. Rev.* **56:** 543–592.

Lebkowski, J.S., S. Clancy, and M.P. Calos. 1985. Simian virus 40 replication in adenovirus-transformed human cells antagonizes gene expression. *Nature* **317:** 169–171.

Lee, D.Y., J.J. Hayes, D. Pruss, and A.P. Wolffe. 1993. A positive role for histone acetylation in transcription factor binding to nucleosomal DNA. *Cell* **72:** 73–84.

Lewis, E.D. and J.L. Manley. 1985. Repression of simian virus 40 early transcription by viral DNA replication in human 293 cells. *Nature* **317:** 172–175.

Louters, K. and R. Chalkley. 1985. Exchange of histones H1, H2A, and H2B *in vivo*. *Biochemistry* **24:** 3080–3085.

Lucchini, R. and J.M. Sogo. 1994. Chromatin structure and transcriptional activity around the replication forks arrested at the 3′ end of the yeast rRNA genes. *Mol. Cell. Biol.* **14:** 318–326.

Miller, A.M. and K.A. Nasmyth. 1984. Role of DNA replication in the repression of silent mating type loci in yeast. *Nature* **312:** 247–251.

Mita-Miyazawa, I., S. Ikegami, and N. Saitoh. 1985. Histospecific acetylcholinesterase development in the presumptive muscle cells isolated from 16 cell stage ascidian embryos with respect to the number of DNA replications. *J. Embryol. Exp. Morphol.* **87:** 1–12.

Paro, R. and D.S. Hogness. 1991. The Polycomb protein shares a homologous domain with a heterochomatin associated protein of *Drosophila*. *Proc. Natl. Acad. Sci.* **88:** 263–267.

Perry, C.A. and A.T. Annunziato. 1989. Influence of histone acetylation on the solubility, H1 content and DNaseI sensitivity of newly replicated chromatin. *Nucleic Acids Res.* **17:** 4275–4291.

Perry, C.A., C.D. Allis, and A.T. Annunziato. 1993. Parental nucleosomes segregated to newly replicated chromatin are underacetylated relative to those assembled *de novo*. *Biochemistry* **32:** 13615–13623.

Philpott, A. and G.H. Leno. 1992. Nucleoplasmin remodels sperm chromatin in *Xenopus* egg extracts. *Cell* **69:** 759–767.

Pruss, D., J.J. Hayes, and A.P. Wolffe. 1995. Nucleosomal anatomy—Where are the histones? *BioEssays* **17:** 161–170.

Randall, S.K. and T.J. Kelly. 1992. The fate of parental nucleosomes during SV40 DNA replication. *J. Biol. Chem.* **267:** 14259-14265.

Ruiz-Carrillo, A., L.J. Wangh, and V.G. Allfrey. 1975. Processing of newly synthesized histone molecules. *Science* **190:** 117-128.

Saffer, L.D. and O.L. Miller, Jr. 1986. Electron microscopic study of *Saccharomyces cerevisiae* rDNA chromatin replication. *Mol. Cell. Biol.* **6:** 1147-1157.

Simpson, R.T. 1991. Nucleosome positioning: Occurrence, mechanisms and functional consequences. *Prog. Nucleic Acids Res. Mol. Biol.* **40:** 143-184.

Smith, P.A., V. Jackson, and R. Chalkley. 1984. Two stage maturation process for newly replicated chromatin. *Biochemistry* **23:** 1576-1581.

Smith, S. and B.W. Stillman. 1989. Purificiation and characterization of CAF1, a human cell factor required for chromatin assembly during DNA replication in vitro. *Cell* **58:** 15-25.

———. 1991. Stepwise assembly of chromatin during DNA replication *in vitro*. *EMBO J.* **10:** 971-980.

Sogo, J.M., H. Stahl, T. Koller, and R. Knippers. 1986. Structure of the replicating SV40 minichromosomes: The replication fork, core histone segregation and terminal structures. *J. Mol. Biol.* **189:** 189-204.

Stillman, B.W. 1986. Chromatin assembly during SV40 DNA replication in vitro. *Cell* **45:** 555-565.

Sugasawa, K., Y. Ishimi, T. Eki, J. Hurwitz, A. Kikuchi, and F. Hanaoka. 1992. Nonconservative segregation of parental nucleosomes during SV40 chromosome replication in vitro. *Proc. Natl. Acad. Sci.* **89:** 1055-1059.

Takata, C., J.F. Albright, and T. Yomada. 1964. Lens antigens in a lens regenerating system studied by the immunofluorescent technique. *Dev. Biol.* **9:** 385-397.

Tjian, R. and T. Maniatis. 1994. Transcriptional activation: A complex puzzle with few easy pieces. *Cell* **77:** 5-8.

Tremethick, D., K. Zucker, and A. Worcel. 1990. The transcription complex of the 5S RNA gene, but not transcription factor TFIIIA alone, prevents nucleosomal repression of transcription. *J. Biol. Chem.* **265:** 5014-5023.

Ura, K., A.P. Wolffe, and J.J. Hayes. 1994. Core histone acetylation does not block linker histone binding to a nucleosome including a *Xenopus borealis* 5S rRNA gene. *J. Biol. Chem.* **269:** 27171-27174.

Wilson, A.C. and R.K. Patient. 1993. DNA replication facilitates the action of transcriptional enhancers in transient expression assays. *Nucleic Acids Res.* **21:** 4296-4304.

Wolffe, A.P. 1989. Transcriptional activation of *Xenopus* class III genes in chromatin isolated from sperm and somatic nuclei. *Nucleic Acids Res.* **17:** 767-780.

———. 1990. Transcription complexes. *Prog. Clin. Biol. Res.* **322:** 171-186.

———. 1991. Implications of DNA replication for eukaryotic gene expression. *J. Cell Sci.* **99:** 201-206.

———. 1993. Replication timing and *Xenopus* 5S RNA gene transcription *in vitro*. *Dev. Biol.* **157:** 224-231.

———. 1994a. Transcription: In tune with the histone. *Cell* **77:** 13-16.

———. 1994b. Nucleosome positioning and modification: Chromatin structures that potentiate transcription. *Trends Biochem. Sci.* **19:** 240-244.

Wolffe, A.P. and D.D. Brown. 1986. DNA replication *in vitro* erases a *Xenopus* 5S RNA gene transcription complex. *Cell* **47:** 217-227.

———. 1988. Developmental regulation of two 5S ribosomal RNA genes. *Science* **241:**

1626–1632.
Worcel, A., S. Han, and M.L. Wong. 1978. Assembly of newly replicated chromatin. *Cell* **15:** 969–977.
Wormington, W.M., M. Schlissel, and D.D. Brown. 1982. Developmental regulation of *Xenopus* 5S RNA genes. *Cold Spring Harbor Symp. Quant. Biol.* **47:** 879–884.

10
Roles of Phosphorylation in DNA Replication

Klaus Weisshart[1] and Ellen Fanning[1,2]
[1]Institute for Biochemistry
81375 Munich, Germany

[2]Department of Molecular Biology
Vanderbilt University
Nashville, Tennessee 37235

Phosphorylation of proteins by protein kinases and dephosphorylation by protein phosphatases represents one of the most common, versatile, and perhaps confusing regulatory mechanisms in eukaryotic cells. Many, perhaps most, proteins in the eukaryotic nucleus are phosphoproteins, among them the proteins involved in DNA replication and its control. The importance of protein phosphorylation as a regulatory mechanism lies in its ready response to intracellular or extracellular signaling, its reversibility, and its ability to act as a measuring device to translate gradual changes into a molecular switch thrown at a threshold level of phosphorylation of a key target protein. The net level of protein phosphorylation is determined by the balance between the activities of protein kinases and those of protein phosphatases (for review, see Cohen 1989; Hubbard and Cohen 1993). The activities of kinases and phosphatases themselves are regulated by their own phosphorylation state. To complicate matters further, the number of known protein kinases, phosphatases, and regulatory subunits is growing rapidly, and their specificity is not always predictable from the primary sequence of the substrate protein (see Moreno and Nurse 1990; Cegielska et al. 1994a). Moreover, the effect of phosphorylation on the activity of a given target protein usually depends on the exact sites that are modified and unmodified. Thus, to understand how protein phosphorylation regulates the activity of replication proteins, one must know which sites in a target protein are phosphorylated, how physiological signals affect phosphorylation at each site, and how this phosphorylation influences the protein's activity. Knowledge of protein kinases or phosphatases that act at each phosphorylation site is also useful in elucidating the role of protein phosphorylation in regulating the target protein.

The role of protein phosphorylation in control of the timing of DNA replication in the cell cycle has been studied extensively (for review, see Norbury and Nurse 1992; Coverley and Laskey 1994), and these phosphorylation events, as well as the cyclins, cyclin-dependent kinases, and their effects on substrates such as the retinoblastoma tumor suppressor protein are discussed by Nasmyth (this volume). Here we discuss specific examples of DNA replication proteins whose regulation by phosphorylation has been described in detail, summarize our current understanding of the regulation of several cellular replication proteins by phosphorylation, and, finally, speculate how phosphorylation could control the activity of some DNA replication proteins in the cell cycle.

PROTEINS THAT RECOGNIZE AND ACTIVATE REPLICATION ORIGINS

Simian Virus 40 Large Tumor Antigen

The replication of SV40 DNA is a useful model for eukaryotic DNA replication (see Brush and Kelly; Hassell and Brinton; both this volume). SV40 DNA is double-stranded, circular, and supercoiled. In infected monkey cells and in the virus particle, viral DNA is complexed with cellular histones, forming nucleosomes that resemble those in cellular chromatin (for review, see Fanning and Knippers 1992). A set of ten human replication proteins that is sufficient to replicate SV40 DNA in vitro has been purified and characterized (Waga et al. 1994; see Brush and Kelly; Hassell and Brinton; both this volume). Many of these mammalian proteins are functionally conserved in yeast, and the genes encoding many of them have been shown to be essential for viability in yeast (for discussion of these proteins, see Stillman; Wang; Coen; Hübscher et al.; Borowiec; Nash and Lindahl; all this volume), providing strong support for the validity of the SV40 model system and arguing that mammalian chromosomes are probably replicated by the same set of cellular proteins. Like mammalian chromosomal DNA, SV40 DNA replicates in the nucleus during the S phase in the cell cycle. It is not clear how SV40 DNA replication is limited to the S phase, but it is evident that cell-cycle-dependent protein kinases, phosphatases, and their regulatory subunits play an important role. However, unlike mammalian chromosomes, SV40 chromatin undergoes multiple rounds of replication during S phase, indicating that SV40 somehow circumvents the cellular mechanisms that limit chromosomal replication to a single round per S phase. Thus, viral replication is subject to some of the cellular controls on DNA replication, but not all of them.

The only viral protein needed for SV40 DNA replication in vitro and in cultured monkey cells is the large tumor antigen (T antigen). T antigen is a multifunctional phosphoprotein whose intrinsic biochemical activities include sequence-specific binding to the SV40 origin of DNA replication, ATP binding and hydrolysis, and $3'-5'$ DNA helicase activity (for review, see Fanning 1992; Fanning and Knippers 1992). T antigen resides primarily in the nucleus of infected cells, where it serves to stimulate the cell to progress to the S phase prior to viral DNA replication and controls expression of viral and cellular genes and virion assembly, in addition to participating directly in viral DNA replication.

A basic understanding of the process of SV40 DNA replication and the action of the proteins involved has been developed through work performed using cell-free systems (for review, see Kelly 1988; Stillman 1989; Borowiec et al. 1990; Hurwitz et al. 1990). In the current model, SV40 T antigen binds specifically to sequences in duplex origin DNA and assembles in the presence of ATP into a double hexamer on the origin, causing local distortion or melting in the origin DNA sequences flanking the T-antigen-binding site (see Borowiec, this volume). In the next step of the reaction, the DNA helicase activity of the double hexamers, driven by ATP hydrolysis, bidirectionally unwinds the two parental strands by reeling the DNA through the complex (Dean et al. 1992; Wessel et al. 1992). The single-stranded regions are stabilized by the single-stranded DNA-binding protein RP-A, and the enzyme responsible for initiation of synthesis, DNA polymerase-α:primase (pol-α:primase), joins the protein-DNA complex. The assembly of this complex is directed by specific protein-protein interactions that occur among T antigen, RP-A, and pol-α:primase (Dornreiter et al. 1990, 1992, 1993; Melendy and Stillman 1993; Murakami and Hurwitz 1993; for review, see Diffley 1992). After synthesis of the first primer and Okazaki fragment, replication factor C (RF-C) displaces pol-α:primase from the primer-template junction and recruits proliferating cell nuclear antigen (PCNA). The RF-C/PCNA complex then targets DNA polymerase-δ to the junction, where it takes over leading-strand DNA synthesis (Lee et al. 1991a; Tsurimoto and Stillman 1991a, b, c). On the lagging strand, pol-α:primase initiates a new RNA primer, extends it into a short DNA primer, dissociates and then repeats this process to generate the next DNA primer (see Salas et al., this volume). To extend the DNA primers on the lagging strand into Okazaki fragments, polymerase δ, again with the help of RF-C and PCNA, binds to the termini and extends the fragments until it encounters the previous Okazaki fragment. RNase H, together with maturation factor I (MFI), removes the RNA primer, and ligase I joins the two fragments (Turchi et al. 1994; Waga and Stillman 1994; Waga et al. 1994; see Brush and Kelly, this volume). Topoisomerases I and II release the torsional stress caused by the progression of the replication fork, and topoisomerase II also functions in the resolution of catenated replication products (Yang et al. 1987; Ishimi et al. 1988).

The replication functions of SV40 T antigen are tightly controlled by its phosphorylation state (for review, see Prives 1990; Fanning 1992, 1994). T antigen expressed in mammalian cells is phosphorylated on multiple serine (S) and threonine (T) residues clustered at the carboxyl and amino termini of the protein (Fig. 1A) (Scheidtmann et al. 1991). Mutational analysis of the T antigen phosphorylation sites in cell culture demonstrated that phosphorylation at six of these sites (S106, S112, S639, S677, S679, and T701) was dispensable, whereas T-antigen mutants bearing alanine in place of S120, S123, or T124 were defective

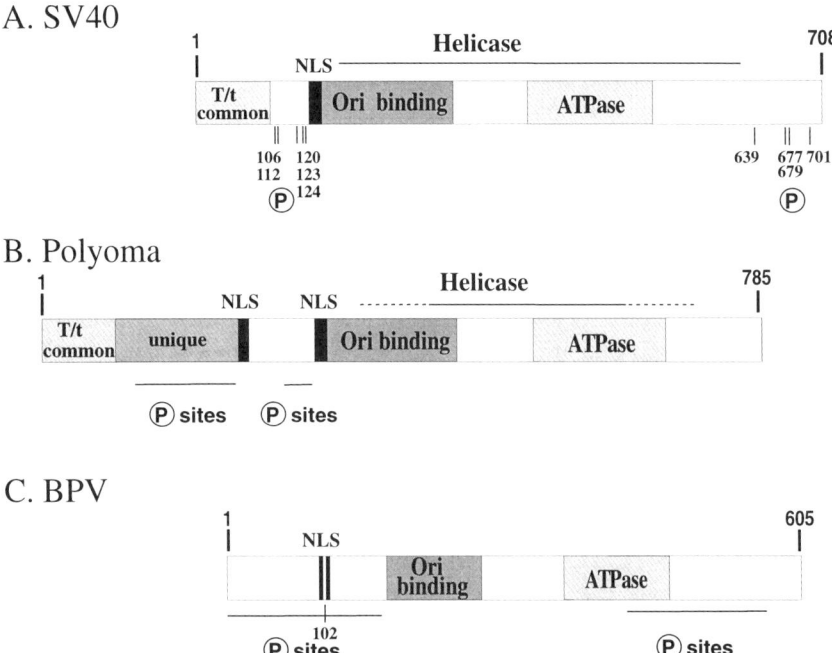

Figure 1 Functional domains and known phosphorylation sites of viral origin recognition proteins. (*A*) SV40 T antigen; (*B*) polyoma T antigen; (*C*) bovine papillomavirus E1 protein. The data were compiled from Fanning (1994); E.H. Wang et al. (1993); Lambert (1991); and unpublished data kindly provided by Mike Lentz. Phosphorylation sites (P) are indicated below each polypeptide by residue numbers (*A*), and regions containing phosphorylation sites by lines (*B, C*).

in viral replication in vivo (Schneider and Fanning 1988), suggesting that modification of these sites is required for SV40 DNA replication in monkey cells.

The role of phosphorylation of T antigen at these three sites in the control of viral DNA replication has been further investigated in cell-free reactions using purified proteins. These studies confirm the requirement for phosphorylation of T124 that was observed in vivo but suggest a quite different role for modification of S120 and S123, namely, that it inhibits viral DNA replication in vitro. For example, the bulk of the T antigen isolated from SV40-infected mammalian cells is hyperphosphorylated and inactive in SV40 DNA replication in vitro (Fig. 2) (Virshup et al. 1989). T antigen expressed in bacteria is not modified at these sites and is also unable to initiate SV40 DNA replication in vitro (Fig. 2) (McVey et al. 1989). However, T antigen expressed in recombinant baculovirus systems has an intermediate level of phosphorylation, with T124 modified in nearly all molecules, but with most serine sites remaining unmodified (Höss et al. 1990); this T antigen is highly active in SV40 replication in vitro (Fig. 2).

In agreement with these findings, treatment of T antigen with protein kinases and phosphatases that alter the phosphorylation state of T124, S120, or S123 activates or inhibits its replication activity in vitro. Phosphorylation of T124 by cdc2-cyclin B activated the replication activity of unphosphorylated bacterially expressed T antigen (McVey et al. 1989). The positive effect of phosphorylation of T124 on the replication activity of T antigen was reversed by treating it with isoforms of protein phosphatase PP2A that specifically targeted T124, i.e., PP2A-D and PP2A-T55 (Cegielska et al. 1994b). Treatment of hyperphosphorylated T antigen from mammalian cells with the catalytic subunit of PP2A (C36, $PP2A_c$) or with an isoform of PP2A (PP2A-T72) stimulated SV40 replication by dephosphorylating S120 and S123 (Virshup et al. 1992; Cegielska et al. 1994b). More recently, an isoform of casein kinase I was shown to phosphorylate these serines in vitro (Cegielska and Virshup 1993; Cegielska et al. 1994a). Kinase treatment of hypophosphorylated replication-active T antigen isolated from baculovirus-infected insect cells led to inactivation of SV40 replication activity. Taken together, these results indicate that SV40 DNA replication in vitro is regulated positively and negatively by the phosphorylation state of T antigen (Fig. 2).

The biochemical mechanism by which phosphorylation of T antigen regulated its replication activity was elucidated by comparing the activities of baculovirus-expressed wild-type and mutant T antigens.

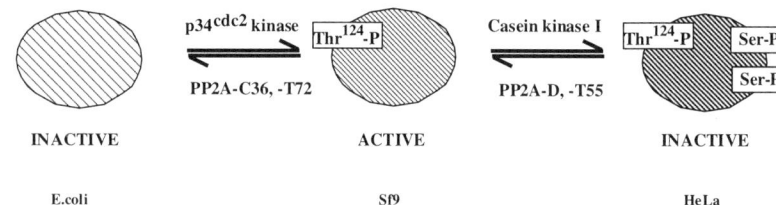

Figure 2 Phosphorylation controls SV40 T-antigen replication activity in vitro. Unphosphorylated T antigen synthesized in bacteria is inactive in replication unless it is first phosphorylated on Thr-124. Hypophosphorylated T antigen expressed in insect cells using baculovirus vectors is active, whereas hyperphosphorylated T antigen expressed in mammalian cells is inactive. Protein kinases and phosphatases (Cegielska et al. 1994a,b) that act on the crucial Thr-124, Ser-120, and Ser-123 sites are indicated. (Reprinted, with permission, from Fanning 1994.)

Phosphorylation at T124 is necessary for the bidirectional unwinding of duplex SV40 origin DNA by T antigen in the initiation reaction (McVey et al. 1993; Moarefi et al. 1993b). In contrast, other enzymatic activities like hexamer assembly, unidirectional DNA helicase, DNA binding, and origin DNA distortion were not markedly influenced by phosphorylation of T124. The specific effect of phosphorylation on bidirectional unwinding appears to result from its ability to stabilize hexamer-hexamer contacts that are necessary for bidirectional, but not unidirectional, DNA unwinding (K. Weisshart and E. Fanning, unpubl.; T. Kelly, pers. comm.). Phosphorylation of T124 also results in cooperative assembly of the two hexamers on the SV40 origin (K. Weisshart and E. Fanning, unpubl.; T. Kelly, pers. comm.).

Interestingly, phosphorylation of T antigen on the inhibitory serine residues also decreased the cooperative assembly of T antigen double hexamers on SV40 origin DNA, blocking bidirectional origin-unwinding activity and viral DNA replication in vitro (Virshup et al. 1992; Cegielska and Virshup 1993; Cegielska et al. 1994a). Thus, T antigen unphosphorylated on T124 or phosphorylated on S120 and S123 could potentially bind to the origin as a double hexamer and distort it locally without initiating bidirectional unwinding and replication until it became phosphorylated on T124 and dephosphorylated on S120 or 123, thereby facilitating interactions between the two hexamers that are required for bidirectional unwinding (Fig. 3). We suggest that the interactions between the hexamers are dynamic, undergoing cycles of association and dissociation as unwinding proceeds (Fig. 3). Although this mechanism

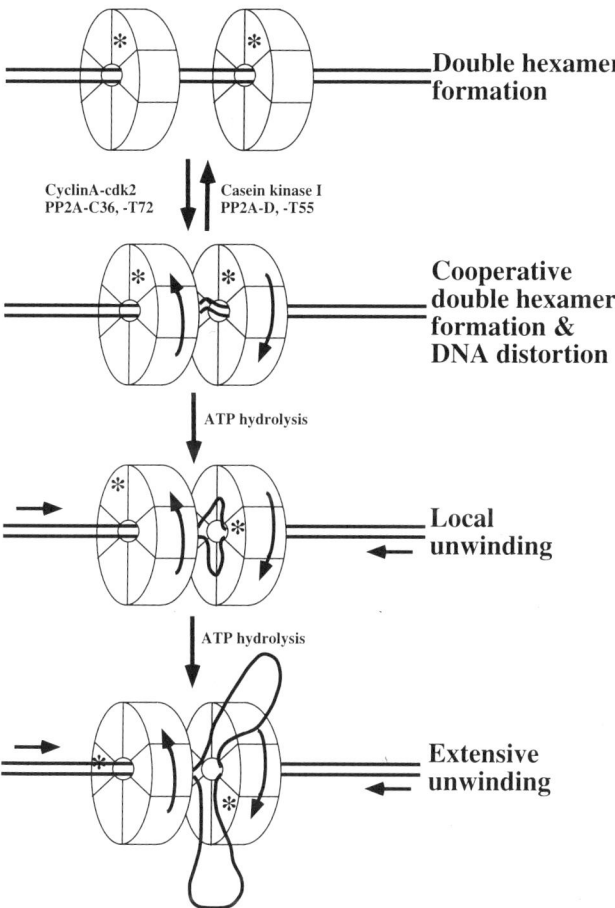

Figure 3 The phosphorylation state of SV40 T antigen governs hexamer-hexamer interactions in bidirectional origin unwinding. Prior to viral DNA replication in infected mammalian cells, T antigen accumulates primarily in hyperphosphorylated forms that can assemble as a double hexamer on the SV40 origin, but the hexamers cannot interact properly with each other to unwind the origin and permit initiation. Replication-active hexamers can be formed upon dephosphorylation of Ser-120 or Ser-123, or by phosphorylation of newly synthesized unphosphorylated T antigen on Thr-124. When a sufficient concentration of this hypophosphorylated form of T antigen is reached in the infected cell, ATP hydrolysis and dynamic hexamer-hexamer interactions lead to bidirectional unwinding at the origin and initiation of replication. Subsequent phosphorylation of Ser-120 and Ser-123 or dephosphorylation of Thr-124 would block reinitiation of replication. (Reprinted, with permission, from Fanning 1994.)

remains speculative, the data demonstrate clearly that the phosphorylation state of T antigen controls SV40 DNA replication both negatively and positively at the level of unwinding of origin DNA. It will be interesting to see whether phosphorylation of cellular origin recognition or unwinding proteins regulates their function in a similar manner.

The significance of T antigen phosphorylation in viral DNA replication in vivo remains for the most part unresolved. Although it is well established that phosphorylation of T124 is essential for SV40 DNA replication in vivo (Schneider and Fanning 1988), this modification may play additional roles in regulation of replication. One attractive model proposes that phosphorylation of T antigen at T124 could link the replication activity of T antigen to the cell cycle (McVey et al. 1989, 1993). T antigen is associated with cdk2-cyclin A in the cell (Adamczewski et al. 1993) and is a much better substrate for this kinase than for cdc2-cyclin B in vitro (C. Voitenleitner et al., unpubl.). cdk2-cyclin A is activated at the G_1-S transition and is found associated with replicating SV40 DNA (Fotedar and Roberts 1991). Together these data suggest a role for cdk2-cyclin A as the best candidate for phosphorylation of T124 in vivo and support the notion that this modification event could link SV40 DNA replication to S phase. However, the phosphate turnover on this site is relatively slow in SV40-infected cells, and most T-antigen molecules carry the modification (Scheidtmann 1986; for review, see Fanning and Knippers 1992). Furthermore, there is no direct evidence to suggest that modification of T124 in SV40-infected or -transformed cells is cell-cycle-dependent. Thus, it remains an open question whether the cell-cycle-dependence of SV40 DNA replication is controlled by this phosphorylation event.

The physiological role of phosphorylation of T-antigen residues S120 and S123 in vivo is even more puzzling. Whereas phosphorylation of S120 and S123 inhibits viral DNA replication in vitro, phosphorylation at these sites is required in vivo (Schneider and Fanning 1988). Reexamination of the kinetics of SV40 DNA replication in cultured cells with the S120A and S123A mutants yielded preliminary evidence that the mutant T antigens did begin viral DNA replication, but only a small amount of DNA accumulated (Moarefi et al. 1993a). To resolve the apparent discrepancy between the in vivo and in vitro results, it has been suggested that phosphorylation of these serine residues is necessary for regulatory functions not detectable in the in vitro replication system (Fanning 1994). Support for this hypothesis comes from the finding that phosphate turnover on these serines is rapid (Scheidtmann 1986), so that the balance between phosphorylation and dephosphorylation on these two serines could determine the fraction of T antigen active in initiation of DNA replication (Fanning 1992). Phosphorylation at these serines may be required in the termination of replication, or it could influence the association of viral DNA with replication foci (Hozak and Cook 1994). These or other mechanisms might reduce the ability of T antigen to sustain multiple rounds of viral DNA replication in vivo, resulting in the defect observed with the S120 and S123 mutants in cell culture but not in vitro (Schneider and Fanning 1988; Moarefi et al. 1993a). Further work is required to resolve this paradox.

The physiological relevance of PP2A and casein kinase I action on T antigen in vivo also remains obscure. Casein kinase I is ubiquitously expressed and not regulated in the cell cycle. In addition, PP2A is found primarily in the cytoplasm rather than in the nucleus. However, one cannot exclude the possibility that the different isoforms of kinase and phosphatase that affect T antigen have different subcellular localizations. Indeed, there is evidence that the activity of PP2A on T antigen is restricted to S phase (Ludlow 1992). Other protein phosphatases that act at the G_1-S border may also target T antigen, such as a mammalian homolog of SIT4 from budding yeast and *Drosophila* (Fernandez-Sarabia et al. 1992; Mann et al. 1993) or a mammalian homolog of the yeast cdc25A phosphatase (Hoffmann et al. 1994; Jinno et al. 1994).

Polyomavirus Large Tumor Antigen

The initiator protein for polyomavirus (PyV) replication is the polyoma large T antigen, Py T, which interacts with specific sequences within the PyV origin of replication (Cowie and Kamen 1984; Hassell and Brinton, this volume). Like its SV40 counterpart, Py T antigen unwinds duplex DNA (Wang and Prives 1991) using its ATPase/helicase function (Seki et al. 1990). Py T antigen is a phosphoprotein with two clusters of phosphorylation sites located at the amino terminus (Fig. 1B) (Hassauer et al. 1986; Bockus and Schaffhausen 1987). No assessment of the importance of single phosphorylation sites has been undertaken, but treatment with calf intestinal phosphatase had differential effects on Py T antigen, depending on the degree of phosphate removal (E.H. Wang et al. 1993). Treatment with limiting amounts of phosphatase stimulated origin binding, whereas incubation with high concentrations of phosphatase had the opposite effect. These data are reminiscent of early results obtained with partially dephosphorylated SV40 T antigen (Grässer et al. 1987) and suggest that the replication activity of PyV T antigen might be regulated through differential phosphorylation. It will be interesting to test whether phosphorylation of Py T regulates initiation of viral DNA replication by controlling the ability of Py T to catalyze bidirectional unwinding of the origin DNA as in the SV40 system.

Papillomavirus E1 Protein

Two bovine papillomavirus (BPV)-encoded proteins are necessary and sufficient for BPV DNA replication in vivo, the 68-kD E1 protein and the 48-kD E2 protein (for review, see Lambert 1991; Ustav and Stenlund 1991; Ustav et al. 1993; Stenlund, this volume). The E2 protein can be dispensable in vitro when high levels of E1 are used (Seo et al. 1993b; Spalholz et al. 1993). All other replication proteins are supplied by the host cell. E1, like SV40 and Py T antigens, is a phosphoprotein that possesses several activities necessary for replication function (Ustav and Stenlund 1991; Seo et al. 1993a; Thorner et al. 1993; Yang et al. 1993; Park et al. 1994; Bonne-Andrea et al. 1995): origin binding activity, polymerase-α binding, ATPase activity, and $3'-5'$ helicase activity (Fig. 1C). E2 protein is a transcriptional activator (Spalholz et al. 1993) and also a phosphoprotein (Bream et al. 1993) that stimulates the replication activity of E1 protein in vitro, at least in part by binding to it (Seo et al. 1993b; Spalholz et al. 1993; Benson and Howley 1995). It has been proposed that the role of E2 is to tether E1 more tightly to the origin and to recruit cellular replication factors, e.g., RP-A and pol-α:primase (Li and Botchan 1993), to the BPV origin.

The facts that BPV DNA is maintained as an episome at a constant copy number in latently transformed cells (Law et al. 1981) and that it replicates during S phase along with the host chromosomal DNA (Gilbert and Cohen 1987) might suggest that the activities of E1 and E2 are regulated in the cell cycle. Since both proteins are phosphoproteins, this regulation could depend on phosphorylation events. The E1 residue T102 near the nuclear localization signal is a substrate in vitro for cdc2-cyclin kinase (Fig. 1C). However, replacing that threonine by isoleucine showed no observable effect on either the nuclear localization of E1 or its efficiency in DNA replication (Lentz et al. 1993). Thus, the functional role, if any, of phosphorylation of the BPV replication proteins remains unknown.

Yeast Origin Recognition Complex

The most likely candidate for the replication initiator in yeast is the origin recognition complex (ORC) (for review, see Bell and Stillman 1992; Diffley 1992; Huberman 1992; Li and Alberts 1992; see Newlon;

Nasmyth; Stillman; all this volume). It consists of six polypeptides with molecular masses of 50, 53, 56, 62, 72, and 120 kD, respectively, that have been termed ORC-1 to ORC-6 according to the decrease in molecular weight. The ORC complex binds specifically to a bipartite site in yeast origins of replication in an ATP-dependent manner (Rao and Stillman 1995; Rowley et al. 1995). Mutations in the gene coding for ORC-2 have shown that ORC is necessary for the initiation of DNA replication and transcriptional silencing at the mating-type locus (Bell et al. 1992; Foss et al. 1993; Micklem et al. 1993; for review, see Newlon 1993; Huberman 1994; Kelly et al. 1994). ORC-6 has been shown to be important for the initiation of DNA replication (Li and Herskowitz 1993).

Evidence has been obtained by in vivo footprinting that ORC is bound to ARS-1 sequences in yeast cells in the same manner as it is in vitro (Diffley and Cocker 1992). Interestingly, ORC remains bound to the origin throughout the cell cycle. These data suggested that ORC could repress replication except in S phase, when it could be activated for replication. ORC-2 and ORC-6 contain multiple potential phosphorylation sites for cyclin-dependent kinases and other kinases (Bell et al. 1992; Li and Herskowitz 1993; Micklem et al. 1993), so that phosphorylation by cell-cycle-regulated kinases may influence ORC activity. However, mutations in the potential sites of phosphorylation by cyclin-dependent kinases have not yet revealed any loss of replication function (J. Diffley, pers. comm.). Interestingly, DBF4, the CDC7 protein kinase regulatory subunit, interacts with the yeast replication origin (Dowell et al. 1994), and activation of CDC7 by DBF4 at the G_1/S-phase transition is important for the initiation of S phase (Jackson et al. 1993; for review, see Newlon 1993; Nasmyth, this volume). Thus, the interaction of DBF4 with ORC-2 may target the CDC7 catalytic subunit to the ORC complex, enabling it to phosphorylate the ORC-2 polypeptide (for review, see Baringa 1994). However, the functional effects of this phosphorylation have not yet been described.

Although the footprint of ARS sequences remains detectable throughout the cell cycle, there is an increase in the protected area during G_1 phase, which is indicative of prereplicative complexes binding in the vicinity of ORC (Diffley et al. 1994). The enlarged footprint may arise through the association of additional proteins with the ORC complex. There is evidence that ORC-6 interacts with CDC6 and CDC46/MCM5 proteins (Li and Herskowitz 1993), which are good candidates for components of the prereplicative complex. CDC6 of budding yeast and its counterpart Cdc18 from fission yeast are important for entry into S phase

(Kelly et al. 1993). The CDC6 gene product shows ATPase activity, but no detectable DNA unwinding activity (Zwerschke et al. 1994). CDC46 also acts at an early stage in DNA replication. The subcellular localization of CDC46 and other members of the MCM 2-3-5 family is regulated in the cell cycle (Hennesey et al. 1991; Y. Chen et al. 1992; Yan et al. 1993; for review, see Tye 1994). These proteins are nuclear only between M and S phase, consistent with the notion that they could participate in the G_1-specific enlargement of the footprint on the origin of replication, but no data are available on this point. One might expect that cell-cycle-dependent phosphorylation events would regulate the functions of the MCM proteins, but again, mutations in the potential sites of phosphorylation of MCM proteins by cyclin-dependent kinases have not affected their function (B. Tye, pers. comm.) (but see section below). It has been speculated based on their amino acid sequences that the MCM proteins may function as ATP-dependent helicases (Koonin 1993).

Xenopus and Mammalian Initiation Proteins

The MCM proteins are found not only in yeast, but also in frogs, mice, and humans (for review, see Tye 1994). Their importance in control of DNA replication is underscored by their recent identification (Chong et al. 1995; Kubota et al. 1995; Madine et al. 1995) as components of a replication licensing factor that was postulated by Blow and Laskey (1988) to limit eukaryotic chromosomal DNA replication to one round per cell cycle (for review, see Coverley and Laskey 1994). Taken together, the data strongly support the notion that a complex of different MCM proteins is required for chromatin replication in the *Xenopus* in vitro system, that their presence in the nucleus before replication is necessary, and that upon replication, they are lost or become easily extractable from the nucleus. Interestingly, replication licensing factor activity is blocked by treatment of *Xenopus* extracts with protein kinase inhibitors, suggesting that phosphorylation is directly or indirectly involved in their function (for review, see Coverley and Laskey 1994; Tye 1994).

A mammalian homolog of MCM3, the P1 protein, was found associated with DNA pol-α:primase (Thömmes et al. 1992b). The nuclear localization of P1 is regulated during the cell cycle like that of its yeast and frog counterparts (Kimura et al. 1994). Nuclear P1 appears to exist in distinct subpopulations: one that is hyperphosphorylated and loosely bound in the nucleus, and another that is underphosphorylated and tightly bound in the nucleus. In early S phase, the tightly bound form of P1 was predominant, but as S phase progressed, the loosely bound P1 gradually increased and became the major form by the end of S phase (Kimura et al. 1994). This behavior would be consistent with a role for P1 in prereplicative chromatin complexes and disassembly of these complexes after cell-cycle-dependent phosphorylation.

OTHER PROTEINS THAT ARE REQUIRED FOR DNA REPLICATION

Pol-α:Primase

Pol-α:primase is composed of four subunits with molecular masses of 180, 68, 58, and 48 kD (p180, p68, p58, and p48) (Thömmes and Hübscher 1990; Wang 1991 and this volume). All the cDNAs encoding the human, mouse, and yeast subunits have been cloned (Lucchini et al. 1987; Wong et al. 1988; Collins et al. 1993; Miyazawa et al. 1993; Stadlbauer et al. 1994). The structure and function of pol-α:primase are highly conserved in eukaryotic organisms (Wang 1991). Pol-α:primase is an essential component of the cellular replication machinery (Pizzagalli et al. 1988; Francesconi et al. 1991), where it functions both in initiation at origins of replication and in the synthesis of Okazaki fragments on the lagging strand of the replication fork (Focher et al. 1988; Prelich and Stillman 1988; Thömmes and Hübscher 1990; Wang 1991). Pol-α:primase also initiates the leading strand at the origin of SV40 replication (Tsurimoto et al. 1990). The p180 subunit harbors the DNA polymerase activity (Pizzagalli et al. 1988; Wong et al. 1988; Nasheuer et al. 1991), the p68 subunit is thought to play a role in the initiation process (Collins et al. 1993; Foiani et al. 1994), and the p58 and p48 subunits comprise the primase activity (Santocanale et al. 1993).

During activation of quiescent mammalian cells, the expression of all subunits is up-regulated. However, in actively cycling cells, gene expression is constitutive and mRNA and protein levels are only marginally enhanced prior to S phase (Wahl et al. 1988; Miyazawa et al. 1993). Thus, if pol-α:primase function is regulated during the cell cycle, it is likely to involve a posttranslational mechanism. Indeed, both the p180 and the p68 subunits are phosphorylated on serine and threonine residues in a cell-cycle-dependent manner (Wong et al. 1986). The p180 subunit is phosphorylated throughout the cell cycle but becomes hyperphosphorylated in the G_2-M transition, whereas the p68 subunit is phosphorylated beginning in S and more heavily at the G_2-M transition (Nasheuer et al. 1991).

Phosphorylation of DNA polymerase-α:primase at different points in the cell cycle has also been studied in budding yeast (Foiani et al. 1995). Both the p180 and p86 (B) subunits are phosphorylated in yeast, but only the B subunit showed a reduced electrophoretic migration upon modification. The kinetics of phosphorylation during the cell cycle revealed that newly synthesized B subunit occurred in an underphosphorylated form during S phase and became modified during G_2. Maternal B subunit was stable and became phosphorylated in early S phase. Both populations of B subunit were dephosphorylated during exit from mitosis (Foiani et al. 1995). The observation that the B subunit is required very early in S phase for initiation, but has no effect on the enzymatic functions of yeast pol-α:primase, suggested that phosphorylation may regulate the activity of the B subunit in initiation of replication (Fig. 4) (Foiani et al. 1995).

The functional relevance of phosphorylation of pol-α:primase has been investigated in vitro. Treatment of pol-α:primase with casein kinase II in vitro phosphorylated the p180 and p58 subunits, but had no influence on either polymerase or primase activity (Podust et al. 1990). Some of the in vivo cell-cycle-dependent phosphorylation sites on the p180 and p68 subunits of mammalian pol-α:primase were shown to be phosphorylated by p34^{cdc2} kinases in vitro, implying these kinases as possible regulators. However, these kinases also had little effect on the enzymatic activities of pol-α:primase; only a slight reduction in affinity for single-stranded DNA was found (Nasheuer et al. 1991).

The functional effect of phosphorylation of pol-α:primase by cyclin-dependent kinases has recently been reexamined using purified proteins. Pol-α:primase expressed in the baculovirus system was modified on the p180 and p68 subunits by cyclin E-, A-, and B-dependent kinases (C. Voitenleitner et al., unpubl.). All of these kinases utilized pol-α:primase approximately equally well as a substrate. Phosphorylation of the enzyme did not affect its polymerase activity, in agreement with published data, although it stimulated primase activity severalfold. In contrast, the ability of pol-α:primase to initiate SV40 DNA replication in vitro in a reaction performed with purified proteins was markedly inhibited after phosphorylation by cyclin-A-dependent kinases. Treatment with cyclin-B-dependent kinase reduced initiation activity more modestly. Cyclin-E-dependent kinase caused no reduction in initiation activity; indeed, a small stimulation was observed in some experiments (C. Voitenleitner et al., unpubl.).

These results raise the question whether phosphorylation of pol-α:primase may influence its interaction with other initiation proteins. Both the p180 and the p68 subunits interact with SV40 T antigen (Dornreiter et al. 1990, 1992, 1993; Collins and Kelly 1991; Collins et al. 1993; Murakami and Hurwitz 1993), and these interactions play a sub-

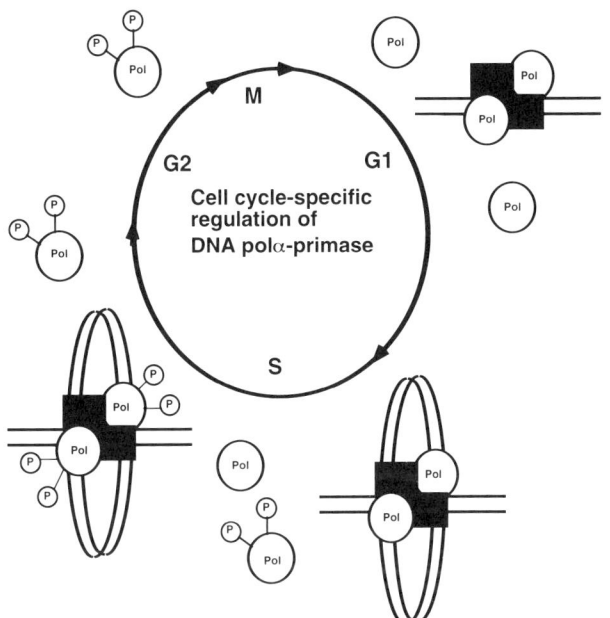

Figure 4 Model for regulation of origin-specific initiation activity of pol-α:primase by cell-cycle-dependent phosphorylation. Hypophosphorylated pol-α:primase that appears in late M or early G_1 (Foiani et al. 1995) is suggested to assemble in a prereplicative complex together with cellular initiation proteins on the cognate origin of DNA replication (Diffley et al. 1994). The complex would remain dormant until activation at the G_1/S border, an event that probably requires cyclin-E-dependent kinase directly or indirectly (Knoblich et al.1994; Jackson et al. 1995; Sauer et al. 1995). Upon initiation, phosphorylation of pol-α:primase is postulated to inhibit its ability to initiate new rounds of replication at an origin. Perhaps phosphorylation of pol-α:primase at the origin is even involved in the transition from initiation to elongation. If cyclin A-dependent kinases, which accumulate as the cell progresses from S through M phase, were responsible for these modifications, the fraction of initiation-active pol-α:primase would be predicted to decline from S to M phase without a concomitant reduction in lagging-strand DNA replication activity. A similar mechanism may regulate the SV40 initiation activity of pol-α:primase in virus-infected cells, thereby facilitating assembly of new virus particles in late S phase and G_2.

stantial role in the initiation of SV40 replication. The ability of phosphorylated pol-α:primase to undergo these interactions has not yet been assayed directly. Intriguingly, however, only a subpopulation of mouse pol-α:primase was associated with Py T antigen, and only this population was able to stimulate viral DNA replication (Moses and

Prives 1994). It will be interesting to investigate the phosphorylation state of these different fractions of pol-α:primase.

Taken together, the results of these studies in different systems with different sets of reagents argue that cell-cycle-dependent phosphorylation of pol-α:primase regulates its function in initiation of DNA replication. Figure 4 depicts a speculative model for this regulation. Since the model makes several predictions that can be tested experimentally, it may be useful in gaining a better understanding of the role of protein phosphorylation in replication.

Replication Protein A

Replication protein A (RP-A), also known as replication factor A (RF-A) and human single-stranded DNA-binding protein (HSSB), was identified as an essential factor for SV40 DNA replication in vitro (Wobbe et al. 1987; Fairman and Stillman 1988; Wold and Kelly 1988). It is the major single-stranded DNA-binding protein in mammalian cells (Seroussi and Lavi 1993) and is highly conserved in all eukaryotes (Wold and Kelly 1988; Brill and Stillman 1989; Mitsis et al. 1993). RP-A is a heterotrimeric protein with subunits of 14, 32, and 70 kD (p14, p32, and p70) (Fairman and Stillman 1988; Wold and Kelly 1988). The cDNAs encoding all three subunits have been cloned from human (Erdile et al. 1990, 1991; Umbricht et al. 1993) and yeast (Heyer et al. 1990; Brill and Stillman 1991) and are essential for cell viability in yeast (Heyer et al. 1990; Brill and Stillman 1991; Longhese et al. 1994).

Human RP-A binds tightly to single-stranded DNA (Wobbe et al. 1987; Fairman and Stillman 1988; Wold and Kelly 1988) with low cooperativity (Kim and Wold 1995) in two distinct complexes, one occupying 8–10 nucleotides, the other 20–30 nucleotides (Blackwell and Borowiec 1994; Kim et al. 1994). Binding affinity to RNA and double-stranded DNA is much lower (Brill and Stillman 1989; Wold et al. 1989; Kim et al. 1992). RP-A has been shown to be associated with an intrinsic DNA-unwinding activity (Georgaki et al. 1992). The nucleic acid-binding properties of the homolog from *Drosophila melanogaster* (dRPA) were reported to be quite similar to those of human RP-A (Mitsis et al. 1993). However, RP-A from *Saccharomyces cerevisiae* (yRPA) was found to be quite different, binding to a site of about 90 nucleotides with high cooperativity (Alani et al. 1992). Yeast RP-A, however, cannot substitute for its human counterpart in SV40 DNA replication in vitro (Brill and Stillman 1991).

The p70 subunit of RP-A contains the intrinsic DNA-binding activity (Wold et al. 1989; Kenny et al. 1990). The p32 and p14 subunits form a stable subcomplex (Henricksen et al. 1994; Stigger et al. 1994) that is important for proper formation of the heterotrimeric complex (Henricksen et al. 1994). The biological functions of the 32-kD and 14-kD subunits are not known. However, monoclonal antibodies directed against the 70-kD, 32-kD, or 14-kD subunits block SV40 DNA replication, suggesting a role for all three subunits in this process (Erdile et al. 1990; Kenny et al. 1990; Umbricht et al. 1993).

RP-A has multiple functions in the replication process (Kenny et al. 1989). During bidirectional unwinding of the SV40 origin of replication by T antigen, it stabilizes the created single-stranded regions. Moreover, it is essential during primosome assembly (Melendy and Stillman 1993) and primer synthesis on the SV40 origin (Schneider et al. 1994), reflecting the specific protein-protein interactions that occur between RP-A, T antigen, and pol-α:primase (Dornreiter et al. 1992, 1993; Nasheuer et al. 1992; Murakami and Hurwitz 1993; K. Weisshart and E. Fanning, unpubl.). RP-A also plays a role in the elongation reaction, where it has been shown to stimulate DNA polymerases α (Kenny et al. 1989; Erdile et al. 1991; Brown et al. 1992), δ (Kenny et al. 1989; Tsurimoto and Stillman 1989; Lee et al. 1991a), and ε (Lee et al. 1991b). In addition, it increases the fidelity of DNA synthesis by pol-α:primase (Carty et al. 1992; Roberts et al. 1993; Suzuki et al. 1994) and stimulates copurifying helicases (Seo et al. 1991; Thömmes et al. 1992a; Georgaki et al. 1994).

Besides its essential functions in DNA replication, RP-A is also required for DNA repair and recombination in vivo (Longhese et al. 1994) and nucleotide excision repair of UV-damaged DNA in vitro (Coverley et al. 1991). It also stimulates in vitro strand-exchange proteins involved in homologous recombination (Heyer et al. 1990; Moore et al. 1991; Alani et al. 1992). RP-A has also been shown to interact with the acidic domains of transcription factors VP16, GAL4, and p53 (Dutta et al. 1993; He et al. 1993; Li and Botchan 1993). The significance of these interactions is still unknown. On the one hand, these factors stimulated BPV replication (Li and Botchan 1993); on the other hand, p53 inhibited SV40 replication (Dutta et al. 1993). It was found recently that yRP-A binds specifically to sequence elements that are involved in the regulation of transcription (Luche et al. 1993; Singh and Samson 1995), so that the interaction of RP-A with transcription factors could play a role in transcriptional regulation.

Several posttranslational modifications of the RP-A subunits have been observed. Both the p70 and p32 subunits are pol-ADP-ribosylated

(Eki and Hurwitz 1991) and phosphorylated (Din et al. 1990; Fang and Newport 1993). The abundance of RP-A remains constant throughout the cell cycle, but the p32 subunit is phosphorylated in a cell-cycle-specific manner on multiple serines (Din et al. 1990). In cycling cells, p32 becomes extensively phosphorylated at the G_1-S transition and remains phosphorylated until G_2-M, when dephosphorylation occurs (Din et al. 1990). The hyperphosphorylated form of p32 shows decreased mobility during gel electrophoresis and runs as a 36-kD species.

On the basis of amino acid sequence, human RP-A p32 has two potential sites for cyclin-dependent kinases, five for DNA-PK, and five for casein kinase II. There is good evidence that cyclin-dependent kinases and DNA-PK phosphorylate RP-A p32 in vitro and in vivo. In vitro, cdc2-cyclin B kinase phosphorylates p32 at Ser-23 and Ser-29, sites that are also phosphorylated in vivo (Dutta and Stillman 1992), concomitant with a modest stimulation of SV40 DNA replication and origin unwinding in G_1 extracts. The same sites were also targeted specifically within the replication initiation complex by cdk2-cyclin A kinase (Fotedar and Roberts 1991, 1992; Dutta and Stillman 1992; Elledge et al. 1992; Pan et al. 1993a). However, mutant RP-A lacking the two cyclin-dependent kinase sites in the p32 subunit supported SV40 DNA replication in vitro as well as wild-type RP-A, demonstrating that phosphorylation of RP-A by cyclin-dependent kinases is dispensable in this system (Henricksen and Wold 1994). Consistent with these results, phosphorylated and unphosphorylated RP-A were equally active in SV40 DNA replication in vitro (Pan et al. 1995).

Under the conditions of the in vitro SV40 replication system, underphosphorylated RP-A becomes hyperphosphorylated (Fotedar and Roberts 1992; Henricksen and Wold 1994). The kinase involved appears to be DNA-PK, but cyclin A-cdk2 may also play a role. RP-A p32 is predominantly phosphorylated when bound to DNA, and the kinase that phosphorylates it is itself bound to single-stranded DNA (Fotedar and Roberts 1992). The p32 subunit in RP-A from cycling embryonic *Xenopus* extracts is also phosphorylated in S phase only in nuclear-localized RP-A bound to single-stranded DNA (Fang and Newport 1993). Hyperphosphorylation of RP-A p32 by DNA-PK in a cyclin-A-activated G_1 extract was reported to require prephosphorylation by cyclin A-cdk2 (Pan et al. 1994). However, mutant RP-A lacking the two potential cyclin-dependent kinase sites in p32 still becomes hyperphosphorylated (Henricksen and Wold 1994; Lee and Kim 1995). Moreover, replication-competent extracts depleted of cyclin-dependent kinases still hyperphosphorylate DNA-bound RP-A in vitro (Dutta and Stillman

1992), whereas extracts depleted of DNA-PK do not (Brush et al. 1994). Human RP-A is also found hyperphosphorylated in G_1 cells that have been exposed to ionizing radiation (Liu and Weaver 1993) or UV light (Carty et al. 1994). A good candidate for the kinase that phosphorylates RP-A during the damage response is the DNA-PK (Lees-Miller and Anderson 1991), a kinase that is known to be involved in DNA repair and is stimulated by double-strand breaks (for review, see Gottlieb and Jackson 1994). Purified DNA-PK phosphorylates RP-A p32 in vitro (Brush et al. 1994; Pan et al. 1994), and cells that are deficient in DNA-PK activity fail to hyperphosphorylate RP-A after irradiation (Boubnov and Weaver 1995). Hyperphosphorylated RP-A in extracts from irradiated cells appeared to be inactive in SV40 DNA replication in vitro, and additional RP-A purified from unirradiated cells relieved that block (Carty et al. 1994), suggesting that phosphorylation on the potential DNA-PK sites might activate RP-A for its tasks in repair while blocking its function in the replication process. However, other workers have found that phosphorylation of RP-A by DNA-PK did not affect its activity in SV40 DNA replication in vitro (Brush et al. 1994).

Based on the evidence from in vitro SV40 DNA replication, phosphorylation of RP-A appears to have little or no functional importance. However, it may play a role in vivo, such as correct localization to replication centers (for review, see Hozak and Cook 1994). Chromatin replication in *Xenopus* egg extracts is dependent on RP-A (Adachi and Laemmli 1994). Before being assembled into replication centers, RP-A is targeted to the prereplicative sites that form after mitosis and prior to initiation (Adachi and Laemmli 1994). This association is loose due to the absence of single-stranded DNA and is likely to be mediated by protein-protein interactions. Assembly of RP-A into prereplicative complexes does not occur in the presence of active cyclin B-cdc2 kinase, suggesting that only the underphosphorylated form of RP-A found after mitosis can be targeted to the complexes. Active cyclin A-cdk2 kinase did not facilitate or inhibit the assembly of prereplicative complexes (Adachi and Laemmli 1994). However, it colocalized with RP-A, and presumably many other replication proteins, to replication foci (Adachi and Laemmli 1992, 1994; Cardoso et al. 1993), suggesting a possible function in this process.

In summary, it seems likely that phosphorylation of RP-A regulates its activity in vivo, but the mechanism of this regulation remains poorly understood. Much of the confusion may stem from the multiple functions of RP-A in the cell, but it should be noted that the assay system used most frequently to assess the effects of phosphorylation, in vitro SV40

DNA replication, probably does not reveal all of the relevant steps in in vivo regulation of DNA replication.

DNA Polymerases δ and ε

Polymerases δ and/or ε are thought to play a pivotal role in leading-strand DNA synthesis as well as lagging-strand synthesis (Burgers 1991; Nethanel et al. 1992; Podust and Hübscher 1993; see Wang, this volume). Polymerase-δ is a two-subunit complex with polypeptides of approximately 124 kD and 55 kD (Chung et al. 1991) whose activity and processivity are stimulated by complexing with PCNA (Bauer and Burgers 1988). Polymerase-ε consists of a catalytic subunit of 258 kD (Kesti et al. 1993) that copurifies with a smaller 55-kD polypeptide in human (Syväoja and Linn 1989; Lee et al 1991b) and with polypeptides of 80, 34, 30, and 29 kD in yeast extracts (Hamatake et al. 1990). RF-C and PCNA also function to stimulate polymerase-ε (Burgers 1991; Lee et al. 1991b). Besides their involvement in DNA replication, both polymerase-δ and polymerase-ε are implicated in DNA repair (Z. Wang et al. 1993; Blank et al. 1994; Zeng et al. 1994b; Budd and Campbell 1995).

Polymerase-δ mRNA and protein expression is somewhat increased at the G_1-S border, but more significantly, the protein is most actively phosphorylated during S phase (Zeng et al. 1994a). It will be important to assess the enzymatic activities of hyper- and hypophosphorylated polymerase-δ and to look for differences in its interactions with other replication proteins like PCNA (Bauer and Burgers 1988) or RP-A (Longhese et al. 1994).

Polymerase-ε mRNA expression and protein levels are strongly dependent on cell proliferation (Kesti et al. 1993; Tuusa et al. 1995), but their abundance fluctuated less dramatically in actively cycling cells. The phosphorylation state of polymerase-ε has not been investigated.

PCNA and RF-C

PCNA is a phosphoprotein and its phosphorylation state appears to vary in a cell cycle-dependent fashion (Prosperi et al. 1993, 1994). Interestingly, highly phosphorylated PCNA associated with insoluble nuclear structures primarily during S phase, whereas soluble PCNA was found in early G_1 and G_2/M in a weakly phosphorylated form. Expression of PCNA appears to be independent of the cell cycle in actively growing cells. On the other hand, its expression declines in quiescent cells and is strongly induced by mitogens (Mathews et al. 1984). It is also subject to regulation by binding to cdk-cyclin kinases and their inhibitors (for review, see Heichman and Roberts 1994; Pines 1994).

RF-C is a multisubunit protein consisting of five polypeptides (p36.5, p37, p38, p40, and p140; also designated RFC1–5) (Lee et al. 1991b; Tsurimoto and Stillman 1991c). The mammalian and yeast genes encoding the RF-C subunits have been cloned (M. Chen et al. 1992a,b; Bunz et al. 1993; Pan et al. 1993b; Li and Burgers 1994a,b; Luckow et al. 1994; Noskov et al. 1994; Cullman et al. 1995; Hübscher et al., this volume). The p140 subunit has an ATP-binding domain and binds unspecifically to DNA. Its transcripts are most abundant in strongly proliferative tissues but are also relatively high in brain (Luckow et al. 1994). The p40 polypeptide has an ATP-binding motif and shows ATPase activity. It seems to interact with PCNA, polymerase-δ, and the p37 subunit of RF-C (Pan et al. 1993b). The steady-state levels of the mRNA fluctuate only slightly during the cell cycle (Noskov et al. 1994). The p38 subunit binds ATP, and its ATPase activity is stimulated by single-stranded DNA (Li and Burgers 1994a). The p37 subunit shows specific affinity to polymerase-ϵ and binds specifically to primer-template termini (Pan et al. 1993b). The p37 mRNA levels do not change significantly during the cell cycle (Li and Burgers 1994b). Since RNA levels of two of the subunits are constant during the cell cycle, protein modifications like phosphorylation might contribute to the regulation of RF-C function. A thorough investigation of RF-C expression and modification, however, must await the availability of suitable immunological reagents.

Topoisomerases

Topoisomerases are involved in the maintenance of chromatin and DNA structure (see Hangaard-Andersen et al., this volume). Topoisomerase I relaxes supercoiled DNA, whereas topoisomerase II catalyzes decatenation and unknotting of topologically linked DNA circles, as well as the relaxation of supercoiled DNA (Wang 1985). Both topoisomerases have been shown to be phosphorylated.

Dephosphorylation of serine/threonine residues in topoisomerase I isolated from different sources reduced or abolished its enzymatic activity (Durban et al. 1983; Kaiserman et al. 1988; Pommier et al. 1990; Tournier et al. 1992). Phosphorylation of the protein on tyrosine decreased activity as well (Tse-Dinh et al. 1984). Conversely, phosphorylation of topoisomerase I by casein kinase II and related kinases (Durban et al. 1983; Kaiserman et al. 1988) or protein kinase C (Pommier et al. 1990; Samuels and Shimizu 1992) stimulated activity.

The decatenation activity of topoisomerase II was shown to be stimulated by phosphorylation by casein kinase II, protein kinase C, cdc2-cyclin kinases, and Ca^{++}/calmodulin-dependent protein kinase (Sahyoun et al. 1986; Rottman et al. 1987; Ackerman et al. 1988; Cardenas and Gasser 1993; for review, see Poljak and Käs 1995). The major in vivo kinase acting on topoisomerase II in *S. cerevisiae* has been identified as casein kinase II (Cardenas et al. 1992). Topoisomerase II is hyperphosphorylated during G_2/M phase (Cardenas et al. 1992), consistent with the finding that this phosphorylation is required for activity of topoisomerase II during chromosome segregation. The stimulation of decatenation induced by phosphorylation was found to be at least partly due to an increase in the DNA-binding affinity of topoisomerase II (Dang et al. 1994) and an increased rate of ATP hydrolysis (Corbett et al. 1993).

SUMMARY AND OUTLOOK

Our understanding of the role of protein phosphorylation in regulation of DNA replication remains fragmentary. All of the proteins that recognize and activate eukaryotic origins of replication are phosphorylated, but with the exception of SV40 T antigen, there is no clear picture of how these modifications affect the function of the proteins. Similarly, the key replication enzymes in eukaryotic cells are phosphoproteins, but the functional importance of phosphorylation is established only for topoisomerase II, and in preliminary fashion for pol-α:primase. The lack of data to support a role for phosphorylation in regulation of DNA replication is not due to lack of effort. The mutational analysis of potential phosphorylation sites that was successful in investigating SV40 T antigen has now been extended to many of the cellular origin recognition proteins and several replication enzymes, including pol-α:primase and RP-A, without conspicuous success. Does this mean that these phosphorylation events are fortuitous? The cell-cycle-dependence of phosphorylation of replication proteins, the correlation between phosphorylation of MCM proteins with changes in their subcellular localization, and the importance of the cyclin-dependent kinases, other cell-cycle-dependent kinases, and the DNA-dependent protein kinase suggest that phosphorylation of most replication proteins is probably physiologically relevant. How then can one demonstrate this importance more directly? It should be noted that the current understanding of phosphorylation in control of SV40 T antigen's replication activity was not achieved solely through mutagenesis. Not only were the in vivo phosphorylation sites mapped, but several overexpression systems and a wide variety of biochemical assays were available in many laboratories to probe the functions of T antigen and its mutants before and after treatment with purified kinases and phosphatases over the course of the past decade. As outlined in the introduction, such a broad approach will likely be required to understand the roles of phosphorylation in the control of DNA replication.

ACKNOWLEDGMENTS

We thank Andreas Zeitvogel and Achim Dickmanns for assistance with the figures; Mel DePamphilis, Vladimir Podust, and Christoph Rehfuess for helpful criticism of the manuscript; and M. Lentz, M. Foiani, C. Lehner, T. Kelly, B. Tye, and J. Diffley for communication of results before publication. Work in the authors' laboratories was supported by the Deutsche Forschungsgemeinschaft, Fonds der Chemischen Industrie, BMFT-Genzentrum, the European Community (CHRX-CT93-0248 DG12), and the National Institutes of Health (1 RO1 GM-52948-01).

REFERENCES

Ackerman, P., C.V. Glover, and N. Osherhoff. 1988. Phosphorylation of DNA topoisomerase *in vivo* and in total homogenates of *Drosophila* Kc cells. *J. Biol. Chem.* **263:** 12653–12660.

Adachi Y. and U. Laemmli. 1992. Identification of nuclear pre-replication centers poised for DNA synthesis in *Xenopus* egg extracts: Immunolocalization study of replication protein A. *J. Cell Biol.* **119:** 1–15.

———. 1994. Study of the cell cycle dependent assembly of DNA pre-replication centers in *Xenopus* egg extracts. *EMBO J.* **13:** 4153–4164.

Adamczewski, J.P., J.V. Gannon, and T. Hunt. 1993. Simian virus 40 large T-antigen associates with cyclin A and p33^{cdk2}. *J. Virol.* **67:** 6551–6557.

Alani, E., R. Thresher, J.D. Griffith, and R.D. Kolodner. 1992. DNA binding properties of yRP-A and interactions with yeast and *E. coli* strand-exchange proteins. *J. Mol. Biol.* **227:** 54–71.

Baringa, M. 1994. Yeast enzyme finds fame in link to DNA replication. *Science* **265:** 1175–1176.

Bauer, G.A. and P.M.J. Burgers. 1988. The yeast analog of mammalian cyclin/proliferating cell nuclear antigen interacts with mammalian DNA polymerase δ. *Proc. Natl. Acad. Sci.* **85:** 7506–7510.

Bell, S.P. and B. Stillman. 1992. ATP-dependent recognition of eukaryotic origins of DNA replication by a multiprotein complex. *Nature* **357:** 128–134.

Bell, S.P., R. Kobayashi, and B. Stillman. 1993. Yeast origin recognition complex functions in transcription silencing and DNA replication. *Science* **262:** 1844–1849.

Benson, J.D. and P.M. Howley. 1995. Amino-terminal domains of the bovine papil-

lomavirus type 1 E1 and E2 proteins participate in complex formation. *J. Virol.* **69**: 4364-4372.
Blackwell, L.J. and J.A. Borowiec. 1994. Human replication protein A binds single-stranded DNA in two distinct complexes. *Mol. Cell. Biol.* **14**: 3993-4001.
Blank, A., B. Kim, and L.A. Loeb. 1994. Polymerase δ is required for base excision repair of DNA methylation damage in *Saccharomyces cerevisiae*. *Proc. Natl. Acad. Sci.* **91**: 9047-9051.
Blow, J.J. and R.A. Laskey. 1988. A role for the nuclear envelope in controlling DNA replication within the cell cycle. *Nature* **332**: 546-548.
Bockus, B.J. and B. Schaffhausen. 1987. Localization of the phosphorylations of polyomavirus large T antigen. *J. Virol.* **61**: 1155-1163.
Bonne-Andrea, C., S. Santuccii, P. Clertant, and F. Tillier. 1995. Bovine papillomavirus E1 protein binds specifically DNA polymerase α but not replication protein A. *J. Virol.* **69**: 2341-2350.
Borowiec, J.A., F.B. Dean, P.A. Bullock, and J. Hurwitz. 1990. Binding and unwinding—How T antigen engages the SV40 origin of DNA replication. *Cell* **60**: 181-184.
Boubnov, N.V. and D.T. Weaver. 1995. scid cells are deficient in Ku and replication protein A phosphorylation by the DNA-dependent protein kinase. *Mol. Cell. Biol.* **15**: 5700-5706.
Bream, G.L., C.-A. Ohmstede, and W.C. Phelps. 1993. Characterization of human papillomavirus type 11 E1 and E2 proteins expressed in insect cells. *J. Virol.* **67**: 2655-2663.
Brill, S.J. and B. Stillman. 1989. Replication factor-A from *Saccharomyces cerevisiae* functions in the unwinding of SV40 origin of replication. *Nature* **342**: 92-95.
———. 1991. Replication factor-A from *Saccharomyces cerevisiae* is encoded by three essential genes coordinately expressed at S-phase. *Genes Dev.* **5**: 1589-1600.
Brown, G.W., T.E. Melendy, and D.S. Ray. 1992. Conservation of structure and function of DNA replication protein A in the trypanosomatid *Crithidia fasciculata*. *Proc. Natl. Acad. Sci.* **89**: 10227-10231.
Brush, G.S., C.W. Anderson, and T.J. Kelly. 1994. The DNA-activated protein kinase is required for the phosphorylation of replication protein A during simian virus 40 DNA replication. *Proc. Natl. Acad. Sci.* **91**: 12520-12524.
Budd, M.E. and J.L. Campbell. 1995. DNA polymerases required for repair of UV-induced damage in *Saccharomyces cerevisiae*. *Mol. Cell. Biol.* **15**: 2173-2179.
Bunz, F., K. Kobayashi, and B. Stillman. 1993. cDNAs encoding the large subunit of human replication factor C. *Proc. Natl. Acad. Sci.* **90**: 11014-11018.
Burgers, P. 1991. *Saccharomyces cerevisiae* replication factor C. II. Formation and activity of complexes with the proliferating cell nuclear antigen and with DNA polymerases δ and ε. *J. Biol. Chem.* **266**: 22698-22706.
Cardenas, M.E. and S.M. Gasser. 1993. Casein kinase II copurifies with yeast topoisomerase II and reactivates the dephosphorylated enzyme. *J. Cell Sci.* **104**: 533-543.
Cardenas, M.E., Q. Dang, C.V.C. Glover, and S.M. Gasser. 1992. Casein kinase II phosphorylates the eukaryote-specific C-terminal domain of topoisomerase II *in vivo*. *EMBO J.* **11**: 1785-1795.
Cardoso, M.C., H. Leonhardt, and B. Nadal-Ginard. 1993. Reversal of terminal differentiation and control of DNA replication: Cyclin A and cdk2 specifically localize at subnuclear sites of DNA replication. *Cell* **74**: 979-992.

Carty, M.P., A.S. Levine, and K. Dixon. 1992. HeLa cell single-stranded DNA-binding protein increases the accuracy of DNA synthesis by DNA polymerase α in vitro. *Mutat. Res.* **274:** 29–43.

Carty, M.P., M. Zernik-Kobak, S. McGrath, and K. Dixon. 1994. UV light-induced DNA synthesis arrest in HeLa cells is associated with changes in phosphorylation of human single-stranded DNA-binding protein. *EMBO J.* **13:** 2114–2123.

Cegielska, A. and D.M. Virshup. 1993. Control of simian virus 40 replication by the HeLa nuclear kinase casein kinase I. *Mol. Cell. Biol.* **13:** 1202–1211.

Cegielska, A., I. Moarefi, E. Fanning, and D.M. Virshup. 1994a. T-antigen kinase inhibits simian virus 40 DNA replication by phosphorylation of intact T antigen on serines 120 and 123. *J. Virol.* **68:** 269–275.

Cegielska, A., S. Shaffer, R. Derua, J. Goris, and D.M. Virshup. 1994b. Different oligomeric forms of protein phosphatase 2A activate and inhibit simian virus 40 DNA replication. *Mol. Cell. Biol.* **14:** 4616–4623.

Chen, M., Z.-Q. Pan, and J. Hurwitz. 1992a. Sequence and expression in *Escherichia coli* of the 40-kDa subunit of activator 1 (replication factor C) of HeLa cells. *Proc. Natl. Acad. Sci.* **89:** 2516–2552.

———. 1992b. Studies of the cloned 37-kDa subunit of activator 1 (replication factor C) of HeLa cells. *Proc. Natl. Acad. Sci.* **89:** 5211–5215.

Chen, Y., K. Hennessy, D. Botstein, and B.-K. Tye. 1992. CDC46/MCM5, a yeast protein whose subcellular localization is cell-cycle regulated, is involved in DNA replication at autonomously replicating sequences. *Proc. Natl. Acad. Sci.* **89:** 10459–10463.

Chong, J.P.J., H.M. Mahbubani, C.-Y. Khoo, and J.J. Blow. 1995. Purification of an MCM-containing complex as a component of the DNA replication licensing system. *Nature* **375:** 418–421.

Chung, D.W., J. Zhang, C.-K. Tan, E.W. Davie, A.G. So, and K.M. Downey. 1991. Primary structure of the catalytic subunit of human DNA polymerase δ and chromosomal location of the gene. *Proc. Natl. Acad. Sci.* **88:** 11197–11201.

Cohen, P. 1989. The structure and regulation of protein phosphatases. *Annu. Rev. Biochem.* **58:** 453–508.

Collins, K.L. and T.J. Kelly. 1991. The effects of T antigen and replication protein A on the initiation of DNA synthesis by polymerase α:primase. *Mol. Cell. Biol.* **11:** 2108–2115.

Collins, K.L., A.A.R. Russo, B.Y. Tseng, and T.J. Kelly. 1993. The role of the 70 kDa subunit of human DNA polymerase α in DNA replication. *EMBO J.* **12:** 4555–4566.

Corbett, A.H., A.W. Fernald, and N. Osterhoff. 1993. Protein kinase C modulates the catalytic activity of topoisomerase II by enhancing the rate of ATP hydrolysis: Evidence for a common mechanism of regulation by phosphorylation. *Biochemistry* **32:** 2090–2097.

Coverley, D. and R.A. Laskey. 1994. Regulation of eukaryotic DNA replication. *Annu. Rev. Biochem.* **63:** 745–776.

Coverley, D., M.K. Kenny, M. Munn, W.D. Rupp, D.P. Lane, and R.D. Wood. 1991. Requirement for the replication protein SSB in human excision repair. *Nature* **349:** 538–541.

Cowie, A. and R. Kamen. 1984. Multiple binding sites for polyomavirus large T antigen within regulatory sequences of polyomavirus DNA. *J. Virol.* **52:** 750–760.

Cullman, G., K. Fien, R. Kobayashi, and B. Stillman. 1995. Characterization of the five replication factor C genes of *Saccharomyces cerevisiae*. *Mol. Cell. Biol.* **15**: 4661–4671.

Dang, Q., G.-C. Alghisi, and S.M. Gasser. 1994. Phosphorylation of the C-terminal domain of yeast topoisomerase II by casein kinase II affects DNA-protein interaction. *J. Mol. Biol.* **243**: 10–24.

Dean, F.B., J.A. Borowiec, T. Eki, and J. Hurwitz. 1992. The simian virus 40 T antigen double hexamer assembles around the DNA at the replication origin. *J. Biol. Chem.* **267**: 14129–14137.

Diffley, J.F.X. 1992. Early events in eukaryotic DNA replication. *Trends Cell Biol.* **2**: 298–303.

Diffley, J.F.X. and J.H. Cocker. 1992. Protein-DNA interactions at a yeast replication origin. *Nature* **357**: 169–172.

Diffley, J.F.X., J.H. Cocker, J. Dowell, and A. Rowley. 1994. Two steps in the assembly of complexes at yeast replication origins *in vivo*. *Cell* **78**: 303–316.

Din, S.U., S.J. Brill, M.P. Fairman, and B. Stillman. 1990. Cell cycle regulated phosphorylation of DNA replication factor A from human and yeast cells. *Genes Dev.* **4**: 968–977.

Dornreiter, I., W.C. Copeland, and T.S.F. Wang. 1993. Initiation of simian virus 40 DNA replication requires the interaction of a specific domain of human DNA polymerase α with large T antigen. *Mol. Cell. Biol.* **13**: 809–820.

Dornreiter, I., A. Höss, A.K. Arthur, and E. Fanning. 1990. SV40 T antigen binds directly to the catalytic subunit of DNA polymerase α. *EMBO J.* **9**: 3329–3336.

Dornreiter, I., L.F. Erdile, U. Gilbert, D. von Winkler, T.J. Kelly, and E. Fanning. 1992. Interaction of DNA polymerase α-primase with replication protein A and SV40 T antigen. *EMBO J.* **11**: 769–776.

Dowell, S.J., P. Romanowski, and J.F.X. Diffley. 1994. Interaction of Dbf4, the Cdc7 protein kinase regulatory subunit, with yeast replication origins *in vivo*. *Science* **265**: 1243–1246.

Durban, E., J.S. Mills, D. Roll, and H. Busch. 1983. Phosphorylation of purified Novikoff hepatoma topoisomerase I. *Biochem. Biophys. Res. Commun.* **111**: 897–905.

Dutta, A. and B. Stillman. 1992. cdc2 family kinases phosphorylate a human cell DNA replication factor, RPA, and activate DNA replication. *EMBO J.* **11**: 2189–2199.

Dutta, A., J.M. Ruppert, J.C. Aster, and E. Winchester. 1993. Inhibition of DNA replication by p53. *Nature* **365**: 79–82.

Eki, T. and J. Hurwitz. 1991. Influence of poly (ADP-ribose) polymerase on the enzymatic synthesis of SV40 DNA. *J. Biol. Chem.* **266**: 3087–3100.

Elledge, S.J., R. Richman, F.L. Hall, R.T. Williams, N. Lodgson, and J.W. Harper. 1992. *CDK2* encodes a 33-kDa cyclin A-associated protein kinase and is expressed before *CDC2* in the cell cycle. *Proc. Natl. Acad. Sci.* **89**: 2907–2911.

Erdile, L.F., M.S. Wold, and T.J. Kelly. 1990. The primary structure of the 32-kDa subunit of human replication protein A. *J. Biol. Chem.* **265**: 3177–3182.

Erdile, L.F., W.-D. Heyer, R.D. Kolodner, and T.J. Kelly. 1991. Characterization of a cDNA encoding the 70-kDa single-stranded DNA-binding subunit of human replication protein A and its role in DNA replication. *J. Biol. Chem.* **266**: 12090–12098.

Fairman M.P. and B. Stillman. 1988. Cellular factors required for multiple stages of SV40 DNA replication *in vitro*. *EMBO J.* **7**: 1211–1218.

Fang, F. and J.W. Newport. 1993. Distinct roles of cdk2 and cdc2 in RP-A phosphorylation during the cell-cycle. *J. Cell Sci.* **106**: 983–994.

Fanning, E. 1992. Simian virus 40 large T antigen: The puzzle, the pieces, and the emerging picture. *J. Virol.* **66:** 1289–1293.

———. 1994. Control of SV40 DNA replication by protein phosphorylation: A model for cellular DNA replication. *Trends Cell. Biol.* **4:** 250–255.

Fanning, E. and R. Knippers. 1992. Structure and function of simian virus 40 large T antigen. *Annu. Rev. Biochem.* **61:** 55–85.

Fernandez-Sarabia, M.J., A. Sutton, T. Zhong, and K.T. Arndt. 1992. SIT4 protein phosphatase is required for the normal accumulation of SWI4, CLN1, CLN2, and HCS26 RNAs during late G1. *Genes Dev.* **2:** 2417–2428.

Focher, F., E. Ferrari, S. Spadari, and U. Hübscher. 1988. Do DNA polymerases δ and α act coordinately as leading and lagging strand replicases. *FEBS Lett.* **229:** 6–10.

Foiani, M., G. Liberi, G. Lucchini, and P. Plevani. 1995. Cell cycle-dependent phosphorylation and dephosphorylation of the yeast DNA polymerase α-primase B subunit. *Mol. Cell. Biol.* **15:** 883–891.

Foiani, M., F. Marini, D. Gamba, G. Lucchini, and P. Plevani. 1994. The B subunit of the DNA polymerase α-primase complex in *Saccharomyces cerevisiae* executes an essential function at the initial stage of DNA replication. *Mol. Cell. Biol.* **14:** 923–933.

Foss, M., F.J. McNally, P. Laurenson, and J. Rine. 1993. Origin recognition complex (ORC) in transcriptional silencing and DNA replication in *S. cerevisiae*. *Science* **262:** 1838–1844.

Fotedar, R. and J.M. Roberts. 1991. Association of p34^{cdc2} with replicating DNA. *Cold Spring Harbor Symp. Quant. Biol.* **56:** 325–333.

———. 1992. Cell cycle regulated phosphorylation of RPA-32 occurs within the replication initiation complex. *EMBO J.* **11:** 2177–2187.

Francesconi, S., M.P. Longhese, A. Piseri, C. Santocanale, G. Lucchini, and P. Plevani. 1991. Mutations in the conserved yeast DNA primase domains impair DNA replication in vivo. *Proc. Natl. Acad. Sci.* **88:** 3877–3881.

Georgaki, A., B. Strack, V. Podust, and U. Hübscher. 1992. DNA unwinding activity of replication protein A. *FEBS Lett.* **308:** 240–244.

Georgaki, A., N. Tuteja, B. Sturzenegger, and U. Hübscher. 1994. Calf thymus DNA helicase F, a replication protein A copurifying enzyme. *Nucleic Acid Res.* **22:** 1128–1134.

Gilbert, D.M. and S.N. Cohen. 1987. Bovine papillomavirus plasmids replicate randomly in mouse fibroblasts throughout S-phase of the cell cycle. *Cell* **50:** 59–68.

Gottlieb, T.M. and S.P. Jackson. 1994. Protein kinases and DNA damage. *Trends Biochem. Sci.* **19:** 500–503.

Grässer, F.A., K. Mann, and G. Walter. 1987. Removal of serine phosphates from simian virus 40 T antigen increases its ability to stimulate DNA replication *in vitro* but has no effect on ATPase and DNA binding. *J. Virol.* **61:** 3373–3380.

Hamatake, R.K., H. Hasegawa, A.B. Clark, K. Bebenek, T. Kunkel, and A. Sugino. 1990. Purification and characterization of DNA polymerase II from the yeast *Saccharomyces cerevisiae*. *J. Biol. Chem* **265:** 4072–4083.

Hassauer, M., K.H. Scheidtman, and G. Walter. 1996. Mapping of phosphorylation sites in polyomavirus large T antigen. *J. Virol.* **58:** 805–816.

He, Z., B.T. Brinton, J. Greenblatt, J.A. Hassell, and C.J. Ingles. 1993. The transactivator proteins VP16 and GAL4 bind replication factor A. *Cell* **73:** 1223–1232.

Heichman, K.A. and J.M. Roberts. 1994. Rules to replicate by. *Cell* **79:** 557–562.

Hennesey, K., C.D. Clarek, and D. Botstein. 1991. Subcellular localisation of yeast *CDC*

46 varies with the cell cycle. *Genes Dev.* **4:** 2252–2263.
Henricksen, L.A. and M.S. Wold. 1994. Replication protein A mutants lacking phosphorylation sites for p34cdc2 kinase support DNA replication. *J. Biol. Chem.* **269:** 24203–24208.
Henricksen, L.A., C.B. Umbricht, and M.S. Wold. 1994. Recombinant replication protein A: Expression, complex formation, and functional characterization. *J. Biol. Chem.* **269:** 11121–11132.
Heyer, W.-D., M.R.S. Rao, L.F. Erdile, T.J. Kelly, and R.D. Kolodner. 1990. An essential *Saccharomyces cerevisiae* single-stranded DNA binding protein is homologous to the large subunit of RP-A. *EMBO J.* **9:** 2321–2329.
Hoffmann, I., G. Draetta, and E. Karsenti. 1994. Activation of the phosphatase activity of human cdc25A by a cdk2-cyclin E dependent phosphorylation at the G1/S transition. *EMBO J.* **13:** 4302–4310.
Höss, A., I. Moarefi, K.-H. Scheidtmann, L.J. Cisek, J.L. Corden, I. Dornreiter, A.K. Arthur, and E. Fanning. 1990. Altered phosphorylation pattern of simian virus 40 T antigen expressed in insect cells by using a baculovirus vector. *J. Virol.* **64:** 4799–4807.
Hozak, P. and P.R. Cook. 1994. Replication factories. *Trends Cell Biol.* **4:** 48–52.
Hubbard, M.J. and P. Cohen. 1993. On target with a new mechanism for the regulation of protein phosphorylation. *Trends Biochem Sci.* **18:** 172–177.
Huberman, J.A. 1992. Quest's end and question's begin. *Curr. Biol.* **2:** 351–352.
———. 1994. A tale of two functions. *Nature* **367:** 20–21.
Hurwitz, J., F.B. Dean, A.D. Kwong, and S.H. Lee. 1990. The *in vitro* replication of DNA containing the SV40 origin. *J. Biol. Chem.* **265:** 18043–18046.
Ishimi, Y., A. Claude, P. Bullock, and J. Hurwitz. 1988. Complete enzymatic synthesis of DNA containing the SV40 origin of replication. *J. Biol. Chem.* **263:** 19723–19733.
Jackson, A.L., P.M.B. Pahl, K. Harrison, J. Rosamond, and R.A. Sclafani. 1993. Cell cycle regulation of the yeast Cdc7 protein kinase by association with Dbf4 protein. *Mol. Cell. Biol.* **13:** 2899–2908.
Jackson, P.K., S. Chevalier, M. Philippe, and M.W. Kirschner. 1995. Early events in DNA require cyclin E and are blocked by p21[CIP1]. *J. Cell Biol.* **130:** 755–769.
Jinno, S., K. Suto, A. Nagata, M. Igarashi, Y. Kanaoka, H. Nojima, and H. Okayama. 1994. Cdc25A is a novel phosphatase functioning early in the cell cycle. *EMBO J.* **13:** 1549–1556.
Kaiserman, H.B., T.S. Ingebritsen, and R.M. Benbow. 1988. Regulation of *Xenopus laevis* DNA topoisomerase I activity by phosphorylation *in vitro*. *Biochemistry* **27:** 3216–3222.
Kelly, T.J. 1988. SV40 DNA replication. *J. Biol. Chem.* **263:** 17889–17892.
Kelly, T.J., P.V. Jallepalli, and R.K. Clyne. 1994. Silence of the ORCs. *Curr. Biol.* **4:** 238–241.
Kelly, T.J., G.S. Martin, S.L. Forsburg, R.J. Stephen, A. Russo, and P. Nurse. 1993. The fission yeast cdc18+ gene product couples S phase to START and mitosis. *Cell* **74:** 371–382.
Kenny, M.K., S.-H. Lee, and J. Hurwitz. 1989. Multiple functions of human single-stranded-DNA binding protein in simian virus 40 DNA replication: Single-strand stabilization and stimulation of DNA polymerases α and δ. *Proc. Natl. Acad. Sci.* **86:** 9757–9761.
Kenny, M., U. Schlegel, H. Furneaux, and J. Hurwitz. 1990. The role of human single-

stranded DNA binding protein and its individual subunits in simian virus 40 DNA replication. *J. Biol. Chem.* **265:** 7693–7700.

Kesti, T., H. Frantti, and J.E. Syväoja. 1993. Molecular cloning of the cDNA for the catalytic subunit of human DNA polymerase ε. *J. Biol. Chem.* **268:** 10238–10245.

Kim, C. and M.S. Wold. 1995. Recombinant human replication protein A binds to polynucleotides with low cooperativity. *Biochemistry* **34:** 2058–2064.

Kim, C., B.F. Paulus, and M.S. Wold. 1994. Interactions of human replication protein A with oligonucleotides. *Biochemistry* **33:** 14197–14206.

Kim, C., R.O. Snyder, and M.S. Wold. 1992. Binding properties of replication protein A from human and yeast cells. *Mol. Cell. Biol.* **12:** 3050–3059.

Kimura, H., N. Nozaki, and K. Sugimoto. 1994. DNA polymerase α associated protein P1, a murine homolog of yeast MCM3, changes its intranuclear distribution during the DNA synthetic period. *EMBO J.* **13:** 4311–4320.

Knoblich, J.A., K. Sauer, L. Jones, H. Richardson, R. Saint, and C.F. Lehner. 1994. Cyclin E controls S phase progression and its down-regulation during *Drosophila* embryogenesis is required for the arrest of cell proliferation. *Cell* **77:** 107–120.

Koonin, E.V. 1993. A common set of conserved motifs in a vast variety of putative nucleic acid-dependent ATPases including MCM proteins involved in the initiation of eukaryotic DNA replication. *Nucleic Acid. Res.* **21:** 2541–2547.

Kubota, Y., S. Mimura, S.-I. Nishimoto, H. Takisawa, and H. Nojima. 1995. Identification of the yeast MCM3-related protein as a component of *Xenopus* DNA replication licensing factor. *Cell* **81:** 601–609.

Lambert, P.F. 1991. Papillomavirus DNA replication. *J. Virol.* **65:** 341–342.

Law, M.-F., D.R. Lowy, I. Dvoretzky, and P.M. Howley. 1981. Mouse cell transformed by bovine papillomavirus contain only extrachromosomal viral DNA sequences. *Proc. Natl. Acad. Sci.* **78:** 2727–2731.

Lee, S.-H. and D.K. Kim. 1995. The role of the 34-kDa subunit of human replication protein A in simian virus 40 DNA replication *in vitro*. *J. Biol. Chem.* **270:** 12801–12807.

Lee, S.-H., A.D. Kwong, Z.Q. Pan, and J. Hurwitz. 1991a. Studies on the activator 1 protein complex, an accessory factor for proliferating cell nuclear antigen-dependent DNA polymerase δ. *J. Biol. Chem.* **266:** 594–602.

Lee, S.-H., Z.-Q. Pan, A.D. Kwong, P.M.J. Burgers, and J. Hurwitz. 1991b. Synthesis of DNA by DNA polymerase ε *in vitro*. *J. Biol. Chem.* **266:** 22707–22717.

Lees-Miller, S.P. and C.W. Anderson. 1991. The DNA-activated protein kinase DNA-PK: A potential coordinator of nuclear events. *Cancer Cells* **3:** 341–346.

Lentz, M.R., D. Pak, I. Mohr, and M.R. Botchan. 1993. The E1 replication protein of bovine papillomavirus type 1 contains an extended nuclear localization signal that includes a p34^{cdc2} phosphorylation site. *J. Virol.* **67:** 1414–1423.

Li, J.J. and B.M. Alberts. 1992. Eukaryotic initiation rites. *Nature* **357:** 114–115.

Li, J.J. and I. Herskowitz. 1993. Isolation of ORC6, a component of the yeast origin recognition complex by a one-hybrid system. *Science* **262:** 1870–1874.

Li, R. and M.R. Botchan. 1993. The acidic transcriptional activation domains of VP16 and p53 bind the cellular replication protein A and stimulate *in vitro* BPV-1 DNA replication. *Cell* **73:** 1207–1221.

Li, X. and P.M.J. Burgers. 1994a. Molecular cloning and expression of the *Saccharomyces cerevisiae RFC3* gene, an essential component of replication factor C. *Proc. Natl. Acad. Sci.* **91:** 868–872.

———. 1994b. Cloning and characterization of the essential *Saccharomyces cerevisiae RFC4* gene encoding the 37-kDa subunit of replication factor C. *J. Biol. Chem.* **269:** 21880-21884.
Liu, V.F. and D.T. Weaver. 1993. The ionizing-radiation induced replication protein A phosphorylation response differs between ataxia telangiectasia and normal human cells. *Mol. Cell. Biol.* **13:** 7222-7231.
Longhese, M.P., P. Plevani, and G. Lucchini. 1994. Replication factor A is required *in vivo* for DNA replication, repair and recombination. *Mol. Cell. Biol.* **14:** 7884-7890.
Lucchini, G., S. Francesconi, M. Foiani, G. Badaracco, and P. Plevani. 1987. Yeast DNA polymerase-primase complex: Cloning of *PRI1*, a single essential gene related to DNA primase activity. *EMBO J.* **6:** 737-742.
Luche, R.M., W.C. Smart, T. Marion, M. Tillman, R.A. Sumrada, and T.G. Cooper. 1993. *Saccharomyces cerevisiae* BUF protein binds to sequences participating in DNA replication in addition to those mediating transcriptional repressing (URS1) and activation. *Mol. Cell. Biol.* **13:** 5749-5761.
Luckow, B., F. Bunz, B. Stillman, P. Lichter, and G. Schütz. 1994. Cloning, expression and chromosomal localization of the 140-kDa subunit of replication factor C from mice and humans. *Mol. Cell. Biol.* **14:** 1626-1634.
Ludlow, J.W. 1992. Selective ability of S-phase cell extracts to dephosphorylate large T antigen in vitro. *Oncogene* **7:** 1011-1014.
Madine, M.A., C.-Y. Khoo, A.D. Mills, and R.A. Laskey. 1995. MCM3 complex required for cell cycle regulation of DNA replication in vertebrate cells. *Nature* **375:** 421-424.
Mann, D.J., V. Dombradi, and P.-T.W. Cohen. 1993. *Drosophila* protein phosphatase V functionally complements a SIT4 mutant in *Saccharomyces cerevisiae* and its amino-terminal region can confer this complementation to a heterologous phosphatase catalytic domain. *EMBO J.* **12:** 4833-4842.
Mathews, M.B., R.M. Bernstein, B.R. Franza, and J.T. Garrels. 1984. Identity of the proliferating cell nuclear antigen and cyclin. *Nature* **309:** 374-376.
McVey, D., L. Brizuela, I. Mohr, D.R. Marshak, Y. Gluzman, and D. Beach. 1989. Phosphorylation of large tumor antigen by cdc2 stimulates SV40 DNA replication. *Nature* **341:** 503-507.
McVey, D., S. Ray, Y. Gluzman, L. Berger, A.G. Wildeman, D.J. Marshak, and P. Tegtmeyer. 1993. Cdc2 phosphorylation of threonine 124 activates the origin-unwinding functions of simian virus 40 T antigen. *J. Virol.* **67:** 5206-5215.
Melendy, T. and B. Stillman. 1993. An interaction between replication protein A and SV40 T antigen appears essential for primosome assembly during SV40 DNA replication. *J. Biol. Chem.* **268:** 3389-3395.
Micklem, G., A. Rowley, J. Harwood, K. Nasmyth, and J.F.X. Diffley. 1993. Yeast origin recognition complex is involved in DNA replication and transcriptional silencing. *Nature* **366:** 87-89.
Mitsis, P.G., S.C. Kowalczykowski, and I.R. Lehman. 1993. A single-stranded DNA binding protein from *Drosophila melanogaster*: Characterization of the heterotrimeric protein and its interaction with single-stranded DNA. *Biochemistry* **32:** 5257-5266.
Miyazawa, H., M. Izumi, S. Tada, R. Takada, M. Masutani, M. Ui, and F. Hanaoka. 1993. Molecular cloning of the cDNAs for the four subunits of mouse DNA polymerase α-primase complex and their gene expression during cell proliferation and the cell cycle. *J. Biol. Chem.* **268:** 8111-8122.

Moarefi, I., C. Schneider, K. van Zee, A. Höss, A.K. Arthur, and E. Fanning. 1993a. Control of SV40 DNA replication by protein phosphorylation. In *DNA replication and the cell cycle* (ed. E. Fanning et al.), pp. 157–169. Springer Verlag, Berlin.

Moarefi, I., D. Small, I. Gilbert, M. Höffner, S.K. Randall, C. Schneider, A.A.R. Russo, U. Ramsperger, A.K. Arthur, H. Stahl, T.J. Kelly, and E. Fanning. 1993b. Mutation of cyclin-dependent kinase phosphorylation site in simian virus 40 (SV40) large T antigen specifically blocks SV40 origin DNA unwinding. *J. Virol.* **67:** 4992–5002.

Moore, S.P., L. Erdile, T. Kelly, and R. Fishel. 1991. The human homologous pairing protein HPP-1 is specifically stimulated by the cognate single-stranded binding protein hRP-A. *Proc. Natl. Acad. Sci.* **88:** 9067–9071.

Moreno, S. and P. Nurse. 1990. Substrates for p34^{cdc2}: In vivo veritas? *Cell* **61:** 549–551.

Moses, K. and C. Prives. 1994. A unique subpopulation of murine DNA polymerase α/primase specifically interacts with polyomavirus T antigen and stimulates DNA replication. *Mol. Cell. Biol.* **14:** 2767–2776.

Murakami, Y. and J. Hurwitz. 1993. DNA polymerase α stimulates the ATP-dependent binding of simian virus tumor T antigen to the SV40 origin of replication. *J. Biol. Chem.* **268:** 11018–11027.

Nasheuer, H.-P., A. Moore, A.F. Wahl, and T.S.-F. Wang. 1991. Cell cycle-dependent phosphorylation of human DNA polymerase α. *J. Biol. Chem.* **266:** 7893–7903.

Nasheuer, H.-P., D. von Winkler, C. Schneider, I. Dornreiter, I. Gilbert, and E. Fanning. 1992. Purification and functional characterization of bovine RP-A in an *in vitro* replication system. *Chromosoma* **102:** S52–S59.

Nethanel, T., T. Zlotkin, and G. Kaufmann. 1992. Assembly of simian virus 40 Okazaki pieces from DNA primers is reversibly arrested by ATP depletion. *J. Virol.* **66:** 6634–6640.

Newlon, C.S. 1993. Two jobs for the origin replication complex. *Science* **262:** 1830–1831.

Norbury, C. and P. Nurse. 1992. Animal cell cycles and their control. *Annu. Rev. Biochem.* **61:** 441–470.

Noskov, V., S. Maki, Y. Kawasaki, S.-H. Leem, B.-I. Ono, H. Araki, Y. Pavlov, and A. Sugino. 1994. The *RFC2* gene encoding a subunit of replication factor C of *Saccharomyces cerevisiae*. *Nucleic Acids Res.* **22:** 1527–1535.

Pan, Z.-Q., A. Amin, and J. Hurwitz. 1993a. Characterization of the *in vitro* reconstituted cyclin A or B1 dependent cdk2 and cdc2 kinase activities. *J. Biol. Chem.* **268:** 20443–20451.

Pan, Z.-Q., M. Chen, and J. Hurwitz. 1993b. The subunits of activator 1 (replication factor C) carry out multiple functions for proliferating-cell nuclear antigen-dependent DNA synthesis. *Proc. Natl. Acad. Sci.* **90:** 6–10.

Pan, Z.-Q., A.A. Amin, E. Gibbs, H. Niu, and J. Hurwitz. 1994. Phosphorylation of p34 subunit of human single-stranded-DNA-binding protein in cyclin A-activated G_1 extracts is catalyzed by cdk-cyclin A complex and DNA-dependent protein kinase. *Proc. Natl. Acad. Sci.* **91:** 8343–8347.

Pan, Z.-Q., C.-H. Park, A.A. Amin, J. Hurwitz, and A. Sancar. 1995. Phosphorylated and unphosphorylated forms of human single-stranded DNA-bindng protein are equally active in simian virus 40 DNA replication and in nucleotide excision repair. *Proc. Natl. Acad. Sci.* **92:** 4636–4640.

Park, P., W. Copeland, L. Yang, T. Wang, M.R. Botchan. and I.J. Mohr. 1994. The cellular DNA polymerase α-primase is required for papillomavirus DNA replication and

associates with the viral E1 helicase. *Proc. Natl. Acad. Sci.* **91:** 8700–8704.
Pines, J. 1994. p21 inhibits cyclin shock. *Nature* **369:** 520–521.
Pizzagalli, A., P. Valsasnini, P. Plevan, and G. Lucchini. 1988. DNA polymerase I gene of *Saccharomyces cerevisiae*: Nucleotide sequence, mapping of temperature sensitive mutation, and protein homology with other DNA polymerases. *Proc. Natl. Acad. Sci.* **85:** 3772–3776.
Podust, V. and U. Hübscher. 1993. Lagging strand DNA synthesis by calf thymus DNA polymerases α, β, δ, and ε in the presence of auxiliary proteins. *Nucleic Acids Res.* **21:** 841–846.
Podust, V., G. Bialek, H. Sternbach, and F. Grosse. 1990. Casein kinase II phosphorylates DNA-polymerase-α-DNA-primase without affecting its basic enzymatic properties. *Eur. J. Biochem.* **193:** 189–193.
Poljak, L. and E. Käs. 1995. Resolving the role of topoisomerase II in chromatin structure and function. *Trends Cell Biol.* **5:** 348–354.
Pommier, Y., D. Kerrigan, K.J.D. Hartman, and R.I. Glatzer. 1990. Phosphorylation of mammalian DNA topoisomerase I and activation by protein kinase C. *J. Biol. Chem.* **265:** 9418–9422.
Prelich, G. and B. Stillman. 1988. Coordinated leading and lagging strand synthesis duing SV40 DNA replication *in vitro* requires PCNA. *Cell* **53:** 117–126.
Prives, C. 1990. The replication functions of SV40 T antigen are regulated by phosphorylation. *Cell* **61:** 735–738.
Prosperi, E., A.I. Scovassi, L.A. Stivala, and L. Bianchi. 1994. Proliferating cell nuclear antigen bound to DNA synthesis sites: Phosphorylation and association with cyclin D1 and cyclin A. *Exp. Cell Res.* **215:** 257–262.
Prosperi, E., L.A. Stivala, E. Sala, I.A. Scovassi, and L. Bianchi. 1993. Proliferating cell nuclear antigen complex formation induced by ultraviolet irradiation in human quiescent fibroblasts as detected by immunostaining and flow cytometry. *Exp. Cell Res.* **205:** 320–325.
Rao, H. and B. Stillman. 1995. The origin recognition complex interacts with a bipartite DNA binding site within yeast replicators. *Proc. Natl. Acad. Sci.* **92:** 2224–2228.
Roberts, J.D., D. Nguyen, and T.A. Kunkel. 1993. Frameshift fidelity during replication of double-stranded DNA in HeLa cell extracts. *Biochemistry* **32:** 4083–4089.
Rottman, M., H.C. Schroder, M. Gramzow, K. Reinneisen, B. Kurelec, A. Dorn, U. Friese, and W.E. Muller. 1987. Specific phosphorylation of proteins in pore complex-laminae from sponge *Geodia cydonium* by the homologous aggregation factor and phorbol ester. Role of protein kinase C in phosphorylation of DNA topoisomerase II. *EMBO J.* **6:** 3939–3944.
Rowley, A., J.H. Cocker, J. Harwood, and J.F.X. Diffley. 1995. Initiation complex assembly at budding yeast replication origins begins with the recognition of a bipartite sequence by limiting amounts of the initiator, ORC. *EMBO J.* **14:** 2631–2641.
Sahyoun, N., M. Wolf, J. Besterman, T. Hsieh, M. Sander, H. LeVine, K.J. Chang, and P. Cuatrecasas. 1986. Protein kinase C phosphorylates topoisomerase II: Topoisomerase activation and its possible role in phorbol ester-induced differentiation of HL-60 cells. *Proc. Natl. Acad. Sci.* **83:** 1603–1607.
Samuels, D.S. and N. Shimizu. 1992. DNA topoisomerase I phosphorylation in mouse fibroblasts treated with 12-O-tetradecanoylphorbol-13-acetate *in vitro* by protein kinase C. *J. Biol. Chem.* **267:** 11156–11162.
Santocanale, C., M. Foiani, G. Lucchini, and P. Plevani. 1993. The isolated 48,000-dalton

subunit of yeast DNA primase is sufficient for RNA primer synthesis. *J. Biol. Chem.* **268:** 1343–1348.

Sauer, K., J.A. Knoblich, H. Richardson, and C.F. Lehner. 1995. Distinct modes of cyclin E/cdc2c kinase regulation and S-phase control in mitotic and endoreduplication cycles of *Drosophila* embryogenesis. *Genes Dev.* **9:** 1327–1339.

Scheidtmann, K.-H. 1986. Phosphorylation of simian virus 40 large T antigen: Cytoplasmic and nuclear phosphorylation sites differ in their metabolic stability. *Virology* **150:** 85–95.

Scheidtmann, K.-H., M. Buck, J. Schneider, D. Kalderon, E. Fanning, and A.E. Smith. 1991. Biochemical characterization of phosphorylation site mutants of simian virus 40 large T antigen: Evidence for interactions between amino- and carboxy-terminal domains. *J. Virol.* **65:** 1479–1490.

Schneider, C., K. Weisshart, L.A. Guarino, I. Dornreiter, and E. Fanning. 1994. Species-specific functional interactions of DNA polymerase α-primase with simian virus 40 (SV 40) T antigen require SV 40 origin DNA. *Mol. Cell. Biol.* **14:** 3176–3185.

Schneider, J. and E. Fanning. 1988. Mutations in the phosphorylation sites of simian virus 40 (SV40) T antigen alter its origin DNA-binding specificity for site I or II and affect SV40 DNA replication activity. *J. Virol.* **62:** 1598–1605.

Seki, M., T. Enomoto, T. Eki, A. Miyahima, Y. Murakami, R. Hanaoka, and M. Ui. 1990. DNA helicase and nucleotide 5′ triphosphate activities of polyoma virus large tumor antigen. *Biochemistry* **29:** 1003–1009.

Seo, Y.-S., S.-H. Lee, and J. Hurwitz. 1991. Isolation of a DNA helicase from HeLa cells requiring the multisubunit human single-stranded DNA-binding protein for activity. *J. Biol. Chem.* **266:** 13161–13170.

Seo, Y.-S., F. Müller, M. Lusky, and J. Hurwitz. 1993a. Bovine papillomavirus (BPV)-encoded E1 protein contains multiple activities required for BPV DNA replication. *Proc. Natl. Acad. Sci.* **90:** 702–706.

Seo, Y.-S., F. Müller, M. Lusky, E. Gibbs, H.-Y. Kim, B. Phillips, and J. Hurwitz. 1993b. Bovine papillomavirus (BPV)-encoded E2 protein enhances binding of E1 protein to the BPV replication origin. *Proc. Natl. Acad. Sci.* **90:** 2865–2869.

Seroussi, E. and S. Lavi. 1993. Replication protein A is the major single-stranded DNA binding protein detected in mammalian cell extracts by gel retardation assays and UV-crosslinking of long and short single-stranded DNA molecules. *J. Biol. Chem.* **268:** 7147–7154.

Singh, K.K. and L. Samson. 1995. Replication protein A binds to regulatory elements in yeast DNA repair and DNA metabolism genes. *Proc. Natl. Acad. Sci.* **92:** 4907–4911.

Spalholz, B.A., A.A. McBride, T. Sarafi, and J. Quintero. 1993. Binding of bovine papillomavirus E1 to the origin is not sufficient for DNA replication. *Virology* **193:** 201–212.

Stadlbauer, F., A. Brueckner, C. Rehfuess, C. Eckerskorn, F. Lottspeich, V. Förster, B.Y. Tseng, and H.-P. Nasheuer. 1994. DNA replication *in vitro* by recombinant DNA-polymerase α-primase. *Eur. J. Biochem.* **222:** 781–793.

Stigger, E., F.B. Dean, J. Hurwitz, and S.-H. Lee. 1994. Reconstitution of functional single-stranded DNA-binding protein from individual subunits expressed by recombinant baculoviruses. *Proc. Natl. Acad. Sci.* **91:** 579–583.

Stillman, B. 1989. Initiation of eukaryotic DNA replication *in vitro*. *Annu. Rev. Cell Biol.* **5:** 197–245.

Suzuki, M., S. Izuta, and S. Yoshida. 1994. DNA polymerase α overcomes an error-prone pause site in the presence of replication protein-A. *J. Biol. Chem.* **269**: 10225–10228.

Syväoja, J. and S. Linn. 1989. Characterization of a large form of DNA polymerase δ from HeLa cells that is insensitive to proliferating cell nuclear antigen. *J. Biol. Chem.* **264**: 2489–2497.

Thömmes P. and Hübscher, U. 1990. Eukaryotic DNA replication: Enzymes and proteins acting at the fork. *Eur. J. Biochem.* **194**: 699–712.

Thömmes, P., E. Ferrari, R. Jessberger, and U. Hübscher. 1992a. Four different DNA helicases from calf thymus. *J. Biol. Chem.* **267**: 6063–6073.

Thömmes, P., R. Fett, B. Schray, R. Burkhart, M. Barnes, C. Kennedy, N.C. Brown, and R. Knippers. 1992b. Properties of the nuclear P1 protein, a mammalian homolog of the yeast Mcm3 replication protein. *Nucleic Acids Res.* **20**: 1069–1074.

Thorner, L.K., D.A. Lim, and M.R. Botchan. 1993. DNA-binding domain of papillomavirus type 1 E1 helicase: Structural and functional aspects. *J. Virol.* **67**: 6000–6014.

Tournier, M.-F., J. Sobczak, B. Nechaud, and M. Duguet. 1992. Comparison of biochemical properties of DNA-topoisomerase I from normal and proliferating liver. *Eur. J. Biochem.* **210**: 359–364.

Tse-Dinh, W.-C., T.W. Wong, and A.R. Goldberg. 1984. Virus and cell-encoded protein tyrosine kinases inactivate DNA topoisomerases *in vitro*. *Nature* **312**: 785–786.

Tsurimoto, T. and B. Stillman. 1989. Multiple replication factors augment DNA synthesis by two eukaryotic DNA polymerases, α and δ. *EMBO J.* **8**: 3883–3889.

———. 1991a. Replication factors required for SV40 DNA replication *in vitro*. I. DNA structure specific recognition of a primer-template junction by eukaryotic DNA polymerases and their accessory proteins. *J. Biol. Chem.* **266**: 1950–1960.

———. 1991b. Replication factors required for SV40 DNA replication *in vitro*. II. Switching of DNA polymerase α and δ during initiation of leading and lagging strand synthesis. *J. Biol. Chem.* **266**: 1961–1968.

———. 1991c. Purification of cellular replication factor, RF-C, that is required for coordinated leading and lagging strands during simian virus 40 DNA replication *in vitro*. *Mol. Cell. Biol.* **9**: 609–619.

Tsurimoto, T., T. Melendy, and B. Stillman. 1990. Sequential initiation of lagging and leading strand synthesis by two different polymerase complexes at the SV40 DNA replication origin. *Nature* **346**: 534–539.

Turchi, J.J., L. Huang, R.S. Murante, Y. Kim, and R.A. Bambara. 1994. Enzymatic completion of mammalian lagging-strand DNA replication. *Proc. Natl. Acad. Sci.* **91**: 9803–9807.

Tuusa, J., L. Uitto, and J.E. Syväoja. 1995. Human DNA polymerase ε is expressed during cell proliferation in a manner characteristic of replicative DNA polymerases. *Nucleic Acids Res.* **23**: 2178–2183.

Tye, B.-K. 1994. The MCM2-3-5 proteins: Are they replication licensing factors? *Trends Cell Biol.* **4**: 160–166.

Umbricht, C.B., L.F. Erdile, E.W. Jabs, and T.J. Kelly. 1993. Cloning, overexpression, and genomic mapping of the 14-kDa subunit of replication protein A. *J. Biol. Chem.* **268**: 6131–6138.

Ustav, E. and A. Stenlund. 1991. Transient replication of BPV-1 requires two viral polypeptides encoded by the E1 and E2 open reading frames. *EMBO J.* **10**: 449–457.

Ustav, E., M. Ustav, P. Szymanski, and A. Stenlund. 1993. The bovine papillomavirus origin of replication requires a binding site for the E2 transcriptional activator. *Proc. Natl. Acad. Sci.* **90:** 898–902.

Virshup, D.M., M.G. Kaufmann, and T.J. Kelly. 1989. Activation of SV40 DNA replication *in vitro* by cellular protein phosphatase 2A. *EMBO J.* **8:** 3891–3898.

Virshup, D.M., A.A.R. Russo, and T.J. Kelly. 1992. Mechanism of activation of simian virus 40 DNA replication by protein phosphatase 2A. *Mol. Cell. Biol.* **12:** 4883–4895.

Waga, S. and B. Stillman. 1994. Anatomy of a DNA replication fork revealed by reconstitution of SV40 DNA replication *in vitro*. *Nature* **369:** 207–212.

Waga, S., G. Bauer, and B. Stillman. 1994. Reconstitution of complete SV40 DNA replication with purified DNA replication factors. *J. Biol. Chem.* **269:** 10923–10934.

Wahl, A.F., A.M. Geis, B.H. Spain, S.W. Wong, D. Korn, and T.S.-F. Wang. 1988. Gene expression of human DNA polymerase α during cell proliferation and the cell cycle. *Mol. Cell. Biol.* **8:** 5016–5025.

Wang, E.H. and C. Prives. 1991. DNA helicase and duplex DNA unwinding activities of polyoma and SV40 large T antigen display similarities and differences. *J. Biol. Chem.* **266:** 12668–12675.

Wang, E.H., S. Bhattacharyya, and C. Prives. 1993. The replication functions of polyomavirus large tumor antigen are regulated by phosphorylation. *J. Virol.* **67:** 6788–6796.

Wang, J.C. 1985. DNA topoisomerases. *Annu. Rev. Biochem.* **54:** 665–697.

Wang, T.S.-F. 1991. Eukaryotic DNA polymerases. *Annu. Rev. Biochem.* **60:** 513–552.

Wang, Z., X. Wu, and E.C. Friedberg. 1993. DNA repair synthesis during base excision repair *in vitro* is catalyzed by polymerase ε and is influenced by polymerase α and δ in *Saccharomyces cerevisiae*. *Mol. Cell. Biol.* **13:** 1051–1058.

Wessel, R., J. Schweizer, and H. Stahl. 1992. Simian virus 40 T-antigen DNA helicase is a hexamer which forms a binary complex during bidirectional unwinding from the viral origin of DNA replication. *J. Virol.* **66:** 804–815.

Wobbe, C.R., L. Weissbach, J.A. Borowiec, F.B. Dean, Y. Murakami, P. Bullock, and J. Hurwitz. 1987. Replication of simian virus 40 origin-containing DNA with purified components. *Proc. Natl. Acad. Sci.* **84:** 1834–1838.

Wold, M.S. and T. Kelly. 1988. Purification and characterization of replication protein A, a cellular protein required for *in vitro* replication of simian virus 40 DNA. *Proc. Natl. Acad. Sci.* **85:** 2523–2527.

Wold, M.S., D.H. Weinberg, D.M. Virshup, J.J. Li, and T.J. Kelly. 1989. Identification of cellular proteins required for simian virus 40 DNA replication. *J. Biol. Chem.* **264:** 2801–2809.

Wong, S.W., L.R. Paborski, P.A. Fisher, T.S.-F. Wang, and D. Korn. 1986. Structural and enzymological characterization of immunoaffinity-purified DNA polymerase α-primase complex from KB cells. *J. Biol. Chem.* **261:** 7958–7968.

Wong, S.W., A.F. Wahl, P.-M. Yuan, N. Arai, B.E. Pearson, K.-I. Arai, D. Korn, M.W. Hunkapiller, and T.S-F. Wang. 1988. Human DNA polymerase α gene expression is cell proliferation dependent and its primary structure is similar to both prokaryotic and eukaryotic replicative DNA polymerases. *EMBO J.* **7:** 37–47.

Yan, H., A.M. Merchant, and B.K. Tye. 1993. Cell cycle-regulated nuclear localization of MCM2 and MCM3, which are required for the initiation of DNA synthesis at chromosomal replication origins in yeast. *Genes Dev.* **7:** 2149–2160.

Yang, L., M.S. Wold, J.J. Li, T.J. Kelly, and L.F. Liu. 1987. Roles of DNA topoisomerases in simian virus 40 DNA replication *in vitro*. *Proc. Natl. Acad. Sci.* **84:** 950–954.

Yang, L., I. Mohr, E. Fouts, D.A. Lim, M. Nohaile, and M. Botchan. 1993. The E1 protein of bovine papillomavirus 1 is an ATP-dependent DNA helicase. *Proc. Natl. Acad. Sci.* **90:** 5086–5090.

Zeng, X.-R., H. Hao, Y. Jiang, and M.Y.W.T. Lee. 1994a. Regulation of human polymerase δ during the cell cycle. *J. Biol. Chem.* **269:** 24027–24033.

Zeng, X.-R., Y. Jiang, S.J. Zhang, H. Hao, and M.Y.W.T. Lee. 1994b. Polymerase δ is involved in the cellular response to UV damage in human cells. *J. Biol. Chem.* **269:** 1348–1351.

Zwerschke, W., H.-W. Rottjakob, and H. Küntzel. 1994. The *Saccharomyces cerevisiae* CDC6 gene is transcribed at late mitosis and encodes an ATP/GTPase controlling S phase initiation. *J. Biol. Chem.* **269:** 23351–23356.

11
Control of S Phase

Kim Nasmyth
Institute for Molecular Pathology
A1030, Vienna, Austria

Sustained cell proliferation requires the duplication and segregation between daughter cells of every cell constituent. Because chromosomes are present in only one or two copies per cell, their duplication and segregation are particularly tightly regulated. To ensure that daughter cells inherit at least one copy of each of their genes and to maintain an appropriate balance in their relative numbers, the vast majority of eukaryotic cells replicate each sequence in their genomes once and only once per cell cycle. To ensure that daughter cells inherit sufficient but not excessive cytoplasm for the execution of their genetic programs, i.e., for protein synthesis, the duplication and segregation of chromosomes must occur no more or less frequently than cells duplicate the rest of their constituents. Finally, to ensure that cells maintain their ploidy, which is essential if they are to contribute to the germ line, reduplication must not recur until sister chromatids have been segregated at anaphase.

STARTING DNA REPLICATION IN G_1

Coordinating DNA Replication and Cell Growth

Pulse labeling with radioactive tracers showed that DNA replication occupies a defined window within the interdivision period, which is separated from the previous and preceding M phases by two gap periods, called G_1 and G_2, respectively (Howard and Pelc 1951). Chromosome duplication and segregation are therefore periodic processes. Unlike chromosomes, most cellular constituents like ribosomes and enzymes are present in large numbers and are synthesized fairly continuously during the cell cycle (Mitchison 1970). During sustained cell proliferation, successive rounds of DNA replication and chromosome segregation (known as the chromosome cycle) must occur with the same frequency as cells double the number of their ribosomes and other cell constituents, known as cell growth (see Fig. 1). Coordination between these two processes is

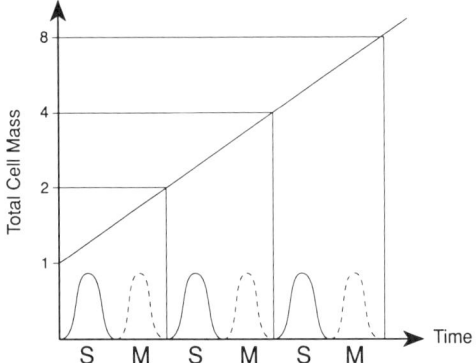

Figure 1 Coordinating the chromosome cycle and cell growth. Total cell mass increases exponentially (for a short while even in the absence of progression through the chromosome cycle), and the signals that trigger S and M phases (shown as oscillations on the abcissa) must occur with the same frequency as cytoplasmic mass doubling if nuclear/cytoplasmic ratios are to be kept within narrow bounds.

needed to maintain the ratio between nucleus and cytoplasm within certain bounds. It may be achieved either by a dependence of the chromosome cycle on the growth cycle or vice versa.

Independence of Cell Growth from the Chromosome Cycle

Dependence of cell growth on the chromosome cycle has been most carefully analyzed in the the fission yeast *Schizosaccharomyces pombe*, where increases in the rates of total protein or RNA synthesis are largely unaffected for several hours after inactivation of the Cdk1 (cdc2) protein kinase, which specifically blocks the chromosome cycle (Creanor and Mitchison 1984). To a first approximation, therefore, DNA replication does not limit the rate of cell growth. A similar conclusion can be drawn from the behavior of chromosome cycle mutants of the budding yeast *Saccharomyces cerevisiae* (Hartwell et al. 1974). A lack of DNA replication does eventually cause cell growth to slow down, after two to three generation times. Presumably, templates for mRNA synthesis eventually become limiting, at least for a few key growth genes.

Mutants in key cell cycle regulators do not exist to test the dependence of growth on the chromosome cycle in mammalian cells, so inhibitors have instead been employed. Some cell lines, such as HeLa, appear to shut down protein synthesis in the presence of DNA synthesis inhibitors, whereas others, such as CHO, continue cell growth like yeast *cdk1* mutants (Kung et al. 1993). Whether the growth cycle is largely independent of the chromosome cycle in animal or plant cells is therefore an unresolved but important issue that would best be investigated using primary cell lines, or better still, analyzing cells in their natural environment in vivo. Arrest in late G_1 using newly identified cyclin kinase inhibitors would be preferable to replication inhibitors, because the shutdown in protein synthesis seen in HeLa cells could be a specific response to arrest in S phase itself. Anecdotal evidence on the behavior of senescent cells, whose cell cycles are blocked in G_1, suggests that they become very large, indicating that their cell cycle arrest does not interfere with their growth (Lucibello et al. 1993). Many differentiated cells, for example, neurons, must continue growth long after they have ceased proliferating. An intrinsic independence of growth from the chromosome cycle would therefore not only favor microorganisms whose growth rates would be restricted by a tight linkage, but would also facilitate the formation of very differently sized differentiated cells in metazoa.

Dependence of S Phase on Growth

In the absence of any tight dependence of growth on the chromosome cycle, nuclear/cytoplasmic ratios are largely maintained by a dependence of the chromosome cycle on growth. The chromosome cycles of most eukaryotic cells are, with few exceptions, tightly dependent on RNA and protein synthesis. The dependence of DNA replication on cell growth was first noticed by Killander and Zetterberg, who observed that there was much greater variation in the size of fibroblast cells at birth than at the onset of S phase and proposed that DNA replication depends on growth to a critical cell size (Killander and Zetterberg 1965). A particularly clear example of this size control is found in the yeast *S. cerevisiae*, whose asymmetric division by budding gives rise to mother and daughter cells with very different sizes. Mother cells are born equal to or larger than the critical size and have very short G_1 periods, whereas the daughter cells are born well below the critical size and must spend long periods growing before they can enter S phase (Hartwell and Unger 1977). Many eukaryotic cells like amoeba, *Physarum*, and the fission yeast *S. pombe* have very short G_1 periods and appear to break this rule. Size controls over S-phase onset nevertheless exist, even in these organisms. In *S. pombe*, for example, cells are usually born larger than the size needed for S-phase onset due to size controls over entry into mitosis. Abolition of this mitotic control creates cells smaller than the critical size, which, like fibroblasts or budding yeast daughter cells, must grow considerably before they can undergo DNA replication (Nurse and Thuriaux 1977). Thus, size controls over the onset of S phase can be obscured by controls that restrict passage through other cell cycle transitions. An extreme example is the repeated rounds of replication without any growth during embryonic cleavage divisions, which are possible only because prior growth of the egg ensures that it can undergo many divisions before cells below a critical size are produced.

One consequence of the asymmetric interdependence of cell growth and the chromosome cycle is that sustained increases in cellular proliferation cannot be driven simply by accelerating the chromosome cycle but must instead be driven by the forces that cause cells to grow. This is well established for microorganisms, where accelerating the chromosome cycle merely leads to reductions in cell size (Nurse 1975), but it is not widely appreciated that it might be valid also for metazoan cell proliferation. It is often assumed that deregulation of chromosome cycle activators, for example Cdk1, -2, and -4 cyclin-dependent kinases, should accelerate cell division (Quelle et al. 1993) and thereby contribute to transformation. This can only be true if the chromosome cycle or regulators concerned with it also affect the rate of a cell's growth, which does not occur in microorganisms and remains to be established for mammalian cells.

Growth-related Protein Synthesis Needed for S-phase Onset

How cell growth to a critical size triggers S phase is an important but still largely unsolved problem. Size per se is not the critical variable but rather the nuclear/cytoplasmic ratio, because cell size is directly proportional to ploidy. One corollary of the observation that increases in the rate of protein synthesis are largely independent of the chromosome cycle is that a cell's rate of protein synthesis is largely proportional to cell size in growing G_1 cells. Thus, the total rate of protein synthesis in a cell may be a good measure of its size, and a critical level may be required for S-phase onset. The widespread observations that passage from G_1 to S phase is extremely sensitive to inhibitors of protein synthesis are consistent with this notion (Cross et al. 1989).

In yeast, there are at least two broadly different programs of protein synthesis, one occurring in quiescent stationary-phase cells, which includes synthesis of several heat shock and polyubiquitin proteins (Iida and Yahara 1984), and another occurring in growing cells, which includes synthesis of growth-related proteins like ribosomal proteins and presumably those needed specifically for the chromosome cycle. These two programs of protein synthesis are distinguished by their different dependence on the initiation factor eIF-4e, which binds to mRNA caps and is encoded by *CDC33* (Brenner et al. 1988). Growth-related but not stationary-phase-related protein synthesis depends on eIF-4e. The G_1 arrest of temperature-sensitive *cdc33* mutations at the restrictive temperature implies, not surprisingly, that S-phase onset depends on proteins synthesized as part of the growth-related program. How cells turn on the eIF-4e dependent and growth-related program of protein synthesis in response to nutrients or growth factors is therefore a very important aspect of S-phase control but is outside the realm of this review. Growth-factor- or nutrient-induced transcription of growth-related genes clearly also has an important part in the stimulation of growth-related protein synthesis, but changes in the translational apparatus that allow these mRNAs to be translated may be equally, if not more, important.

Two Parables about Growth and Division

Certain phenomena concerning the control of S phase cannot be understood without appreciating the distinction between cell growth and the chromosome cycle. A good example is the part played by cAMP in *S. cerevisiae*. Mutants defective in adenyl cyclase or cAMP-dependent protein kinases cannot sustain growth-related protein synthesis, and they arrest in G_1 (Matsumoto et al. 1982), from which it has been concluded that cAMP must be a positive effector of S phase. More recently, it has been found that cAMP is involved in modulating cell size in response to nutrients. Cells growing in poor media are smaller than those growing in rich media, principally because of a reduction in the critical size needed for entry into S phase and budding (Lorincz and Carter 1979). Mutants whose cAMP-dependent protein kinases cannot respond to cAMP cannot alter their size in response to nutrients (Tokiwa et al. 1994). Furthermore, addition of cAMP alone to the medium mimics the effects of transferring cells to a richer medium; that is, it increases the cell size needed for S phase. It is thought to do this by inhibiting the accumulation of G_1 cyclins needed for S phase (Baroni et al. 1994; Tokiwa et al. 1994). cAMP must therefore be considered as an inhibitor of S phase; but how can it be both an inhibitor and a promoter of S phase? This paradox is resolved only if one appreciates that cAMP is regulating two very different types of physiological processes, both of which are needed for S phase. On the one hand, it is involved (directly or indirectly) in signaling activation of growth-related protein synthesis, whereas, on the other hand, it inhibits the accumulation of mRNAs for G_1 cyclins, whose translation could be part of the growth-related protein synthesis program and which have highly specific roles in triggering S phase. In having these two effects, cAMP helps to increase the rate of protein synthesis in cells transferred to rich media while simultaneously increasing their cell size. The function of this size control is not clear, but it could be beneficial for the rapid execution of the chromosome cycle due to a cell size-related increase in protein synthetic capacity.

A factor that reduces growth-related protein synthesis while accelerating the onset of S phase (i.e., the opposite effect to cAMP) may have an important role in the differentiation of erythrocytes (Dolznig et al. 1995). Undifferentiated precursor cells whose proliferation is supported by mixture of growth factors can be induced to differentiate synchronously into mature nondividing erythrocytes by transfer to medium containing erythropoetin. The rate of protein synthesis remains fairly constant during the first 72 hours but gradually switches from growth-related proteins to proteins specific to erythrocytes like hemoglobin. During much of this period, cell division is temporarily accelerated due to a dramatic reduction in the length of G_1, which is all the more remarkable because the end product is a cell that cannot proliferate at all. Thus, like cAMP, erythropoetin has at least two effects: It promotes the chromosome cycle, presumably by increasing the activity of cyclin-dependent protein kinases (see later), but it also shuts down growth-related protein synthesis. The consequence of accelerating division but not cell growth is the production of erythrocytes that are many times smaller than their precursor cells. A moral to be drawn from these two stories is that changes in cell size are sometimes a better indicator of changes in the regulation of the chromosome cycle than are changes in mass doubling times.

S- and M-phase Promoting Factors

What is synthesized by growing cells that triggers DNA replication and then mitosis? What do oocytes store up that allows fertilized eggs to undergo successive cleavages in the absence of any growth? What do differentiating erythrocytes transiently make more of that reduces their G_1 period? Do cells by mysterious means measure all their constituents, for example, the number of ribosomes or the total rate of growth-related protein synthesis, or do they instead simply measure the quantity of one or a small number of components that are synthesized as part of the general growth program of the cell and whose abundance is proportional to cell size? Evidence for such a "sampling" mechanism has come from two approaches: the isolation of yeast mutants with an altered cell size and the identification of factors and genes needed for the onset of S phase.

Lessons from Cell Fusion Studies

Due to their large size, many multinucleate cells can be readily coalesced by cutting and grafting. This enabled early cell cycle researchers to address what would happen to the chromosome cycle when cells at different stages were combined into one (Johnson and Rao 1971). Experiments with ciliates, amoebae, and, in particular, the myxomycete plasmodia like *Physarum polycephalum* showed that nuclear synchrony was achieved very rapidly. Nuclei derived from "early" cells were accelerated, whereas those from "late" cells were retarded in their progression toward mitosis. The larger the plasmodium from late cells, the more rapidly were the early nuclei accelerated. These studies suggested that a substance capable of triggering mitosis gradually accumulated during G_2.

Most of the cells on which early coalescence experiments were performed have very short G_1 periods, and it was not possible to address the nature of S-phase inducers. These were eventually addressed in experiments with HeLa cells (Rao and Johnson 1970), where it was found that fusion between G_1 and S-phase cells advanced the entry of G_1 nuclei into S phase, but fusion of G_1 with G_2 cells did not. Significantly, DNA replication of S-phase nuclei was barely, if at all, retarded by the addition of G_1 cytoplasm and nuclei. Fusion between S and G_2 cells failed to induce S phase in G_2 nuclei, but again the addition of G_2 cytoplasm and nuclei did not retard DNA replication within S-phase nuclei. These results indicated that S but neither G_1 nor G_2 cells contained a substance (*S*-phase *p*romoting *f*actor or SPF) capable of inducing DNA replication in G_1 but not in G_2 nuclei. Similar experiments suggested that mitotic HeLa cells, like those from *Physarum* (Rusch et al. 1966), contain a substance (now refered to as MPF) that is capable of inducing premature M phase in nuclei at all other stages of the cell cycle (Johnson and Rao 1970). The picture that emerged from these cell fusion experiments is one in which the abundance of S- and M-phase inducers oscillates during the cell cycle with different phases. The former appear in late G_1 and disappear as cells enter G_2, and the latter appear in late G_2 and disappear at the end of mitosis.

An M-phase-promoting factor was eventually purified from mature *Xenopus* eggs (Masui and Markert 1971), which are arrested in metaphase of meiosis II, using an assay that measured its ability to induce chromosome condensation and nuclear membrane breakdown (Lohka et al. 1988). It proved to be identical to the cdc2 protein kinase, which had by this time been implicated in the control of mitosis in *S. pombe* due to the existence of rare dominant alleles that caused cells to divide at half the normal cell size (Nurse and Thuriaux 1980; Nurse 1990). No convenient biochemical assay was ever developed for S-phase-promoting factors, and the present candidates for these factors were identified by genetic studies of the yeast cell cycle.

S-phase-promoting Factors in *S. cerevisiae*

Genes Required for the Onset of S Phase

S. cerevisiae mutants defective in the chromosome cycle fall into two classes: those with primary defects in nuclear division and those with primary defects in entering S phase (Pringle and Hartwell 1981). However, all mutants with S-phase defects eventually also ceased to undergo nuclear division, whereas most mutants with M-phase defects also ceased eventually to replicate DNA, confirming earlier conclusions, drawn for the use of inhibitors, that S and M phases were interdependent. Why cells must normally complete M phase before they can rereplicate their chromosomes is an important question that we address below.

When *S. cerevisiae* cells reach a critical size, they initiate not only DNA replication but also budding, the first step toward cytokinesis, and spindle pole body duplication, the first step toward building a bipolar mitotic spindle. This point in the yeast cell cycle, which is called Start, is also the point at which haploid cells become refractory to cell cycle inhibition by sexual pheromones and the point at which protein synthesis and nutrients become less critical for completing the cell cycle (Pringle and Hartwell 1981). Mutants with primary defects in entering S phase fell into three groups according to their degrees of pleiotropy. The least pleiotropic class included, initially, mutants in only a single gene, *CDC7*, which encodes a protein kinase that is activated in late G_1 (Hollingsworth and Sclafani 1990). *cdc7* mutants fail to initiate DNA replication. Nevertheless, they bud normally, duplicate and separate their spindle pole bodies, form mitotic spindles, and switch off the budding process, with the result that they arrest as large single-budded cells containing an undivided nucleus with a fully formed mitotic spindle (Pringle and Hartwell 1981). Subsequent studies have shown that alleles of *DBF4* (Chapman and Johnston 1989), *CDC6* (Bueno and Russell 1992), *CDC46, CDC47, CDC54* (Hennessy et al. 1991), *MCM2*, and *MCM3* (Gibson et al. 1990) arrest with similar phenotypes. *DBF4* encodes a regulatory subunit of the Cdc7 protein kinase (Jackson et al. 1993), whereas *CDC46*, *MCM3*, and *MCM5* encode members of a family of related proteins with ATPase motifs characteristic of DNA helicases (Tye 1994), homologs of which are needed for the initiation of DNA replication also in fission yeast (Coxon et al. 1992) and in vertebrates (Thommes et al. 1992). The initiation of DNA replication might be the exclusive function of this group of genes.

The next class of mutations mapped to three genes: *CDC4*, *CDC34*, and *CDC53*. These mutants are more pleiotropic than the *CDC7* class, because they fail not only to enter S phase but also to form mitotic spindles (Pringle and Hartwell 1981). However, they duplicate their spindle pole bodies and form buds. Another of their characteristics is that they fail to shut off the budding process after the formation of the first bud, with the result that the mutant cells accumulate with multiple buds (Hartwell et al. 1974). The discovery that *CDC34* encodes an E2 ubiquitin-conjugating enzyme implicated in proteolysis suggests that protein degradation as well as synthesis is needed for S-phase entry and the formation of mitotic spindles in yeast (Goebl et al. 1988).

Cyclin-dependent Kinases

The last, most pleiotropic, class of mutants, represented initially by only a single allele of *CDC28* (Hartwell 1993), failed to undergo any of the four events associated with Start; they failed to enter S phase, to duplicate spindle pole bodies, to form buds, and (in haploids) to shut off sensitivity to cell cycle arrest by mating pheromones (Hartwell et al. 1974; Reid and Hartwell 1977; Pringle and Hartwell 1981). *CDC28* encodes a type of protein kinase (Lorincz and Reed 1984) whose activity is now known to require regulatory subunits called cyclins (Nurse 1990), which had been discovered as proteins whose abundance fluctuated during sea urchin cleavage divisions (Evans et al. 1983). How crucial a role the Cdc28 protein kinase might have in cell cycle control was only fully appreciated upon the subsequent discovery that *cdc2* from *S. pombe*, which had been shown to regulate mitosis, encoded a homologous protein kinase (Beach et al. 1982). Cdc2 from *S. pombe* and Cdc28 from *S. cerevisiae* are the founding members of a large family of similar cyclin-dependent kinases now known as Cdks; they have therefore a second name, Cdk1. Through their association with different cyclin subunits, the two yeast Cdk1 kinases (Forsburg and Nurse 1991a; Nasmyth 1993) have many different cell cycle functions: starting the cell cycle, initiating S phase, orchestrating nuclear division, and ensuring the dependence of S upon M phase, to name but a few. In mammalian cells, the Cdks that participate in these different events differ in both kinase and cyclin subunits (see below). Cyclin-dependent kinases are not exclusively concerned with cell cycle control. For example, distant relatives, cyclin H/Cdk7 from mammals (Fisher and Morgan 1994) and Ccl1/kin28 (Valay et al. 1993) from yeast have been implicated in the phosphorylation of the CTD tail of RNA polymerase II (Kim et al. 1994), whereas the yeast Cdk Pho85 regulates genes involved in phosphate metabolism through its phosphorylation of the Pho4 transcription factor (Kaffman et al. 1994).

Cdks as S-phase-promoting Factors

The first real indication that the onset of S phase might be triggered by cyclin-dependent kinases like Cdk1 (Cdc28) stemmed from the isolation of yeast mutants that started their cell cycles with a smaller than normal cell size. Mutants with a reduced cell size are particularly informative about cell cycle control, because this phenotype can only arise due to hyperactivation of the chromosome cycle (Nurse 1975), whereas mutants with an increased cell size can arise due to defects in any process needed for chromosome duplication or segregation. Dominant alleles of a gene now known as *CLN3* cause yeast to start the cell cycle at half the normal cell size (Sudbery et al. 1980; Cross 1988). *CLN3* encodes a protein distantly related to cyclins A and B, which are subunits of the Cdk1 kinase that promotes M phase in *Xenopus*. The mutant alleles encoded stable variants of Cln3 due to truncation of PEST sequences at its carboxyl terminus (Nash et al. 1988). Drawing on the analogy with mitotic cyclins and the knowledge that Cdk1 is needed for starting the yeast cell cycle, it was proposed that Cln3 might be a partner of Cdk1 and that a Cln3/Cdk1 kinase might trigger early yeast cell cycle events including S phase; i.e., it might correspond to an S-phase-promoting factor.

Deletion of *CLN3* greatly increases the size needed to start the yeast cell cycle, but it is not lethal, suggesting that other cyclins might exist with overlapping functions. Many other genes encoding cyclin partners of Cdk1 have subsequently been isolated, frequently as genes whose overexpression suppresses the temperature-sensitive growth of *cdk1* (*cdc28*) mutants: *CLN1* and *CLN2*, which encode a related pair of cyclins with only distant similarity to mitotic cyclins or to CLN3 (Hadwiger et al. 1989), and *CLB1*, *CLB2*, *CLB3*, *CLB4*, *CLB5*, and *CLB6*, which encode cyclins more homologous to the mitotic B-type cyclins from animals (Ghiara et al. 1991; Surana et al. 1991; Epstein and Cross 1992;

Schwob and Nasmyth 1993). None of these genes had previously been identified by conventional genetic analyses of the cell cycle because none of them are essential genes.

Do Clns and Clbs Correspond to SPFs and MPFs?

Because of the viability of single mutants, much of the phenotypic analysis of cyclin mutants has been performed on strains carrying multiple mutations. The finding that certain combinations of cyclin mutations cause arrest specifically in G_1 or G_2 has usually been taken to mean that the cyclin genes affected had similar functions. For example, inactivation of all three Cln cyclins causes cells to arrest indefinitely as unbudded G_1 cells, whereas all combinations of double mutants are viable, although their cells are much larger than normal (Richardson et al. 1989). The conclusion was that the three Cln cyclins had equivalent functions in promoting budding and S phase. Similar studies indicated, before the discovery of Clb5 and Clb6, that the four B-type cyclins Clb1–Clb4 all participated to a greater or lesser extent in the formation of mitotic spindles but were not needed for the initiation of S phase (Fitch et al. 1992; Richardson et al. 1992).

These two sets of G_1- and G_2-specific cyclins (i.e., Clns and Clbs) are all unstable proteins whose abundance, with the exception of Cln3, fluctuates during the cell cycle. Their periodic accumulation is partly due to transcriptional controls and partly due to fluctuations in their rates of proteolysis. All genes but *CLN3* are transcribed transiently during the cell cycle: *CLN1* and *CLN2* from late G_1 to the middle of S phase (Wittenberg et al. 1990), *CLB3* and *CLB4* from S phase till the end of mitosis (Fitch et al. 1992; Grandin and Reed 1993), and *CLB1* and *CLB2* during G_2 and M phases (Ghiara et al. 1991; Surana et al. 1991). Newly synthesized Cln1, Cln2, and Cln3 proteins have half-lives between 3 and 5 minutes (Deshaies et al. 1995; Yaglom et al. 1995), which have not been reported to vary during the cell cycle. The half-life of Clb2, in contrast, varies dramatically during the cell cycle, switching from 1 minute or less during the interval between the end of mitosis and Start to 60 minutes during the interval between Start and the onset of anaphase (Amon et al. 1994). The periodic accumulation of G_1- and G_2-specific cyclins is consistent with the notion that S phase is triggered by the accumulation or activation of Cln/Cdk1 complexes in late G_1, whereas M phase is triggered by the accumulation or activation of Clb/Cdk1 complexes in G_2. The case for Cln1 and Cln2/Cdk1 kinases being S-phase triggers is particularly strong, not only because these kinases are activated shortly before S phase, but also because the onset of S phase is advanced by premature activation of *CLN2* transcription in early G_1 (see below).

Clns Promote S Phase by Activating Clbs

The simple picture of Clns and Clbs corresponding to S- and M-phase-promoting factors, respectively, was dealt a mortal blow by the discovery of two further members of the B-type class of cyclins, Clb5 and Clb6. *CLB5* and *CLB6* mRNAs are of low abundance in early G_1 cells and, like those for *CLN1* and *CLN2*, accumulate to high levels shortly before S phase (Epstein and Cross 1992; Schwob and Nasmyth 1993). Clb5 associates with Cdk1, and the kinase activity of these complexes fluctuates during the cell cycle, being absent in early G_1 and appearing shortly before S phase, only slowly declining during G_2 and M phases, and finally disappearing as cells enter anaphase (Schwob et al. 1994). Like other cyclins, neither Clb5 nor Clb6 is essential, but the onset of S phase is greatly delayed in their absence (Schwob and Nasmyth 1993). S phase is normally coincident with budding, but in *clb5 clb6* double mutants it starts at least 30 minutes later.

New data show that B-type cyclins do not merely support Cln/Cdk1 kinases in triggering S phase but have a central role in this process. There is an important distinction between the S-phase delay of *cln1 cln2* double mutants and that of *clb5 clb6* double mutants. Both budding and S phase are delayed in the former but only S phase in the latter. The lesser pleiotropy of the Clb defect indicated that Clb5 and Clb6 might have a more direct role in the initiation of S phase than Cln1 and Cln2. However, if activation of the Clb5 or Clb6/Cdk1 kinases were "the" S-phase trigger, why is S phase merely delayed in the *clb5 clb6* double mutant? The surprising answer is that the supposedly "mitotic" B-type cyclins Clb1–Clb4 and not the G_1-specific Cln cyclins assume the S-phase-promoting functions of Clb5 and Clb6 in their absence. S phase is delayed yet longer in *clb3 clb4 clb5 clb6* quadruple mutants and never occurs in sextuple mutants lacking all six Clb cyclins, despite the activity of Cln1 and Cln2/Cdk1 kinases (Schwob et al. 1994). Activation of Cln/Cdk1 kinases is therefore not sufficient for initiating S phase; Clb/Cdk1 kinases are essential. In some circumstances, Clns are not even necessary. Cells arrested in G_1 by being deprived of all three Cln cyclins can be triggered to enter S phase highly synchronously by inducing *CLB5* or even *CLB2* expression from the *GAL1-10* promoter (Schwob and Nasmyth 1993; Amon et al. 1994). None of the *CLB* cyclin genes is efficiently transcribed in the absence of Cln cyclins, and *GAL*-driven expression circumvents this. These data indicate that it is through their activation of Clb cyclins that Cln/Cdk1 kinases promote S phase. An equally important conclusion is that the type of Cdk1 kinase activated in G_1 cells does not determine whether they undergo S phase or M phase (see below).

Regulation of S-phase-promoting Cyclins

Most growth-related mRNAs are absent in quiescent yeast cells and appear rapidly upon the addition of fresh nutrients. The same is true for *CLN3*. Its mRNAs are absent in stationary-phase cells, are induced within minutes of nutrient addition, but do not thereafter fluctuate during the cell cycle (Nash et al. 1988; Hubler et al. 1993). Cln3 protein levels behave similarly; in particular, their relative abundance does not greatly vary between the birth of daughter cells born below the size critical for Start and the onset of S phase (Tyers et al. 1993). Despite this lack of control, it is clear that the amount of Cln3 protein per cell is an important determinant of when yeast cells start the cell cycle; two- or more fold increases in *CLN3* gene dosage or mutations that stabilize Cln3 protein reduce the cell size at which cells bud and enter S phase (Cross 1988; Nash et al. 1988). Because Cln3 protein has a short half-life and its synthesis is not cell-cycle-regulated, its abundance per cell should reflect the general rate of protein synthesis. It is therefore an ideal candidate for a "sample" protein whose abundance is used by the cell to measure its rate of growth. Cln3 associates with Cdk1 (Tyers et al. 1992), and genetic data indicate that its main function is to activate Cdk1 (Cross and Blake 1993). The specific activity values of Cln3/Cdk1 complexes using histone H1 as substrate are, however, so low that it has not been possible to measure them as wild-type cells progress through G_1. It is not therefore clear whether the Cln3/Cdk1 kinase is in any way regulated.

In contrast to *CLN3*, the mRNAs from *CLN1*, *CLN2*, *CLB5*, and *CLB6* are all tightly cell-cycle-regulated, being low or absent in early G_1, accumulating as cells reach a critical size in late G_1, and declining as cells progress through S and G_2 phases (Koch and Nasmyth 1994). This pattern is part of a larger program of transcriptional control exerted by a pair of related transcription factors called SBF and MBF. The mRNAs of most genes involved in DNA replication and many involved in spindle pole body duplication have a similar pattern of accumulation, but the vast majority of these genes encode stable proteins, like DNA polymerase-α, whose de novo synthesis in the preceding G_1 is not needed for S-phase entry (Falconi et al. 1993). In contrast, the short half-lives of Cln1 and Cln2 and Clb5 and Clb6 mean that de novo synthesis of these proteins in late G_1 is an important if not essential step in triggering each and every S phase.

Late G_1-specific Transcription Factors

SBF and MBF bind to SCB and MCB elements found in the promoters of late G_1-specific genes. Both factors are composed of two subunits: a common regulatory subunit Swi6 that is complexed with either Swi4 or Mbp1 DNA-binding proteins (Koch and Nasmyth 1994). Swi4 and Mbp1 have a common architecture, both containing related site-specific DNA-binding domains at their amino termini (Primig et al. 1992; Koch et al. 1993), four related ankyrin motifs within their central regions (Breeden and Nasmyth 1987), and common sequences at their carboxyl termini through which they bind to their common partner Swi6 (Andrews and Moore 1992). Swi6, which lacks a DNA-binding domain, nevertheless has a similar structure, and all three proteins are presumably derived from a common ancestor. Although SBF (Swi4/Swi6) and MBF (Mbp1/Swi6) factors bind preferentially to SCB and MCB elements, respectively, polypeptides containing only their DNA-binding domains bind to both types of sequence. G_1 cyclin genes, such as *CLN1*, *CLN2*, and *HCS26*, are mainly regulated by SBF, whereas most genes encoding DNA replication enzymes are regulated by MBF. In some cases, genes regulated by one factor become in its absence regulated by the other (C. Koch, pers. comm.).

The Yeast Cell Cycle Is Started by the SBF/MBF Transcriptional Program

Several observations suggest that the onset of all events that occur at Start is triggered by the activation of transcription due to SBF and MBF. *CLN1* and *CLN2* mRNAs are lacking in small daughter cells and only appear along with other SBF/MBF-regulated mRNAs shortly before S phase when cells attain a critical size (Koch et al. 1993). Similar levels of a *CLN2* mRNA, albeit with a *MET3* leader sequence, can be induced in early G_1 cells from a *MET3-CLN2* promoter gene fusion, and this advances the onset of S phase and budding (L. Dirick et al., in prep.). More convincing, overexpression of *SWI4* from the *GAL* promoter in cells lacking Mbp1 causes mRNAs from most, if not all, genes normally regulated by SBF and MBF to appear in very small cells; that is, as soon as daughter cells are born (M. Neuberg et al., in prep.). The mRNAs are not overproduced but merely accumulate prematurely to levels found in wild-type cells shortly before S phase. This causes DNA replication and budding in cells whose size is one-third or less than the size needed for these two Start events in wild-type cells. The proliferation of very small cells by virtue of SBF hyperactivation is dependent on the presence of *CLN1* or *CLN2* but is totally independent of *CLN3*, whose activity normally controls the size at which yeast cells start the cell cycle. This suggests that Cln3's main, if not sole, function is to facilitate activation of the SBF/MBF transcriptional program (Tyers et al. 1993) and that the appearance of Cln1 and/or Cln2/Cdk1 kinases due to this program is crucial for the induction of S phase and budding.

Activation of SBF and MBF

How is transcription due to SBF and MBF normally activated only when cells reach a critical size? The transcriptional program fails to be activated or is greatly delayed in *cdk1* (*cdc28*) mutants (Dirick and Nasmyth 1991). It also fails to occur in mutants lacking all three Cln cyclins but is immediately restored by the reactivation of any one of them (Cross and Tinkelenberg 1991). The ability of Cln1 or Cln2/Cdk1 kinases to trigger the transcription of SBF/MBF-regulated genes suggested that the sudden activation of *CLN1* and *CLN2* transcription only when cells reach a certain size could be facilitated by a positive feedback loop by which Cln1 and Cln2 promote their own transcription along with other late G_1-specific genes. This hypothesis predicts that full activation of the transcriptional program should depend on the activities of Cln1 and Cln2, which is not the case. mRNAs known to be regulated by SBF or MBF accumulate with similar kinetics and at similar cell sizes in wild type and in *cln1 cln2* double mutants, but accumulation of the same mRNAs is delayed by several hours in *cln3* mutants (Dirick et al. 1995). These data emphasize that the three Clns do not have equivalent functions. Cln3 is essential for activating SBF/MBF-mediated transcription at normal cell sizes, whereas Cln1 and Cln2 are not.

The mechanisms by which Cln3/Cdk1 activates the SBF and MBF transcription factors only when cells reach a certain size are not understood. Genomic footprinting has shown that functionally important SCBs in the *CLN2* promoter are fully occupied by SBF in cells arrested in G_1 due to a temperature-sensitive *cdk1* mutation. Lowering the temperature of these mutant cells restores Cdk1 kinase and leads to the immediate activation of *CLN2* transcription, suggesting that Cdk1 alters the activity of complexes that are already bound to SCBs (Koch et al. 1996).

Regulation of Clb/Cdk1 Kinases during G_1

Despite their lack of involvement in the activation of gene expression, Cln1 and Cln2 have a crucial role in triggering the onset of both budding and S phase (Hadwiger et al. 1989), which are advanced by premature expression of *CLN2* and which are delayed in *cln1 cln2* double mutants until cells have grown at least three times the normal size (Dirick et al. 1995). Current data are consistent with the notion that Cln1 and Cln2 promote S phase by stimulating activation of Clb/Cdk1 kinases. They do this by facilitating the accumulation of Clb proteins (Amon et al. 1994) and the disappearance of a Cdk1 inhibitory protein p40Sic1 (Schwob et al. 1994).

Rapid proteolysis of Clb2, which is largely shut off during S and G_2 phases, commences during anaphase and persists until activation of Cln1 and Cln2/Cdk1 kinases during the subsequent G_1 period. The extremely short half-life of Clb2 during early G_1 (1 minute or less) ensures that it cannot accumulate in G_1 cells lacking Cln cyclins, even when highly expressed from the *GAL* promoter (Amon et al. 1994). Clb5 protein, with a half-life of 5 minutes (Seufert et al. 1994), is more stable during this period. Cells arrested in G_1, due to deprivation or physiological inactivation of all three Clns, undergo S phase (S. Irniger, pers. comm.) upon inactivation of proteins such as Cdc16, Cdc23, and Cse1, which are required for Clb2 proteolysis (Irniger et al. 1995). This implies that the persistence of rapid Clb proteolysis until the activation of Cln1 and Cln2 could have an important role in preventing premature accumulation of Clb/Cdk1 kinase activity.

Role of a Cdk Inhibitor, p40Sic1

Probably the most important mechanism regulating the onset of S phase in yeast is the late G_1-specific destruction of a Cdk1 inhibitory protein p40Sic1 (Schwob et al. 1994). It had been found that transcription of *CLB5* in small early G_1 cells failed to advance S phase. This was surprising because premature Cln2 expression did advance S phase and yet it was thought that Clb5 had a more direct role than Cln2 in the initiation of DNA replication. The ectopic expression caused Clb5 protein accumulation, but the resulting Clb5/Cdk1 complexes were inactive due to the presence in early G_1 cells of a factor capable of inhibiting the Clb5/Cdk1 kinase. The inhibitory factor is a 40-kD protein encoded by the *SIC1* gene (Nugroho and Mendenhall 1994), which had previously been purified as a potential substrate of Cdk1 and found to be an inhibitor (Mendenhall 1993). p40Sic1 is a potent inhibitor of Clb/Cdk1 but not of Cln1 or Cln2/Cdk1 kinase activities (M. Mendenhall, pers. comm.).

Early G_1 cells contain high levels of p40Sic1 protein, but it disappears shortly before the onset of S phase and does not reappear until cells complete mitosis. This pattern of accumulation is partly due to the periodic accumulation of *SIC1* mRNAs, whose abundance increases transiently from a basal level as cells exit from mitosis (Donovan et al. 1994; Schwob et al. 1994). However, cell-cycle-dependent proteolysis mediated by the Cdc34 ubiquitin-conjugating enzyme is implicated in the sudden disappearance of p40Sic1 in late G_1. The inhibitor accumulates at the end of mitosis to similar levels in wild type and *cdc34* mutants but fails to disappear in the latter when cells ought to be entering S phase (Schwob et al. 1994). This defect is highly specific, because the SBF/MBF transcriptional program, activation of Cln1 and Cln2/Cdk1 kinases, and budding all take place in *cdc34* mutants. The accumulation of p40Sic1 probably inhibits activation of all Clb/Cdk1 kinases in *cdc34* mutants, and this must be responsible for their G_1 arrest because deletion of the *SIC1* gene permits them to undergo DNA replication. The G_1 arrest of *cdc4* and *cdc53* mutants is also due to a failure to destroy p40Sic1.

These data suggest that proteolysis of p40Sic1 mediated by Cdc4, Cdc34, and Cdc53 proteins is an essential step toward the initiation of S phase in *S. cerevisiae*. How then is this process regulated? p40sic1 protein persists in cells arrested in G_1 due to triple *cln* or *cdk1* mutations but not in sextuple *clb1-6* mutants (T. Bohm, pers. comm.), suggesting that Cln but not Clb Cdk1 kinases play a part in its destruction. Unlike SBF/MBF-mediated transcription, the disappearance of p40Sic1 is delayed in *cln1 cln2* double mutants, and this is largely responsible for their delayed S-phase entry (Dirick et al. 1995). Thus, Cln3 is insufficient and Cln1 and Cln2 are necessary to trigger disappearance of p40Sic1 on schedule. p40Sic1 is not an effective inhibitor of the Cln2/Cdk1 kinase either in vivo or in vitro but is an excellent substrate, at least in vitro (M. Mendenhall, pers. comm.). This raises the possibility that phosphorylation of p40Sic1 by Cln1 or Cln2/Cdk1 kinases might be the trigger for its proteolysis via Cdc34. Consistent with this notion, *sic1* mutations that alter potential Cln/Cdk1 phosphorylation sites delay the disappearance of p40Sic1 protein and the onset of S phase (T. Bohm, pers. comm.).

Cdc7

A piece of the S-phase regulatory jigsaw that has not yet been satisfactorily fitted into the above picture is the role and regulation of the Cdc7 kinase. Kinase activity associated with Cdc7 rises in late G_1 along with that of Clb/Cdk1 kinases (Jackson et al. 1993; Yoon et al. 1993). *cdc7* mutants fail to initiate S phase despite possessing active Clb/Cdk1 kinases. Cdc7 is therefore needed in addition to Clb/Cdk1 kinases. Cdc7 associates with the Dbf4 protein, which is also needed for replication and could be a regulatory subunit with properties (but not sequences) similar to cyclins (Jackson et al. 1993). Dbf4 associates in vivo with proteins that bind to replication origins, suggesting that some of these are substrates for the Cdc7 kinase (Dowell et al. 1994). One scenario is that Clbs activate the replication apparatus via their activation of Cdc7; i.e., Clb/Cdk1 kinases are upstream of Cdc7 on a linear pathway leading to replication (Jackson et al. 1993). It is more likely, however, that Clb/Cdk1 kinases have several roles in activating replication, only one of which might be to promote activation of the Cdc7 kinase. The need for Cdc7 and Dbf4 can be bypassed by *bob1* mutations, suggesting that Cdc7's normal function is to relieve a negative control.

Summary: Promoting S Phase in S. cerevisiae

There now exists a detailed, but by no means complete, hypothesis for how S-phase-promoting factors are produced in *S. cerevisiae* (see Fig. 2). Let us start with a small quiescent cell that has just been inoculated into fresh medium. The presence of nutrients induces the appearance, largely due to transcriptional controls, of most, if not all, mRNAs needed for cell growth, whose translation is facilitated by the simultaneous activation of translational initiation factors like eIF-4e. Neither of these events is dependent on the forthcoming chromosome cycle, but both are essential for the eventual production of specific S-phase-promoting factors. One of the many proteins whose synthesis is stimulated in this manner is Cln3, a Cdk1 cyclin, whose abundance per cell, due to its short half-life, reflects the overall rate of translation of "growth" mRNAs and thereby the size and growth rate of the cell. Few, if any, other S-phase-promoting factors are made in these small growing cells until they reach a threshold size, at which Cln3/Cdk1 suddenly activates transcription of a large battery of genes, some of which encode unstable replication proteins like Cdc6 (Piatti et al. 1995), whereas others encode further S-phase-promoting cyclins like Cln1 and Cln2 and Clb5 and Clb6. The size at which this transcriptional explosion occurs determines the size at which cells enter S phase and is regulated by external factors like nutrients, which exert their effect via cAMP; an increase in the activity of cAMP-dependent protein kinases in rich medium increases the size needed to trigger the appearance of *CLN1* and *CLN2* mRNAs. With the activation of the late

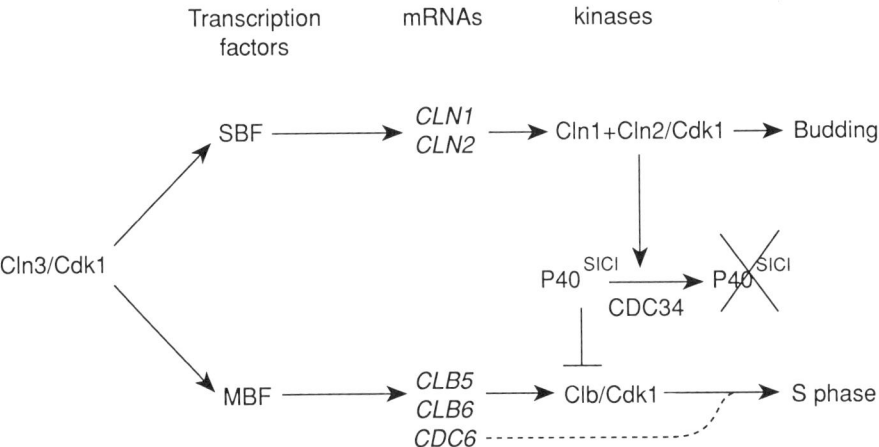

Figure 2 The production of S-phase-promoting factors in *S. cerevisiae*. The process starts with activation of transcription due to SBF and MBF transcription factors when cells reach a critical size. This leads to the synthesis of Cln1/2 and Clb5/6 cyclins and replication proteins like Cdc6 which are needed for the formation of pre-RCs at future replication origins (Cdc6 is also synthesized at the end of mitosis). Clb5 and Clb6/Cdk1 complexes remain inactive until the p40Sic1 inhibitor is destroyed via ubiquitin-mediated proteolysis involving Cdc4, Cdc34, and Cdc53, an event that is thought to be triggered by Cln1 and Cln2/Cdk1 kinases.

G_1-specific transcription program, the "baton," as it were, passes from Cln3 to Cln1 and Cln2. Accumulation of Cln1 or Cln2/Cdk1 kinases, due largely to activation of the SBF transcription factor by Cln3/Cdk1, leads to the next crucial step: activation of Clb/Cdk1 kinases, which are needed to activate the Cdc7 kinase and, presumably, many other replication proteins.

The absence of Clb/Cdk1 kinases in early G_1 cells is due partly to a lack of transcription of *CLB* genes, partly to rapid proteolysis which is started during anaphase but persists during early G_1, and partly to the accumulation at the end of mitosis or in stationary phase of the p40Sic1 protein, which is a potent inhibitor of Clb/Cdk1 kinases. Cln3/Cdk1 induces synthesis of Clbs 5 and 6, whereas Cln1 and Cln2/Cdk1 kinases promote accumulation of active Clb/Cdk1 kinases by drastically reducing proteolysis of cyclins like Clb2 and, more important still, by triggering proteolysis of p40Sic1. How activation of Clb/Cdk1 and Cdc7 kinases triggers the formation of replication forks and why they do this in G_1 but not in G_2 cells will require consideration of the substrates of these kinases.

S-phase-promoting Factors in Other Fungi

A single Cdk kinase related to Cdk1 (known as Cdc2) is needed for the onset of both S and M phases in two other ascomycetes, the fission yeast *S. pombe* (Nurse and Bissett 1981) and the filamentous fungus *Aspergillus nidulans* (Osmani et al. 1994). Less is known, however, about the cyclin partners of Cdk1 (Cdc2) that promote S phase in these two organisms. The *puc1* gene in *S. pombe* encodes a cyclin related in sequence to Cln3 (Forsburg and Nurse 1991b), but there is little or no evidence that *puc1* helps to promote the onset of S phase in vegetative cells (Forsburg and Nurse 1994). This may be less surprising now that we realize that Cln cyclins have only an indirect role in promoting S phase, even in *S. cerevisiae*, and that B-type cyclins are more intimately involved. *S. pombe* contains at least three B-type cyclins encoded by the *cig1* (Bueno et al. 1991), *cig2* (Bueno and Russell 1993; Connolly and Beach 1994; Obara Ishihara and Okayama 1994), and *cdc13* genes (Booher and Beach 1988; Hagan et al. 1988). None is necessary for the G_1-to-S-phase transition in vegetative cells, but *cdc13* is essential for M phase. There are indications that these B-type cyclins are nevertheless important for promoting S phase. For example, *cig2* is important for the terminal rounds of DNA replication that occur as *S. pombe* cells enter stationary phase upon nitrogen starvation (Obara Ishihara and Okayama 1994), and deletion of both *cig1* and *cig2* causes nuclei in vegetatively growing cells to spend longer in G_1 (Connolly and Beach 1994). *S. pombe* might therefore resemble *S. cerevisiae*, in that one of several B-type cyclins is sufficient for triggering S phase.

S phase in *S. pombe* depends on the activation of transcription in late G_1 of key replication genes by a pair of transcription factors Res1/Cdc10 (Tanaka et al. 1992; Caligiuri and Beach 1993) and Res2/Cdc10 (Tanaka et al. 1992; Zhu et al. 1994), which are related to Swi4/Swi6 and Mbp1/Swi6 in *S. cerevisiae*. Cdc10, which resembles Swi6 (Breeden and Nasmyth 1987), is necessary for both factors and is essential for the G_1-to-S-phase transition in *S. pombe* (Nurse et al. 1976). Res1/Cdc10 complexes are more important for S phase in vegetatively growing cells, whereas Res2/Cdc10 complexes are more important for premeiotic DNA replication. Fewer genes involved in DNA replication are regulated by these transcription factors than are regulated by their equivalent factors in *S. cerevisiae*. Nevertheless, transcriptional activation in late G_1 of at least two genes, *cdc18* (a homlog of *CDC6* from *S. cerevisiae* [Kelly et al. 1993]) and *cdt1*, is essential for initiation of the ensuing S phase (Hofmann and Beach 1994). Neither *cig1* nor *cdc13* mRNA are cell-cycle-regulated, but *cig2* mRNAs are periodic (Connolly and Beach 1994), accumulating to maximal levels in late G_1, and could therefore be regulated by one of the two Res/cdc10 factors. As in *S. cerevisiae*, Cdk1 is needed for late G_1-specific gene activation; in this case, there is evidence that Cdk1 is needed for the DNA-binding activity of Res/cdc10 factors (Reymond et al. 1993). A zinc-finger-containing protein, called Rep2, is needed for the activity of Res2/cdc10 but not Res1/cdc10 complexes in mitotic *S. pombe* cells (Nakashima et al. 1995).

Whether changes in the proteolysis of B-type cyclins in *S. pombe* has any role in regulating their activity during G_1 is not known. There exists, however, a Cdk inhibitory protein, encoded by the *rum1* gene, which delays the onset of S phase under certain circumstances (Moreno and Nurse 1994). In fast-growing cells, daughter cells are born larger than the threshold size needed to start the cell cycle, G_1 is extremely short, and deletion of *rum1* has little or no effect. Inactivation of the Wee1 protein kinase eliminates size control over M phase, and daughter cells are now born below the critical size and must spend one-third of their next cycle growing large enough to enter S phase. This longer G_1 period of *wee1* mutants is largely eliminated by deletion of the *rum1* gene. Rum1 is also implicated in delaying or even preventing activation of B-type cyclin/Cdk1 kinases in nitrogen-starved cells and in cells arrested in G_1 due to *cdc10* mutations. The Rum1 protein is a potent inhibitor of Cdc13/Cdk1 kinase (Correa-Bordes and Nurse 1995), but it has little or no sequence similarity to p40Sic1 from *S. cerevisiae* or Cdk inhibitors from mammalian cells, nor is it known whether or how Rum1 abundance is regulated during the cell cycle. There have been no measurements of specific B-type cyclin/Cdk1 kinase activities during G_1, so it is difficult at this stage to paint even a crude picture of how these kinases are regulated during this stage of the *S. pombe* cell cycle.

Although the details differ considerably, the broad outlines of S-phase control seem conserved between *S. pombe* and *S. cerevisiae*; this includes the synthesis of replication proteins like spCdc18 or scCdc6 due to Cdc10/Swi6-like transcription factors, the role of B-type cyclin Cdk kinases, and the control of these kinases by inhibitory proteins.

S-phase-promoting Factors in Animal Cells

It is not possible to assemble as detailed a picture of the events leading to S phase in animal cells as in yeast, not because fewer components have been identified (the advent of two-hybrid screening has largely solved this problem), but because their functional analysis is still much harder (function cannot be as reliably inactivated by antisense RNA or even antibody injections as by gene mutation, which although possible in mice, is still laborious and difficult to make conditional) and because studies are performed on many different types of cells whose behavior can be very different. Nevertheless, one can detect several parallels between animal cells and yeast, which are worth pointing out since they may represent the more highly conserved aspects of S-phase control.

The Restriction Point

It has long been known that de novo protein synthesis during G_1 is needed for S phase in tissue-culture cells. However, entry into S phase becomes less sensitive to inhibitors of protein synthesis about 2 hours before DNA replication commences. It is around this point in late G_1, called the restriction point, that many mammalian tissue-culture cells become capable of entering S phase upon withdrawal of growth factors, which are potent effectors of the cell's rate of protein synthesis (Pardee 1989). G_1-specific cyclins and their catalytic subunits may be some of the proteins whose synthesis preceding the restriction point is stimulated by growth factors and is needed for the initiation of DNA replication.

G_1-specific Cdks

Whereas ascomycetes use the same Cdk subunit to regulate S and M phases, animal cells use at least three different kinase subunits. Cdk1 (cdc2) for M phase (Riabowol et al. 1989; Hamaguchi et al. 1992), Cdk2 for S phase (Fang and Newport 1991; Paris et al. 1991), and Cdk4 or Cdk6 for progression from early to late G_1 (Sherr 1994a). As in yeast, multiple cyclins associate with a given Cdk. Cdk4 and Cdk6 associate with D-type cyclins (Matsushime et al. 1992). Cyclin D1 is absent in quiescent cells and is induced by mitogenic growth factors (Matsushime et al. 1991), reaching maximal levels several hours after their addition but long before the onset of S phase. Cyclin D1 has a half-life of around 30 minutes, which, as a percentage of the doubling time, is similar to that of yeast G_1 cyclins. It accumulates in the nucleus during G_1, but its abundance varies little if at all during the cell cycle of exponentially growing cells (Sewing et al. 1993). Antibody injection experiments had suggested that cyclin D1 is essential for progression from G_1 to S phase in several different types of cells (Baldin et al. 1993; Lukas et al. 1994). However, mouse embryos lacking any cyclin D1 gene can undergo all prenatal development (Sicinski et al. 1995). Whether this means that the effects of antibody injection were less specific than anticipated or that the cell lines used in such studies are more dependent on cyclin D1 than most embryonic cells remains to be determined. It is worth noting that even in yeast few if any cyclin gene deletions are lethal, despite having drastic effects on the kinetics of specific cell-cycle transitions, and the same could be true for mice.

Another means of testing the role of D-type cyclins in promoting S phase has been to analyze the consequences of inducing high levels of synthesis during early G_1. Hyperaccumulation of cyclin D1 caused a modest reduction in G_1 length in cycling cells and a more persuasive reduction during the outgrowth of previously quiescent cells (Quelle et al. 1993; Resnitzky et al. 1994). In a study using regulated cyclin D expression, hyperactivation did not, however, increase the rate of cell proliferation (although one suspects that it reduced cell size), which is consistent with similar findings in yeast and suggests that the G_1-to-S-phase transition is not a rate-determining step for proliferation, at least for the tissue-culture cells used in these studies. Similar experiments, performed with cyclin D transgenes in mice, have little immediate effect on cell proliferation but do eventually give rise to tumors, especially in conjunction with other oncogenes (Bodrug et al. 1994; Lovec et al. 1994).

Cdk2 is found mainly associated with cyclins E and A in mammalian cells. Both proteins oscillate in phase with the chromosome cycle, with cyclin E accumulating to maximal levels in late G_1 (Dulic et al. 1992; Koff et al. 1992), possibly coincident with the restriction point, and cyclin A somewhat later during S and G_2 phases (Pines and Hunter 1990; Tsai et al. 1991). Antibody injection experiments suggest that Cdk2 and cyclin A (Girard et al. 1991; Pagano et al. 1992; Zindy et al. 1992; Tsai et al. 1993) are essential for S phase in mammalian tissue-culture cells. Cyclin A, however, has been found unnecessary for DNA replication in extracts from *Xenopus* eggs and for at least certain divisions during *Drosophila* embryogenesis (Lehner and O'Farrell 1990; Knoblich and Lehner 1993). Cyclin E, in contrast, seems essential for replication in *Drosophila* and *Xenopus*. Maternal cyclin E transcripts disappear suddenly as *Drosophila* cells complete the last rapid mitotic division during embryogenesis (cycle 16) and start to accumulate for the first time in G_1, but the mRNAs reappear due to zygotic gene expression shortly before S phase in the minority of cells that continue cell division at this point; they also reappear during late embryogenesis in tissues that undergo endoduplication cycles. More convincing still, cyclin E mutants cease all DNA replication after cycle 16, and ectopic expression of cyclin E from a heat shock promoter induces S phase in cells that would otherwise have arrested in G_1; however, this only occurs if expression is induced soon after their exit from the previous mitosis (Knoblich et al. 1994). *Drosophila* cyclin E protein associates with a Cdk that resembles the mammalian Cdk2. In mammalian tissue culture, the artificial induction of high levels of cyclin E is insufficient to cause quiescent cells to enter S phase, but it does cause a modest reduction in the length of their G_1 period and thereby enables cells to divide at a smaller cell size (Ohtsubo and Roberts 1993; Resnitzky et al. 1994). It does not, however, alter their rate of proliferation. Thus, the accumulation of cyclin E in late G_1 may have an important role in the onset of S phase in both arthropod and vertebrate cells. Neither mammalian nor *Drosophila* cells seem to require Cdk1 (cdc2) or B-type cyclins for S phase.

Cdk Inhibitors

As in yeast, the activities of D and E type cyclin/Cdk complexes are not just regulated by the pattern of cyclin accumulation. There is a family of small proteins (p15, p16, and p18) composed of ankyrin motifs analogous to those first found in Swi6 (Breeden and Nasmyth 1987), which, it is thought, inactivate Cdk4 and Cdk6/cyclin D complexes by binding to the Cdk subunits and displacing cyclin D (Serrano et al. 1993; Guan et al. 1994; Hannon and Beach 1994). Expression of p15 is induced in keratinocytes by antimitogenic factors like transforming growth factor β (TGF-β). p16INK4, on the other hand, increases in abundance following the stimulation of fibroblasts with serum (Tam et al. 1994); its accumulation to maximal levels during S phase might play a role in inhibiting Cdk4 and Cdk6 kinases as cells enter G_2. Whether any of these inhibitors has a role in preventing premature activation of cyclin D/Cdk kinases during early G_1 is unclear. Transient overproduction of p16INK4 causes cells to arrest in G_1, and mutations in the INK4/MST1 gene for p16 are found in most tumor cell lines and in some primary tumors (Hunter and Pines 1994), suggesting that p16 could have an important role in arresting the proliferation of potential tumor cells.

Other cyclin kinase inhibitors, p21CIP1/WAF1/SDI1 (El-Deiry et al. 1993; Harper et al. 1993) and p27KIP1 (Polyak et al. 1994), bind to and inhibit a wide variety of Cdk kinases including cyclin D/Cdk4, cyclin E/Cdk2, and cyclin A/Cdk2. p27 levels decline modestly as quiescent macrophage cells are stimulated to enter S phase by growth factors (Kato et al. 1994). A proposal that this decline is important for entry into S phase remains to be tested rigorously. p27 abundance declines rapidly in T cells in response to interleukin 2 (IL-2), which could be an important event in the induction of S phase by this factor (Firpo et al. 1994). p21 levels, in contrast, rise upon growth factor stimulation, which might have a role in preventing activation of S-phase-promoting Cdks before cells have made other preparations for S phase. It is still unclear, however, whether inhibitors like p21 and p27 exert a strong effect on the kinetics of S-phase entry in cycling cells. Both certainly have the potential to block cell cycle progression. A good example is the increase in abundance of p21 in response to DNA damage, which is due to activation of p21WAF1 transcription by p53 and has a role in preventing the entry of irradiated cells into S phase (Deng et al. 1995). Other modes of p21 induction, for example, its induction by growth factors, do not involve p53 (Parker et al. 1995), and it is unknown whether p21 levels are sufficient to delay activation of S-phase-promoting Cdks in cells lacking DNA damage and, if so, how cells surmount this barrier.

Phosphorylation

The activities of all Cdks are dependent on phosphorylation of a highly conserved residue (Thr-161 [Solomon et al. 1992]), but the kinase responsible, which is another cyclin-dependent kinase composed of Cdk7(MO15) and cyclin H (Fesquet et al. 1993; Fisher and Morgan 1994; Solomon et al. 1993), and which seems identical to the kinase that phosphorylates the CTD of RNA polymerase II (Roy et al. 1994), is not apparently cell-cycle-regulated (Tassan et al. 1994). Phosphorylation at other conserved sites on Cdk1 (Tyr-15), which inhibits its kinase activity during G_2 in yeast and in animal cells, is reversed by the Cdc25 (*S. pombe*) family of phosphatases. Vertebrates possess at least three members of Cdc25 phosphatases; Cdc25A undergoes phosphorylation and activation in late G_1 and could have a role in the activation of Cdk2 kinases shortly before S phase (Hoffmann et al. 1994).

Late G_1-specific Transcription

As found in yeast, transcripts from many genes implicated in DNA replication or its control oscillate in abundance during the cell cycle, reaching maximal levels in late G_1; examples include mRNAs for thymidine kinase (Dou et al. 1992), dihydrofolate reductase (Bjorklund et al. 1990), cdc2 (Dalton 1992), DNA polymerase-α (Pearson et al. 1991), and cyclin E (Reed et al. 1992). A family of transcription factors collectively called E2F are implicated in this transcriptional program (La Thangue 1994). Most of these factors are heterodimers containing an

E2F and a DP1 subunit. There are several subtypes of each of these subunits, creating many different potential heterodimers, whose regulatory properties almost certainly differ. E2F binds to sequences that resemble the SCBs and MCBs bound by SBF and MBF in yeast, but E2F and DP1 have no sequence similarities with Swi4 or Swi6.

Analyzing patterns of mRNA accumulation in cells lacking particular genes will ultimately be necessary to establish which E2F factors are responsible for regulating which late-G_1-specific genes. Mutation of an E2F gene in *Drosophila* abolishes both S phase and developmentally regulated pulses of DNA replication enzyme mRNAs (Duronio and O'Farrell 1994; Duronio et al. 1995). Analyzing similar mutants in the mouse will undoubtedly be complicated by potential overlaps in the function of different E2F and DP-1 genes, as has been found for SBF and MBF in yeast. Another complication is that genes thought to be activated or controlled by E2F-binding sites do not have a common mode of regulation. The c-*myc* gene, for example, whose promoter contains an E2F site implicated in its transcriptional regulation, is regulated in a very different manner from many other supposedly E2F-regulated genes.

Activation of E2F in Late G_1

At least three different regulatory proteins related to the product of the retinoblastoma gene (Rb itself, p107, and p130) are found associated with different E2Fs during G_1 (Hinds and Weinberg 1994). Rb associates with E2F-1, E2F-2, and E2F-3 and can inhibit their ability to activate transcription in vitro (Dynlacht et al. 1994). Rb is phosphorylated at many residues throughout the cell cycle, but two or three specific residues become phosphorylated only in late G_1 and remain so until the end of mitosis (Sherr 1994b). This hyperphosphorylation prevents Rb from binding to E2F. It is therefore proposed that Rb hyperphosphorylation caused by cyclin D- or cyclin E-associated Cdks that become active in late G_1 enables accumulation of uncomplexed E2F, which is free to activate transcription. Although attractive as a regulatory mechanism, this theory has never been rigorously tested; for example, by characterizing the properties of Rb proteins in which residues whose phosphorylation is cell-cycle-regulated (and not others) have been mutated. The current hypothesis would predict that expression of modest levels of proteins that lack the ability to be hyperphosphorylated should be sufficient to prevent the activation of genes regulated by E2F.

There are two other potential mechanisms for the activation of E2F-regulated genes by G_1-specific Cdks. Phosphorylation of E2F-1's serine residues 332 and 337 hinders its interaction with Rb but facilitates its interaction with the adenovirus E4 protein. Both residues are largely unphosphorylated during early G_1, become phosphorylated during late G_1, and can be phosphorylated by cyclin D/Cdk4 kinase in vitro (Fagan et al. 1994). Cell-cycle-specific phosphorylation of E2F-1 by cyclin D/Cdks could therefore also contribute to the accumulation of "free" E2F in late G_1. Finally, the abundance of E2F mRNAs increases dramatically in late G_1. It has therefore been proposed that activation of Cdks in late G_1 stimulates synthesis of E2F, which could contribute to its accumulation in a form unbound to the Rb protein (Johnson et al. 1994). The effectiveness of this mechanism might depend on whether phosphorylation of Rb or E2F can actually liberate E2F from Rb. If not, then de novo synthesis of E2F in late G_1 could have a vital role and not just a helping hand in the activation of E2F-regulated genes.

The role of Rb as a key cell-cycle regulator has been questioned recently by the discovery that embryonic stem cells homozygous for Rb deletions can contribute to most tissues in the mouse; that is, cell-autonomous Rb function is not essential for the proliferation and differentiation of most types of mammalian cells (Sherr 1994b). It will be important to establish to what extent the normality of Rb-negative cells in vivo is due to the assumption of Rb's regulatory functions by other Rb-related proteins, to E2F remaining adequately regulated in the absence of any such protein, to regulation of E2F-mediated gene expression being less important than hitherto envisaged, or to deregulated entry into S phase being not particularly damaging to embryonic development because cell proliferation can be equally well if not better regulated by growth control or cell death (apoptosis). There are modest increases in the rates of proliferation within certain neuronal tissues in *rb* mutants but these seem to be compensated by increased apoptosis (Morgenbesser et al. 1994).

Comparisons to Yeast: Are Cyclin D and Rb on the Same Pathway?

The similarities between yeast and animal cells are sufficient to encourage a comparison of their factors that promote S phase. Several facts suggest that cyclin D/Cdk4 kinases perform functions that are analogous to Cln3/Cdk1. First, the patterns of synthesis of cyclin D and Cln3 are similar; neither is tightly periodic and both appear to be stimulated primarily by "growth"-promoting signals rather than chromosome cycle signals. Second, both proteins are very different from other cyclins involved later in the cycle, like the E, A, and B types, and differ greatly in their substrate specificity in that neither Cln3/Cdk1 (Tyers et al. 1992) nor cyclin D/Cdk4 readily phosphorylates histone H1 (Kato et al. 1993). Third, although the sequences of Cln3 and cyclin Ds do not greatly resemble each other, there is some evidence that they can perform equivalent tasks when expressed in yeast (Hatakeyama et al. 1994).

One of the more striking similarities between cyclin D and Cln3 is that the main, if not sole, task of both cyclins may be to activate late G_1-specific transcription. The dependence of S phase on Cln3 in yeast (i.e., its occurrence at a normal cell size) is abolished in cells that express late G_1-specific mRNAs constitutively due to hyperactivation of SBF. Likewise, entry into S phase of mammalian cell lines in which E2F cannot be regulated by Rb (due to mutations in the Rb gene) is sensitive neither to the injection of cyclin-D-specific antibodies (Lukas et al. 1994) nor to the overexpression of the cyclin-D-specific inhibitor p16INK4 (Serrano et al. 1995). Thus, both Cln3 and cyclin D appear to be "upstream" effectors of late G_1-specific transcription whose function becomes redundant when the "downstream" transcription factors are activated by other means.

There are indications that the abundance of cyclin D might be regulated by Rb. Cyclin D levels are very low in cell lines lacking Rb (Bates et al. 1994), suggesting that Rb, to which it is capable of binding in vitro, might play some role in stabilizing cyclin D protein. That cells can proliferate with very low levels of cyclin D when Rb is missing is consistent with its function being bypassed in these cells. Whether this is exclusively due to the resulting hyperactivation of E2F or to the deregulation of other S-phase-promoting events in *rb* mutants remains to be established.

What Are the Animal Equivalents of Clb5 and Clb6?

Cyclin E/Cdk2 kinases could perform functions similar to those of Cln1 and Cln2/Cdk1 kinases or those of Clb5 and Clb6/Cdk1 kinases. In contrast to Cln3 and cyclin D, accumulation of all these cyclins is tightly regulated by passage through the chromosome cycle. Certain observations suggest that an analogy between cyclin E and the "early" Clb cyclins Clb5 and Clb6 might be more appropriate. The functions of cyclin E seem highly conserved between vertebrates and arthropods, which is analogous to the conserved role of B-type cyclins (but not Cln cyclins) in regulating S phase in *S. pombe* and *S. cerevisiae*. Furthermore, the sequences of cyclins E resemble B-type cyclins much more than than they do Cln cyclins. The apparent activation of cyclin E/Cdk2 kinases 1 or 2 hours before the onset of S phase (Koff et al. 1992) is, on the other hand, more analogous to the behavior of Cln1 and Cln2, and is at first glance inconsistent with the notion that activation of cyclin E/Cdk2 is the "S phase trigger." If cyclin E/Cdk2 performs functions similar to those of Clb5/Cdk1, then cyclin A/Cdk2 could perform functions similar to those of Clb3 and Clb4/Cdk1 kinases. Cyclins A and Clb3 and Clb4 all accumulate later than cyclins E and Clb5 and Clb6, possibly too late to initiate S phase in wild-type cells, but not too late to facilitate the firing of "late" origins of replication.

Prospects for Genetic Analysis of S Phase in Animals

Analysis of the cell cycle in mammalian cells has until recently been dominated by experiments involving "gains of function"; hyperactivation of cyclins accelerate the cycle, overexpression of Rb suppresses division, cotransfection of transcription factor X can activate reporter gene Y, etc. There are two reasons for this. First, the prime motivation for many who study the mammalian cell cycle has been to understand oncogenesis, which until recently was thought to be largely due to gain-of-function mutations. Second, it is still much easier to study the effects caused by gains of function in mammalian cells than it is to study effects due to loss of function. Antisense has a discredited track record in establishing phenotype, whereas antibody injection is technically demanding and difficult to interpret unambiguously. Gain-of-function approaches only tell us what "can" function, which may be important for oncogenesis, but generally distracts us from understanding how normal cells actually function. The more recent appreciation that many, if not most, mutations that cause cancer are "loss-of-function" mutations and the development of methods to delete genes in mouse ES cells have been a very important step in the right direction. The latter has provided a reliable means of generating either animals or cell lines that are homozygous for mutations in key cell-cycle regulators.

It should not be forgotten, however, that the "reverse genetics" approach pioneered in yeast and now available in the mouse has its limitations, and there will continue to be much scope for the conventional genetic approach of isolating mutants following screening. This is possible in yeast, flies, and (hopefully) in the zebra fish, but is unlikely ever to be practical for mice. In an era when genetic analysis increasingly holds sway in experimental biology, it is seldom appreciated how difficult it is to determine gene function merely by analyzing phenotype. Gene knockouts, whether they are performed in mice or yeast, appear either to have little effect or to cause lethality. In the first case, it is difficult to ascertain what might have gone wrong without previous intuitions about the gene's function, whereas in the second case, it is difficult to distinguish primary from secondary phenotypes. Conventional mutant screens, when successful, are highly selective on both counts: Mutants are sought with specific defects in an area of biology in which the investigator is an expert and pleiotropic mutants are usually avoided. Reverse genetics will be most valuable when we have some idea where to look for defects.

STOPPING DNA REPLICATION IN G_2

Could Cell-cycle State Be Determined Solely by Cdks?

How do cells ensure that no DNA sequence is replicated more than once during S phase and that mitosis precedes the next round of replication? There is ample evidence in yeast and animal cells that the timing of S and M phases is determined by the pattern of activation of G_1- and G_2-specific Cdks. Might not just the timing but also whether a cell undergoes S or M phase be determined by the state of Cdk-cyclin complexes? Let us create an imaginary cell that possesses only two types of Cdks, an S-phase- and an M-phase-promoting form (SPF and MPF), which oscillate in abundance with identical frequencies but out of phase with each other. Let us further specify the shape of each wave such that the S-phase-promoting Cdk is inactive during the first quarter of the cycle, active during the second, but inactive again during the third and fourth quarters and such that the M-phase-promoting Cdk is inactive throughout first and second quarters, partially active during the third, and fully active during the fourth. Let us further assume that whether or not the cell undergoes chromosome duplication (S phase) or segregation (M phase) is determined entirely by the state of the two Cdks: If SPF is active and MPF inactive, then cells replicate DNA, whereas if MPF is fully active and SPF is inactive, then cells undergo mitosis. Our imaginary cells will undergo alternate S and M phases with intervening gaps equivalent to G_1 and G_2. This hypothesis, that cell-cycle state is defined solely by the state of S- and M-phase-specific Cdks (Nurse 1994), predicts that activation of G_1-specific Cdks in the absence of G_2-specific forms should trigger S phase, whereas activation of G_2-specific Cdks in the absence of G_1-specific forms should trigger M phase, irrespective of a cell's position in the cycle at the time. In other words, it should be possible, by artificial manipulation of SPF and MPF activities, to cause cells to jump between states that are not normally adjacent; that is, it should be possible to override the cell's history. Two observations in *S. pombe* are consistent with this view (Hayles et al. 1994). First, *S. pombe* mutants that lack the "mitotic" cdc13/Cdk1 kinase fail to enter M phase and instead undergo multiple rounds of DNA replication. Thus, preventing activation of an MPF allows cells to reenter the S-phase state without proceeding through the M-phase state. Second, *cdc10* mutants, which normally arrest in G_1 at their restrictive temperature due to a failure to activate late-G_1-specific transcription, can be induced instead to enter mitosis merely by overproducing the two components of the mitotic Cdk: cdc13 and Cdk1. Thus, premature activation of MPF can induce cells lacking SPF and MPF to enter M phase without any prior S phase.

If, as indicated by these experiments, the "Cdk state" model is correct, then understanding how the cell cycle runs in the correct direction (i.e., how cells ensure that chromosome duplication precedes their segregation) should be simply a question of understanding the dynamics or logic of SPF and MPF oscillations. Much progress has been made in understanding how different cyclin-dependent kinases regulate each other in *S. cerevisiae*. Consider the S-phase-promoting Cln1 and Cln2/Cdk1 kinases and the M-phase-promoting Clb1-Clb4/Cdk1 kinases. Cln1 and Cln2 promote activation of Clbs by turning off their proteolysis and by triggering destruction of p40Sic1. Clbs then repress transcription of *CLN1* and *CLN2*, promote mitosis and finally their own destruction, leading back to a state in which *CLN1* and *CLN2* genes can be activated by Cln3 in the absence of any interference by Clbs. But are such regulatory networks really solely responsible for imposing the procession of S and M phases?

The State of Nuclei Determines the Response to a Given Cdk

A major problem with the Cdk state hypothesis is that it provides no explanation for how cells avoid the repetitive firing of origins during S phase. S-phase-promoting Cdks must presumably be active throughout S phase to fire both early and late origins of replication. How do cells prevent the refiring of early origins as they fire old ones? It can be debated whether this remarkable property is relevant to the problem of how S and M phases alternate (Roberts 1993), but there is no denying that it would be nice to have a theory that could explain both.

Doubts as to whether cell-cycle state could solely be determined by the state of S- and M-phase-promoting factors were raised long before the discovery of Cdks even, when it was found that G_1 but not G_2 nuclei can be triggered into S phase by fusion of G_1 or G_2 cells with S-phase cells (Johnson and Rao 1971). However, this could still be explained by the "Cdk state" hypothesis if G_2 nuclei continued to harbor G_2-specific Cdks even after fusion to S-phase cells. The Cdk state hypothesis cannot, however, explain the recent discoveries that Clb/Cdk1 and not Cln/Cdk1 kinases are essential for S phase in *S. cerevisiae* (Schwob et al. 1994), that supposedly mitotic Clb/Cdk1 kinases can trigger S phase in G_1 cells (Amon et al. 1994), or that mutants deprived of the mitotic Clb1–Clb4/Cdk1 kinases possess high levels of the S-phase-promoting Clb5/Cdk1 kinase but arrest permanently in G_2 (Amon et al. 1993). These data show that the distinction between S- and M-phase-promoting Cdks is not a clear one. Mitotic Clb/Cdk1 kinases can also promote S phase! Cell-cycle state cannot therefore be determined solely by the presence or absence of specific S- or M-phase-promoting Cdks. It must also be influenced by a cell's history; that is, whether it has recently responded to these Cdks by undergoing S phase. Put in molecular terms, whether or not a cell responds to an S-phase-promoting Cdk by undergoing DNA replication must be determined by the state of Cdk substrates at the time of this Cdk's activation. G_1 nuclei, but not G_2 nuclei, presumably contain substrates whose phosphorylation by S-phase-promoting Cdks triggers DNA replication.

A Two-step Mechanism for S-phase Onset

Licensing Factor

What the crucial difference between a G_1 and a G_2 nucleus might be has until very recently been largely investigated using extracts from activated *Xenopus* eggs (Lohka and Masui 1984) in which exogenous DNA can undergo multiple rounds of DNA replication and where the distinction between G_1 and G_2 nuclei can be measured in vitro (see Laskey and Madine, this volume). *Xenopus* eggs are arrested in metaphase of meiosis II and contain high levels of MPFs like cyclin B/Cdk1. Fertilization, or artificial "activation," triggers destruction of cyclin B, and the eggs exit from mitosis and enter interphase. Extracts made from these "activated" eggs support replication of sperm chromatin following its assembly into nuclei (Blow and Sleeman 1990); the extracts presumably contain SPFs like cyclin E/Cdk2. Shortly after DNA replication, cyclin B/Cdk1 activity accumulates, which causes nuclear envelope breakdown and chromosome condensation, but the kinase later disappears due to cyclin B destruction, which leads to nuclear reformation and, in some cases, a second round of DNA replication (Hutchinson et al. 1987; Blow and Laskey 1988). The extracts thus undergo, at least partly, the natural sequence of events that occurs during cleavage divisions, in that cyclin B levels oscillate and DNA replication occurs only once per cell cycle (Murray and Kirschner 1989). The system's resemblance to the in vivo situation is striking in that no DNA molecules replicate more than once per cell cycle (Blow and Laskey 1986).

Addition of cycloheximide to the extracts prevents the appearance of cyclin B/Cdk1 kinase activity but presumably does not block activation of S-phase-promoting Cdks, because a single round of replication still occurs within the freshly assembled nuclei. Addition of an MPF activity to cycloheximide-treated extracts causes nuclear envelope breakdown and allows at least partial rereplication. The cycloheximide-treated extracts also cause G_1 but not G_2 nuclei from HeLa cells to enter S phase. The cycloheximide extracts therefore partly mimic the cytoplasm of S-phase HeLa cells, in that they can trigger S phase in G_1 but not in G_2 nuclei (Leno et al. 1992).

To explain these observations, it has been suggested (Blow and Laskey 1988) that initiation of DNA replication might be a two-step process. The first step would be the binding to origins of a factor needed for the initiation of DNA replication whose entry into the nucleus is restricted to M phase, when nuclear envelopes are disassembled. This factor would license origins to fire upon the subsequent arrival of SPF in the next cell cycle. To explain the lack of rereplication in the absence of MPF, it was envisaged that initiation of DNA replication destroys all licensing factor within the nucleus and that the nuclear membrane must be disassembled before licensing factor residing in the cytoplasm can regain access to origins. In an experiment that is widely regarded as an important confirmation of this hypothesis, it was found that permeabilization of the nuclear envelope permits rereplication of G_2 nuclei in the absence of MPF activation (Blow and Laskey 1988; Coverley et al. 1993). Purification of the putative licensing factor has been thwarted until now due to the lack of a good assay (but see Laskey and Madine, this volume).

Preparations for S Phase as Cells Exit from Mitosis

The *CDC46*, *MCM2*, and *MCM3* gene products in yeast have certain properties reminiscent of the putative licensing factor (Hennessy and Botstein 1991; Hennessy et al. 1991; Tye 1994). They are necessary for the initiation of DNA replication in yeast, and their access to chromatin is under cell cycle control. Cdc46 and Mcm3 proteins accumulate in the cytoplasm during G_2 and only enter nuclei as cells undergo anaphase. Vertebrate homologs of these proteins are not like their yeast counterparts excluded from nuclei during G_2, but their association with chromatin may nevertheless be under cell cycle control (Kimura et al.

1994). The P1 protein, a homolog of Mcm3 that binds to DNA polymerase-α, is distributed in the cytoplasm during metaphase and associates with chromatin only as chromosomes decondense during telophase. The behavior of Cdc46/Mcm proteins and that of the putative licensing factor differ drastically in one important respect. The Cdc46/Mcm proteins enter nuclei or associate with chromatin as cells exit from M phase (Hennessy et al. 1990), as cyclin B/Cdk1 kinase is destroyed (Surana et al. 1993), and not during M phase, when cyclin B/Cdk1 kinase is high, as originally proposed for licensing factor. This difference could be vital because, as we shall see, the cyclin B/Cdk1 kinase might be an important regulator of DNA rereplication (Hayles et al. 1994; Dahmann et al. 1995). Another example of preparations for S phase occurring at the end of M phase concerns the 70-kD subunit of RP-A, a three-subunit single-stranded DNA-binding protein needed for prepriming, whose association with subnuclear foci proposed to be prereplication centers (pre-RCs) as chromosomes decondense following exit from M phase (Adachi and Laemmli 1994) can be inhibited by cyclin B/Cdk1 kinase.

Chromatin Structure of Yeast Origins during the Cell Cycle

Yet more persuasive evidence of preparations for replication occurring at the end of mitosis have come from analyses of chromatin structure at DNA replication origins during the cell cycle (Diffley et al. 1994). So far, discrete replication origins have been well defined only in *S. cerevisiae*, where a 200-bp DNA sequence (ARS) containing a core consensus sequence called A and flanking sequences B1, B2, and B3 is sufficient for origin function on episomes and within chromosomes (Marahrens and Stillman 1992). A 250-kD protein complex called the origin recognition complex (ORC) binds to sequences A and B1 (Bell and Stillman 1992; Diffley and Cocker 1992), whereas a separate protein, Abf1, binds to the B3 element (Diffley and Stillman 1989). The state of these origins during the cell cycle has been analyzed by genomic footprinting. Both ORC and Abf1 appear to be bound throughout the cell cycle, but there are important changes in the neighboring chromatin (Diffley et al. 1994). From S phase until late anaphase, the pattern of cleavage by DNase I in vivo resembles that induced by ORC binding in vitro, which causes an adjacent DNA sequence to become hypersensitive. As cells exit from mitosis, this DNase I hypersensitive site disappears and does not reappear until cells enter S phase in the next cycle. The data indicate that a second factor binds next to ORC during G_1 but not at other stages of the cell cycle.

CDC6 Is Necessary for Pre-RCs

A key question is whether the cell-cycle-regulated factor that binds to yeast origins is required for DNA replication or for preventing its premature onset. If the former were true, then disappearance of DNase I hypersensitivity at the end of mitosis should depend on proteins needed for origin firing. A good candidate for such a protein is that encoded by *CDC6*, which is required for the initiation of DNA replication in *S. cerevisiae* (Bueno and Russell 1992), as is a homologous gene, *cdc18*, in *S. pombe* (Kelly et al. 1993). Minichromosomes whose replication is dependent on a single ARS sequence are much more frequently lost in *cdc6* mutants than in wild-type cells, but this defect is suppressed when the minichromosome carries multiple ARS sequences, implying that *cdc6* mutants are not defective in DNA replication per se but rather in the function of origins (Hogan and Koshland 1992). The Cdc6 protein is unstable and is synthesized in a burst at the end of mitosis (as well as in late G_1) simultaneous with the change in chromatin structure that occurs at origins (Piatti et al. 1995). More important, the disappearance of ORC-induced DNase I hypersensitivity at origins does not take place when cells exit from mitosis in the absence of Cdc6 synthesis (Cocker et al. 1996). Furthermore, the "protected" state of origins characteristic of cells arrested in G_1 (by pheromone) is temperature sensitive in *ts cdc6* mutants. Cdc6 is therefore necessary for the formation and maintenance of a chromatin configuration at origins that precedes DNA replication and is presumably essential for its initiation. The observation that Cdc6 and ORC can form complexes in vitro (Liang et al. 1995) is consistent with this view.

S-phase onset in yeast might therefore involve two temporarily distinct events at replication origins. The first would be the formation of pre-RCs, whose appearance during telophase depends on *CDC6* and coincides with the destruction of Clb/Cdk1 kinases (MPFs). The second event would be the activation of S-phase-promoting Clb/Cdk1 kinases (Schwob et al. 1994) in late G_1 (SPFs), which would trigger pre-RCs to initiate DNA replication. There is a pleasing congruence between this scheme and the arrival on chromatin at the end of mitosis of RP-A subnuclear foci and of Cdc46/Mcm proteins, which might also be components of yeast prereplicative complexes.

Control of pre-RC Formation

If the formation of pre-RCs at future origins is a prerequisite for initiating DNA replication, then control of pre-RC formation could be as vital an aspect of S-phase control as the production of S-phase-promoting factors (Cdks). In *S. cerevisiae*, activation of cyclin B/Cdk1 kinases during G_1 causes the onset of S phase, but the very same kinases are active during G_2 and yet cells do not reenter S phase. This can now be explained by the absence in G_2 cells of pre-RCs that are, as it were, the key substrates for S-phase-promoting Cdks. What then excludes the formation of pre-RCs during S, G_2, and M phases?

In yeast, the adoption by origins of a prereplicative state indicative of pre-RCs occurs with very similar kinetics at the end of mitosis as the destruction of cyclin B/Cdk1 kinases (Surana et al. 1993; Diffley et al. 1994). This suggests that cyclin B/Cdk1 kinases could have a role in preventing the de novo assembly of pre-RCs. Direct evidence that destruction of Cdks in G_2 or M phase is sufficient to trigger pre-RCs and thereby creates the potential to rereplicate upon activation of S-phase-promoting Cdks has been obtained using an inducible *SIC1* gene in *S. cerevisiae* (Dahmann et al. 1995). The microtubule-disrupting drug nocodazole prevents nuclear division in *S. cerevisiae* and causes cells to arrest with duplicated DNA, origins in a postreplicative state, and high levels of all six Clb kinases. Induction of high levels of the Clb/Cdk1 inhibitor p40Sic1 from a *GAL-SIC1* gene fusion inhibits completely both Clb2 and Clb5/Cdk1 kinases (and presumably all other Clb kinases as well), causes cells to rebud without nuclear division, causes origins to adopt a prereplicative chromatin structure, but does not induce rereplication. However, reactivation of Clb/Cdk1 kinases by subsequently shutting off p40Sic1 synthesis causes all cells to rereplicate their DNA.

Mutants That Rereplicate

Quite independent evidence implicating cyclin B/Cdk1 kinases in blocking rereplication in G_2 cells stems from the characterization of yeast mutants that undergo successive S phases in the absence of an intervening mitosis. Such mutants have been isolated on the basis of their ability to diploidize and thereby sporulate. Budding yeast mutant cells with a temperature-sensitive allele of the *ESP1* gene replicate normally at the restrictive temperature, but they subsequently fail to undergo anaphase due to an unspecified mitotic spindle defect. Despite their mitotic failure, the mutant cells proceed as usual with destruction of Clb/Cdk1 kinase, and they subsequently embark on a new cell cycle, during which they rereplicate their chromosomes (McGrew et al. 1992; Surana et al. 1993). The behavior of *esp1* mutants shows that Clb destruction is accompanied by replication even when mitosis has not occurred. Thus, mitosis per se is not necessary.

Many more diploidizing mutants have subsequently been analyzed in *S. pombe*. The vast majority fall into the *esp1* class, in that they attempt to enter mitosis, fail, and proceed with the next cycle. In rare mutants, however, cells rereplicate DNA without attempting mitosis (although nitrogen starvation is also required). Upon return to the permissive temperature, some of these "G_2" cells rereplicate their DNA before entering mitosis. Mutations in two genes give rise to this phenotype. Remarkably, they encode *cdc2* (*cdk1*) and *cdc13*, the two components of the cyclin B/Cdk needed for mitosis in fission yeast (Broek et al. 1991; Hayles et al. 1994). The behavior of cells completely lacking the *cdc13* gene product is even more striking. Upon inoculation into fresh medium, wild-type spores germinate, replicate their DNA, spend the next 90 minutes in G_2, and then undergo mitosis. The second round of DNA replication follows almost immediately after completion of the first mitosis. Spores lacking *cdc13* resemble wild-type cells up to the point at which the Cdc13/Cdc2 kinase should have become active. Due to its absence, the mutant cells never enter mitosis; G_2 is prolonged, but not indefinitely, because 5 hours after the first S phase, cells rereplicate their DNA and this process recurs thereafter, roughly every generation time. As a consequence, some cells eventually accumulate up to 32 copies of their genome. Therefore, it seems that Cdc13 is necessary to prevent *S. pombe* cells from entering a G_1-like state after they have spent some time in G_2. Why might this be so? One possibility is that loss of Cdc13/Cdk1 kinase encourages the formation of pre-RCs.

Cyclin B/Cdk1 kinases have also been implicated in preventing DNA replication between meiosis I and II. During maturation of *Xenopus* oocytes, cyclin B/Cdk1 is inactivated after meiosis I but, due to the action of c-*mos*, is promptly reactivated prior to meiosis II. Ablation of c-*mos* product causes maturing oocytes instead to reform nuclei and replicate DNA after the first meiotic division (Furuno et al. 1994). The same effect is observed with dominant-negative alleles of Cdk1. Thus, cyclin B/Cdk1 kinases inhibit DNA replication between meioses I and II as well as between S and M phases during vegetative cell cycles. Whether they do so by inhibiting pre-RC formation is yet to be determined.

Preventing Rereplication during S and G_2 Phases: Do S-phase-promoting Cdks Have a Dual Function?

We now have a coherent explanation for the behavior of origins during mitosis and G_1. Assembly of pre-RCs requires destruction of mitotic Cdks and the synthesis of pre-RC components, either at the end of mitosis or in late G_1. Pre-RCs are then transformed into replication forks by activation of S-phase-promoting Cdks. Thus, the lack of DNA replication during G_1 and M phases has very different causes. G_1 cells that have formed pre-RCs have the potential to enter S phase but do not do so for lack of S-phase-promoting Cdks, whereas M-phase cells could not enter S phase even if they were to express S-phase-promoting Cdks (and some types of cell indeed do) because assembly of pre-RCs is inhibited by mitotic cyclin B/Cdk1 kinases. This model cannot yet account for the behavior of S and G_2 cells. During DNA replication, S-phase-promoting Cdks are active whereas mitotic cyclin B/Cdk1 kinases are inactive, and cells should therefore be able both to form pre-RCs and to turn them into replication forks. What then prevents origins from firing more than once? A similar problem could afflict G_2 cells that have not yet activated cyclin B/Cdk1 kinases but contain Cdks like cyclin A/Cdk2 in animals or Clb1–Clb4/Cdk1 in yeast, which are capable of promoting S phase.

The phenotype of *S. cerevisiae* mutants lacking G_2-specific B-type cyclins Clb1–Clb4 may provide an important clue to this mystery. Strains lacking Clb1, Clb3, and Clb4 and kept alive with a temperature-sensitive allele of *CLB2* (conditional *clb1–clb4* mutants) can replicate normally but fail to build a mitotic spindle at the restrictive temperature. In this regard, their phenotype resembles that of *S. pombe cdc13* deletion mutants. Their phenotype differs greatly in another regard, for they arrest in G_2 indefinitely without rereplicating. Clb5/Cdk1 kinase remains active during this arrest. Is it possible that this kinase also inhibits assembly of pre-RCs and thereby prevents rereplication? Mutations in at least two genes (*SIM1* and *SIM2*) allow *clb1-4* mutants to rereplicate (C. Dahmann et al., in prep.) Remarkably, the mutants are all defective in maintaining high levels of Clb5/Cdk1 kinase and rebud at the same time as they rereplicate, which is a sign of Clb kinase decline in vivo. Moreover, they fail to rereplicate if *CLB5* gene dosage is increased. The behavior of these mutants shows that Clb5/Cdk1 kinase acts to suppress replication in *clb1-4* mutants, which is all the more remarkable because Clb5/Cdk1 normally promotes S phase.

The apparent contradiction that Clb5/Cdk1 both promotes and inhibits S phase could be explained if the kinase had different replication functions in G_1 and G_2: promoting it during G_1 but blocking it during G_2. Might the very same kinase that promotes transformation of pre-RCs into replication forks also inhibit de novo assembly of pre-RCs (see Fig. 3)? Indeed, it seems likely that all of the six Clb/Cdk1 kinases that are capable of promoting S phase in *S. cerevisiae* also inhibit formation of pre-RCs. Clb5 and Clb6 might be primarily responsible for inhibiting

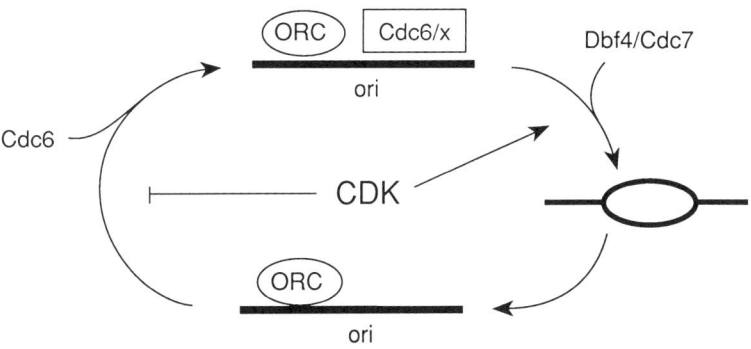

Figure 3 A dual function for Cdks: promoting initiation of DNA replication and inhibiting pre-RC assembly. Initiation is a two-step process, the first step of which is the formation of pre-RCs at future origins. This process depends on the synthesis of the unstable replication protein, Cdc6. It usually occurs at the end of mitosis when there is a burst of Cdc6 synthesis but can also occur in late G_1, when there is a second burst of Cdc6 synthesis, at least in daughter cells. The second step is activation in late G_1 of cyclin B/Cdk1 kinases along with the Cdc7 kinase. It seems that activity of the very same set of cyclin B/Cdk1 kinases prevents de novo assembly of pre-RCs. Each round of initiation therefore depends on a cycle of cyclin activation/destruction (see Fig. 4).

pre-RC assembly during S phase, Clb3 and Clb4 during early G_2, and Clb1 and Clb2 during late G_2 and during M phase, when the respective kinases are most active. If this model were correct, the different behavior of *S. pombe cdc13* mutants, which undergo multiple rounds of replication, and *S. cerevisiae clb1-4* quadruple mutants, which arrest indefinitely in G_2, might be simply one of degree. *cdc13* mutants also do not re-replicate for many hours after their first S phase, during which time other B-type cyclins, e.g., Cig1 and Cig2, might prevent the assembly of pre-RCs. The Cig1 and Cig2 B-type cyclins might either be somewhat less effective than Clb5 and Clb6 and eventually be overcome by the forces driving the assembly of pre-RCs or, for reasons not yet understood, they may simply not remain active during G_2 for as long as Clb5 and Clb6.

Pre-RCs Can Form Either at the End of M Phase or in Late G_1

At first sight, a mechanism whereby S-phase initiation depends on events that occurred at the end of the previous mitosis seems somewhat precarious. It is hard to imagine how cells ensure that prereplication complexes formed at the end of mitosis could be relied on to survive an extended arrest in G_1, as occurs during stationary phase in yeast. The same problem would afflict quiescent mammalian cells, which can spend months, if not years, arrested in G_1 before they are stimulated to reenter S phase. In fact, the chromatin structure thought to represent pre-RCs at yeast origins can form not only at the end of mitosis, as occurs in cycling cells, but also during G_1 phase when inoculation into fresh medium stimulates stationary-phase yeast cells to reenter the cell cycle (Diffley et al. 1994). This observation is not at odds with the notion that destruction of cyclin B/Cdk1 kinases is necessary for the formation of pre-RCs, according to which pre-RC assembly should be possible throughout G_1, for as long as the inhibitory cyclin B/Cdk1 kinases remain dormant. Thus, DNA replication could be seeded predominantly from pre-RCs formed at the end of mitosis in cells with short G_1 periods, as in yeast mother cells and embryonic cells undergoing cleavage divisions, whereas it could be seeded mainly by pre-RCs formed during G_1 in cells with long G_1 periods.

Democracy of the Genome

The notion that assembly of new pre-RCs might be prevented by the very same set of Cdks that promote their transition into replication forks is an interesting idea, because it explains how cells ensure that there is no stage during the cell cycle in which pre-RCs can both be assembled and be transformed into replication forks (see Fig. 4). S-phase-promoting Cdks with dual functions would preclude any possibility of origins firing more than once during the cell cycle. Furthermore, if we assume that all Cdk1 (or Cdk2) forms can hinder pre-RC formation, we could explain why G_2 nuclei cannot enter S phase even when they are placed in a cytoplasm containing S-phase-promoting Cdks (Johnson and Rao 1971). We presume, by analogy with *S. cerevisiae*, that from S phase until the end of mitosis, animal cells also continuously maintain an environment that is hostile to the formation of pre-RCs. S-phase-promoting Cdks like cyclin E/Cdk2 or cyclin A/Cdk2 might prevent their formation during S phase, whereas cyclin A/Cdk1 and cyclin B/Cdk1 assume this function during G_2 and M phase.

Primordial Cell Cycles

Although multiple Cdks may participate in the exclusion of pre-RC assembly during the S, G_2, and M phases of most present-day eukaryotic

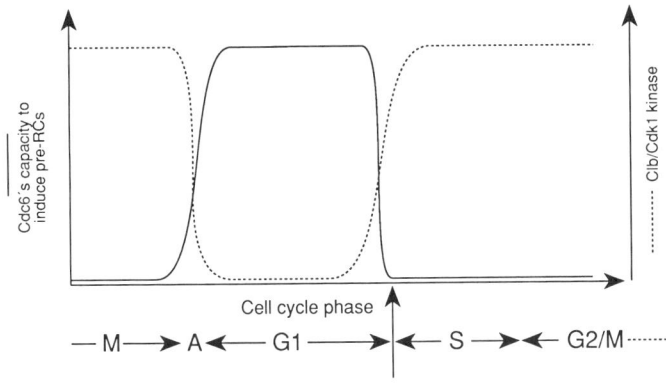

POINT OF NO RETURN

Figure 4 The pre-RC/Cdk cycle in yeast. Cyclin B/Cdk1 kinases prevent the formation of pre-RCs during S, G_2, and M phases. Pre-RCs therefore only form upon proteolysis of cyclin B molecules during anaphase. Reactivation of cyclin B/Cdk1 kinases during the subsequent G_1 period triggers initiation at origins that had formed pre-RCs during G_1, when cyclin B/Cdk1 kinases were inactive.

cells, and specialized Cdks may be employed to transform pre-RCs into replication forks, it is worth considering that discrete rounds of replication could theoretically be driven by the alternate activation and inactivation of a single "dual function" Cdk. In *S. cerevisiae*, where mitosis does not entail extensive chromosome condensation, entry into mitosis and DNA replication can co-exist. It is therefore conceivable that in the primordial eukaryotic cell, activation of a single Cdk might have triggered simultaneously transformation of pre-RCs into replication forks (i.e., S phase) and attachment of chromosomes to mitotic spindles (i.e., M phase). Subsequent inactivation of this same Cdk could have signaled both the assembly of pre-RCs and exit from mitosis. It is therefore possible to envisage how the oscillation of a single Cdk could have formed the heart of our distant ancestors' cell cycle. To function with high fidelity, such a system might have required a surveillance mechanism that ensured that destruction of Cdks was never initiated before DNA replication was complete. This might then have been the cell cycle's oldest checkpoint control.

Dependence of S on Anaphase

The hypothesis that mitotic Cdks prevent pre-RC assembly also explains why sister chromatid separation always precedes a second round of DNA replication. S-phase-promoting Cdks decline during G_2, G_2-specific Cdks like cyclin A/Cdk1 decline during metaphase, but the destruction of the last remaining Cdks capable of inhibiting the assembly of pre-RCs does not occur until mitosis-specific Cdks like cyclin B/Cdk1 are inactivated due to cyclin B proteolysis at anaphase. The latter depends on activation of a specialized multisubunit ubiquitin ligase, which is necessary also for the separation of sister chromatids and is therefore called the anaphase-promoting complex (APC) (see Fig. 5). The APC is thought to catalyze proteolysis not only of B-type cyclins but also of proteins that inhibit sister chromatid separation (Holloway et al. 1993; Irniger et al. 1995; King et al. 1995). Thus, cells cannot begin to assemble pre-RCs (a pre-condition for chromosome duplication) without having also triggered anaphase.

A Three-parameter Model for the Cell Cycle

We started this section with the notion that cell-cycle state might be determined by the activity of two key parameters: the level of S- and M-phase-promoting Cdks. This idea turns out to be untenable. At least three parameters are required to define the major cell-cycle states of most existing eukaryotic cells: the levels of S- and M-phase-promoting Cdks (SPF and MPF) and the presence or absence of pre-RCs. Early G_1 cells lack SPF and MPF but usually (although not always) contain pre-RCs; cells embarking on S phase contain SPF and pre-RCs; cells in S phase contain SPF and are in the process of converting their pre-RCs into replication forks; G_2 cells have finished this process and may or may not contain SPF but have not fully activated MPF; M-phase cells have fully activated MPF, and anaphase cells are in the process of destroying their MPF and other proteins whose destruction is needed for chromosome segregation and are about to form pre-RCs. Only when we take pre-RCs into account can we explain why G_1 and G_2 nuclei respond differently to activation of S-phase-promoting Cdks. The three-parameter model of the cell cycle also allows us to explain the very different effects of activating mitotic Cdks in G_1-arrested *S. pombe* and *S. cerevisiae* cells; it induced replication in *S. cerevisiae* cells deprived of Clns but failed to do so in *S. pombe cdc10* mutants (Amon et al. 1994; Hayles et al. 1994). *cdc10* is needed for the synthesis of Cdc18, which is homologous to Cdc6 from *S. cerevisiae* and might therefore be necessary for the formation of pre-RCs. The presence of pre-RCs in *cln*-arrested *S. cerevisiae* cells and their absence in *S. pombe cdc10* mutants would explain their different responses.

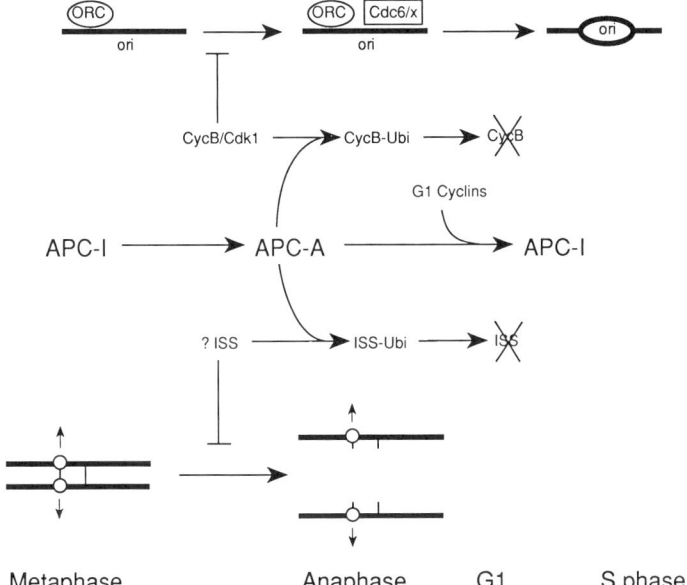

Figure 5 The APC could link S phase to anaphase. The APC is a 20S particle that contains Cdc16, Cdc23, and Cdc27 proteins (among others) and catalyzes the ligation of multiple ubiquitin molecules to cyclin B proteins containing a destruction box, targeting them for destruction by the 26S proteosome. The ubiquitin ligase reaction also requires ATP, ubiquitin, an activating enzyme (E1), and a conjugating enzyme (E2). APC components are needed not only for cyclin B ubiquitination and proteolysis, but also for the onset of anaphase. It is therefore thought that the transition from metaphase to anaphase depends on the destruction of inhibitors of sister chromatid separation (ISS). The APC is thought to be inactive from late G_1 until the onset of anaphase. Its activation then triggers anaphase and the destruction of cyclin B proteins. The latter leads to the formation of pre-RCs and thereby to a new round of DNA replication when cyclin B/Cdk1 kinases are reactivated during the subsequent G_1 period. Activation of Cln1 or Cln2/Cdk1 kinases turns the APC off and allows the accumulation of mitotic cyclin B proteins.

The inclusion of pre-RCs as a major cell-cycle parameter helps to impart a rigid directionality to our cell-cycle model. The presence of these structures is partly determined by the cell's history, but this directionality comes at a price. The timing of cell-cycle events becomes more crucial. For example, activation of S-phase-promoting Cdks before pre-RCs have been assembled, as would be possible in the outgrowth of cells that had long lost pre-RCs inherited from their previous telophase, would completely finesse the replication machinery and lead to an unreplicated state from which a cell could no longer extract itself. Much of the regulation of pre-RC components and S-phase-promoting Cdks, which is more conventionally thought to be contributing to the control of cellular proliferation, may have the more important role of helping to avoid the disastrous consequences of activating S-phase-promoting Cdks before pre-RCs have had time to form.

Why Do Most Quiescent Cells Arrest in G_1?

A final insight that stems from these ideas concerns why so many metazoan cells choose to enter a quiescent state during G_1 and not during S or G_2 phases. Cells at these latter stages must maintain active Cdks to exclude the assembly of pre-RCs. A temporary inactivation of these Cdks due to unspecified physiological disasters could lead to the assembly of pre-RCs and thereby to an unscheduled round of DNA replication if S-phase-promoting Cdks were activated before mitosis could take place. It is interesting in this regard that rereplication of Cdk1 alleles in *S. pombe* was first observed only upon extreme heat shock and nitrogen starvation (Broek et al. 1991); that is, extreme environmental conditions can contribute to the destruction of Cdks. In contrast, destruction of preassembled pre-RCs in G_1 cells should not be disastrous, as long as Cdks capable of inhibiting pre-RC assembly were not yet activated and as long as cells contained pre-RC components or were capable of entering a program that caused their synthesis. The logic of replication control may therefore dictate that the early G_1 state is one from which a cell can more readily recover its true cell-cycle state following an experience that erases its memory of cell-cycle position. It is the one stage of the cell cycle where history does not matter!

Closing the Cycle

The first section of this review dealt with the events needed to drive a G_1 cell into S phase. We treated the resting G_1 cell as a blank slate and considered how cells produce the factors that trigger it to replicate. In the second half, we considered how cells prevent replication during G_2 and M phases. We can now appreciate how events at the end of one cell cycle lead to the start of the next one; that is, we can begin to appreciate its cyclical nature. There is sufficient knowledge about this process in *S. cerevisiae* to piece together the following picture. During G_2 and M phases, Clb/Cdk1 kinases prevent transcription of S-phase-promoting kinases Cln1 and Cln2/Cdk1 by preventing the binding of SBF to SCBs in their promoters. These kinases also prevent the assembly of pre-RCs (how is not understood). Thus, destruction of Clb/Cdk1 kinases during anaphase through the activation of Clb proteolysis and the accumulation of p40Sic1 removes the key barrier to transcriptional activation of *CLN1* and *CLN2* by Cln3/Cdk1 and the formation of pre-RCs at origins (which can form at this stage or later in G_1). In the absence of Clb/Cdk1 kinases, growth of daughter cells to a threshold size facilitates transcription of *CLN1*, *CLN2*, and *CDC6*, which produces active Cln/Cdk1 kinases and yet more pre-RCs. Cln1 and Cln2/Cdk1 kinases then trigger the reactivation of Clb/Cdk1 kinases by promoting proteolysis of p40Sic1 and by turning off proteolysis of Clb cyclins, whose activity now promotes the transition of pre-RCs into replication forks.

ACKNOWLEDGMENTS

I thank Christian Koch, Nick Jones, and Gustav Ammerer for helpful comments on the manuscript.

REFERENCES

Adachi, Y. and U.K. Laemmli. 1994. Study of the cell cycle-dependent assembly of the DNA pre-replication centres in *Xenopus* egg extracts. *EMBO J.* **13:** 4153–4164.

Amon, A., S. Irniger, and K. Nasmyth. 1994. Closing the cell cycle circle in yeast: G2 cyclin proteolysis initiated at mitosis persists until the activation of G1 cyclins in the next cycle. *Cell* **77:** 1037–1050.

Amon, A., M. Tyers, B. Futcher, and K. Nasmyth. 1993. Mechanisms that help the yeast cell cycle clock tick: G2 cyclins transcriptionally activate G2 cyclins and repress G1 cyclins. *Cell* **74:** 993–1007.

Andrews, B.J. and L.A. Moore. 1992. Interaction of the yeast Swi4 and Swi6 cell cycle regulatory proteins in vitro. *Proc. Natl. Acad. Sci.* **89:** 11852–11856.

Baldin, V., J. Lukas, M.J. Marcote, M. Pagano, and G. Draetta. 1993. Cyclin D1 is a nuclear protein required for cell cycle progression in G1. *Genes Dev.* **7:** 812–821.

Baroni, M.D., P. Monti, and L. Alberghina. 1994. Repression of growth-regulated G1 cyclin expression by cyclic AMP in budding yeast (see comments). *Nature* **371:** 339–342.

Bates, S., D. Parry, L. Bonetta, K. Vousden, C. Dickson, and G. Peters. 1994. Absence of cyclin D/cdk complexes in cells lacking functional retinoblastoma protein. *Oncogene* **9:** 1633–1640.

Beach, D., B. Durkacz, and P. Nurse. 1982. Functionally homologous cell cycle control genes in budding and fission yeast. *Nature* **300:** 706–709.

Bell, S.P. and B. Stillman. 1992. ATP-dependent recognition of eukaryotic origins of DNA replication by a multiprotein complex (see comments). *Nature* **357:** 128–134.

Bjorklund, S., S. Skog, B. Tribukait, and L. Thelander. 1990. S-phase-specific expression of mammalian ribonucleotide reductase R1 and R2 subunit mRNAs. *Biochemistry* **29:** 5452-5428.

Blow, J.J. and R.A. Laskey. 1986. Initiation of DNA replication in nuclei and purified DNA by a cell-free extract of *Xenopus* eggs. *Cell* **47:** 577-587.

———. 1988. A role for the nuclear envelope in controlling DNA replication within the cell cycle. *Nature* **332:** 546-547.

Blow, J.J. and A.M. Sleeman. 1990. Replication of purified DNA in *Xenopus* egg extract is dependent on nuclear assembly. *J. Cell Sci.* **95:** 383-391.

Bodrug, S.E., B.J. Warner, M.L. Bath, G.J. Lindeman, A.W. Harris, and J.M. Adams. 1994. Cyclin D1 transgene impedes lymphocyte maturation and collaborates in lymphomagenesis with the *myc* gene. *EMBO J.* **13:** 2124-2130.

Booher, R. and D. Beach. 1988. Involvement of *cdc13*$^+$ in mitotic control in *Schizosaccharomyces pombe*: Possible interaction of the gene product with microtubules. *EMBO J.* **7:** 2321-2328.

Breeden, L. and K. Nasmyth. 1987. Similarity between cell-cycle genes of budding yeast and fission yeast and the *notch* gene of *Drosophila*. *Nature* **329:** 651-654.

Brenner, C., N. Nakayama, M. Goebl, K. Tanaka, A. Toh-e, and K. Matsumoto. 1988. CDC33 encodes mRNA cap-binding protein eIF-4E of *Saccharomyces cerevisiae*. *Mol. Cell. Biol.* **8:** 3556-3559.

Broek, D., R. Bartlett, K. Crawford, and P. Nurse. 1991. Involvement of p34^{cdc2} in establishing the dependency of S phase on mitosis (see comments). *Nature* **349:** 388-393.

Bueno, A. and P. Russell. 1992. Dual functions of CDC6: A yeast protein required for DNA replication also inhibits nuclear division. *EMBO J.* **11:** 2167-2176.

———. 1993. Two fission yeast B-type cyclins, Cig2 and Cdc13, have different functions in mitosis. *Mol. Cell. Biol.* **13:** 2286-2297.

Bueno, A., H. Richardson, S.I. Reed, and P. Russell. 1991. A fission yeast B-type cyclin functioning early in the cell cycle (published erratum in *Cell* 1993 **73:** following 1049). *Cell* **66:** 149-159.

Caligiuri, M. and D. Beach. 1993. Sct1 functions in partnership with Cdc10 in a transcription complex that activates cell cycle START and inhibits differentiation. *Cell* **72:** 607-619.

Chapman, J.W. and L.H. Johnston. 1989. The yeast gene, *DBF4*, essential for entry into S phase is cell cycle regulated. *Exp. Cell Res.* **180:** 419-428.

Cocker, J.H., S. Piatti, C. Santocanale, K. Nasmyth, and J.F.X. Diffley. 1996. An essential role for the Cdc6 protein in forming the pre-replicative complexes of budding yeast. *Nature* **379:** 180-182.

Connolly, T. and D. Beach. 1994. Interaction between the Cig1 and Cig2 B-type cyclins in the fission yeast cell cycle. *Mol. Cell. Biol.* **14:** 768-776.

Correa-Bordes, J. and P. Nurse. 1995. p25^{rum1} orders S phase and mitosis by acting as an inhibitor of the p34^{cdc2} mitotic kinase. *Cell* **83:** 1001-1009.

Coverley, D., C.S. Downes, P. Romanowski, and R.A. Laskey. 1993. Reversible effects of nuclear membrane permeabilization on DNA replication: Evidence for a positive licensing factor. *J. Cell Biol.* **122:** 985-992.

Coxon, A., K. Maundrell, and S.E. Kearsey. 1992. Fission yeast *cdc21*$^+$ belongs to a family of proteins involved in an early step of chromosome replication. *Nucleic Acids Res.* **20:** 5571-5577.

Creanor, J. and J.M. Mitchison. 1984. Protein synthesis and its relation to the DNA division cycle in the fission yeast *Schizosaccharomyces pombe*. *J. Cell Sci.* **69:** 199-210.

Cross, F.R. 1988. *DAF1*, a mutant gene affecting size control, pheromone arrest, and cell cycle kinetics of *Saccharomyces cerevisiae*. *Mol. Cell. Biol.* **18:** 4675-4684.

Cross, F.R. and C.M. Blake. 1993. The yeast Cln3 protein is an unstable activator of Cdc28. *Mol. Cell. Biol.* **13:** 3266-3271.

Cross, F.R. and A.H. Tinkelenberg. 1991. A potential positive feedback loop controlling *CLN1* and *CLN2* gene expression at the start of the yeast cell cycle. *Cell* **65:** 875-883.

Cross, F., J. Roberts, and H. Weintraub. 1989. Simple and complex cell cycles. *Annu. Rev. Cell Biol.* **5:** 341-395.

Dahmann, C., J.F.X. Diffley, and K. Nasmyth. 1995. S-phase-promoting cyclin-dependent kinases prevent re-replication by inhibiting the transition of replication origins to a pre-replicative state. *Curr. Biol.* **5:** 1257-1269.

Dalton, S. 1992. Cell cycle regulation of the human *cdc2* gene. *EMBO J.* **11:** 1797-1804.

Deng, C., P. Zhang, J.W. Harper, S.J. Elledge, and P. Leder. 1995. Mice lacking p21$^{CIP1/WAF1}$ undergo normal development, but are defective in G1 checkpoint control. *Cell* **82:** 675-684.

Deshaies, R., V. Chau, and M. Kirschner. 1995. Ubiquitination of the G1 cyclin Cln2p by a Cdc34p-dependent pathway. *EMBO J.* **14:** 303-312.

Diffley, J.F.X. and J.H. Cocker. 1992. Protein-DNA interactions at a yeast replication origin. *Nature* **357:** 169-172.

Diffley, J.F.X. and B. Stillman. 1989. Similarity between the transcriptional silencer binding proteins ABF1 and RAP1. *Science* **246:** 1034-1038.

Diffley, J.F.X., J.H. Cocker, S.J. Dowell, and A. Rowley. 1994. Two steps in the assembly of complexes at yeast replication origins in vivo. *Cell* **78:** 303-316.

Dirick, L. and K. Nasmyth. 1991. Positive feedback in the activation of G1 cyclins in yeast. *Nature* **351:** 754-757.

Dirick, L., T. Böhm, and K. Nasmyth. 1995. Roles and regulation of Cln-Cdc28 kinases at the start of the cell cycle in *Saccharomyces cerevisiae*. *EMBO J.* **14:** 4803-4813.

Dolznig, H., P. Bartunek, K. Nasmyth, E. Müllner, and H. Beug. 1995. Terminal differentiation of normal chicken, erythroid progenitors: Shortening of G_1 correlates with loss of D-cyclin/cdk4 expression and altered cell size control. *Cell Growth Differ.* **6:** 1341-1352.

Donovan, J.D., J.H. Toyn, A.L. Johnson, and L.H. Johnston. 1994. P40SDB25, a putative CDK inhibitor, has a role in the M/G1 transition in *Saccharomyces cerevisiae*. *Genes Dev.* **8:** 1640-1653.

Dou, Q.P., P.J. Markell, and A.B. Pardee. 1992. Thymidine kinase transcription is regulated at G1/S phase by a complex that contains retinoblastoma-like protein and a *cdc2* kinase. *Proc. Natl. Acad. Sci.* **89:** 3256-60.

Dowell, S.J., P. Romanowski, and J.F. Diffley. 1994. Interaction of Dbf4, the Cdc7 protein kinase regulatory subunit, with yeast replication origins in vivo. *Science* **265:** 1243-1246.

Dulic, V., E. Lees, and S.I. Reed. 1992. Association of human cyclin E with a periodic G1-S phase protein kinase. *Science* **257:** 1958-1961.

Duronio, R.J. and P.H. O'Farrell. 1994. Developmental control of a G1-S transcriptional program in *Drosophila*. *Development* **120:** 1503-1515.

Duronio, R.J., P.H. O'Farrell, J.E. Xie, A. Brook, and N. Dyson. 1995. The transcription factor E2F is required for S phase during *Drosophila* embryogenesis. *Genes Dev.* **9:** 1445-1455.

Dynlacht, B.D., O. Flores, J.A. Lees, and E. Harlow. 1994. Differential regulation of E2F

transactivation by cyclin/cdk2 complexes. *Genes Dev.* **8:** 1772-1786.
El-Deiry, W.D., T. Tokino, V.E. Velcuiescu, D.B. Levy, R. Parsons, J.M. Trent, D. Lin, W.E. Mercer, K.W. Kinzler, and B. Vogelstein. 1993. *WAF1*, a potential mediator of p53 tumor suppression. *Cell* **75:** 817-825.
Epstein, C.B. and F.E. Cross. 1992. CLB5: A novel B cyclin from budding yeast with a role in S phase. *Genes Dev.* **6:** 1695-1706.
Evans, T., E.T. Rosenthal, J. Youngbloom, D. Distel, and T. Hunt. 1983. Cyclin: A protein specified by maternal mRNA in sea urchin eggs that is destroyed at each cleavage division. *Cell* **33:** 389-396.
Fagan, R., K.J. Flint, and N. Jones. 1994. Phosphorylation of E2F-1 modulates its interaction with the retinoblastoma gene product and the adenoviral E4 19 kDa protein. *Cell* **78:** 799-811.
Falconi, M.M., A. Piseri, M. Ferrari, G. Lucchini, P. Plevani, and M. Foiani. 1993. De novo synthesis of budding yeast DNA polymerase α and *POL1* transcription at the G1/S boundary are not required for entrance into S phase. *Proc. Natl. Acad. Sci.* **90:** 10519-10523.
Fang, F. and J.W. Newport. 1991. Evidence that the G1-S and G2-M transitions are controlled by different cdc2 proteins in higher eukaryotes. *Cell* **66:** 731-742.
Fesquet, D., J.C. Labbe, J. Derancourt, J.P. Capony, S. Galas, F. Girard, T. Lorca, J. Shuttleworth, M. Doree, and J.C. Cavadore. 1993. The *MO15* gene encodes the catalytic subunit of a protein kinase that activates cdc2 and other cyclin-dependent kinases (CDKs) through phosphorylation of Thr161 and its homologues. *EMBO J.* **12:** 3111-3121.
Firpo, E.J., A. Koff, M.J. Solomon, and J.M. Roberts. 1994. Inactivation of a Cdk2 inhibitor during interleukin 2-induced proliferation of human T lymphocytes. *Mol. Cell. Biol.* **14:** 4889-4901.
Fisher, R.P. and D.O. Morgan. 1994. A novel cyclin associates with MO15/CDK7 to form the CDK-activating kinase. *Cell* **78:** 713-724.
Fitch, I., C. Dahmann, U. Surana, A. Amon, K. Nasmyth, L. Goetsch, B. Byers, and B. Futcher. 1992. Characterization of four B-type cyclin genes of the budding yeast *Saccharomyces cerevisiae*. *Mol. Biol. Cell.* **3:** 805-818.
Forsburg, S.L. and P. Nurse. 1991a. Cell cycle regulation in the yeasts *Saccharomyces cerevisiae* and *Schizosaccharomyces pombe*. *Annu. Rev. Cell Biol.* **7:** 227-256.
———. 1991b. Identification of a G1-type cyclin puc1$^+$ in the fission yeast *Schizosaccharomyces pombe* (published erratum in *Nature* [1991] **352:** 648). *Nature* **351:** 245-248.
———. 1994. Analysis of the *Schizosaccharomyces pombe* cyclin puc1: Evidence for a role in cell cycle exit. *J. Cell Sci.* **107:** 601-613.
Furuno, N., M. Nishizawa, K. Okazaki, H. Tanaka, J. Iwashita, N. Nakajo, Y. Ogawa, and N. Sagata. 1994. Suppression of DNA replication via Mos function during meiotic divisions in *Xenopus* oocytes. *EMBO J.* **13:** 2399-2410.
Ghiara, J.B., H.E. Richardson, K. Sugimoto, M. Henze, D.J. Lew, C. Wittenberg, and S.I. Reed. 1991. A cyclin B homolog in *S. cerevisiae*: Chronic activation of the Cdc28 protein kinase by cyclin prevents exit from mitosis. *Cell* **65:** 163-174.
Gibson, S.I., R.T. Surosky, and B.K. Tye. 1990. The phenotype of the minichromosome maintenance mutant *mcm3* is characteristic of mutants defective in DNA replication. *Mol. Cell. Biol.* **10:** 5707-5720.
Girard, F., U. Strausfeld, A. Fernandez, and N.J. Lamb. 1991. Cyclin A is required for the

onset of DNA replication in mammalian fibroblasts. *Cell* **67**: 1169–1179.

Goebl, M.G., J. Yochem, S. Jentsch, J.P. McGrath, A. Varshavsky, and B. Byers. 1988. The yeast cell cycle gene *CDC34* encodes a ubiquitin-conjugating enzyme. *Science* **241**: 1331–1335.

Grandin, N. and S.I. Reed. 1993. Differential function and expression of *Saccharomyces cerevisiae* B-type cyclins in mitosis and meiosis. *Mol. Cell. Biol.* **13**: 2113–2125.

Guan, K.-L., C.W. Jenkins, Y. Li, M.A. Nichols, X. Wu, C.L. O'Keefe, A.G. Matera, and Y. Xiong. 1994. Growth suppression by p18, a p16INK4/MTS1-and p14INK4B/MTS2-related CDK6 inhibitor, correlates with wild-type pRb function. *Genes Dev.* **8**: 2939–2952.

Hadwiger, J.A., C. Wittenberg, H.E. Richardson, M. De Barros Lopes, and S.I. Reed. 1989. A family of cyclin homologs that control the G1 phase in yeast. *Proc. Natl. Acad. Sci.* **86**: 6255–6259.

Hagan, I., J. Hayles, and P. Nurse. 1988. Cloning and sequencing the cyclin-related $cdc13^+$ gene and a cytological study of its role in fission yeast mitosis. *J. Cell Sci.* **91**: 587–596.

Hamaguchi, J.R., R.A. Tobey, J. Pines, H.A. Crissman, T. Hunter, and E.M. Bradbury. 1992. Requirement for $p34^{cdc2}$ kinase is restricted to mitosis in the mammalian *cdc2* mutant FT210. *J. Cell Biol.* **117**: 1041–1053.

Hannon, G.J. and D. Beach. 1994. p15INK4B is a potential effector of TGF-β-induced cell cycle arrest. *Nature* **371**: 257–261.

Harper, J.W., G.R. Adami, N. Wei, K. Keyomarsi, and S.J. Elledge. 1993. The p21 Cdk-interacting protein Cip1 is a potent inhibitor of G1 cyclin-dependent kinases. *Cell* **75**: 805–816.

Hartwell, L.H. (1993). Getting started in the cell cycle. In *The early days of yeast genetics* (ed. M.N. Hall and P. Linder), pp. 307–314. Cold Spring Harbor Laboratory Press, Cold Spring Harbor, New York.

Hartwell, L.M. and M.W. Unger. 1977. Unequal division in *Saccharomyces cerevisiae* and its implications for the control of cell division. *J. Cell Biol.* **75**: 422–435.

Hartwell, L., J.Culotti, J.R. Pringle, and B.J. Reid. 1974. Genetic control of the cell division cycle in yeast. *Science* **183**: 46–51.

Hatakeyama, M., J.A. Brill, G.R. Fink, and R.A. Weinberg. 1994. Collaboration of G1 cyclins in the functional inactivation of the retinoblastoma protein. *Genes Dev.* **8**: 1759–1771.

Hayles, J., D. Fisher, A. Woollard, and P. Nurse. 1994. Temporal order of S phase and mitosis in fission yeast is determined by the state of the $p34^{cdc2}$-mitotic B cyclin complex. *Cell* **78**: 813–822.

Hennessy, K.M. and D. Botstein. 1991. Regulation of DNA replication during the yeast cell cycle. *Cold Spring Harbor Symp. Quant. Biol.* **56**: 279–284.

Hennessy, K.M., C.D. Clark, and D. Botstein. 1990. Subcellular localization of yeast CDC46 varies with the cell cycle. *Genes Dev.* **4**: 2252–2263.

Hennessy, K.M., A. Lee, E. Chen, and D. Botstein. 1991. A group of interacting yeast DNA replication genes. *Genes Dev.* **5**: 958–969.

Hinds, P.W. and R.A. Weinberg. 1994. Tumor suppressor genes. *Curr. Opin. Genet. Dev.* **4**: 135–141.

Hoffmann, I., G. Draetta, and E. Karsenti. 1994. Activation of the phosphatase activity of human cdc25A by a cdk2-cyclin E dependent phosphorylation at the G1/S transition. *EMBO J.* **13**: 4302–4310.

Hofmann, J.F. and D. Beach. 1994. *cdt1* is an essential target of the Cdc10/Sct1 transcription factor: Requirement for DNA replication and inhibition of mitosis. *EMBO J.* **13:** 425–434.

Hogan, E. and D. Koshland. 1992. Addition of extra origins of replication to a minichromosome suppresses its mitotic loss in *cdc6* and *cdc14* mutants of *S. cerevisiae. Proc. Natl. Acad. Sci.* **89:** 3098–3102.

Hollingsworth, R.E., Jr. and R.A. Sclafani. 1990. DNA metabolism gene *CDC7* from yeast encodes a serine (threonine) protein kinase. *Proc. Natl. Acad. Sci.* **87:** 6272–6276.

Holloway, S.L., M. Glotzer, R.W. King, and A.W. Murray. 1993. Anaphase is initiated by proteolysis rather than by the inactivation of maturation-promoting factor. *Cell* **73:** 1393–1402.

Howard, A. and S.R. Pelc. 1951. Synthesis of nucleoprotein in bean root cells. *Nature* **167:** 599–600.

Hubler, L., J. Bradshaw Rouse, and W. Heideman. 1993. Connections between the Rascyclic AMP pathway and G1 cyclin expression in the budding yeast *Saccharomyces cerevisiae. Mol. Cell. Biol.* **13:** 6274–6282.

Hunter, T. and J. Pines. 1994. Cyclins and cancer. II. Cyclin D and CDK inhibitors come of age. *Cell* **79:** 573–582.

Hutchinson, C.J., R. Cox, S. Drepaul, M. Gomperts, and C. Ford. 1987. Periodic DNA synthesis in cell-free extracts of *Xenopus* eggs. *EMBO J.* **6:** 2003–2010.

Iida, H. and I. Yahara. 1984. Durable synthesis of high molecular weight heat shock proteins in G0 cells of the yeast and other eukaryotes. *J. Cell Biol.* **99:** 199–207.

Irniger, S., S. Piatti, C. Michaelis, and K. Nasmyth. 1995. Genes involved in sister chromatid separation are needed for B-type cyclin proteolysis in budding yeast. *Cell* **81:** 269–277.

Jackson, A.L., P.M. Pahl, K. Harrison, J. Rosamond, and R.A. Sclafani. 1993. Cell cycle regulation of the yeast Cdc7 protein kinase by association with the Dbf4 protein. *Mol. Cell. Biol.* **13:** 2899–2908.

Johnson, D.G., K. Ohtani, and J.R. Nevins. 1994. Autoregulatory control of E2F1 expression in response to positive and negative regulators of cell cycle progression. *Genes Dev.* **8:** 1514–1525.

Johnson, P.T. and P.N. Rao. 1970. Mammalian cell fusion: Induction of premature chromosome condensation in interphase nuclei. *Nature* **226:** 717–722.

———. 1971. Nucleo-cytoplasmic interactions in the achievement of nuclear synchrony in DNA synthesis and mitosis in multinucleate cells. *Biol. Rev.* **46:** 97–155.

Kaffman, A., I. Herskowitz, R. Tjian, and E.K. O'Shea. 1994. Phosphorylation of the transcription factor PHO4 by a cyclin-CDK complex, PHO80-PHO85 (see comments). *Science* **263:** 1153–1156.

Kato, J.Y., M. Matsuoka, K. Polyak, J. Massagué, and C.J. Sherr. 1994. Cyclic AMP-induced G1 phase arrest mediated by an inhibitor (p27Kip1) of cyclin-dependent kinase 4 activation. *Cell* **79:** 487–496.

Kato, J., H. Matsushime, S.W. Hiebert, M.E. Ewen, and C.J. Sherr. 1993. Direct binding of cyclin D to the retinoblastoma gene product (pRb) and pRb phosphorylation by the cyclin D-dependent kinase CDK4. *Genes Dev.* **7:** 331–342.

Kelly, T.J., G.S. Martin, S.L. Forsburg, R.J. Stephen, A. Russo, and P. Nurse. 1993. The fission yeast *cdc18*+ gene product couples S phase to START and mitosis (see comments). *Cell* **74:** 371–382.

Killander, D. and A. Zetterberg. 1965. A quantitative cytochemical investigation of the relationship between cell mass and the initiation of DNA synthesis in mouse fibroblasts in vitro. *Exp. Cell Res.* **40:** 12–20.

Kim, Y.-J., S. Björklund, Y. Li, M.H. Sayre, and R.D. Kornberg. 1994. A multiprotein mediator of transcriptional activation and its interaction with the C-terminal repeat domain of RNA polymerase II. *Cell* **77:** 599–608.

Kimura, H., N. Nozaki, and K. Sugimoto. 1994. DNA polymerase alpha associated protein P1, a murine homolog of yeast MCM3, changes its intranuclear distribution during the DNA synthetic period. *EMBO J.* **13:** 4311–4320.

King, R.W., J. Peters, S. Tugendreich, M. Rolfe, P. Hieter, and M.W. Kirschner. 1995. A 20S complex containing CDC27 and CDC16 catalyzes the mitosis-specific conjugation of ubiquitin to cyclin B. *Cell* **81:** 279–288.

Knoblich, J.A. and C.F. Lehner. 1993. Synergistic action of *Drosophila* cyclins A and B during the G2-M transition. *EMBO J.* **12:** 65–74.

Knoblich, J.A., K. Sauer, L. Jones, H. Richardson, R. Saint, and C.F. Lehner. 1994. Cyclin E controls S phase progression and its down-regulation during *Drosophila* embryogenesis is required for the arrest of cell proliferation. *Cell* **77:** 107–120.

Koch, C. and K. Nasmyth. 1994. Cell cycle regulated transcription in yeast. *Curr. Opin. Cell Biol.* **6:** 451–459.

Koch, C., A. Schleiffer, G. Ammerer, and K. Nasmyth. 1996. Switching transcription on and off during the yeast cell cycle: Cln/Cdc28 kinases activate bound transcription factor SBF (Swi4/Swi6) at Start, whereas Clb/Cdc28 kinases displace it from the promoter in G_2. *Genes Dev.* **10:** 129–142.

Koch, C., T. Moll, M. Neuberg, H. Ahorn, and K. Nasmyth. 1993. A role for the transcription factors Mbp1 and Swi4 in progression from G1 to S phase. *Science* **261:** 1551–1557.

Koff, A., A. Giordano, D. Desai, K. Yamashita, J.W. Harper, S. Elledge, T. Nishimoto, D.O. Morgan, B.R. Franza, and J.M. Roberts. 1992. Formation and activation of a cyclin E-cdk2 complex during the G1 phase of the human cell cycle. *Science* **257:** 1689–1994.

Kung, A.L., S.W. Sherwood, and R.T. Schimke. 1993. Differences in the regulation of protein synthesis, cyclin B accumulation, and cellular growth in response to the inhibition of DNA synthesis in Chinese hamster ovary and HeLa S3 cells. *J. Biol. Chem.* **268:** 23072–23080.

La Thangue, N.B. 1994. DP and E2F proteins: Components of a heterodimeric transcription factor implicated in cell cycle control. *Curr. Opin. Cell Biol.* **6:** 443–450.

Lehner, C.F. and P.H. O'Farrell. 1990. The roles of *Drosophila* cyclins A and B in mitotic control. *Cell* **61:** 535–547.

Leno, G.H., C.S. Downes, and R.A. Laskey. 1992. The nuclear membrane prevents replication of human G2 nuclei but not G1 nuclei in *Xenopus* egg extract. *Cell* **69:** 151–158.

Liang, C., M. Weinreich, and B. Stillman. 1995. ORC and Cdc6p interact and determine the frequency of initiation of DNA replication in the genome. *Cell* **81:** 667–676.

Lohka, M.J. and Y. Masui. 1984. Roles of cytosol and cytoplasmic particles in nuclear membrane assembly and sperm pronuclear formation in cell-free preparations from amphibian eggs. *J. Cell Biol.* **98:** 1222–1230.

Lohka, M.J., M.K. Hayes, and J.L. Maller. 1988. Purification of maturation-promoting factor, an intracellular regulator of early mitotic events. *Proc. Natl. Acad. Sci.* **85:**

3009-3013.

Lorincz, A. and B.L.A. Carter. 1979. Control of cell size at bud initiation in *Saccharomyces cerevisiae. J. Gen. Microbiol.* **113**: 287-295.

Lorincz, A.T. and S.I. Reed. 1984. Primary structure homology between the product of yeast division control gene *CDC28* and vertebrate oncogenes. *Nature* **307**: 183-185.

Lovec, H., A. Grzeschiczek, M.B. Kowalski, and T. Moroy. 1994. Cyclin D1/bcl-1 cooperates with *myc* genes in the generation of B-cell lymphoma in transgenic mice. *EMBO J.* **13**: 3487-3495.

Lucibello, F.C., A. Sewing, S. Brusselbach, C. Burger, and R. Muller. 1993. Deregulation of cyclins D1 and E and suppression of cdk2 and cdk4 in senescent human fibroblasts. *J. Cell Sci.* **105**: 123-133.

Lukas, J., H. Muller, J. Bartkova, D. Spitkovsky, A.A. Kjerulff, P. Jansen Durr, M. Strauss, and J. Bartek. 1994. DNA tumor virus oncoproteins and retinoblastoma gene mutations share the ability to relieve the cell's requirement for cyclin D1 function in G1. *J. Cell Biol.* **125**: 625-638.

Marahrens, Y. and B. Stillman. 1992. A yeast chromosomal origin of DNA replication defined by multiple functional elements. *Science* **255**: 817-823.

Masui, Y. and C. Markert. 1971. Cytoplasmic control of nuclear behavior during meiotic maturation of frog oocytes. *J. Exp. Zool.* **177**: 129-146.

Matsumoto, K., I. Uno, Y. Oshima, and T. Ishikawa. 1982. Isolation and characterization of yeast mutants deficient in adenyl cyclase and cAMP dependent protein kinase. *Proc. Natl. Acad. Sci.* **79**: 2355-2359.

Matsushime, H., M.F. Roussel, R.A. Ashmun, and C.J. Sherr. 1991. Colony-stimulating factor 1 regulates novel cyclins during the G1 phase of the cell cycle. *Cell* **65**: 701-713.

Matsushime, H., M.W. Ewen, D.K. Strom, J.Y. Kato, S.K. Hanks, M.F. Roussel, and C.J. Sherr. 1992. Identification and properties of an atypical catalytic subunit (p34PSK-J3/cdk4) for mammalian D type G1 cyclins. *Cell* **71**: 323-334.

McGrew, J.T., L. Goetsch, B. Byers, and P. Baum. 1992. Requirement for *ESP1* in the nuclear division of *S. cerevisiae. Mol. Biol. Cell* **3**: 1443-1454.

Mendenhall, M.D. 1993. An inhibitor of p34CDC28 protein kinase activity from *S. cerevisiae. Science* **259**: 216-219.

Mitchison, J.M. 1970. *The biology of the cell cycle.* Cambridge University Press, Cambridge, England.

Moreno, S. and P. Nurse. 1994. Regulation of progression through the G1 phase of the cell cycle by the *rum1*[+] gene. *Nature* **367**: 236-242.

Morgenbesser, S.D., B.O. Williams, T. Jacks, and R.A. DePinho. 1994. p53-dependent apoptosis produced by Rb-deficiency in the developing mouse lens. *Nature* **371**: 72-74.

Murray, A.W. and M.W. Kirschner. 1989. Cyclin synthesis drives the early embryonic cell cycle. *Nature* **339**: 275-286.

Nakashima, N., K. Tanaka, S. Sturm, and M. Okayama. 1995. Fission yeast Rep2 is a putative transcriptional activator subunit for the cell cycle "start" function of Res2-Cdc10. *EMBO J.* **14**: 4794-4802.

Nash, R., G. Tokiwa, S. Anand, K. Erickson, and A.B. Futcher. 1988. The WH11[+] gene of *Saccharomyces cerevisiae* tethers cell division to cell size and is a cyclin homolog. *EMBO J.* **7**: 4335-4346.

Nasmyth, K. 1993. Control of the yeast cell cycle by the Cdc28 protein kinase. *Curr.*

Opin. Cell Biol. **5:** 166-179.
Nugroho, T.T. and M.D. Mendenhall. 1994. An inhibitor of yeast cyclin-dependent protein kinase plays an important role in ensuring the genomic integrity of daughter cells. *Mol. Cell. Biol.* **14:** 3320-3328.
Nurse, P. 1975. Genetic control of cell size at cell division in yeast. *Nature* **256:** 547-551.
———. 1990. Universal control mechanism regulating onset of M-phase. *Nature* **344:** 503-508.
———. 1994. Ordering S phase and M phase in the cell cycle. *Cell* **79:** 547-550.
Nurse, P. and Y. Bissett. 1981. Gene required in G1 for commitment to cell cycle and in G2 for control of mitosis in fission yeast. *Nature* **292:** 558-560.
Nurse, P. and P. Thuriaux. 1977. Controls over the timing of DNA replication during the cell cycle of fission yeast. *Exp. Cell Res.* **107:** 376-385.
———. 1980. Regulatory genes controlling mitosis in the fission yeast *Schizosaccharomyces pombe. Genetics* **96:** 627-637.
Nurse, P., P. Thuriaux, and K. Nasmyth. 1976. Genetic control of the cell division cycle in the fission yeast *Schizosaccharomyces pombe. Mol. Gen. Genet.* **146:** 167-178.
Obara Ishihara, T. and H. Okayama. 1994. A B-type cyclin negatively regulates conjugation via interacting with cell cycle 'start' genes in fission yeast. *EMBO J.* **13:** 1863-1872.
Ohtsubo, M. and J.M. Roberts. 1993. Cyclin-dependent regulation of G1 in mammalian fibroblasts. *Science* **259:** 1908-1912.
Osmani, A.H., N. van Peij, M. Mischke, M.J. O'Connell, and S.A. Osmani. 1994. A single p34cdc2 protein kinase (encoded by nimXcdc2) is required at G1 and G2 in *Aspergillus nidulans. J. Cell Sci.* **107:** 1519-28.
Pagano, M., R. Pepperkok, F. Verde, W. Ansorge, and G. Draetta. 1992. Cyclin A is required at two points in the human cell cycle. *EMBO J.* **11:** 961-971.
Pardee, A.B. 1989. G1 events and regulation of cell proliferation. *Science* **246:** 603-608.
Paris, J., R. Le Guellec, A. Couturier, K. Le Guellec, F. Omilli, J. Camonis, S. MacNeill, and M. Philippe. 1991. Cloning by differential screening of a *Xenopus* cDNA coding for a protein highly homologous to cdc2. *Proc. Natl. Acad. Sci.* **88:** 1039-1043.
Parker, S.B., G. Eichele, P. Zhang, A. Rawls, A.T. Sands, A. Bradley, E.N. Olson, J.W. Harper, and S.J. Elledge. 1995. p53-independent expression of p21Cip1 in muscle and other terminally differentiating cells. *Science* **267:** 1024-1027.
Pearson, B.E., H.P. Nasheuer, and T.S. Wang. 1991. Human DNA polymerase alpha gene: Sequences controlling expression in cycling and serum-stimulated cells. *Mol. Cell. Biol.* **11:** 2081-2095.
Piatti, S., C. Lengauer, and K. Nasmyth. 1995. Cdc6 is an unstable protein whose de novo synthesis in G1 is important for the onset of S phase and for preventing a "reductional" anaphase in the budding yeast *Saccharomyces cerevisiae. EMBO J.* **14:** 3788-3799.
Pines, J. and T. Hunter. 1990. Human cyclin A is adenovirus E1A-associated protein p60 and behaves differently from cyclin B. *Nature* **346:** 760-763.
Polyak, K., M.H. Lee, H. Erdjument Bromage, A. Koff, J.M. Roberts, P. Tempst, and J. Massagué. 1994. Cloning of p27Kip1, a cyclin-dependent kinase inhibitor and a potential mediator of extracellular antimitogenic signals. *Cell* **78:** 59-66.
Primig, M., S. Sockanathan, H. Auer, and K. Nasmyth. 1992. Anatomy of a transcription factor important for the start of the cell cycle in *Saccharomyces cerevisiae. Nature* **358:**

593–597.

Pringle, J.R. and L.H. Hartwell. 1981. The *Saccharomyces cerevisiae* cell cycle. In *The molecular biology of the yeast* Saccharomyces: *Life cycle and inheritance* (ed. J. Strathern et al.), pp. 97–142. Cold Spring Harbor Laboratory Press, Cold Spring Harbor, New York.

Quelle, D.E., R.A. Ashmun, S.A. Shurtleff, J.Y. Kato, D. Bar Sagi, M.F. Roussel, and C.J. Sherr. 1993. Overexpression of mouse D-type cyclins accelerates G1 phase in rodent fibroblasts. *Genes Dev.* **7:** 1559–1571.

Rao, P.N. and R.T. Johnson. 1970. Mammalian cell fusion: studies on the regulation of DNA synthesis and mitosis. *Nature* **225:** 159–164.

Reed, S.I., V. Dulic, D.J. Lew, H.E. Richardson, and C. Wittenberg. 1992. G1 control in yeast and animal cells. *Ciba Found. Symp.* **170:** 7–15 (discussion 15–19).

Reid, B. and L.H. Hartwell. 1977. Regulation of mating in the cell cycle of *S. cerevisiae*. *J. Cell Biol.* **75:** 355–365.

Resnitzky, D., M. Gossen, H. Bujard, and S.I. Reed. 1994. Acceleration of the G1/S phase transition by expression of cyclins D1 and E with an inducible system. *Mol. Cell. Biol.* **14:** 1669–1679.

Reymond, A., J. Marks, and V. Simanis. 1993. The activity of *S. pombe* DSC-1-like factor is cell cycle regulated and dependent on the activity of $p34^{cdc2}$. *EMBO J.* **12:** 4325–4334.

Riabowol, K., G. Draetta, L. Brizuela, D. Vandre, and D. Beach. 1989. The cdc2 kinase is a nuclear protein that is essential for mitosis in mammalian cells. *Cell* **57:** 393–401.

Richardson, H.E., C. Wittenberg, R. Cross, and S.I. Reed. 1989. An essential G1 function for cyclin-like proteins in yeast. *Cell* **59:** 1127–1133.

Richardson, H., D.J. Lew, M. Henze, K. Sugimoto, and S.I. Reed. 1992. Cyclin-B homologs in *Saccharomyces cerevisiae* function in S phase and in G2. *Genes Dev.* **6:** 2021–2034.

Roberts, J.M. 1993. Turning DNA replication on and off. *Curr. Opin. Cell Biol.* **5:** 201–206.

Roy, R., J.P. Adamczewski, T. Seroz, W. Vermeulen, J.-P. Tassan, L. Schaeffer, E.A. Nigg, J.H.J. Hoeijmakers, and J.-M. Egly. 1994. The MO15 cell cycle kinase is associated with the TFIIH transcription-DNA repair factor. *Cell* **79:** 1093–1101.

Rusch, H.P., W. Sachsenmaier, K. Behrens, and V. Gruter. 1966. Synchronization of mitosis by the fusion of the plasmodia of *Physarum polycephalum*. *J. Cell Biol.* **31:** 204–209.

Schwob, E. and K. Nasmyth. 1993. CLB5 and CLB6, a new pair of B cyclins involved in DNA replication in *Saccharomyces cerevisiae*. *Genes Dev.* **7:** 1160–1175.

Schwob, E., T. Bohm, M.D. Mendenhall, and K. Nasmyth. 1994. The B-type cyclin kinase inhibitor p40SIC1 controls the G1 to S transition in *S. cerevisiae*. *Cell* **79:** 233–244.

Serrano, M., G.J. Hannon, and D. Beach. 1993. A new regulatory motif in cell cycle control causing specific inhibition of cyclin D/CDK4. *Nature* **366:** 704–707.

Serrano, M., E. Gomez-Lahoz, R.A. DePinho, D. Beach, and D. Bar-Sagi. 1995. Inhibition of Ras-induced proliferation and cellular transformation by $p16^{INK4}$. *Science* **267:** 249–251.

Seufert, W., B. Futcher, and S. Jentsch. 1994. Role of a ubiquitin-conjugatiing enzyme in degradation of S- and M-phase cyclins. *Nature* **373:** 78–81.

Sewing, A., C. Burger, S. Brusselbach, C. Schalk, F.C. Lucibello, and R. Muller. 1993.

Human cyclin D1 encodes a labile nuclear protein whose synthesis is directly induced by growth factors and suppressed by cyclic AMP. *J. Cell Sci.* **104**: 545–555.

Sherr, C.J. 1994a. G1 phase progression: Cycling on cue. *Cell* **79**: 551–555.

———. 1994b. The ins and outs of RB: Coupling gene expresion to the cell cycle clock. *Trends Cell Biol.* **4**: 15–18.

Sicinski, P., J.L. Donaher, S.B. Parker, T. Li, A. Fazeli, H. Gardner, S.Z. Haslam, R.T. Bronson, S.J. Elledge, and R.A. Weinberg. 1995. Cyclin D1 provides a link between development and oncogenesis in the retina and breast. *Cell* **82**: 621–630.

Solomon, M.J., J.W. Harper, and J. Shuttleworth. 1993. CAK, the p34^{cdc2} activating kinase, contains a protein identical or closely related to p40MO15. *EMBO J.* **12**: 3133–3142.

Solomon, M.J., T. Lee, and M.W. Kirschner. 1992. Role of phosphorylation in p34^{cdc2} activation: Identification of an activating kinase. *Mol. Biol. Cell.* **3**: 13–27.

Sudbery, P.E., A.R. Goodey, and B.L.A. Carter. 1980. Genes which control cell proliferation in the yeast *Saccharomyces cerevisiae*. *Nature* **288**: 401–404.

Surana, U., A. Amon, C. Dowzer, J. McGrew, B. Byers, and K. Nasmyth. 1993. Destruction of the CDC28/CLB mitotic kinase is not required for the metaphase to anaphase transition in budding yeast. *EMBO J.* **12**: 1969–1978.

Surana, U., H. Robitsch, C. Price, T. Schuster, I. Fitch, A.B. Futcher, and K. Nasmyth. 1991. The role of CDC28 and cyclins during mitosis in the budding yeast *S. cerevisiae*. *Cell* **65**: 145–161.

Tam, S.W., J.W. Shay, and M. Pagano. 1994. Differential expression and cell cycle regulation of the cyclin-dependent kinase 4 inhibitor p16Ink4. *Cancer Res* **54**: 5816–5820.

Tanaka, K., K. Okazaki, N. Okazaki, T. Ueda, A. Sugiyama, H. Nojima, and H. Okayama. 1992. A new cdc gene required for S phase entry of *Schizosaccharomyces pombe* encodes a protein similar to the cdc10$^+$ and *SWI4* gene products. *EMBO J.* **11**: 4923–4932.

Tassan, J.P., S.J. Schultz, J. Bartek, and E.A. Nigg. 1994. Cell cycle analysis of the activity, subcellular localization, and subunit composition of human CAK (CDK-activating kinase). *J. Cell Biol.* **127**: 467–478.

Thommes, P., R. Fett, B. Schray, R. Burkhart, M. Barnes, C. Kennedy, N.C. Brown, and R. Knippers. 1992. Properties of the nuclear P1 protein, a mammalian homologue of the yeast Mcm3 replication protein. *Nucleic Acids Res.* **20**: 1069–1074.

Tokiwa, G., M. Tyers, T. Volpe, and B. Futcher. 1994. Inhibition of G1 cyclin activity by the Ras/cAMP pathway in yeast. *Nature* **371**: 342–345.

Tsai, L.H., E. Harlow, and M. Meyerson. 1991. Isolation of the human cdk2 gene that encodes the cyclin A- and adenovirus E1A-associated p33 kinase. *Nature* **353**: 174–177.

Tsai, L.H., E. Lees, B. Faha, E. Harlow, and K. Riabowol. 1993. The cdk2 kinase is required for the G1-to-S transition in mammalian cells. *Oncogene* **8**: 1593–1602.

Tye, B.-K. 1994. The MCM2-3-5 proteins: Are they replication licensing factors? *Trends Cell Biol.* **4**: 160–165.

Tyers, M., G. Tokiwa, and B. Futcher. 1993. Comparison of the *Saccharomyces cerevisiae* G1 cyclins: Cln3 may be an upstream activator of Cln1, Cln2 and other cyclins. *EMBO J.* **12**: 1955–1968.

Tyers, M., G. Tokiwa, R. Nash, and B. Futcher. 1992. The Cln3-Cdc28 kinase complex of *S. cerevisiae* is regulated by proteolysis and phosphorylation. *EMBO J.* **11**: 1773–1784.

Valay, J.G., M. Simon, and G. Faye. 1993. The Kin28 protein kinase is associated with a cyclin in *S. cerevisiae*. *J. Mol. Biol.* **234:** 307–310.

Wittenberg, C., K. Sugimoto, and S.I. Reed. 1990. G1-specific cyclins of *S. cerevisiae*: Cell cycle periodicity, regulation by mating pheromone, and association with the p34CDC28 protein kinase. *Cell* **62:** 225–237.

Yaglom, J., M.H. Linskens, S. Sadis, D.M. Rubin, B. Futcher, and D. Finley. 1995. p34Cdc28-mediated control of Cln3 cyclin degradation. *Mol. Cell. Biol.* **15:** 731–741.

Yoon, H.J., S. Loo, and J.L. Campbell. 1993. Regulation of *Saccharomyces cerevisiae* CDC7 function during the cell cycle. *Mol. Biol. Cell* **4:** 195–208.

Zhu, Y., T. Takeda, K. Nasmyth, and N. Jones. 1994. pct1+, which encodes a new DNA-binding partner of p85cdc10, is required for meiosis in the fission yeast *Schizosaccharomyces pombe*. *Genes Dev.* **8:** 885–898.

Zindy, F., E. Lamas, X. Chenivesse, J. Sobczak, J. Wang, D. Fesquet, B. Henglein, and C. Brechot. 1992. Cyclin A is required in S phase in normal epithelial cells. *Biochem. Biophys. Res. Commun.* **182:** 1144–1154.

12
Temporal Order of DNA Replication

Itamar Simon and Howard Cedar
Hebrew University Medical School
Jerusalem, Israel 91120

THE REPLICATION CLOCK

In every organism, DNA replication takes place in an ordered physical and temporal fashion. Although this process is primarily necessary to carry out the task of copying the genome, it may also indirectly play a role in the regulation of gene expression. In *Escherichia coli*, for example, the entire 4-Mb circular chromosome undergoes bidirectional DNA synthesis that is initiated from a single origin. It takes at least 40 minutes to copy the complete bacterial genome, and as a result, genes located near the origin are replicated early in the cell cycle whereas those positioned at the opposite end of the chromosome are synthesized late in the cycle. One direct consequence of this temporal organization is that in rapidly dividing cells, the early-replicating genes themselves accumulate differentially in high copy number (von Meyenburg and Hansen 1987), and this most certainly has profound effects on the overall expression pattern of the organism.

Compared to prokaryotes, eukaryotic organisms contain a much larger genome. Although the total DNA is divided into separate chromosome molecules, each of these entities in itself carries a considerable amount of genetic information, and DNA synthesis is accomplished through the action of multiple bidirectional origins. If all of these replication units were activated simultaneously, the entire process of replication could be completed very quickly. However, replication starts are actually distributed in a programmed manner throughout the length of S phase, and, as a consequence, each origin is turned on at a specific time. In the yeast *Saccharomyces cerevisiae*, DNA located near telomeres, for example, undergoes replication late in S phase, whereas sequences associated with centromeres are early replicating (Reynolds et al. 1989). Other regions of the genome are replicated at various fixed times within the 30-minute S phase.

Concepts in Eukaryotic DNA Replication
© 1999 Cold Spring Harbor Laboratory Press 0-87969-557-9/99

The process of DNA replication in animal cells takes place over a longer period of time, an average of 8–10 hours. Despite the complexity of these genomes (3000 Mb), a number of related methods have been developed to pinpoint the replication time of individual genes. Basically, replicating DNA is first labeled in vivo with bromodeoxyuridine (BrdU) and then purified by gradient centrifugation from cells at various stages of S phase. This fractionation is accomplished either by prior synchronization (Goldman et al. 1984; Schmidt and Migeon 1990) or by cell-sorting techniques (Braunstein et al. 1982; Gilbert 1986), and the resultant newly synthesized DNA samples are then subjected to blot hybridization with specific probes. Early-replicating genes hybridize to fractions from the beginning of S phase, whereas late-replicating sequences only hybridize with BrdU DNA from cells in the later fractions of S. By increasing the number of interval fractions in S phase, one can obtain better cell-cycle resolution of replication timing. Using this methodology, a large number of specific genes from various organisms have been analyzed, and in general, the results confirm the idea that replication proceeds progressively in an ordered programmed manner throughout S phase (for review, see Holmquist 1987).

REPLICATION TIMING AND GENE EXPRESSION

Although the biological function of replication timing control in the cell cycle is not yet clear, studies in yeast and in animal cells have revealed a striking association between replication timing and gene expression. In *S. cerevisiae*, for example, the unexpressed *HML* and *HMR* loci are located in subtelomeric regions that replicate late in S phase. When, as part of the normal process of cell-type-specific switching, they are copied to the *MAT* locus in an early-replicating region of the same chromosome, the promoters are activated (Reynolds et al. 1989). That replication timing is involved in this control mechanism is suggested by the observation that positional silencing at *HMR* requires an autonomously replicating sequence (ARS) element, and mutations in the ARS that abolish origin function also relieve transcriptional silencing (for review, see Laurenson and Rine 1992). In higher organisms, the general relationship between transcription and replication is even more straightforward. Constitutively transcribed housekeeping genes undergo relatively early DNA replication (some time during the first half of S) in a variety of cell types. In contrast, many tissue-specific genes are late-replicating (the last 50% of S) in their repressed state, but are early-replicating in cells in which they are expressed (for review, see Holmquist 1987). It thus appears that in

Table 1 Replication times for genes in vertebrate cells

Gene	Replication time
Housekeeping	
HGRPT	E
APRT	E
CAD	E
DHER	E
arginine succinate synthetase	E
G6PD	E
β-actin	E
β-tubulins	E
metallothionein-1	E
PDG X linked	E
Tyr aminotransferase	E
c-ras	E
c-myc	E
all histones tested	E
Xenopus somatic 5s	E
Tissue-specific	
apolipoprotein	E
placental lactogen	E
α_1-antitrypsin	L
β-casein	L
Phe hydroxylase	L
α-globin	E
fibronectin	L
β-globin	L*
immunoglobulin V_H	L*
Xenopus oocyte 5s	L*
cystic fibrosis	L*

Data on replication timing were either derived from standard cell-culture procedures (adapted from Holmquist 1987) or obtained by FISH analysis (e.g., cystic fibrosis genes). It has already been demonstrated that these two different techniques yield similar results (Selig et al. 1992). As shown, all housekeeping genes replicate in the first half of S phase (E), whereas many tissue-specific genes replicate in the last half (L) of S in cell types that do not express the gene. In some cases, these same gene sequences were found to replicate early in expressing cells (*). This table only includes representative examples and is not exhaustive.

these organisms, replication timing is a developmentally regulated process that is closely associated with gene expression (Table 1).

Unlike some organisms, such as *Xenopus* or *Drosophila*, that are characterized by extremely short replication cycles during early-cleavage stages of development, in mammals, programmed, cell-cycle-controlled replication is present even at the earliest cell divisions (Takagi 1974). To obtain a better picture of how the replication timing pattern of the genome is established during development, replication was analyzed in F9 teratocarcinoma cells (S. Selig, unpubl.). These studies suggest that tissue-specific genes, such as β-globin or albumin, are initially late-replicating in cells of the early embryo prior to organogenesis. It is likely that these same genes then remain late-replicating in most cell types, but shift to early replication as a part of the differentiation process that brings about gene activation. In parallel, housekeeping genes remain early-replicating and active throughout development. Almost all studies on replication timing have been carried out in rapidly growing tissue-culture cells, and much work must still be done to evaluate this process in vivo.

Tissue-specific genes, which are initially late-replicating in the early embryo, undergo conversion to early replication in specific cell types. However, other DNA sequences in the genome have an opposite pattern and become late-replicating in a developmentally regulated manner. Genes on the X chromosome, for example, shift dramatically from early to late replication in association with the inactivation process that occurs at about the time of implantation in female eutherans (Takagi 1974). At the same stage in development, centromeric mouse satellite sequences also undergo a shift to a later time of replication (Selig et al. 1988). Although the role of replication timing in this case is not known, this change probably reflects the overall stage-specific structural alterations that take place on these chromosomal regions.

REPLICATION TIME ZONES IN MAMMALIAN CELLS

Since in each cell type, individual sectors of DNA are programmed to undergo replication at specific times in S phase, the genome must be organized into replication time units. By following BrdU labeling, these can be visualized cytogenetically as an alternating pattern of early- and late-replicating regions on each chromosome (Fig. 1). In this procedure, nonsynchronized cell cultures are incubated with BrdU for fixed times, and the resulting incorporated label is then detected in metaphase chromosomes. After 4 hours of labeling, for example, all of the BrdU visualized in metaphase represents sequences that replicate late in S phase just prior to condensation in G_2 (~2 hours), and this produces a striking pattern of highly ordered late-replication bands on each chromosome. Labeling with BrdU for longer intervals reveals a continuous series of chromosomal bands that replicate in a programmed manner throughout S (Latt 1973; Wolf and Perry 1974; Vogel et al. 1986). The very existence of this visible banding pattern immediately

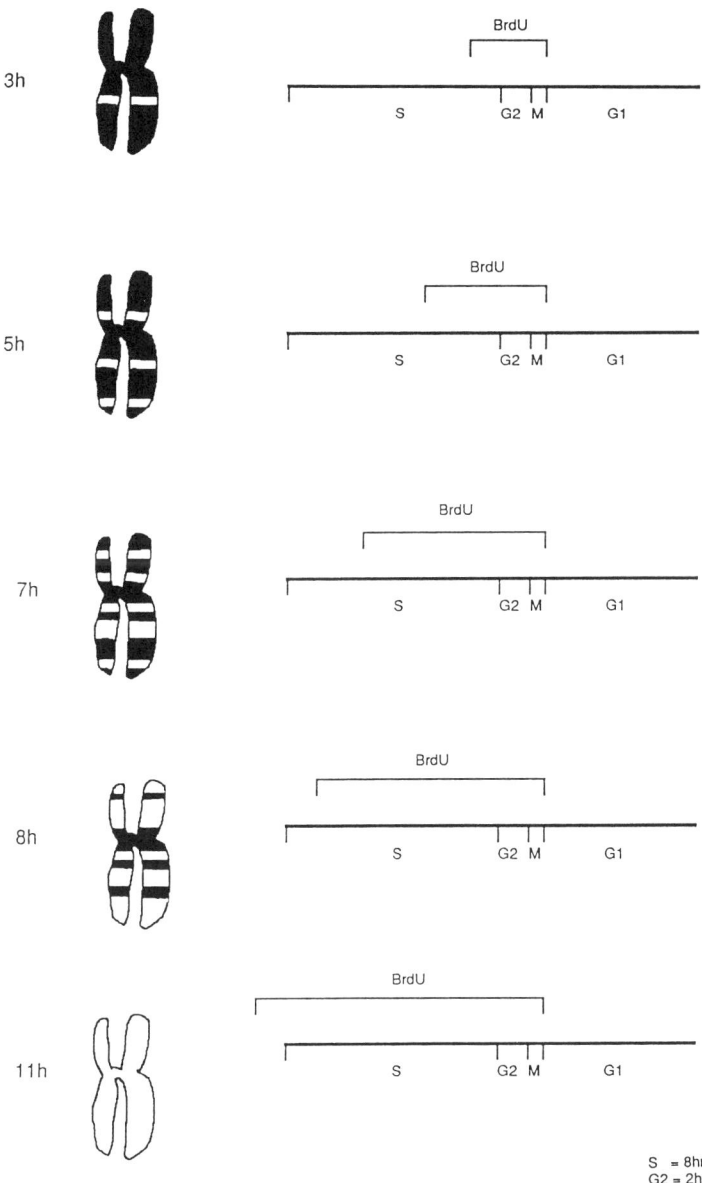

Figure 1 Replication banding. Nonsynchronous cells are cultured in the presence of BrdU for various fixed times, and nuclei are then prepared and stained (see Drouin et al. 1990). The locations of BrdU incorporation are visualized as light (quenched) bands. Since chromosomes are only observed in metaphase cells (M), the 3-hr incubation labels only those regions that replicate in the last hour of S, and the 5-hr incubation labels those that replicate in the last 3 hr of S. After 11 hr, the entire chromosome is labeled with BrdU.

suggests that DNA replication timing units are not only quite large, but also spatially well delineated. Replication bands are an integral part of the basic chromosome unit structure, and these same early-replicating DNA regions appear to comap quite precisely with Giemsa-stained light G bands (Hand 1978). Using DNase I sensitivity as a molecular probe for genome accessibility, it has been possible to map "active" regions (D bands) on metaphase chromosomes, and these also correlate with early-replicating time zones (Kerem et al. 1984). The fact that these banding patterns do not vary appreciably from cell type to cell type suggests that the early-replicating active regions may represent genomic loci rich in housekeeping genes. Taken together, these morphological observations indicate that the entire genome may be subdivided into a series of basic modules which represent units of common structure, function, and replication timing.

Original estimates for the size of each replication band suggest that these domains may cover 5–10 Mb and contain multiple replicons. High-resolution mapping of both replication bands (Drouin et al. 1990) and G bands (Yunis 1981) on extended pro-metaphase chromosomes, however, has delineated at least 2000 individual stripes, suggesting that each chromosomal unit is actually much smaller, containing, on average, l.5 Mb of DNA (Kitsberg et al. 1991). Due to their relatively large size, most of these replication domains have not yet been mapped at the molecular level, but the availability of extensive cloned contigs, as well as recent advances in technology, should now make it possible to examine these units in detail.

The replication time of a genomic DNA fragment can be measured in almost any cell type by using fluorescence in situ hybridization (FISH) to interphase nuclei. In this method, unreplicated DNA is visualized as a single hybridization dot, and replicated DNA appears as a double dot. In a nonsynchronized population of dividing cells, a high percentage of nuclei with double hybridization signals indicates that this particular gene replicates relatively early in S phase, whereas a low count is obtained for gene sequences that replicate late in the cycle (Selig et al. 1992). Furthermore, by labeling cells with BrdU prior to harvesting, it becomes possible to specifically identify nuclei in S phase and thus to disregard single-dot nuclei from G_1 and double-dot nuclei from G_2 (Kitsberg et al. 1993b). This method is particularly accurate for measuring the replication time of one zone relative to another. In this case, each probe is labeled with a different fluorescent dye, making it possible to detect nuclei where one gene has already replicated while the other is still uncopied in the same cell (Selig et al. 1992). Using this approach, replication time zones can probably be resolved within minutes on the cell-cycle clock.

A 1.5-Mb region of DNA containing the cystic fibrosis (CF) gene has been studied using this method. By assaying a contiguous set of phage and cosmid clones in fibroblasts or lymphocytes, cells that do not express this gene, one observes two distinct, but internally uniform, replication time bands: an early/middle zone located 5' to the gene and an extensive late zone that includes the 250-kb domain that carries the CF coding sequence. In the intestinal epithelial-derived cell line, Caco-2, in which the CF gene is transcriptionally active, the basic replication structure is quite similar, except for a distinct region of 500–700 kb around the CF gene that replicates very early (Selig et al. 1992). Thus, in this region, replication timing is organized into defined spatial domains and is developmentally regulated (Fig. 2). The human β-globin gene region has also been mapped for replication timing at the molecular level. In this case, as well, a relatively large domain containing the β-like genes is late-replicating in a number of different cell types, but early-replicating in erythroleukemia cells (Epner et al. 1988; Hatton et al. 1988). Because only 250 kb of sequence surrounding these genes has been analyzed, it has not been possible to map the full extent and outer boundaries of this time zone.

REPLICON STRUCTURE

Since animal cell replicons are, on average, 50–300 kb in length (Hand 1978), it is likely that a typical replication band is made up of multiple adjacent replicons that are coordinately regulated to undergo replication at the same time in S. This was originally derived from autoradiograph studies of DNA synthesis, which clearly reveal replicating regions containing strings of multiple pulse-labeled origins (Edenberg and Huberman 1975), and this observation has now been confirmed at the molecular level. The most clear-cut example of this phenomenon is in the mouse major histocompatibility complex (MHC) locus, where a large region of more than 2 Mb contains at least 5–6 distinct replicons that all replicate fairly synchronously as a single time zone (Spack et al. 1992). Several well-mapped replication origins near the DHFR gene also appear to fire in a coordinated manner, all at the beginning of S phase (Ma et al. 1990).

Several lines of evidence indicate that the activation time of any particular origin is not an intrinsic property of the origin sequence itself. The chromosome V origin, ARS501 in yeast, is activated in the second half of S phase and is responsible for the late replication of a 66-kb domain

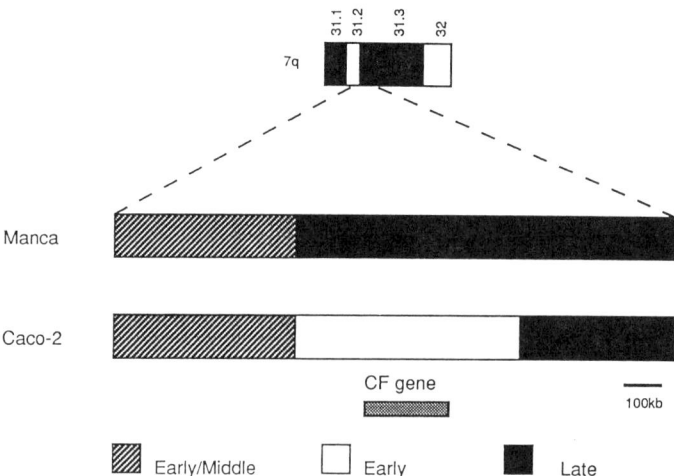

Figure 2 Replication time zones in the cystic fibrosis gene domain. A large number of plasmid, phage, and cosmid clones covering a 1.5-Mb region on human chromosome 7 were analyzed for replication timing in two cell types: Manca lymphocytes, which do not express the CF gene, and Caco-2 cells, which do (see Selig et al. 1992). This map corresponds to chromosome region 7q31.

that includes the right telomere (Ferguson et al. 1991). The time of replication of this origin is regulated by position, since late activation requires the presence of a telomere nearby, and when this origin is transplanted to another site, it replicates early. These observations show that the time at which an origin is activated can depend on its chromosomal context (Ferguson and Fangman 1992). Origins in animal cells apparently behave in a similar manner. In the human β-globin gene domain, mapping has demonstrated that a large region of at least 100 kb is part of a single replicon, with an origin located slightly 5′ to the β-globin gene sequence. Interestingly, this same site serves as the origin both in lymphocytes, where replication occurs late in S, and in erythroleukemia cells, where replication is early (Kitsberg et al. 1993a). This represents clear-cut evidence that the origin sequence serves as a topographical marker for the site of initiating DNA synthesis, whereas temporal activation is controlled by chromosomal context in *cis* and most likely by protein factors in *trans*.

REGULATION OF REPLICATION TIMING

Several experiments indicate that long-range *cis*-acting elements must play a role in the control of replication timing. In animal cells this effect is most obvious in regions that have undergone chromosomal rearrangement. The Xic region on the inactive X, for example, probably plays an important role in controlling replication timing, since its translocation to another chromosome causes adjacent sequences to become late-replicating and, in many cases, inactive (see, e.g., Disteche et al. 1979). Significant changes in replication timing also occur during the normal process of human immunoglobulin heavy chain (IgH) rearrangement in the B-cell lineage. In most somatic cells, the heavy-chain constant region is embedded within a large early-replicating domain (Brown et al. 1987), whereas distal variable sequences replicate late in S. When rearrangement takes place, however, the juxtaposed variable gene falls under control of the constant domain and assumes an early replication time (Calza et al. 1984). This shift could be part of the variable gene promoter activation process, which is ultimately responsible for antibody production in B cells. Similar changes in replication timing also occur at the site of the IgC_H-*myc* translocation in MPC11 plasmacytoma cells (Calza et al. 1984) or at other translocated loci in human leukemias (Karube and Watanabe 1988).

Genomic mapping studies in yeast have succeeded in identifying and characterizing defined *cis*-acting sequences that control replication timing. A telomere itself, for example, can induce late replication on origins located within approximately 30 kb. This effect is undoubtedly mediated through the telomeric DNA sequence itself, since small $C_{1-3}A$ tracts also shift replication timing when placed close to an origin (Ferguson and Fangman 1992). It is likely that *cis*-acting sequences in other parts of the genome can also direct the initiation time of DNA synthesis, and an element of this nature has recently been mapped near the late-replicating KEX2 ARS on yeast chromosome XIV (B.J. Brewer et al., unpubl.).

Replication timing in animal cells is also under the control of *cis*-acting sequences, and there is now good evidence for a defined replication timing regulatory element associated with the locus control region (LCR) in the human β-globin gene domain. This enhancer-like sequence, located about 20 kb upstream of the ε gene, probably plays a major role in the regulation of genes in this cluster, since patients carrying a deletion of this control region (Hispanic thalassemia) do not make any β-globin transcripts (Driscoll et al. 1989). In normal human lymphocytes, this locus is in a repressed configuration, but fusion to mouse erythroleukemia cells induces a striking activation of the domain and a concomitant shift to early replication. In contrast, the β-globin domain from Hispanic thalassemia lymphocytes fails to undergo transcriptional activation and, at the same time, remains late-replicating. It thus appears that seqences associated with, or near, the globin LCR are probably involved in replication timing control (Forrester et al. 1990).

Indirectly, these fusion studies also suggest that the donor erythroleukemia cells contain a set of protein factors which control the replication timing of the globin domain in *trans*, perhaps through specific interactions with LCR elements. Experiments in culture have shown that treatment of cells with the demethylating agent, 5-azacytidine (5azaC), dramatically shifts the replication time of both the inactive X chromosome (Shafer and Priest 1984; Jablonka et al. 1985) and mouse satellite sequences (Selig et al. 1988) from late to early in the cycle. The highly coordinated and quantum nature of this response suggests that this drug does not act independently on each of the multiple chromosomal regions themselves. Rather, 5azaC mediates the changes in replication timing either through an effect on a master *cis*-acting control element or by inducing *trans*-acting protein factors. This latter possibility is supported by cell fusion experiments. Thus, for example, when embryonic carcinoma cells are fused with female-derived thymocytes, this brings about a dramatic early shift in X-chromosome replication timing (Takagi et al. 1983). The abnormal early replication time of trisomic chromosomal region 15E seen in cases of T-cell leukemia represents another case where protein factors in the nucleus must play some role in the control of the replication clock (Somssich et al. 1984). These studies, taken together, thus support a model whereby replication timing is regulated through *cis* elements that are recognized by and interact with specific *trans*-acting factors.

ALLELE-SPECIFIC GENE EXPRESSION

Transcriptional regulation is thought to be carried out in a large part through the interaction of regulatory sequences with combinations of multiple *trans*-acting factors in the nucleus. In parallel, *cis*-acting effects, such as chromatin structure or DNA methylation, also modulate gene expression by affecting the accessibility of gene regions to these same factors. The importance of *cis*-acting mechanisms is nowhere more obvious than in cases where only one allele is expressed in a given nucleus. Thus, for example, genes on both X chromosomes in female cells are potentially exposed to the same *trans*-acting factors in the nucleus, yet structural differences render one X inaccessible and therefore excluded from the normal transcriptional machinery. DNA methylation represents one important mode for mediating allelic selection. Methyl moieties are known to inhibit transcription both by altering specific protein-DNA interactions and by modulating general chromatin structure (Eden and Cedar 1994). When only one allele is modified, this leads to selective inhibition.

A major role of DNA methylation on the X chromosome is to maintain the differential expression pattern from generation to generation (Mohandas et al. 1981). Once established, methyl moieties are conservatively preserved in daughter cells due to the action of a maintenance methylase that copies the modification pattern of the previous generation (Cedar 1988). Another parameter that differentiates between the active and inactive X is their time of replication in S phase; this may represent yet another *cis*-acting structural marker for distinguishing between two otherwise equivalent alleles in the same cell. Assuming that replication timing is set by cell-cycle-controlled factors, it is easy to see that this parameter, much like methylation, can be propagated from generation to generation and can thus be used to maintain allelic identity.

Replication timing may also be involved in the allele-specific regulation of imprinted genes. A number of these genes have now been identified and characterized in the mouse. Insulin-like growth factor 2 (*Igf2*), U2af-binding protein-related sequence, and small nuclear ribonucleoprotein polypeptide n (*Snrpn*), for example, are expressed exclusively from the paternal allele, whereas insulin-like growth factor 2 receptor (*Igf2r*) and *H19* are transcribed only from their maternal copy in somatic cells (for review, see Razin and Cedar 1994; see also Villar and Pedersen 1994). Imprinting has also been observed for some of these same genes in humans. The Prader-Willi and Angelman syndromes (PWS/AS), for example, come about as a result of either uniparental disomy or monoparental deletions on chromosome 15, and it is likely that imprinting of *SNRPN* or other gene sequences located in this region may explain the variety of symptoms seen in these diseases (Nicholls 1993).

IMPRINTED REPLICATION TIME ZONES

Much like the genes on the X chromosome, these sequences have a genomic structure that distinguishes between the two parental alleles. Thus, as might be expected, almost all of the loci that have been analyzed show differential methylation (Razin and Cedar 1994). In situ hybridization has been employed to evaluate the replication timing properties of these genes. This method is especially appropriate for detecting asynchronous replication. In any cell population, FISH analysis generally reveals mainly nuclei with either two single or two double hybridization dots, indicating that both homologs of most genes replicate

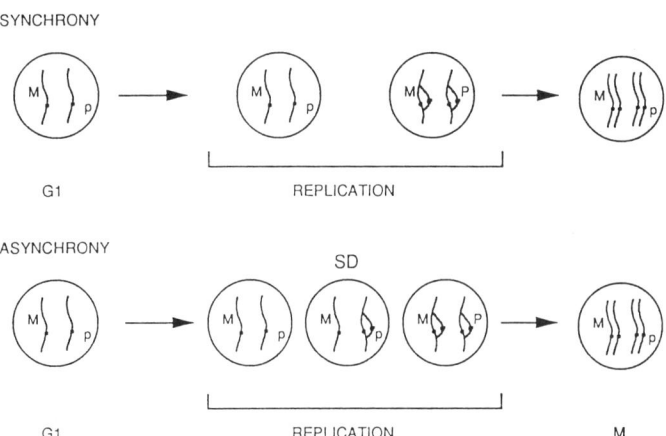

Figure 3 Allele-specific replication timing. By using FISH analysis, all gene regions are unreplicated on both alleles in G_1 and therefore show two single hybridization dots. For most sequences, both homologs replicate relatively synchronously, so nuclei in S phase either have two single dots or two double dots. For imprinted genes, each allele replicates at a different time in S, so a high percentage of nuclei show both an unreplicated (single dot) and a replicated (double dot) allele. At metaphase, all of the DNA has already replicated.

synchronously during the cell cycle. In contrast, for imprinted genes, a large number (~30%) of nuclei contain one replicated and one unreplicated gene (Fig. 3), suggesting that these alleles undergo DNA synthesis asynchronously in the cell cycle (Kitsberg et al. 1993b; Knoll et al. 1994). It appears from these studies that many imprinted genes are imbedded within large differentially replicating domains which, in some cases, can even be visualized by standard replication banding techniques (Izumikawa et al. 1991). By using polymorphic markers to identify each parental allele individually, it has been shown that most of these regions replicate coordinately in an allele-specific manner, with the paternal allele generally replicating earlier in the cell cycle (Kitsberg et al. 1993b). As a result of this regional pattern, the active allele of some imprinted genes actually replicates late in the S phase. This clearly represents an exception to the general correlation between early replication timing and gene expression and suggests that this standard structure/function relationship may be altered for imprinted genes.

REGIONAL REGULATION

The finding of domain-wide replication-timing control strongly suggests that imprinting itself may be regulated at the regional level. This idea is supported by the observation that imprinted genes are clustered. Experiments with mono-parental disomic mice initially indicated the existence of a limited number of relatively well defined chromosomal regions that may be involved in the generation of parent-of-origin defects during development (for review, see Cattanach 1986), and more recently, genomic mapping has shown that imprinted genes are indeed clustered. *Igf2* and *H19*, for example, are positioned within 90 kb of each other and are imprinted in both mouse and man (Razin and Cedar 1994). Furthermore, the nearby insulin gene has been found to be imprinted in extraembryonic tissues (Giddings et al. 1994), and *Mash-2*, a newly identified paternally transcribed gene, is located in the same chromosomal region about 500 kb upstream of the *Igf2* locus (Guillemot et al. 1995).

Even more striking is the PWS region on human chromosome 15, where at least four individual genes are transcribed exclusively from the paternal allele (Sutcliffe et al. 1994; Wevrick et al. 1994). Further evidence suggesting regional control at this locus comes from cases of PWS carrying small deletions around the *Snrpn* gene on the paternal chromosome (Buiting et al. 1995). Although these mutations are relatively small, several nearby intact imprinted genes on this chromosome become transcriptionally inactive (Sutcliffe et al. 1994), selective adjacent sites adopt a maternal pattern of methylation (Reis et al. 1994), and for at least one case, an entire 600-kb domain undergoes a dramatic change in replication timing on the paternal allele (Gunaratne et al. 1995). It is thus likely that these individuals have a defect in *cis*-acting DNA sequences directly involved in the regionwide regulation of imprinting. In general, these imprinted domains include gene sequences that are not necessarily expressed in an allele-specific manner. It thus seems likely that the regulation of imprinting is a complicated process that utilizes multiple control mechanisms and most certainly involves both local and regional effectors of expression.

The organization and control of asynchronous replication, both on the X chromosome and in imprinted domains, are not well understood. Although the two X chromosomes in female cells generally replicate asynchronously (Takagi 1974), the inactive X still retains a detailed multireplication-band structure, with each individual domain replicating at a fixed time in S, usually later than the corresponding region on the active X (Wahrman et al. 1983). Because of this structure, most genes on the active X replicate in the first half of S, but for some genes, despite their asynchrony, both alleles actually replicate in late S phase (Hansen et al. 1993; Boggs and Chinault 1994). Whereas replication timing on the X chromosome appears to be coordinated, individual replication bands may also be regulated autonomously. In the fragile X syndrome, for example, a single large domain of over 150 kb is shifted to a very late replication time while the rest of the chromosome replicates with normal timing properties (Hansen et al. 1993). Overall replication timing asynchrony on the X is most likely centrally regulated by *cis*-acting sequences at *Xic*, a locus that may itself be essential for the inactivation process (Brown et al. 1991b). Further studies have highlighted the possible role of the *Xist* gene which maps to this region and codes for an untranslated nuclear RNA species transcribed exclusively from the inactive chromosome (Brown et al. 1991a, 1992). Interestingly, this gene is embedded in a unique domain that replicates asynchronously, but unlike the rest of the chomosome, is copied earlier on the inactive X (Boggs and Chinault 1994; Torchia et al. 1994). Recent studies have shown that this gene is not required for the maintenance of X inactivation, making it unlikely that it itself plays a role in the control of replication timing (Brown and Willard 1994).

Complex patterns of replication time zones may also be characteristic of imprinted regions. Both in the Igf2r and Igf2 domains, asynchronous replication appears to be uniform and coordinated in such a way that large regions on each parental chromosome replicate together (Kitsberg et al. 1993b). In contrast, it is likely that the asynchronously replicating 3- to 4-Mb PWS/AS domain is actually made up of smaller individual subunits, each subject to its own replication timing control. Most DNA from this region clearly replicates with the paternal allele early, but it has also been demonstrated that sequences near the GABA receptor genes replicate maternally early (LaSalle and Lalande 1995), and at least one probe near the P locus replicates asynchronously, but without any allele specificity (Knoll et al. 1994).

ALLELIC EXCLUSION

The most striking aspect of asynchronous replication timing is its consistent association with the phenomenon of domain-wide allelic inactivation. Olfactory receptor gene loci may represent yet another example of this unique type of regulatory structure. The sense of smell in a number of different organisms is mediated through olfactory neurons in the nasal organ (Buck and Axel 1991). Each of these neurons evidently expresses one, or a small number, out of a repertoire of about 1000 odorant receptor genes (Vassar et al. 1993). The problem of distinguishing which receptors have been activated therefore reduces to a problem of identifying which neurons have been activated. In such a model, perception of an odor encountered by the organism would result from the identification of the subset of cells responsive to that odorant.

Recent evidence suggests that in an individual neuron the receptor gene is expressed from only one of the two parental alleles (Chess et al. 1994). In some cells it is the paternal gene which is active, whereas in others the maternal gene undergoes transcription. Since olfactory receptor genes are clustered at a number of different chromosomal sites, it has been suggested that a single gene in an array is turned on stochastically by means of a *cis*-acting locus control element. If this is the case, there must be a mechanism of allelic exclusion to prevent the second allele from activating another gene. Although there is, as yet, no direct experimental proof for this model, it has been shown that these loci indeed replicate asynchronously (Table 2), and this suggests indirectly that only one allele in each cell may be in an active conformation. Unlike imprinted gene domains, the replication pattern of the olfactory receptor loci is not allele-specific, and this is in keeping with the observation that either allele for any given receptor gene can be active. The asynchronous pattern is probably generated in the early embryo, since it has been observed in a variety of different cell types and in F9 teratocarcinoma cells (Chess et al. 1994). In a sense, this structure would thus represent a form of autosomal allelic exclusion analogous to X-chromosome inactivation. There may very well be additional regions of the genome that behave in this manner, and the replication assay provides a convenient method for detecting these.

REGULATION OF GENE EXPRESSION

Although there is a correlation between replication timing and gene expression, the cause and effect relationship between these phenomena has not yet been explained. One possibility is that shifts in replication timing are secondary to changes which occur in gene expression or chromatin structure. Since replication time zones are so large, however, this model would predict that a local modulation in the transcription of one particular gene could bring about a domain-wide shift in replication timing. It is much more likely that replication time is controlled independently by long-range *cis*-acting elements (e.g., X-inactivation center or LCRs) and the resulting domain-wide structure then influences specific genes at the local level.

Studies on *Xenopus* 5S RNA genes provide a good model for understanding how replication timing may modulate transcription. Two different sequence types for this gene are carried in the genome; one pro-

Table 2 Synchronous and asynchronous replication timing

Locus	Single-double (%)
Control loci	
c-*myc*	9
H-*ras*-1	12
IgH	10
albumin	9
p53	12
C48	9
Pfk-L	11
Imprinted gene	
Igf2	23
Igf2R	35
H19	30
Snrpn	34
Olfactory receptors	
102	34
107	35
I7	38
I28	38
Y22(MEL)	32
Y22(F9)	39
Y22(PEF)	31

A number of gene regions were tested for replication timing using FISH analysis and the number of nuclei showing asynchronous replication (see Fig. 3) was scored (% single-double). Most loci replicate relatively synchronously with approximately 10% single-double nuclei. Imprinted genes and olfactory receptor genes have a high percentage of nuclei showing asynchronous replication. In the case of imprinted genes, replication is allele-specific (Kitsberg et al. 1993b), whereas for olfactory receptor genes either allele can replicate early (Chess et al. 1994). Olfactory receptor gene replication was analyzed in mouse erythroleukemia cells (MEL), embryonal carcinoma (F9), or primary embryonic fibroblasts (PEF).

vides the major transcript in the oocyte and another is expressed mainly in somatic tissues. In the oocyte, high concentrations of transcription factors activate both gene types equally, but since the oocyte 5S RNA genes are more numerous, this species provides the overwhelming majority of RNA molecules. In somatic cells, on the other hand, the oocyte genes are in a repressed chromatin state which prevents their transcription (Wolffe and Brown 1988).

In contrast to the somatic 5S RNA genes, the oocyte genes replicate late in the cell cycle (Gilbert 1986). It has been suggested that this may put them at a disadvantage for binding the critical transcription factors in somatic cells, and through this simple mechanism, they would be maintained in an inactive state even following replication. The feasibility of this mechanism is, of course, based on the assumption that certain protein factors can only interact with the gene while it is undergoing replication and the basic chromatin structure is thus disrupted. In keeping with this prediction, transcription factors indeed stimulate oocyte 5S RNA synthesis, but only after histone H1 is removed from the chromatin (Wolffe and Brown 1988). Replication timing appears to be an important part of this model, since even the oocyte 5S RNA gene is activated in a cell line carrying a translocation that forces this sequence to undergo early replication (Guinta et al. 1986). There is no question that the control of somatic and ooctye 5S RNA gene expression is both complex and multifaceted, but it certainly appears that replication timing is an important element in this process.

Direct evidence for an effect of replication timing on gene expression has been obtained from an analysis of the *URA3* gene in the yeast *S. cerevisiae*. At its normal locus in the genome, this gene is constitutively expressed at a basal level, but it can be induced to higher levels of expression by the binding of the *trans*-activator PPR1 to the gene regulatory domain (Roy et al. 1990). When *URA3* is located in a late-replicating telomeric region, however, its basal level of expression is markedly repressed and, as a result, the cells become phenotypically ura3⁻, unless they are rescued by the presence of *PPR1*. In an elegant series of experiments, the *PPR1* gene was put under the regulation of a galactose-inducible promoter. Cells were then arrested at the G_1/S interphase in the absence of *PPR1* expression, a condition that prevents *URA3* transcription. Surprisingly, the induction of *PPR1* did not bring about *URA3* activation, and only when the cells were allowed to continue through the later part of the cell cycle did *URA3* become transcriptionally competent (Aparicio and Gottschling 1994). These studies clearly show that the induction of this gene is dependent on cell-cycle events and strongly suggest that replication is required for the activation process.

Two models have evolved for the effect of chromatin structure on gene regulation. In one case, chromatin prevents *trans*-activating proteins from gaining access to the DNA, thus keeping the gene irreversibly repressed. However, during DNA replication the chromatin structure of the gene is perturbed, and activation factors have the opportunity to gain access and establish transcription before reassembly of the chromatin is completed. In an alternate model, the *trans*-acting proteins can induce gene transcription at any time in a replication-independent manner, effectively disrupting the repressive nature of the chromatin. Examples of both forms of regulation have been observed (Felsenfeld 1992).

Assuming that at least some genes require a replication event in order to set up an active transcription complex, the time at which replication takes place could play a big role in determining the fate of each particular sequence. Since dynamic changes take place during the cell cycle, the levels of pertinent factors may fluctuate and thus be present in the nucleus only during specific intervals of S. In this way, the activation of a target gene may only be possible if it is replicated in this specified "window of opportunity." Replication at an alternate time in S could lead to an inactive configuration. E2F proteins may represent an example of this type of transcription factor, since their availability is carefully monitored in a cell-cycle-dependent manner (Shirodkar et al. 1992).

CONCLUSIONS

DNA replication takes place in an ordered and programmed manner during S phase. In animal cells, the genome is divided into distinct Mb-size time zones whose replication timing is probably controlled by *trans*-acting factors and central *cis*-acting elements. Housekeeping genes generally replicate during the first half of S phase, whereas many tissue-specific genes are late-replicating, yet developmentally regulated to undergo early replication in their tissue of expression. Replication time is apparently associated with chromosome structure, but it is not yet clear whether it is a consequence of chromatin interactions or whether it actually plays a role in the regulation of gene expression. Alternate replication time structures can also serve to distinguish between two allelic regions in the same nucleus, as is observed for imprinted genes, for the X chromosome in female cells, and in cases of allelic exclusion.

ACKNOWLEDGMENTS

Some of the research described in this review was supported by the National Institutes of Health, the Israel Academy of Sciences, and the Israel Cancer Research Fund.

REFERENCES

Aparicio, O.M. and D.E. Gottschling. 1994. Overcoming telomeric silencing: A *trans*-activator competes to establish gene expression in a cell cycle-dependent way. *Genes Dev.* **8:** 1133–1146.

Boggs, B.A. and A.C. Chinault. 1994. Analysis of replication timing properties of human X-chromosomal loci by fluorescence in situ hybridization. *Proc. Natl. Acad. Sci.* **91:** 6083–6087.

Braunstein, D., D. Schulze, T. DelGiudice, A. Furst, and C.L. Schildkraut. 1982. The temporal order of replication of murine immunoglobulin heavy chain constant region

sequences corresponds to their linear order in the genome. *Nucleic Acids Res.* **10:** 6887–6902.
Brown, C.J. and H.F. Willard. 1994. The human X-inactivation centre is not required for maintenance of X-chromosome inactivation. *Nature* **368:** 154–156.
Brown, C.J., A. Ballabio, J.L. Rupert, R.G. Lafreniere, M. Groumpe, R. Tonlorenzi, and H.F. Willard. 1991a. A gene from the region of the human X inactivation centre is expressed exclusively from the inactive X chromosome. *Nature* **349:** 38–44.
Brown, C.J., B.D. Hendrich, J.L. Rupert, R.G. Lafreniere, Y. Xing, J. Lawrence, and H.F. Willard. 1992. The human Xist gene: Analysis of 17 kb inactive X-specific RNA that contains conserved repeats and is highly localized within the nucleus. *Cell* **71:** 527–542.
Brown, C.J., R.G. Lafreniere, V.E. Powers, G. Sebastio, A. Ballabio, A.L. Pettigrew, D.H. Ledbetter, E. Levy, I.W. Craig, and H.F. Willard. 1991b. Localization of the X inactivation centre on the human X chromosome in Xq13. *Nature* **349:** 82–84.
Brown, E.H., M.A. Iqbal, S. Stuart, K.S. Hatton, J. Valinsky, and C.L. Schildkraut. 1987. Rate of replication of the murine immunoglobulin heavy-chain locus: Evidence that the region is part of a single replicon. *Mol. Cell. Biol.* **7:** 450–457.
Buck, L. and R. Axel. 1991. A novel multigene family may encode odorant receptors: A molecular basis for odor recognition. *Cell* **65:** 175–187.
Buiting, K., S. Saitoh, S. Gross, B. Dittrich, S. Schwartz, R.D. Nicholls, and B. Horsthemke. 1995. Inherited microdeletions in the Angelman and Prader-Willi syndromes define an imprinting centre on human chromosome 15. *Nat. Genet.* **9:** 395–400.
Calza, R.E., L.A. Eckardt, T. DelGiudice, and C.L. Schildkraut. 1984. Changes in gene position are accompanied by a change in time of replication. *Cell* **36:** 689–696.
Cattanach, B.M. 1986. Parental origin effects in mice. *J. Embryol. Exp. Morphol.* (suppl.) **97:** 137–150.
Cedar, H. 1988. DNA methylation and gene activity. *Cell* **53:** 3–4.
Chess, A., I. Simon, H. Cedar, and R. Axel. 1994. Allelic inactivation regulates olfactory receptor gene expression. *Cell* **78:** 823–834.
Disteche, C.M., E.M. Eicher, and S.A. Latt. 1979. Late replication in an X-autosome translocation in the mouse: Correlation with genetic inactivation and evidence for selective effects during embryogenesis. *Proc. Natl. Acad. Sci.* **76:** 5234–5238.
Driscoll, M.C., C.S. Dobkin, and B.P. Alter. 1989. γδβ-Thalassemia due to a de novo mutation deleting the 5′ β-globin gene activation-region hypersensitive sites. *Proc. Natl. Acad. Sci.* **86:** 7470–7474.
Drouin, R., N. Lemieux, and C.-L. Richer. 1990. Analysis of DNA replication during S phase by means of dynamic chromosome banding at high resolution. *Chromosoma* **99:** 273–280.
Eden, S. and H. Cedar. 1994. Role of DNA methylation in the regulation of transcription. *Curr. Opin. Genet. Dev.* **4:** 255–259.
Edenberg, H.J. and J.A. Huberman. 1975. Eukaryotic chromosome replication. *Annu. Rev. Genet.* **9:** 245–285.
Epner, E., W.C. Forrester, and M. Groudine. 1988. Asynchronous DNA replication within the human β-globin gene locus. *Proc. Natl. Acad. Sci.* **85:** 8081–8085.
Felsenfeld, G. 1992. Chromatin as an essential part of the transcriptional mechanism. *Nature* **355:** 219–224.
Ferguson, B.M. and W.L. Fangman. 1992. A position effect on the time of replication origin activation in yeast. *Cell* **68:** 333–339.

Ferguson, B.M., B.J. Brewer, A.E. Reynolds, and W.L. Fangman. 1991. A yeast origin of replication is activated late in S phase. *Cell* **65:** 507-515.

Forrester, W.C., E. Epner, M.C. Driscoll, T. Enver, M. Brice, T. Papayannopoulou, and M. Groudine. 1990. A deletion of the human β-globin locus activation region causes a major alteration in chromatin structure and replication across the entire β-globin locus. *Genes Dev.* **4:** 1637-1649.

Giddings, S.J., C.D. King, K.W. Harman, J.F. Flood, and L.R. Carnaghi. 1994. Allele specific inactivation of insulin 1 and 2, in the mouse yolk sac, indicates imprinting. *Nat. Genet.* **6:** 310-313.

Gilbert, D.M. 1986. Temporal order of replication of *Xenopus laevis* 5S ribosomal RNA genes in somatic cells. *Proc. Natl. Acad. Sci.* **83:** 2924-2928.

Goldman, M.A., G.P. Holmquist, L.A. Caston, and A. Nag. 1984. Replication timing of genes and middle repetitive sequences. *Science* **224:** 686-692.

Guillemot, F., T. Caspary, S.M. Tilghman, N.G. Copeland, D.J. Gilbert, N.A. Jenkins, D.J. Anderson, A.L. Joyner, J. Rossand, and A. Nagy. 1995. Genomic imprinting of *Mas2*, a mouse gene required for trophoblast development. *Nat. Genet.* **9:** 235-242.

Guinta, D.R., J. Yun Tso, S. Narayanswami, B.A. Hamkalo, and L.J. Korn. 1986. Early replication and expression of oocyte-type 5S RNA genes in a *Xenopus* somatic cell line carrying a translocation. *Proc. Natl. Acad. Sci.* **83:** 5150-5154.

Gunaratne, P.H., M. Nakao, D.H. Ledbetter, J.S. Sutcliffe, and A.C. Chinault. 1995. Tissue-specific and allele-specific replication timing control in the imprinted human Prader-Willi syndrome region. *Genes Dev.* **9:** 808-920.

Hand, R. 1978. Eucaryotic DNA: Organization of the genome for replication. *Cell* **15:** 317-325.

Hansen, R.S., T.K. Canfield, M.M. Lamb, S.M. Gartier, and C.D. Laird. 1993. Association of fragile X. syndrome with delayed replication of the *FMR1* gene. *Cell* **73:** 1403-1409.

Hatton, K.A., V. Dhar, E.H. Brown, M.A. Iqbal, S. Stuart, V.T. Didamo, and C.L. Schildkraut. 1988. Replication program of active and inactive multigene families in mammalian cells. *Mol. Cell. Biol.* **8:** 2149-2158.

Holmquist, G.P. 1987. Role of replication time in the control of tissue-specific gene expression. *J. Hum. Genet.* **40:** 151-173.

Izumikawa, Y., K. Naritomi, and K. Hirayama. 1991. Replication asynchrony between homologs 15q11.2: Cytogenetic evidence for genomic imprinting. *Hum. Genet.* **87:** 1-5.

Jablonka, E., R. Goitein, M. Marcus, and H. Cedar. 1985. DNA hypomethylation causes an increase in DNase-I sensitivity and an advance in the time of replication of the entire inactive X chromosome. *Chromosoma* **93:** 152-156.

Karube, T. and S. Watanabe. 1988. Analysis of the chromosomal DNA replication pattern using the bromodeoxyuridine labeling method. *Cancer Res.* **48:** 219-222.

Kerem, B.-S., R. Goitein, G. Diamond, H. Cedar, and M. Marcus. 1984. Mapping of DNAase I sensitive regions of mitotic chromosomes. *Cell* **38:** 493-499.

Kitsberg, D., S. Selig, and H. Cedar. 1991. Chromosome structure and eukaryotic gene organization. *Curr. Opin. Genet. Dev.* **1:** 534-537.

Kitsberg, D., S. Selig, I. Keshet, and H. Cedar. 1993a. Replication structure of the human β-globin gene domain. *Nature* **366:** 588-590.

Kitsberg, D., S. Selig, M. Brandeis, I. Simon, I. Keshet, D.J. Driscoll, R.D. Nicholls, and H. Cedar. 1993b. Allele-specific replication timing of imprinted gene regions. *Nature*

364: 459-463.
Knoll, J.H.M., S.-D. Cheng, and M. Lalande. 1994. Allele specificity of DNA replication timing in the Angelman/Prader-Willi syndrome imprinted chromosomal region. *Nat. Genet.* **6:** 41-46.
LaSalle, J.M. and M. Lalande. 1995. Domain organization of allele-specific replication within the GABRB3 gene cluster requires a biparental 15q11-13 contribution. *Nat. Genet.* **9:** 386-394.
Latt, S.A. 1973. Microfluorometric detection of deoxyribonucleic acid replication in human metaphase chromosomes. *Proc. Natl. Acad. Sci.* **70:** 3395-3399.
Laurenson, P. and J. Rine. 1992. Silencers, silencing, and heritable transcriptional states. *Microbiol. Rev.* **56:** 543-560.
Ma, C., T.H. Leu, and J.L. Hamlin. 1990. Multiple origins of replication in the dihydrofolate reductase amplicons of a methotrexate-resistant Chinese hamster cell line. *Mol. Cell. Biol.* **10:** 1338-1346.
Mohandas, T., R.S. Sparker, and L.J. Shapiro. 1981. Reactivation of an inactive human X chromosome: Evidence for X inactivation by DNA methylation. *Science* **211:** 393-396.
Nicholls, R.D. 1993. Genomic imprinting and candidate genes in the Prader-Willi and Angelman syndromes. *Curr. Opin. Genet. Dev.* **3:** 445-456.
Razin, A. and H. Cedar. 1994. DNA methylation and genomic imprinting. *Cell* **77:** 473-476.
Reis, A., B. Dittrich, V. Greger, K. Buiting, M. Lalande, G. Gillessen-Kaesbach, M. Anvret, and B. Horsthemke. 1994. Imprinting mutations suggested by abnormal DNA methylation patterns in familial Angelman and Prader-Willi syndromes. *Am. J. Hum. Genet.* **54:** 741-747.
Reynolds, A.E., R.M. McCarroll, C.S. Newlon, and W.L. Fangman. 1989. Time of replication of ARS elements along yeast chromosome III. *Mol. Cell. Biol.* **9:** 4488-4494.
Roy, A., F. Exinger, and R. Losson. 1990. cis- and trans-acting regulatory elements of the yeast URA 3 promoter. *Mol. Cell. Biol.* **10:** 5257-5270.
Schmidt, M. and B.R. Migeon. 1990. Asynchronous replication of homologous loci on human active and inactive X chromosome. *Proc. Natl. Acad. Sci.* **87:** 3685-3689.
Selig, S., K. Okumura, D.C. Ward, and H. Cedar. 1992. Delineation of DNA replication time zones by fluorescence in situ hybridization. *EMBO J.* **11:** 1217-1225.
Selig, S., M. Ariel, R. Goitein, M. Marcus, and H. Cedar. 1988. Regulation of mouse satellite DNA replication time. *EMBO J.* **7:** 419-426.
Shafer, D.A. and J.H. Priest. 1984. Reversal of DNA methylation with 5-aza-C alters chromosome replication patterns in human lymphocyte and fibroblast cultures. *Am. J. Hum. Genet.* **36:** 534-545.
Shirodkar, S., M. Ewen, J.A. DeCaprio, J. Morgan, D.M. Livingston, and T. Chittenden. 1992. The transcription factor E2F interacts with the retinoblastoma product and a p107-cyclin A complex in a cell cycle-regulated manner. *Cell* **68:** 157-166.
Somssich, I.E., J. Spira, H. Hameister, and G. Klein. 1984. Correlation between tumorigenicity and banding pattern of chromosome 15 in murine T-cell leukemia cells. *Chromosoma* **91:** 39-45.
Spack, E.G., E.D. Lewis, B. Paradowski, R.T. Schimke, and P.P. Jones 1992. Temporal order of DNA replication in the H-2 major histocompatibility complex of the mouse. *Mol. Cell. Biol.* **12:** 5174-5188.
Sutcliffe, J.S., M. Nakao, S. Christian, K.H. Orstavik, N. Tommerup, D.H. Ledbetter, and

A.L. Beaudet. 1994. Deletions of a differentially methylated CpG island at the SNRPN gene define a putative imprinting control region. *Nat. Genet.* **8:** 52–58.

Takagi, N. 1974. Differentiation of X chromosomes in early female mouse embryos. *Exp. Cell Res.* **86:** 127–135.

Takagi, N., M.A. Yoshida, O. Sugawara, and M. Sasaki. 1983. Reversal of X-inactivation in female mouse somatic cells hybridized with murine teratocarcinoma stem cells in vitro. *Cell* **34:** 1053–1062.

Torchia, B.S., L.M. Call, and B.R. Migeon. 1994. DNA replication analysis of FMR1, XIST, and factor 8C loci by FISH shows nontranscribed X-linked genes replicate late. *Am. J. Hum. Genet.* **55:** 96–104.

Vassar, R., J. Ngai, and R. Axel. 1993. Spatial segregation of odorant receptor expression in the mammalian olfactory epithelium. *Cell* **74:** 309–318.

Villar, A.J. and R.A. Pedersen. 1994. Parental imprinting of the Mas protooncogene in mouse. *Nat. Genet.* **8:** 373–379.

Vogel, W., M. Autenrieth, and G. Speit. 1986. Detection of bromodeoxyuridine-incorporation in mammalian chromosomes by a bromodeoxyuridine-antibody. 1. Demonstration of replication patterns. *Hum. Genet.* **72:** 129–132.

von Meyenburg, K. and F.G. Hansen. 1987. Regulation of Chromosome Replication. In Escherichia coli *and* Salmonella typhimurium: *Cellular and molecular biology* (ed. F.C. Neidhardt), pp. 1555–1577. American Society for Microbiology, Washington, D.C.

Wahrman, J., C. Richler, E. Neufeld, and A. Friedman. 1983. The origin of multiple sex chromosomes in the gerbil *Gerbillus gerbillus* (Rodentia: Gerbillinae). *Cytogenet. Cell Genet.* **35:** 161–180.

Wevrick, R., J.A. Kerns, and U. Francke. 1994. Identification of a novel paternally expressed gene in the Prader-Willi syndrome region. *Hum. Mol. Genet.* **3:** 1877–1882.

Wolf, S. and P. Perry. 1974. Differential staining of sister chromatids and the study of sister chromatid exchanges without autoradiography. *Chromosoma* **48:** 341–353.

Wolffe, A.P. and D.D. Brown. 1988. Developmental regulation of two 5S ribosomal RNA genes. *Science* **241:** 1626–1632.

Yunis, J.J. 1981. Mid prophase human chromosomes. The attainment of 2000 bands. *Hum. Genet.* **56:** 293–298.

13
Changes in DNA Replication during Animal Development

Janet L. Carminati[1] and Terry L. Orr-Weaver
Department of Biology, Massachusetts Institute of Technology
and Whitehead Institute
Cambridge, Massachusetts 02142

Throughout animal growth and development, the cell cycle is modified in response to developmental signals. These alterations in the cell cycle influence the control of S phase and the properties of DNA replication (outlined in Table 1). The production of haploid gametes requires a modified cell cycle, meiosis, in which two rounds of chromosome segregation follow a single S phase. Thus, S phase must be prevented between the two meiotic divisions. In the early embryos of many animals, a rapid cell cycle occurs in which S phase oscillates with mitosis without gap phases. This S-M cycle necessitates unique controls for the entry into S phase. Later in embryogenesis in *Drosophila* and *Xenopus*, a G_1 phase is added to the cell cycle, resulting in another developmental alteration of the onset of S phase. There are numerous examples of tissues that become polytene as a consequence of a modified cell cycle with only an S phase and a gap phase.

These developmental changes in the cell cycle require special controls for entry into S phase. The regulation of the onset of S phase has been extensively investigated in the normal cell cycle with G_1-S-G_2-M phases and is reviewed by Nasymth (this volume) and Weisshart and Fanning (this volume). Cyclin-dependent kinases (cyclin-cdk complexes), including cyclins E, D, and A complexed with cdk2, 4, and 6, are all known to play a role in S-phase regulation of higher eukaryotes (Sherr 1993, 1994). G_1 cyclin kinases are thought to phosphorylate the retinoblastoma gene product, pRb, releasing it from the transcription factor E2F, thus allowing the S-phase transcriptional program (Sherr 1994). Other S-phase kinases, perhaps cyclin A kinase, might then inactivate E2F once its transcriptional program is complete (Dynlacht et al. 1994; Krek et al. 1994). In the yeast *Saccharomyces cerevisiae*, three functionally redundant

---
[1]Present address: Department of Biological Sciences, Stanford University, Stanford, California 94305.

Table 1 Alterations in regulation of DNA replication during development

Developmental change	S-phase control	Replication modifications
Meiosis	distinct entry into S phase	slower S phase *S. cerevisiae*: same origins used, asynchronous activation mouse: slower fork rate newt: fewer origins activated
Oocyte activation	*Xenopus*: Mos inhibition of S phase during meiosis	sea urchin: inactive initiation factors in the unfertilized egg
Restart of S phase at fertilization	*Drosophila*: inhibition of S phase prior to fertilization by *plu*, *png*, and *gnu*	
Early embryonic cycles	S/M cycle posttranscriptional control of S-phase genes *Drosophila*: genes that couple S and M phases	*Xenopus* and *Drosophila*: periodic spacing of origins; sequence independent and synchronous activation
Onset of transcription and addition of G_1 phase (MBT)	G_1 phase added; transcriptional control of S-phase genes	longer S phase; asynchronous origin activation and late-replicating heterochromatin
Polyploidy/polyteny	*Drosophila*: endo cell cycle (S/G); spatial and temporal pattern	*Drosophila*: removal of block to rereplication; late replication and underrepresentation of heterochromatin
Amplification		tissue and temporal control; removal of block to rereplication

cyclins, CLN 1, 2, and 3, control the G_1/S transition, and two cyclin-B homologs, CLB5 and 6, also act to control S phase (Richardson et al. 1989; Epstein and Cross 1992; Schwob and Nasmyth 1993). These cyclins all interact with the CDC28 kinase. In addition to these positive regulators, in the normal cell cycle, inhibitors of cyclin kinases exist that are able to inhibit progression of the cell cycle in response to various environmental cues (Sherr 1994; Sherr and Roberts 1995).

In the developmentally modified cell cycles discussed here, entry into S phase is regulated by different mechanisms from those of the normal cell cycle (Table 1). In yeast meiosis, the CDC28 kinase appears to act later than S phase. During the embryonic S-M cycles, entry into S phase must be controlled posttranscriptionally and must be regulated in the presence of constitutive cyclin E kinase activity. The addition of a G_1 phase during embryogenesis requires that extrinsic developmental cues influence the onset of S phase. In *Drosophila*, this is mediated at least in part through cyclin E. During the S-G cycle that produces polytene cells, entry into S phase can no longer have the completion of mitosis as a prerequisite.

In addition to differential control of the onset of S phase, the actual parameters of S phase are altered during development. By parameters of S phase, we mean the intrinsic properties of DNA replication. The parameters of DNA replication changed in modified cell cycles include replication origin usage and activation, the rate of replication fork movement, and the block to rereplication (Table 1). We discuss how these parameters are changed in each of the variant cell cycles. For a background on the normal intrinsic regulation of DNA replication, we refer the reader to DePamphilis; Heintz; Newlon; Simon and Cedar; and Blow and Chong (all this volume).

S PHASE OF THE MEIOTIC CELL CYCLE

Meiosis is a modified cell cycle in which two rounds of chromosome segregation follow a single S phase. This results in the chromosome number being reduced by half. This is essential so that the chromosome number is restored when the sperm and egg fuse at fertilization. The S phase that precedes chromosome segregation in meiosis has been termed "premeiotic S." The use of the term premeiotic S implies that the term meiosis refers only to the actual segregation of chromosomes. We refer to the S phase of the meiotic cell cycle as premeiotic S, but the reader should realize that premeiotic S phase occurs as part of the same cell cycle in which the two meiotic divisions occur.

Regulation of Entry into S Phase

Studies addressing the regulation of premeiotic S phase have been investigated mainly in the yeast, *S. cerevisiae*. Many genes that govern S phase during the mitotic cell cycle have been characterized and are covered by Nasymth (this volume). However, several genes that are required for entry into S phase during the mitotic cell cycle are not required for the control of premeiotic S phase. Interestingly, some of these genes that are involved in S phase during the mitotic cell cycle do have a role in meiosis, yet are required after replication and before the meiosis I division. One example is the mutation *cdc7*, in which mitotic cells are blocked prior to replication, yet meiotic cells are arrested following premeiotic DNA synthesis (Schild and Byers 1978; Buck et al. 1991). Recently, Cdc7 has been linked to replication origins, in that a protein that binds to and activates the Cdc7 kinase, Dbf4, also binds to *ARS* elements (Jackson et al. 1993; Dowell et al. 1994). If Cdc7 does prove to have a role at the replication origin, this might then suggest that premeiotic replication is differentially regulated. Other similar examples include the *S. cerevisiae* mutations, *cdc28* and *cdc4*, in which DNA replication is blocked in the mitotic cell cycle, yet premeiotic synthesis occurs (Simchen and Hirschberg 1977; Piggott et al. 1982; Shuster and Byers 1989; Reed and Wittenberg 1990). Cdc28 is a component of maturation promoting factor (MPF) and is required both at the G_1/S and G_2/M transitions during the mitotic cell cycle yet does not appear to be required for the G_1/S transition of premeiotic S phase. It should be noted, however, that the *Schizosaccharomyces pombe* homolog of *cdc28*, *cdc2*, is required for premeiotic DNA replication (Iino et al. 1995).

Spatial Control of Meiotic Replication Origins

Comparisons of S phase during the meiotic and mitotic cell cycles using fiber autoradiography indicated that a similar spacing of origins is present and that replication forks travel at a similar rate (Johnston et al. 1982; Newlon 1988). However, the resolution of fiber autoradiography was not precise enough to determine whether the specific origins used were the same origins used during the mitotic cycle. Given the fact that not all *ARS* elements identified by the plasmid assay are active chromosomal origins during the mitotic cell cycle, it is possible that the other *ARS* elements were specifically used during premeiotic S phase. In one study designed to examine *ARS1* function in premeiotic S phase, plasmid loss occurred in both mitosis and meiosis upon induction of transcription through the *ARS1* element (Hollingsworth and Sclafani 1993). Although this study used an indirect assay for *ARS* function, it suggests that the same origin can function in S phase in both the meiotic and mitotic cycles. In another study to identify replication origins active during premeiotic S phase, two-dimensional gel analysis was used to examine origins on chromosome III (Collins and Newlon 1994). The five premeiotic S-phase origins map to the same *ARS* elements as the origins from the mitotic cell cycle. One possible exception was an origin located at *CEN3* that had weak activity in the mitotic cycle and did not appear to have any activity in the meiotic cycle. No additional premeiotic S-phase origins were found that did not correspond to origins used in the mitotic cycle. Further analysis will be needed to determine if this conclusion is true for all chromosomes, but these studies strongly suggest that meiotic origins are indeed the same as mitotic origins.

In higher eukaryotes, perhaps this is not the case. Using fiber autoradiography in the newt *Triturus*, the spacing between origins used in meiosis is larger than that seen during the mitotic cycle, suggesting that either different origins are used or a subset of mitotic origins are activated (Callan 1974).

Temporal Control of Meiotic Origins

S phase in eukaryotes can vary widely in length, from extremely short S phases during embryonic development (several minutes) to longer S phases in somatic cells (hours) and often very prolonged S phases during premeiotic S phase (at least 24 hours in the *Drosophila* ovary) (Chandley 1966; Grell 1973). S phase seems to lengthen as development places increasingly complex controls on the cell cycle. In *S. cerevisiae*, premeiotic S phase is at least twice as long as S phase in the mitotic cycle (65 min versus 30 min) (Williamson et al. 1983). Despite this lengthening of S phase, origin usage and the rate of fork movement are the same as during mitotic S phase (Johnston et al. 1982).

Two possibilities for the longer premeiotic S phase are either that origins within a chromosome are fired asynchronously or that different chromosomes are replicated at different times. In the study of origins in the meiotic cell cycle on chromosome III mentioned above, the efficiency of usage of specific *ARS* elements and characteristic termination patterns were similar between meiosis and mitosis (Collins and Newlon 1994). This implies that within a single chromosome the kinetics of initiation and termination are roughly the same in meiosis and mitosis. Further studies are needed to determine if this is true for all chromosomes. Collins and Newlon therefore suggest that the longer premeiotic S phase is due to different initiation times for different chromosomes.

In the premeiotic S phase of mouse spermatogonia, the lengthening of S phase was studied by fiber autoradiography (Jagiello et al. 1983). As in yeast, origin spacing during S phase in the meiotic and mitotic cycle is similar. However, the rate of fork movement is slower during meiosis and might account for the longer S phase.

S PHASE DURING EARLY EMBRYOGENESIS

Inhibition of DNA Replication in the Developing Oocyte and Unfertilized Egg

Many animals undergo rapid cell divisions following fertilization, including *Xenopus*, *Drosophila*, sea urchin, starfish, and clam. DNA replication factors are stockpiled in the egg to prepare for the high demand during the rapid embryonic divisions. However, DNA replication must be repressed in the oocyte so as not to occur during meiosis or prior to fertilization. This repression could be at the level of S-phase control such that entry into S phase is blocked during meiosis, or alternatively, at the level of replication parameters, such as initiation factors that might be kept in an inactive state. Two well-characterized mechanisms of replication control in the developing oocyte and unfertilized egg have been found in *Xenopus* and sea urchin.

The inability of *Xenopus* oocytes to replicate DNA is mediated by Mos activity blocking entry into S phase during meiosis (Furuno et al. 1994). Mos is a serine/threonine kinase that is expressed at high levels in germ cells of vertebrates and is maternally loaded into the *Xenopus* oocyte (Sagata et al. 1988). Mos acts at three steps in *Xenopus* oocyte maturation: (1) It activates meiosis; (2) it represses DNA replication between the meiotic divisions; and (3) it helps maintain the metaphase II arrest. The first activity of Mos is that of a meiotic initiator. *Xenopus* immature oocytes (stage VI) are arrested at G_2/M of meiosis I. Following hormone stimulation, germinal vesicle breakdown (GVBD) occurs, and oocytes progress to metaphase II of meiosis. This maturation is dependent on the translation of Mos mRNA following hormone stimulation, after which mature oocytes remain arrested at metaphase II until fertilization.

The second Mos activity, the ability of Mos to block DNA replication during meiosis, is relevant for this review. By examining a precise time course of MPF activity throughout meiosis, it was found that MPF is inactivated early in meiosis I (early metaphase I) and is then reactivated during late metaphase I, well before meiosis II (Furuno et al. 1994; Ohsumi et al. 1994). Mos mediates this reactivation of MPF, which then suppresses DNA replication between the two meiotic divisions. Ablation studies using c-*mos* antisense RNA or Mos antibodies showed that upon inactivation during GVBD, mature oocytes enter S phase inappropriately following meiosis I (Furuno et al. 1994). The role of MPF in suppressing an intervening S phase was also confirmed by studies that block MPF activation at meiosis II, resulting in DNA replication (Furuno et al. 1994; Ohsumi et al. 1994). Once MPF is reactivated, it is stabilized by the action of cytostatic factor (CSF) at the metaphase II arrest (Minshull 1993). This is the third meiotic activity of Mos, because CSF is composed of both Mos and an unknown factor. Thus, Mos is needed to maintain the metaphase II arrest in the mature oocyte.

Mos homologs have not been identified in invertebrates, suggesting that different mechanisms might act to control replication during oocyte maturation. In contrast to the control of S-phase entry seen in *Xenopus*, the mechanism that inhibits DNA replication in the unfertilized sea urchin egg is the absence of active initiation factors responsible for DNA synthesis. Following oocyte maturation, sea urchin eggs complete meiosis and arrest at G_1 of the first mitotic cycle. Fertilization then releases this arrest, and replication factors are posttranslationally activated within a short 3 minutes following fertilization (Zhang and Ruderman 1993). The first S phase occurs 20 minutes following fertilization, even in the absence of protein synthesis.

The unfertilized egg of sea urchin is not permissive for DNA replication, as assayed by the inability of egg extracts to replicate added sperm nuclei or double-stranded DNA templates (Zhang and Ruderman 1993). Conversely, embryonic interphase extracts are capable of supporting replication of added templates. Mixing experiments of egg and embryonic extracts showed no evidence of negative factors in the egg capable of repressing replication; the egg extract did not inhibit the ability of the embryonic extract to replicate DNA. Thus, the unfertilized egg does not contain inhibitors of replication, and instead, the inhibition of replication seen is due to inactive initiation factors present at this stage of development.

Restart of S Phase at Fertilization

In many organisms, the unfertilized egg is arrested during or following meiosis, and fertilization then releases the egg from this arrest. This coupling of fertilization with the resumption of the cell cycle ensures that the female and male pronuclei can then enter the first S phase and subsequent cell cycles with the proper timing. In *Drosophila*, starfish, and sea urchin, the completion of meiosis is not coupled to fertilization. In *Drosophila*, the mature oocyte is arrested at metaphase I in the ovary. Upon passage of the egg through the uterus, the egg becomes activated, and meiosis is completed regardless of whether fertilization occurs. Fertilization is necessary to restart the cell cycle in the embryo following the completion of meiosis, and unique regulators may be needed at this developmental step.

In *Drosophila*, three genes that act at this point to couple fertilization with DNA replication are the maternal-effect genes *giant nuclei* (*gnu*), *pan gu* (*png*), and *plutonium* (*plu*). Unfertilized mutant eggs complete meiosis resulting in four meiotic products, yet then undergo improper DNA replication, resulting in large polyploid nuclei (Freeman and Glover 1987; Shamanski and Orr-Weaver 1991). These genes normally act as negative regulators of DNA replication to make the restart of S phase dependent on fertilization. Fertilization must overcome the action of these genes in order to resume the cell cycle.

In many organisms, including starfish, *Xenopus*, sea urchin, and mouse, a transient increase in intracellular calcium (Ca^{++}) occurs at fertilization and is associated with the onset of development including meiotic maturation, pronuclear migration, DNA replication, and nuclear envelope breakdown. Time-lapse confocal imaging was used to study calcium dynamics throughout starfish early development (Stricker 1995). No transient increases in intracellular calcium were observed during oocyte maturation, whereas a single prolonged transient coincided with fertilization, and repetitive calcium oscillations occurred during the early cleavage divisions. The levels of inositol triphosphate also fluctuate during fertilization and might act to trigger the calcium transients seen during sea urchin development (Ciapa et al. 1994). However, the cause and effect relationship between calcium and the second messengers in other systems remains less clear (Whitaker and Swann 1993).

Following the calcium burst in *Xenopus* mature oocytes, the calmodulin-dependent kinase, CaMKII, is responsible for the inactivation of both MPF and CSF, thus releasing the oocyte from a metaphase II arrest (Lorca et al. 1993). How this then regulates the resumption of DNA replication following meiosis is less clear. The role of calcium bursts in controlling DNA replication can be seen in starfish, where the calcium ionophore, A 23187, acts as a parthenogenic agent capable of inducing several rounds of replication in mature oocytes (Picard et al. 1987). The calcium ionophore also triggers the onset of the cell cycle in unfertilized sea urchin eggs (Whitaker and Patel 1990). Whether transient fluxes in calcium levels play a role in *Drosophila* fertilization and the restart of S phase is unclear.

Rapid Early Cycles in *Xenopus* and Flies

Many organisms undergo rapid embryonic cycles following fertilization. The early embryonic cycles of *Xenopus* and *Drosophila* have been well characterized and consist of rapid cycles of alternating S phase and M phase that are controlled by maternally supplied products present in the egg. DNA replication is differentially regulated during this time in development. Unique regulation occurs at both the entry into S phase and in the replication parameters used. S-phase entry is regulated posttranscriptionally due to the absence of zygotic transcription. DNA replication parameters are also developmentally controlled such that many synchronous origins are activated to ensure complete replication within the very rapid 3- to 10-minute S phases.

Posttranscriptional Control of S Phase

The rapid early cycles of *Xenopus* and *Drosophila* are controlled by posttranscriptional modifications of regulators during S phase and M phase. In *Drosophila*, the first 13 divisions in the fertilized embryo are rapid, synchronous nuclear divisions within a common shared cytoplasm, where S phase occurs in an extremely short 3- to 4-minute period. The early cycles in *Xenopus* are similar in that the first 12 divisions consist of rapid, synchronous cycles with S phase occurring in a brief 10 minutes. During these early embryonic cycles of both *Xenopus* and *Drosophila*, transcription of the zygotic nucleus does not occur. In *Drosophila*, maximal zygotic transcription occurs following cellularization at cycle 14, whereas in *Xenopus*, zygotic transcription occurs at the midblastula transition (Newport and Kirschner 1982b; Edgar and Schubiger 1986).

Known regulators of S phase are present at high levels in *Drosophila* and *Xenopus* embryos. In *Drosophila*, the S-phase cyclin E-cdk2 kinase is present throughout early embryonic development (Richardson et al. 1993; Knoblich et al. 1994). Similarly, a large pool of maternal cyclin E is present during the early cleavage divisions of *Xenopus* (Jackson et al. 1995). The constitutive high levels of these proteins indicate that control of entry into S phase during the early division cycles must be regulated differently than during the later cycles when S phase is preceded by a G_1 phase.

In addition to the posttranscriptional control of known S-phase regulators, unique regulators might also be used early in development for the rapid cell cycle. In *Drosophila*, three maternal-effect genes, *giant nuclei* (*gnu*), *pan gu* (*png*), and *plutonium* (*plu*), appear to be novel cell cycle regulators. As discussed above, these genes are needed to couple S phase with fertilization such that mutant unfertilized eggs undergo inappropriate DNA replication. When fertilized, these embryos also give rise to giant polyploid nuclei, suggesting that these genes control S phase during the early cycles. Although fertilized, these embryos fail to properly couple S phase and M phase, and some aspects of mitosis such as centrosome duplication continue to cycle independently from nuclear division (Freeman et al. 1986; Freeman and Glover 1987; Shamanski and Orr-Weaver 1991). *png* is unique in that several presumably leaky alleles transiently couple S phase and M phase, resulting in embryos containing many more giant nuclei. However, defects in mitotic figures and DNA condensation can be seen as early as the first division, and the uncoupling of replication and mitosis progresses as development proceeds (J. Carminati, unpubl.). The unfertilized and fertilized phenotypes can be explained by the proposal that these genes inhibit DNA replication. In unfertilized eggs they function to make S phase dependent on fertilization, whereas in fertilized embryos these gene products make S phase dependent on the proper completion of mitosis. However, at what level these genes act is unclear. They might control either the entry into S phase, the block to rereplication, or other aspects of the cell cycle, such as chromosome condensation, that link mitosis to replication.

Molecular data confirm that *plu* is a unique regulator that acts solely during the early *Drosophila* divisions and not in later canonical cell cycles. RNA null alleles are maternal-effect alleles, and expression of the *plu* transcript is not present during later stages of development (Axton et al. 1994). Plu encodes a 19-kD protein consisting of three ankyrin repeats. Interestingly, another small ankyrin repeat protein is the cdk4 inhibitor, p16^{INK4}, which acts to inhibit cdk4, thus disrupting its association with the S-phase cyclin D protein (Serrano et al. 1993). Plu might act analogously in its role in repressing DNA replication, possibly by inhibiting S-phase cyclin kinases until their proper time of action.

Replication Origins

Studies in both *Xenopus* and *Drosophila* suggest that in the extremely rapid cycles of the early embryo, origins are controlled by a unique mechanism perhaps involving chromosome folding or attachment to the nuclear envelope. This could ensure the complete replication of the genome during the brief S phase. This mechanism acts temporally and spatially, such that periodically spaced origins are activated synchronously at the beginning of each S phase.

Experiments using *Xenopus* eggs showed that plasmid replication upon injection into the egg is under cell cycle control. Replication is not dependent on specific sequences, but instead depends on the size of the plasmid injected (Harland and Laskey 1980; Méchali and Kearsey 1984). Two-dimensional gel analysis of replication intermediates revealed that in both *Xenopus* eggs and extracts, plasmids containing either rDNA repeats or single-copy sequences initiated and terminated replication at random sites throughout the plasmid (Hyrien and Méchali 1992; Mahbubani et al. 1992). It was also determined that although initiation could occur at random sites on the plasmid, a single initiation event gave rise to complete replication of each plasmid molecule.

Similar conclusions were reached upon examination of the replication of chromosomal rDNA repeats in early embryos prior to the midblastula transition (Hyrien and Méchali 1993). Initiation occurred at random positions, and the estimated replicon size was 9–12 kb. To complete replication in the rapid S phase, all origins were presumed to be activated synchronously at the beginning of S phase. Whereas replication initiation is random with respect to sequence, the periodic spacing of replicons suggests that initiation is not random with respect to higher-order chromatin folding. If initiation were entirely random, there could be instances where some replicons were too far apart to finish replication in the short S phase. Thus, the authors suggest that chromosomal folding might specify a periodic spacing of origins, guaranteeing complete replication within the 10-minute S phase.

A similar periodic spacing of replicons was seen in the *Drosophila* early cleavage nuclei in which replication occurs in a 3- to 4-minute S phase. By electron microscopy (EM) studies, the average replicon size was 7.9 kb with a preferred periodicity of 3.4 kb and a maximum size of 19 kb (Blumenthal et al. 1974). An estimated 20,000 bidirectional origins must be activated nearly synchronously to finish S phase in the extremely short time. The maximum replicon size correlates with the amount of DNA that can be replicated in 3–4 minutes, given the rate of fork movement observed. However, these studies did not resolve whether initiation occurs at defined or random sequences. Using two-dimensional gel analysis of replication intermediates from early embryos, it was found that random initiation occurs both within the histone repeats and within a 40-kb single-copy sequence (Shinomiya and Ina 1991). Therefore, the periodicity of replicon spacing and the sequence-independent nature of replication are similar to that found in early *Xenopus* embryos. Thus, there appears to be a specific control of origin usage in the early rapid cycles.

S-phase Slowdown

Following the rapid embryonic cell cycles, the cell cycle lengthens to allow certain developmental processes to occur, such as the onset of zygotic gene transcription and gastrulation. In both *Xenopus* and *Drosoph-*

ila, the increase of the nuclear-to-cytoplasmic ratio results in the slowing of the cell cycle (Newport and Kirschner 1982a,b; Edgar and Schubiger 1986). During this time in development, both the control of replication origins and the entry into S phase are altered. In *Xenopus*, this is a one-step process that occurs at the midblastula transition (MBT), whereas in *Drosophila* these events occur in two distinct steps. First, following cellularization at cycle 13, alterations in the temporal control of origins occur. At this time, a G_2 phase is added to the cell cycle, and zygotic transcription commences. Following three more divisions, a G_1 phase is added, permitting transcriptional control of S phase. This transcriptional control in *Drosophila* and the MBT in *Xenopus* are discussed below.

In *Drosophila*, replication proceeds from the fast synchronous S phase of the early cycles to a prolonged S phase in which origin activation becomes asynchronous. A change in condensation of both euchromatin and heterochromatin occurs during this lengthening of S phase (Foe et al. 1993). During the first 13 cycles of the *Drosophila* embryo, the euchromatin remains decondensed throughout interphase. The heterochromatic regions decondense for a short time in S phase, decondensing progressively later in interphase as the cycle slows during cycles 11 to 13. By cycle 13, S phase has lengthened from 3–4 minutes to 13 minutes. During the three post-blastoderm cycles 14 to 16, S phase is eight times as long as the early S phases, taking 35–45 minutes.

In contrast to the early cycles, at least 200 particles of highly condensed euchromatin can be seen during interphase of cycles 14, 15, and 16 (Foe et al. 1993). EM studies show that new forks appear throughout the first 20 minutes of S phase during cycle 14, indicating that origin activation becomes asynchronous during interphase of these later cycles (McKnight and Miller 1977). Heterochromatin remains condensed throughout most of interphase and is not replicated until euchromatic replication has finished (Edgar and O'Farrell 1990; Foe et al. 1993). This altered regulation of origins and concurrent lengthening of S phase correlates with the onset of zygotic transcription and tissue-specific expression of certain genes. Replication must be coordinated with other processes now occurring during the cell cycle. In the post-blastoderm cycles, replication is followed by a G_2 phase of varying length, whereas mitosis occurs in distinct temporal and spatial domains.

Mouse

Similar to *Xenopus*, mouse embryos also are able to support the replication of injected double-stranded DNA templates, whereas injected plasmids are not replicated in oocytes (Wirak et al. 1985). Replication in embryos is dependent on specific *cis*-acting origin sequences of polyomavirus (PyV) or simian virus 40 (SV40) present on the plasmids. This is in contrast to results in *Xenopus* eggs, in which replication is independent of *cis*-acting sites and initiates at random sequences.

Using mouse embryos, an added layer of control is seen when comparing arrested one-cell embryos containing the unfused male and female pronuclei with two-cell embryos containing zygotic nuclei (Martinez-Salas et al. 1988, 1989). One-cell embryos are able to replicate DNA containing a minimal origin core sequence from PyV, whereas the two-cell embryo requires enhancer sequences present in *cis* to the PyV origin. Enhancers are also required for gene expression in the zygote as opposed to the one-cell embryo (Martinez-Salas et al. 1989; Majumder et al. 1993; Melin et al. 1993). The cytoplasmic factors that act on promoters and enhancers are not present in the one-cell embryo, but rather appear during the formation of the two-cell embryo (Henery et al. 1995). Enhancers are postulated to prevent the repression of origins and promoters by altered chromatin structure that is thought to occur upon zygote formation.

Replication in the mouse embryo is distinct from that occurring in the rapid early embryonic cycles of *Xenopus* and *Drosophila*, possibly due to the difference seen in zygotic transcription in these organisms. *Xenopus* and *Drosophila* have an early period lacking zygotic transcription, so are able to support fast cycles of alternating S and M phase. In contrast, transcription occurs in the early mouse embryo following zygotic formation, and thus replication must be coordinated with transcription. Enhancers might be associated with origins to overcome chromatin effects due to a transcriptionally active genome.

DEVELOPMENTAL SHIFT TO TRANSCRIPTIONAL REGULATION OF S PHASE

Following the rapid cycles of *Xenopus* and *Drosophila* that are controlled by posttranscriptional modifications of cell cycle regulators, a developmental shift occurs and transcription of the zygotic genome becomes active. At this time in development, S-phase regulators can be controlled at a transcriptional level, a mode of regulation characteristic of the somatic cell cycle.

MBT in *Xenopus*

The cell cycle of *Xenopus* is modified during the MBT, which occurs after the first 6 hours of development. The early rapid cycles of S and M phase slow, giving rise to a longer cell cycle in which both G_1 and G_2 gap phases are added to the cell cycle. This process then allows for the resumption of zygotic transcription (Kimelman et al. 1987). This slowing of the cell cycle is thought to occur due to a mitotic initiation factor that becomes rate-limiting at the MBT (Kirschner et al. 1985; Newport et al. 1985). Blastula cleavage becomes less synchronous, cells become motile, and zygotic transcription turns on. Recent results determined that the excess of histones present in the early embryo is responsible for the repression of transcription prior to the MBT (Prioleau et al. 1994).

As mentioned previously, the developmental regulation of S phase is altered at two different levels during the MBT in *Xenopus*. First, similar to *Drosophila*, origin activation presumably becomes asynchronous, giving rise to a lengthened S phase. DNA replication becomes asynchronous between cells, as evidenced by a variation of proliferating cell nuclear antigen (PCNA) staining, with some nuclei showing a peripheral staining and others showing a homogeneous staining (Leibovici et al. 1992). This is in contrast to the early cycles in which PCNA staining is homogeneous throughout S phase. Second, the entry into S phase is altered such that a G_1 phase is added to the cell cycle. Zygotic transcription resumes, permitting S-phase regulators to become transcriptionally controlled.

Addition of G_1 in *Drosophila*

The developmental shift to asynchronous origin activation has already occurred by cycle 14 of the *Drosophila* embryo. The second developmental control placed on S phase occurs following cycle 16 when a G_1 phase is added to the cell cycle. Following cycle 16, mitotic embryonic cells either arrest in G_1 and divide later in development (imaginal cells) or continue to divide (neuronal cells). The cells giving rise to most of the larval tissues become polytene, and as described below, these cells go through additional cycles that have only S and gap phases, but no mitosis. Thus, cycle 17 is the first time during embryogenesis there is a G_1 phase and, with the exception of a small group of cells on the ventral epidermis, it occurs in cells that are becoming polytene. At this point in embryogenesis, the cyclin E gene changes from being constitutively transcribed to a dynamic pattern of transcription in which the appearance of cyclin E transcripts precedes S phase. The down-regulation of cyclin E is needed for the arrest of cells in G_1, and cyclin E is then necessary for the G_1/S transition (Richardson et al. 1993; Knoblich et al. 1994). Both the cyclin E transcript and protein are down-regulated following the last mitotic division of epidermal cells during cycle 16 and remain off in these G_1-arrested cells (Knoblich et al. 1994). During *Drosophila* development, the transcriptional control of cyclin E during the added G_1 phase is a new mode of S-phase regulation that occurs specifically following embryonic cell cycles 1–16. Mutations in the E2F transcription factor block DNA replication beginning at cycle 17, correlating with the first appearance of G_1 (Duronio et al. 1995). In embryonic cells that have a G_1 cell cycle phase, cyclin E is the downstream target of E2F, as evidenced by the observation that ectopic expression of cyclin E can overcome the requirement for E2F for S phase (Duronio and O'Farrell 1995). Thus, E2F appears to control the developmental pattern of cyclin E transcription that is responsible for the G_1 to S transition. Cyclin E in turn activates the transcription of a set of genes encoding essential replication factors, including polymerase-α, *PCNA*, and ribonuclease reductase 1 and 2 (*RNR1, 2*) (Duronio and O'Farrell 1994). E2F and cyclin E exert a different role in the cells becoming polytene than in the neuronal cells that lack a G_1 phase (Duronio and O'Farrell 1995). In the nervous system, cyclin E continues to be constitutively expressed, and this expression is independent of E2F. Rather, in the nervous system, cyclin E appears to regulate E2F.

POLYPLOIDY/POLYTENY

Changes in S-phase Regulation

Polyploid cells exist in a number of organisms including plants, ciliates, and dipteran insects, as well as in some mammalian cell types such as the trophoblasts that give rise to the mammalian placenta. Polyploidy is often associated with cells or tissues in which a requirement for increased protein production is needed; multiple chromosome copies is one way in which evolution has met that demand. An area of current research concerns the identification of regulators that govern the developmental transition leading to polyploidy. Similar to the developmental alterations of DNA replication already discussed, polyploid replication is controlled at two basic levels. These include changes in regulation of the cell cycle as well as alterations at the level of replication origins and other parameters used during DNA synthesis. In *Drosophila*, the transition to polyteny results in an altered cell cycle, termed the endo cell cycle, which consists of an alternating S phase and gap phase (Smith and Orr-Weaver 1991). Parameters of replication also become altered, and the block to rereplication is overcome during the endo cell cycle.

Polyploid and polytene cells are defined as those in which DNA replication has become uncoupled from mitosis, giving rise to cells with greater than diploid content of DNA. The degree to which replication is uncoupled from mitotic aspects of the cell cycle can vary. Polytene cells uncouple replication from all aspects of mitosis; 1000 or more chromosome copies remain synapsed, forming the large polytene chromosomes characteristic of the *Drosophila* salivary gland. Some polyploid cells do not uncouple replication from all aspects of mitosis, and chromosome segregation or cycles of chromosome condensation still occur. This is referred to as endopolyploidy or endomitosis. Finally, multinucleate cells have been referred to as polyploid cells. The *Drosophila* larval polytene cells are among the best characterized in terms of alterations in S phase that occur during this developmental transition.

Regulators of the Endo Cell Cycle

One of the best-understood examples of polyteny is in the *Drosophila* larval tissues (Smith and Orr-Weaver 1991). Following embryogenesis, *Drosophila* larval growth is due to an increase in cell size upon polytenization, because only cells in the nervous system and imaginal tissues undergo mitosis during larval development. In *Drosophila*, most tissues enter the endo cell cycle during late embryogenesis, and this transition is temporally and spatially regulated (Smith and Orr-Weaver 1991). The first transitions to the endo cell cycle occur in tissue-specific domains that replicate at characteristic times after cycle 16, with the salivary gland being the first tissue to enter the endo cell cycle, followed by the midgut, hindgut, and Malpighian tubules. It was determined for the hindgut that cells enter the endo cell cycle from the G_1 phase of the cell cycle, whereas salivary gland cells may enter the endo cell cycle following G_2. Gap phases of the endo cell cycle can vary widely in length, from 3 hours for the midgut to 18 hours for the salivary gland (Smith and Orr-Weaver 1991).

The spatially and temporally regulated pattern seen in *Drosophila* polytene tissues argues that a novel factor controls these cycles. However, regulators of the endo cell cycle could in theory be known mitotic cell cycle regulators that also act to control the endo cell cycle. In this case, the dependency between S phase and M phase must be disrupted in this altered cell cycle. In *Drosophila*, the mitotic cell cycle regulators, cyclin A, cdc2, and the cdc25 phosphatase homolog, *string*, are not needed for the endo cell cycle. Endoreplication proceeds in embryos lacking any of these regulators (Smith and Orr-Weaver 1991; Smith et al. 1993; Stern et al. 1993).

The cyclin E gene is essential for polytenization. The endo cell cycle does not occur in embryos lacking cyclin E or E2F, and cyclin E expression parallels S phase in endoreplicating tissues (Knoblich et al. 1994; Duronio et al. 1995). Thus, developmental regulation of the transcription of the cyclin E gene can account for the pattern of polytene S phases. Cyclin E transcription in endoreplicating cells is subject to a negative feedback, because ectopic expression of cyclin E down-regulates endogenous cyclin E transcription in these cells (Sauer et al. 1995). This negative feedback loop may be responsible for the transient appearance of cyclin E transcripts in the endo cell cycle and could be necessary for cyclic, rather than continuous, DNA replication in these cells. It may simply be that periodic transcription of cyclin E in the absence of cyclins A and B results in a cycle of S phase in the absence of mitosis (Sauer et al. 1995). However, it is possible that novel regulators couple developmental signals to the regulation of the endo cell cycle.

An interesting gene, *escargot*, is a transcription factor that appears to maintain the diploid state of arrested imaginal cells in the *Drosophila* larva (Hayashi et al. 1993; Fuse et al. 1994). In certain allelic combinations, a group of imaginal cells known as the histoblast nests overreplicate. Inappropriate expression of *escargot* in the polytene salivary gland represses endoreplication. A model has been proposed by which *escargot* maintains diploidy via transcriptional repression of regulators of the endo cell cycle. However, an alternate model might be that *escargot* plays a more direct role in controlling the cell fate of imaginal cells.

In the fission yeast, *S. pombe*, the absence of cyclin B due to a *cdc13* deletion causes cells to undergo multiple rounds of S phase, resulting in high levels of polyploidy (Hayles et al. 1994). These results have led to a model whereby high levels of cyclin B-cdc2 kinase promote entry into M phase and, conversely, low levels cause the entry into S phase. Thus, by simply disrupting the cyclin B kinase activity, cells are able to reset to a G_1 phase and enter S phase. In this example, *S. pombe* has created a cell cycle leading to polyploidy solely by altering the control of mitotic cell cycle regulators. The cell cycle of *S. pombe*, however, represents a very simplified cell cycle. In higher eukaryotes a more complex set of regulators controls the cell cycle and ensures the proper coupling of replication and mitosis. Therefore, in most organisms, the transition to the endo cell cycle most likely requires more than simply inactivating G_2/M regulators, although this might be a necessary step for the transition to polyteny. Other regulatory changes must also occur, such as the alteration of checkpoints that act to couple S phase and mitosis, as well as the block to rereplication which must be removed.

Removal of the Block to Rereplication

A major alteration in the parameters of replication that must occur during the endo cell cycle is the removal of the block to rereplication. One postulated mechanism that normally limits DNA replication to once per cell cycle is the existence of licensing factor. This factor is proposed to permit one round of DNA replication and to then become inactivated following replication (Blow and Laskey 1988). Active licensing factor is then excluded from the nucleus, and reentry into the nucleus is only permitted following nuclear envelope breakdown in mitosis (see Blow and Chong, this volume). Perhaps the actions of licensing factor or other factors involved in replication initiation are altered in the endo cell cycle, but the nature of this alteration is unclear. A mutation has been identified in the *Drosophila MCM2* gene, a proposed component of licensing factor (Treisman et al. 1995). It is interesting that this mutation inhibits proliferation of the diploid imaginal and neuronal cells but has almost no effect on the polytene cells. The polytene cells grow and undergo DNA replication, but the chromosomes may be more fragile in the mutant. Thus, licensing factor may become dispensable in the endo cell cycle. In the case of the multiple rounds of replication that occur in *S. pombe* as described above, the block to rereplication was removed solely by the disruption of cyclin B without a direct effect on licensing factor.

Late-replicating Heterochromatin/Underrepresentation

Another aspect of altered replication during the endo cell cycle is the temporal control of polytene replication. Fiber autoradiography of polytene chromosomes shows a pattern of late-replicating heterochromatic regions (Spradling and Orr-Weaver 1987). In *Drosophila virilus* the replicon size in polytene chromosomes is similar to diploid brain cells, yet the rate of fork movement is three times slower (Steinemann 1981a,b). Similar studies of *Drosophila nasuta* show that shorter and slower replicons are seen in late-replicating regions (Lakhotia and Sinha 1983). Perhaps the chromatin organization of polytene heterochromatin inhibits replication fork movement. This slower replication might account for the late replication of the heterochromatin.

A second characteristic alteration of DNA replication in the endo cell cycle is that 20–30% of the genome is underrepresented, including the centric heterochromatin and rDNA and histone repeats. Less underrepresentation of rDNA is seen, however, in the polyploid nurse cells, most likely due to the function of the nurse cells in producing the rRNA for the developing oocyte. Regions of euchromatin are generally replicated to the same extent during polyploidization. In *Drosophila* polytene chromosomes, both bands and interbands also replicate to the same extent, as confirmed by quantitative Southern blots (Spierer and Spierer 1984).

Heterochromatic underrepresentation might be caused by incomplete replication of specific sequences during polytenization, or alternatively, by elimination of sequences from the chromosome (Karpen and Spradling 1990; Glaser et al. 1992). One hypothesis is that elimination of specific regions is caused by the excision of transposable elements. Interestingly, many transposable elements are found solely in heterochromatic regions. A testable prediction of the elimination model is that novel DNA junctions should be formed upon elimination. The role of underrepresentation during the endo cell cycle is unclear; perhaps these sequences are not needed for the high level of protein expression characteristic of most polytene tissues.

Amplification

Amplification of specific genomic sequences is a developmentally regulated mechanism that allows the production of large amounts of protein products in a short developmental time frame. Amplification control occurs at the level of the block to rereplication of specific sequences, whereby reinitiation of replication leads to multiple copies of genomic sequences. Amplification provides a model system to study the developmental regulation of a eukaryotic replicon. Two well-characterized examples of developmental amplification are the *Drosophila* chorion genes and the *Sciara* DNA puffs. This is reviewed more extensively by Gerbi and Urnov (this volume).

Drosophila Chorion Genes

During oogenesis, the somatic follicle cells surrounding the egg chamber are responsible for the secretion of the chorionic eggshell layers encompassing the developing oocyte. Following polyploidization of these cells, a further tissue-specific mechanism ensures an increased copy number of the chorion genes, so that proteins can be made in a rapid developmental time window. The major chorion genes are organized into two chromosomal clusters present on the X and third chromosomes, which amplify to levels of 15-fold and 60-fold, respectively (Orr-Weaver 1991). Amplification within these clusters occurs by repeated reinitiation of an origin as shown by multiple eye forms (bubbles within bubbles) in EM spreads (Osheim et al. 1988).

Studies of both clusters have identified *cis*-acting regions responsible for amplification, termed the amplification control element (ACE). ACE3 of the third chromosome cluster has been delineated to a 320-bp region that acts in a distance- and orientation-independent manner (Orr-Weaver et al. 1989). Replication intermediates in this region have been analyzed, and a predominant replication origin has been mapped 1.5 kb downstream from ACE3. This lies in a region important for high levels of amplification known as amplification enhancing region-d, (AER-d) (Delidakis and Kafatos 1989; Heck and Spradling 1990). However, initiation events also occur throughout a 12-kb region surrounding both ACE3 and AER-d.

ACE3 is able to direct the autonomous amplification of sequences when inserted throughout the genome, albeit at lower amplification levels (Carminati et al. 1992). These studies suggest a model by which ACE3 controls the reinitiation of nearby origins, perhaps by capturing limiting replication factors and overcoming a block to rereplication.

Sciara DNA Puffs

Amplification also occurs in the fungus fly *Sciara coprophila* within puff regions of the larval salivary gland polytene chromosomes. Similar to the *Drosophila* chorion genes, amplification is temporally and tissue-regulated. Amplification as well as transcription of the *Sciara* DNA puffs is developmentally regulated by the steroid hormone, ecdysone (Bienz-Tadmor et al. 1991; Gerbi et al. 1993). Amplification presumably allows the rapid production of proteins needed during late larval development such as those for the formation of the pupal case. Similarly, in *Rhynchosciara*, puffs encode polypeptides necessary for the production of the pupal cocoon and undergo amplification (Glover et al. 1982).

Amplification occurs in several major puff regions in *Sciara*, and puff expansion is due to a burst of transcription following amplification of these sequences to approximately 20-fold levels (DiBartolomeis and Gerbi 1989; Wu et al. 1993). DNA amplification within one of the major puffs, II/9A, has been well characterized and is consistent with an onion-skin mechanism, similar to the *Drosophila* chorion genes. Puff II/9A encodes two genes that share 85% sequence similarity, and a major amplification origin has been mapped to a 1-kb region lying upstream of the two genes (Liang et al. 1993; Liang and Gerbi 1994). Replication from this origin occurs bidirectionally.

SUMMARY

DNA replication is regulated by a wide variety of mechanisms that act throughout development to coordinate S phase and replication with developmental transitions. Replication can be controlled both at the level of key S-phase regulators and at the level of parameters of DNA synthesis, such as origin usage. By studying the unique ways that different developmental events control and alter S phase, we will broaden our understanding of the regulators and mechanisms involved in DNA replication.

ACKNOWLEDGMENTS

We thank Tim Hunt, Peter Jackson, Marc Kirschner, Helena Richardson, and Rob Saint for providing unpublished information. This work was supported by National Institutes of Health grant GM-39341. J.C. was supported by National Institutes of Health predoctoral training grant CA-09541.

REFERENCES

Axton, J.M., F.L. Shamanski, L.M. Young, D.S. Henderson, J.B. Boyd, and T.L. Orr-Weaver. 1994. The inhibitor of DNA replication encoded by the *Drosophila* gene *plutonium* is a small, ankyrin repeat protein. *EMBO J.* **13:** 462–470.

Bienz-Tadmor, B., H.S. Smith, and S.A. Gerbi. 1991. The promoter of DNA puff gene II/9-1 of *Sciara coprophila* is inducible by ecdysone in late prepupal salivary glands of *Drosophila melanogaster*. *Cell Regul.* **2:** 875–878.

Blow, J.J. and R.A. Laskey. 1988. A role for the nuclear envelope in controlling DNA replication within the cell cycle. *Nature* **332:** 546–548.

Blumenthal, A.B., H.J. Kriegstein, and D.S. Hogness. 1974. The units of DNA replication in *Drosophila melanogaster* chromosomes. *Cold Spring Harbor Symp. Quant. Biol.* **38:** 205–223.

Buck, V., A. White, and J. Rosamond. 1991. CDC7 protein kinase activity is required for mitosis and meiosis in *Saccharomyces cerevisiae*. *Mol. Gen. Genet.* **227:** 452–457.

Callan, H.G. 1974. DNA replication in the chromosomes of eukaryotes. *Cold Spring Harbor Symp. Quant. Biol.* **38:** 195–203.

Carminati, J.L., C.G. Johnston, and T.L. Orr-Weaver. 1992. The *Drosophila* ACE3 chorion element autonomously induces amplification. *Mol. Cell. Biol.* **12:** 2444–2453.

Chandley, A.C. 1966. Studies on oogenesis in *Drosophila melanogaster* with 3H-thymidine label. *Exp. Cell Res.* **44:** 201–215.

Ciapa, B., D. Pesando, M. Wilding, and M. Whitaker. 1994. Cell-cycle calcium transients driven by cyclic changes in inositol trisphosphate levels. *Nature* **368:** 875–878.

Collins, I. and C.S. Newlon. 1994. Chromosomal DNA replication initiates at the same origins in meiosis and mitosis. *Mol. Cell. Biol.* **14:** 3524–3534.

Delidakis, C. and F.C. Kafatos. 1989. Amplification enhancers and replication origins in the autosomal chorion gene cluster of *Drosophila*. *EMBO J.* **8:** 891–901.

DiBartolomeis, S.M. and S.A. Gerbi. 1989. Molecular characterization of DNA puff

II/9A genes in *Sciara coprophila. J. Mol. Biol.* **210**: 531-543.
Dowell, S.J., P. Romanowski, and J.F. Diffley. 1994. Interaction of Dbf4, the Cdc7 protein kinase regulatory subunit, with yeast replication origins in vivo. *Science* **265**: 1243-1246.
Duronio, R.J. and P.H. O'Farrell. 1994. Developmental control of a G_1-S transcriptional program in *Drosophila. Development* **120**: 1503-1515.
―――. 1995. Developmental control of the G1 to S transition in *Drosophila*: Cyclin E is a limiting downstream target of E2F. *Genes Dev.* **9**: 1456-1468.
Duronio, R.J., P.H. O'Farrell, J.-E. Xie, A. Brook, and N. Dyson. 1995. The transcription factor E2F is required for S phase during *Drosophila* embryogenesis. *Genes Dev.* **9**: 1445-1455.
Dynlacht, B.D., O. Flores, J.A. Lees, and E. Harlow. 1994. Differential regulation of E2F *trans*-activation by cyclin/cdk2 complexes. *Genes Dev.* **8**: 1772-1786.
Edgar, B.A. and P.H. O'Farrell. 1990. The three postblastoderm cell cycles of *Drosophila* embryogenesis are regulated in G2 by *string. Cell* **62**: 469-480.
Edgar, B.A. and G. Schubiger. 1986. Parameters controlling transcriptional activation during early *Drosophila* development. *Cell* **44**: 871-877.
Epstein, C.B. and F.R. Cross. 1992. CLB5: A novel B cyclin from budding yeast with a role in S phase. *Genes Dev.* **6**: 1695-1706.
Foe, V.E., G.M. Odell, and B.A. Edgar. 1993. Mitosis and morphogenesis in the *Drosophila* embryo: Point and counterpoint. In *The development of* Drosophila melanogaster (ed. M. Bate and A. Martinez Arias), pp. 149-300. Cold Spring Harbor Laboratory Press. Cold Spring Harbor, New York.
Freeman, M. and D.M. Glover. 1987. The *gnu* mutation of *Drosophila* causes inappropriate DNA synthesis in unfertilized and fertilized eggs. *Genes Dev.* **1**: 924-930.
Freeman, M., C. Nüsslein-Volhard, and D.M. Glover. 1986. The dissociation of nuclear and centrosomal division in *gnu*, a mutation causing giant nuclei in *Drosophila. Cell* **46**: 457-468.
Furuno, N., M. Nishizawa, K. Okazaki, H. Tanaka, J. Iwashita, N. Nakajo, Y. Ogawa, and N. Sagata. 1994. Suppression of DNA replication via Mos function during meiotic divisions in *Xenopus* oocytes. *EMBO J.* **13**: 2399-2410.
Fuse, N., S. Hirose, and S. Hayashi. 1994. Diploidy of *Drosophila* imaginal cells is maintained by a transcriptional repressor encoded by *escargot. Genes Dev.* **8**: 2270-2281.
Gerbi, S.A., C. Liang, N. Wu, S.M. DiBartolomeis, B. Bienz-Tadmor, H.S. Smith, and F.D. Urnov. 1993. DNA amplification in DNA puff II/9A of *Sciara coprophila. Cold Spring Harbor Symp. Quant. Biol.* **58**: 487-493.
Glaser, R.L., G.H. Karpen, and A.C. Spradling. 1992. Replication forks are not found in a *Drosophila* minichromosome demonstrating a gradient of polytenization. *Chromosoma* **102**: 15-29.
Glover, D.M., A. Zaha, A.J. Stocker, R.V. Santelli, M.T. Pueyo, S.M. De Toledo, and F.J. Lara. 1982. Gene amplification in *Rhynchosciara* salivary gland chromosomes. *Proc. Natl. Acad. Sci.* **79**: 2947-2951.
Grell, R.F. 1973. Recombination and DNA replication in the *Drosophila melanogaster* oocyte. *Genetics* **73**: 87-108.
Harland, R.M. and R.A. Laskey. 1980. Regulated replication of DNA microinjected into eggs of *Xenopus laevis. Cell* **21**: 761-771.
Hayashi, S., S. Hirose, T. Metcalfe, and A.D. Shirras. 1993. Control of imaginal cell de-

velopment by the *escargot* gene of *Drosophila*. *Development* **118**: 105–115.
Hayles, J., D. Fisher, A. Woollard, and P. Nurse. 1994. Temporal order of S phase and mitosis in fission yeast is determined by the state of the p34cdc2-mitotic B cyclin complex. *Cell* **78**: 813–822.
Heck, M.M. and A.C. Spradling. 1990. Multiple replication origins are used during *Drosophila* chorion gene amplification. *J. Cell Biol.* **110**: 903–914.
Henery, C.C., M. Miranda, M. Wiekowski, I. Wilmut, and M.L. DePamphilis. 1995. Repression of gene expression at the beginning of mouse development. *Dev. Biol.* **169**: 448–460.
Hollingsworth, R., Jr. and R.A. Sclafani. 1993. Yeast pre-meiotic DNA replication utilizes mitotic origin ARS1 independently of *CDC7* function. *Chromosoma* **102**: 415–420.
Hyrien, O. and M. Méchali. 1992. Plasmid replication in *Xenopus* eggs and egg extracts: A 2D gel electrophoretic analysis. *Nucleic Acids Res.* **20**: 1463–1469.
―――――. 1993. Chromosomal replication initiates and terminates at random sequences but at regular intervals in the ribosomal DNA of *Xenopus* early embryos. *EMBO J.* **12**: 4511–4520.
Iino, Y., Y. Hiramine, and M. Yamamoto. 1995. The role of *cdc2* and other genes in meiosis in *Schizosaccharomyces pombe*. *Genetics* **140**: 1235–1245.
Jackson, A.L., P.M. Pahl, K. Harrison, J. Rosamond, and R.A. Sclafani. 1993. Cell cycle regulation of the yeast Cdc7 protein kinase by association with the Dbf4 protein. *Mol. Cell. Biol.* **13**: 2899–2908.
Jackson, P.K., S. Chevalier, M. Philippe, and M.W. Kirschner. 1995. Early events in DNA replication require cyclin E and are blocked by p21^{CIP1}. *J. Cell Biol.* **130**: 755–769.
Jagiello, G., W.K. Sung, and J. Van't Hof. 1983. Fiber DNA studies of premeiotic mouse spermatogenesis. *Exp. Cell Res.* **146**: 281–287.
Johnston, L.H., D.H. Williamson, A.L. Johnson, and D.J. Fennell. 1982. On the mechanism of premeiotic DNA synthesis in the yeast *Saccharomyces cerevisiae*. *Exp. Cell Res.* **141**: 53–62.
Karpen, G.H. and A.C. Spradling. 1990. Reduced DNA polytenization of a minichromosome region undergoing position-effect variegation in *Drosophila*. *Cell* **63**: 97–107.
Kimelman, D., M. Kirschner, and T. Scherson. 1987. The events of the midblastula transition in *Xenopus* are regulated by changes in the cell cycle. *Cell* **48**: 399–407.
Kirschner, M., J. Newport, and J. Gerhart. 1985. The timing of early developmental events in *Xenopus*. *Trends Genet.* **1**: 41–47.
Knoblich, J.A., K. Sauer, L. Jones, H. Richardson, R. Saint, and C.F. Lehner. 1994. Cyclin E controls S phase progression and its down-regulation during *Drosophila* embryogenesis is required for the arrest of cell proliferation. *Cell* **77**: 107–120.
Krek, W., M.E. Ewen, S. Shirodkar, Z. Arany, W. Kaelin, Jr., and D.M. Livingston. 1994. Negative regulation of the growth-promoting transcription factor E2F-1 by a stably bound cyclin A-dependent protein kinase. *Cell* **78**: 161–172.
Lakhotia, S.C. and P. Sinha. 1983. Replication in *Drosophila* chromosomes. X. Two kinds of active replicons in salivary gland polytene nuclei and their relation to chromosomal replication patterns. *Chromosoma* **88**: 265–276.
Leibovici, M., G. Monod, J. Geraudie, R. Bravo, and M. Méchali. 1992. Nuclear distribution of PCNA during embryonic development in *Xenopus laevis*: A reinvestigation of

early cell cycles. *J. Cell Sci.* **102:** 63-69.
Liang, C. and S.A. Gerbi. 1994. Analysis of an origin of DNA amplification in *Sciara coprophila* by a novel three-dimensional gel method. *Mol. Cell. Biol.* **14:** 1520-1529.
Liang, C., J.D. Spitzer, H.S. Smith, and S.A. Gerbi. 1993. Replication initiates at a confined region during DNA amplification in *Sciara* DNA puff II/9A. *Genes Dev.* **7:** 1072-1084.
Lorca, T., F.H. Cruzalegui, D. Fesquet, J.C. Cavadore, J. Mery, A. Means, and M. Doree. 1993. Calmodulin-dependent protein kinase II mediates inactivation of MPF and CSF upon fertilization of *Xenopus* eggs. *Nature* **366:** 270-273.
Mahbubani, H.M., T. Paull, J.K. Elder, and J.J. Blow. 1992. DNA replication initiates at multiple sites on plasmid DNA in *Xenopus* egg extracts. *Nucleic Acids Res.* **20:** 1457-1462.
Majumder, S., M. Miranda, and M.L. De Pamphilis. 1993. Analysis of gene expression in mouse preimplantation embryos demonstrates that the primary role of enhancers is to relieve repression of promoters. *EMBO J.* **12:** 1131-1140.
Martinez-Salas, E., D.Y. Cupo, and M.L. DePamphilis. 1988. The need for enhancers is acquired upon formation of a diploid nucleus during early mouse development. *Genes Dev.* **2:** 1115-1126.
Martinez-Salas, E., E. Linney, J. Hassell, and M.L. DePamphilis. 1989. The need for enhancers in gene expression first appears during mouse development with formation of the zygotic nucleus. *Genes Dev.* **3:** 1493-1506.
McKnight, S.L. and O.L.J. Miller. 1977. Electron microscopic analysis of chromatin replication in the cellular blastoderm *Drosophila melanogaster* embryo. *Cell* **12:** 795-804.
Méchali, M. and S. Kearsey. 1984. Lack of specific sequence requirement for DNA replication in *Xenopus* eggs compared with high sequence specificity in yeast. *Cell* **38:** 55-64.
Melin, F., M. Miranda, N. Montreau, M.L. De Pamphilis, and D. Blangy. 1993. Transcription enhancer factor-1 (TEF-1) DNA binding sites can specifically enhance gene expression at the beginning of mouse development. *EMBO J.* **12:** 4657-4666.
Minshull, J. 1993. Cyclin synthesis: Who needs it? *BioEssays* **15:** 149-155.
Newlon, C.S. 1988. Yeast chromosome replication and segregation. *Microbiol. Rev.* **52:** 568-601.
Newport, J. and M. Kirschner. 1982a. A major developmental transition in early *Xenopus* embryos. I. Characterization and timing of cellular changes at the midblastula stage. *Cell* **30:** 675-686.
―――. 1982b. A major developmental transition in early *Xenopus* embryos. II. Control of the onset of transcription. *Cell* **30:** 687-696.
Newport, J., T. Spann, J. Kanki, and D. Forbes. 1985. The role of mitotic factors in regulating the timing of the midblastula transition in *Xenopus*. *Cold Spring Harbor Symp. Quant. Biol.* **50:** 651-656.
Ohsumi, K., W. Sawada, and T. Kishimoto. 1994. Meiosis-specific cell cycle regulation in maturing *Xenopus* oocytes. *J. Cell Sci.* **107:** 3005-3013.
Orr-Weaver, T.L. 1991. *Drosophila* chorion genes: Cracking the eggshell's secrets. *BioEssays* **13:** 97-105.
Orr-Weaver, T.L., C.G. Johnston, and A.C. Spradling. 1989. The role of ACE3 in *Drosophila* chorion gene amplification. *EMBO J.* **8:** 4153-4162.
Osheim, Y.N., O. Miller, Jr., and A.L. Beyer. 1988. Visualization of *Drosophila*

melanogaster chorion genes undergoing amplification. *Mol. Cell. Biol.* **8:** 2811-2821.

Picard, A., E. Karsenti, M.C. Dabauvalle, and M. Doree. 1987. Release of mature starfish oocytes from interphase arrest by microinjection of human centrosomes. *Nature* **327:** 170-172.

Piggott, J.R., R. Rai, and B.L.A. Carter. 1982. A bifunctional gene product involved in two phases of the yeast cell cycle. *Nature* **298:** 391-393.

Prioleau, M.N., J. Huet, A. Sentenac, and M. Méchali. 1994. Competition between chromatin and transcription complex assembly regulates gene expression during early development. *Cell* **77:** 439-449.

Reed, S.I. and C. Wittenberg. 1990. Mitotic role for the *Cdc28* protein kinase of *Saccharomyces cerevisiae*. *Proc. Natl. Acad. Sci.* **87:** 5697-5701.

Richardson, H.E., L.V. O'Keefe, S.I. Reed, and R. Saint. 1993. A *Drosophila* G1-specific cyclin E homolog exhibits different modes of expression during embryogenesis. *Development* **119:** 673-690.

Richardson, H.E., C. Wittenberg, F. Cross, and S.I. Reed. 1989. An essential G_1 function for cyclin-like proteins in yeast. *Cell* **59:** 1127-1133.

Sagata, N., M. Oskarsson, T. Copeland, J. Brumbaugh, and G.F. Vande Woude. 1988. Function of *c-mos* proto-oncogene product in meiotic maturation in *Xenopus* oocytes. *Nature* **335:** 519-525.

Sauer, K., J.A. Knoblich, H. Richardson, and C.F. Lehner. 1995. Distinct modes of cyclin E/cdc2c kinase regulation and S-phase control in mitotic and endoreduplication cycles of *Drosophila* embryogenesis. *Genes Dev.* **9:** 1327-1339.

Schild, D. and B. Byers. 1978. Meiotic effects of DNA-defective cell division cycle mutations of *Saccharomyces cerevisiae*. *Chromosoma* **70:** 109-130.

Schwob, E. and K. Nasmyth. 1993. *CLB5* and *CLB6*, a new pair of B cyclins involved in DNA replication in *Saccharomyces cerevisiae*. *Genes Dev.* **7:** 1160-1175.

Serrano, M., G.J. Hannon, and D. Beach. 1993. A new regulatory motif in cell-cycle control causing specific inhibition of cyclin D/CDK4. *Nature* **366:** 704-707.

Shamanski, F.L. and T.L. Orr-Weaver. 1991. The *Drosophila plutonium* and *pan gu* genes regulate entry into S phase at fertilization. *Cell* **66:** 1289-1300.

Sherr, C.J. 1993. Mammalian G1 cyclins. *Cell* **73:** 1059-1065.

———. 1994. G1 phase progression: Cycling on cue. *Cell* **79:** 551-555.

Sherr, C.J. and J.M. Roberts. 1995. Inhibitors of mammalian G1 cyclin-dependent kinases. *Genes Dev.* **9:** 1149-1163.

Shinomiya, T. and S. Ina. 1991. Analysis of chromosomal replicons in early embryos of *Drosophila melanogaster* by two-dimensional gel electrophoresis. *Nucleic Acids Res.* **19:** 3935-3941.

Shuster, E.O. and B. Byers. 1989. Pachytene arrest and other meiotic effects of the start mutations in *Saccharomyces cerevisiae*. *Genetics* **123:** 29-43.

Simchen, G. and J. Hirschberg. 1977. Effects of the mitotic cell-cycle mutation *cdc4* on yeast meiosis. *Genetics* **86:** 57-72.

Smith, A.V. and T.L. Orr-Weaver. 1991. The regulation of the cell cycle during *Drosophila* embryogenesis: The transition to polyteny. *Development* **112:** 997-1008.

Smith, A.V., J.A. King, and T.L. Orr-Weaver. 1993. Identification of genomic regions required for DNA replication during *Drosophila* embryogenesis. *Genetics* **135:** 817-829.

Spierer, A. and P. Spierer. 1984. Similar level of polyteny in bands and interbands of *Drosophila* giant chromosomes. *Nature* **307:** 176-178.

Spradling, A. and T. Orr-Weaver. 1987. Regulation of DNA replication during

Drosophila development. *Annu. Rev. Genet.* **21:** 373-403.
Steinemann, M. 1981a. Chromosomal replication in *Drosophila virilis*. II. Organization of active origins in diploid brain cells. *Chromosoma* **82:** 267-288.
―――. 1981b. Chromosomal replication in *Drosophila virilis*. III. Organization of active origins in the highly polytene salivary gland cells. *Chromosoma* **82:** 289-307.
Stern, B., G. Ried, N.J. Clegg, T.A. Grigliatti, and C.F. Lehner. 1993. Genetic analysis of the *Drosophila cdc2* homolog. *Development* **117:** 219-232.
Stricker, S.A. 1995. Time-lapse confocal imaging of calcium dynamics in starfish embryos. *Dev. Biol.* **170:** 496-518.
Treisman, J.E., P.J. Follette, P.H. O'Farrell, and G.M. Rubin. 1995. Cell proliferation and DNA replication defects in a *Drosophila MCM2* mutant. *Genes Dev.* **9:** 1709-1715.
Whitaker, M. and R. Patel. 1990. Calcium and cell cycle control. *Development* **108:** 525-542.
Whitaker, M. and K. Swann. 1993. Lighting the fuse at fertilization. *Development* **117:** 1-12.
Williamson, D.H., L.H. Johnston, D.J. Fennell, and G. Simchen. 1983. The timing of the S phase and other nuclear events in yeast meiosis. *Exp. Cell Res.* **145:** 209-217.
Wirak, D.O., L.E. Chalifour, P.M. Wassarman, W.J. Muller, J.A. Hassell, and M.L. DePamphilis. 1985. Sequence-dependent DNA replication in preimplantation mouse embryos. *Mol. Cell. Biol.* **5:** 2924-2935.
Wu, N., C. Liang, S.M. DiBartolomeis, H.S. Smith, and S.A. Gerbi. 1993. Developmental progression of DNA puffs in *Sciara coprophila*: Amplification and transcription. *Dev. Biol.* **160:** 73-84.
Zhang, H. and J.V. Ruderman. 1993. Differential replication capacities of G1 and S-phase extracts from sea urchin eggs. *J. Cell Sci.* **104:** 565-572.

14
Comparison of DNA Replication in Cells from Prokarya and Eukarya

Bruce Stillman
Cold Spring Harbor Laboratory
Cold Spring Harbor, New York 11724

It is now supposed that utilization of DNA as genetic material emerged after an RNA world existed (Gesteland and Atkins 1993) and that all contemporary organisms evolved from a common ancestor related to primitive photosynthetic bacteria (Woese and Pace 1993). It is therefore not surprising that cells in the three major kingdoms, Bacteria, Archaea, and Eukarya, use roughly similar strategies and mechanisms for genome duplication. Nevertheless, the diversity of DNA replication is evident when the varied strategies used for replication of bacteriophage, plasmid, and virus genomes in both prokaryotes and eukaryotes are considered. This diversity becomes obvious when scanning the chapters that summarize the replication of virus genomes later in this book (Hassell and Brinton; Stenlund; Hay; Challberg; Yates; Traktman; Cotmore and Tattersall; Seeger and Mason; Bisaro; Ahrens et al.). I do not attempt to review these varied mechanisms here but rather focus on comparing the replication of cellular DNA in bacteria and eukaryotes. Unfortunately, except for some studies on DNA polymerases, little is known about the process of DNA replication in Archaea, and therefore inclusion of this kingdom adds little to the discussion.

In organisms from bacteria to the multicellular eukaryotes, DNA replication is intimately coupled to the physiology of the individual cell and to global growth controls imposed on a population of cells, whether in a colony of bacteria or in a subset of cells within a developing organism. These regulatory pathways are different in bacteria and eukaryotes and also vary between different cell types within a single organism. Nevertheless, the actual process of DNA replication has striking parallels between the two groups of organisms. A universal challenge for the replication machinery is that the process must be precise and tightly controlled to ensure faithful transmission of the genetic heritage to the next generation of cells. If this goal is not achieved, disastrous consequences await the organism or its descendants.

In general, the replication of chromosomes in both groups can be separated into multiple stages. First is the recognition of specific DNA sequences within the genome by initiator proteins. This interaction is the primary determinant of the location of the origin of DNA replication. At this location, a pre-replication complex of proteins is established that renders the genome competent for replication. The two important enzyme activities that must be attracted to the origin are a helicase that will eventually unwind the template DNA and a primase that will provide the necessary primer for initiation of DNA synthesis by DNA polymerases. Activation of the pre-replicative complex is a key step that eventually leads to the unwinding of the DNA and DNA synthesis. The establishment of a multiprotein complex called a replisome at the replication fork involves the assembly of the DNA polymerases and accessory proteins, followed by fork movement. Finally, termination of DNA replication occurs prior to separation of the daughter chromosomes. It is in each of these areas that the processes in bacteria and eukaryotes will be compared. This discussion relies heavily on the reviews presented in this volume on the replication mechanisms in eukaryotes for specific examples, and much of the primary literature is cited in these other chapters.

CHROMOSOME STRUCTURE AND REPLICATION

Perhaps the most obvious difference between the two classes of cells is how the genetic material is organized into chromosomes. Bacteria have a circular chromosome that contains one principal origin of DNA replication located at the *OriC* locus in *Escherichia coli*. The rate at which the DNA polymerases move along the *E. coli* chromosome is approximately 100 kb per minute, and thus the entire chromosome can be replicated from a single origin in about 40 minutes (Kornberg and Baker 1992). The genomes of eukaryotes are much larger than the bacterial chromosomes, and the rate of replication fork movement averages about 2 kb per minute, a rate that is prohibitive for even the smallest genome to replicate from a single origin (Fangman and Brewer 1992). Moreover, DNA in eukaryotes is invariably sequestered into multiple chromosomes, which demands more than one origin per genome. In certain artificial chromosomes in *Saccharomyces cerevisiae*, one origin is sufficient to replicate a chromosome of about 100 kb in size, but if the chromosome is larger than this limit, it becomes very unstable with only one origin (Wellinger and Zakian 1989; Dershowitz and Newlon 1993; Newlon, this volume).

It has been known for more than three decades that different regions of individual eukaryotic chromosomes replicate at distinct times during S phase, demonstrating that multiple origins of DNA replication must exist (Hand 1978; Fangman and Brewer 1992; DePamphilis, this volume). Unlike bacteria, eukaryotes initiate chromosome replication in a temporally regulated manner that can change during development (Simon and Cedar, this volume). The existence of multiple origins affords considerable flexibility in the timing of replication.

In addition to the relative timing of replication of different regions of the genome, the total time taken to replicate the entire genome can vary considerably during development of an organism (Blumenthal et al. 1974; Callan 1974). This appears to be due to a change in the frequency of active origins of DNA replication distributed along the chromosome, rather than a change in the rate of replication fork movement. For example, in the rapid replication cycles of nuclei in the early *Drosophila* embryo, origins of DNA replication are located on average 3–7 kb apart, but later during embryogenesis, the frequency of replication origins changes to about 40–100 kb as cells begin to differentiate (Blumenthal et al. 1974). Interestingly, if mammalian cells are arrested for a long period of time at the G_1 to S phase, the frequency along the genome of active origins of DNA replication increases following removal of the block, and the total time taken to pass through S phase is much shorter than the time taken in the absence of the inhibitor (Taylor 1977). This suggests that a factor may accumulate in the nucleus of cells that determines the frequency of active origins of DNA replication. A good candidate protein is the mammalian homolog of the $p65^{cdc18}$ protein from *Schizosaccharomyces pombe* and the related protein Cdc6p from *S. cerevisiae* (Kelly et al. 1993; Liang et al. 1995).

A major difference between the two groups of cells emerges from these observations. The existence of a single origin in the bacterial chromosome fixes the relative order of gene duplication, except perhaps for occasional selective gene amplification. In contrast, eukaryotes have enormous flexibility in the relative time during S phase when specific genes replicate. This permits the replication of specific regions of the genome to be linked to gene expression, a topic discussed in an accompanying chapter (Simon and Cedar, this volume). This flexibility may be an important component of developmental patterning in cells of metazoan origin. Even though the number of active origins in a single bacterial chromosome does not vary, as described below in the section on control of DNA replication, the frequency of active origins within a cell can vary considerably in bacteria due to reinitiation in a single cell-division cycle.

In bacteria that have distinct cell types, developmentally imposed control of the initiation of DNA replication can occur. In *Caulobacter crescentus,* cell division can produce two morphologically different cell types that have different replicative fates (Marczynski et al. 1995). One cell type, called the stalked cell, replicates the entire chromosome after cell division, whereas the other cell type, the swarmer cell, only replicates its DNA after a delay. In the stalked cell, a single chromosomal origin of DNA replication is utilized. The replicator element, *Cori,* has recently been characterized, and in addition to the initiator protein-binding sites that are analogous to the DnaA initiator protein-binding sites, within the *E. coli OriC* replicator, *Cori* contains essential transcriptional promoter elements that are only active in the stalked cell (Marczynski et al. 1995). Thus, cell-type-specific transcription can influence replication.

It has been known for some years that transcription can facilitate initiation of DNA replication from the bacteriophage λ origin as well as the *E. coli OriC* (Kornberg and Baker 1992). The *C. crescentus* case is an example in bacteria where transcription may effect cell-type-specific initiation of DNA replication. Transcription factors in eukaryotic cells can also be intimately involved in the initiation of DNA replication and influence origin firing in a cell-type- and temporal-specific manner (van der Vliet, this volume; for review, see DePamphilis 1993). A recent example in mammalian cells is the locus controlling region (LCR) that regulates the developmentally controlled expression of the human β-globin gene. The LCR is required for the initiation of DNA replication from an origin located near the β-globin gene (Aladjem et al. 1995). These selected examples demonstrate that despite the quite different organization of the chromosomes, and the number of replication origins, bacteria and eukaryotes can utilize similar strategies to control whether or not initiation of DNA replication will occur.

Consideration of the influence of chromosome structure on DNA replication in bacteria and eukaryotes must also take into account the different organization of DNA in the cell. The bacterial chromosome is associated with the cell membrane but otherwise is exposed to the entire intracellular environment. In contrast, eukaryotic DNA replication occurs in a distinct compartment in the cell, affording the separation of proteins that may influence the initiation of DNA replication. Within the nucleus, initiation of eukaryotic DNA replication occurs at pre-replicative complexes that are established prior to the beginning of S phase, and these presumably contain the initiator and other replication proteins (Adachi and Laemmli 1994; Diffley et al. 1994). A similar initiation complex may exist at the membrane-bound bacterial replicator, since DnaA protein from *E. coli* is a lipid-binding protein and is associated with the membrane (Sekimizu and Kornberg 1988; Sekimizu et al. 1988a,b). Thus, in both cell types, initiation may actually occur on a solid-state support, albeit that the supports and environment may be quite different.

REPLICATORS

The mechanism of initiation was first proposed for replication of the bacterial chromosome in the early 1960s (Jacob et al. 1964). In this classic paper, a *cis*-acting DNA element called the replicator was proposed to be a primary determinant of the location in the genome where initiation of DNA replication was to occur. Furthermore, it was posited that initiation of replication of the bacterial genome also required a *trans*-acting factor called the initiator that interacted with the replicator. The replicator can be defined by genetic means, and, indeed, the *E. coli* replicator *OriC* was characterized by the isolation of fragments from the *E. coli* genome that could support autonomous replication of recombinant plasmids (Hirota et al. 1979; Messer et al. 1979; von Meyenburg et al. 1979). These plasmids behaved in a manner similar to the complete chromosome.

Fine-structure analysis by mutagenesis and evolutionary comparison of replicators from five enteric bacteria revealed that the bacterial replicator is highly conserved between species and is composed of multiple essential DNA sequence elements (Fig. 1A). The *E. coli* replicator has four recognition sites for the initiator protein DnaA that are flanked on one side by an A·T-rich sequence and on the other by a series of repeats that can be easily unwound in negatively supercoiled DNA (Kornberg and Baker 1992). These three repeats, or 13-mers, are locally unwound by DnaA once it is bound to *OriC* (Bramhill and Kornberg 1988). This general structure for bacterial replicators is also found in the origins of DNA replication from the lambdoid phages. Interspersed within the *E. coli* replicator are ten recognition sites for the DNA adenine methyltransferase (dam methylase), and methylation of these sites plays an important role in the control of initiation (see below).

The replicators present in eukaryotic viruses have been characterized by mutational analysis. One of the best understood of these is the simian virus 40 (SV40) replicator called the SV40 *ori* (Stillman 1989; Hassell and Brinton, this volume). Like the bacterial replicators, the SV40 *ori* contains multiple binding sites for the SV40 initiator protein, T antigen, and these binding sites are flanked by an A·T-rich sequence on one side and a partial palindromic repeated sequence (Fig. 1B) (Parsons et al.

A. OriC

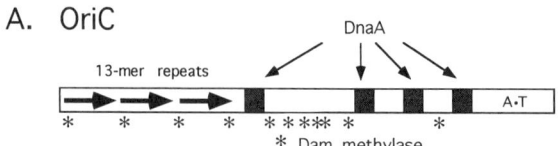

B. SV40

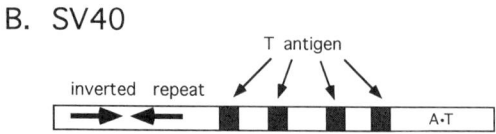

C. S. cerevisiae

```
                                    ORC
ABF1                               ↙  ↘
▓▓▓▓▓▓░░░░▓▓▓░░░░▓▓▓▓▓░░░░▓▓▓▓▓░░░░█████
   B3      B2      B1            A
```

Figure 1 Comparison of the replicators from *E. coli*, SV40, and *S. cerevisiae*. (A) *E. coli* OriC. (B) The SV40 *ori*. (C) *S. cerevisiae* ARS1.

1990). The latter repeat also interacts with T antigen, resulting in the unwinding of 8 bp of DNA. Thus, the SV40 *ori* is remarkably similar in organization to the *E. coli* replicator.

Understanding the nature of replicators in eukaryotic cell chromosomes has been more problematic, particularly in multicellular eukaryotes. Only in the yeasts have replicators been isolated by selection of fragments that permit autonomous replication of plasmids (Campbell and Newlon 1991; DePamphilis; Newlon; both this volume). Attempts to isolate such autonomous replicating sequences (ARSs) in other eukaryotes have not been successful, and thus the replicators from metazoan cells remain poorly defined. In *S. cerevisiae*, a number of replicators have been characterized extensively, and the general picture that has emerged is that the replicators have little similarity to the bacterial and viral replicators (Fig. 1C). The best-characterized replicator is ARS1 from chromosome IV, and it consists of four DNA sequence elements, only one of which (A) is essential (Marahrens and Stillman 1992). The initiator protein for eukaryotes is the origin recognition complex (ORC), and this six-subunit protein binds to the A and B1 elements in an

ATP-dependent manner (Bell and Stillman 1992; Rao and Stillman 1995; Rowley et al. 1995). Another element, which is not essential and is only present in some replicators in the *S. cerevisiae* chromosomes, is a binding site for the transcription factor ABF1 (Marahrens and Stillman 1992). Interestingly, other transcription factors can substitute for ABF1 and activate initiation of DNA replication, similar to the stimulation of initiation of replication at virus origins by transcription factors (DePamphilis 1993; Hassell and Brinton; van der Vliet; both this volume).

In general, the yeast replicators contain DNA sequences that are easily unwound when a plasmid in which they are cloned is in a highly negatively supercoiled state (Natale et al. 1993). A similar phenomenon has been observed with the *E. coli OriC* and the SV40 *ori* (Kowalski and Eddy 1989; Lin and Kowalski 1994), and in these cases, the region that unwinds first corresponds to the region that is unwound by the binding of DnaA or SV40 T antigen, respectively. This suggests that all origins of DNA replication have a tendency to unwind, and the energy for separation of the double helix either comes from torsional strain in the DNA, binding of the initiator protein, or both. For replication of plasmid DNAs containing the phage λ or *OriC* replicators in vitro, torsional strain provided by a balance of gyrase, topoisomerase I, and the HU protein greatly facilitates origin unwinding and initiation of DNA replication (Sekimizu et al. 1988a,b; Mensa-Wilmot et al. 1989; Hwang and Kornberg 1992). Furthermore, transcription of the template DNA in the vicinity of the origin is necessary for the earliest stages of duplex strand unwinding during the initiation of DNA replication (Baker and Kornberg 1988; Learn et al. 1993). Whether transcription per se or the action of transcriptional activation domains present on sequence-specific DNA-binding proteins activates initiation at eukaryotic virus and cell origins remains to be determined.

INITIATORS

As indicated above, initiator proteins bind in a sequence-specific manner to the replicator, and together they function as the primary determinants of the location of the origin of DNA replication. The primary function of the initiator-replicator complex is as a landing pad for other replication proteins such as the helicase and the primase. Initiators can come in many different versions. All function as DNA-binding proteins, and all interact with other proteins that are required for the initiation of DNA replication. Some, such as the phage λ O protein, function only in this capacity (Dodson et al. 1985), whereas others have additional activities.

For example, the DnaA protein from *E. coli* is an ATP/ADP-binding protein with the ability to slowly hydrolyze the ATP to ADP, whereas the SV40 T antigen, the papillomavirus E1 protein, and the herpes simplex virus (HSV) UL9 initiators bind ATP, have a potent intrinsic ATPase activity, and in addition, contain DNA helicase function (Stahl et al. 1986; Sekimizu et al. 1988a,b; Bruckner et al. 1991; see Boroweic; Challberg; Hassell and Brinton; Stenlund; all this volume). One intriguing initiator from bacteriophage P4 is a sequence DNA-binding protein, an ATPase, a DNA helicase, and a DNA primase, and thus fulfills many of the functions required for initiation of DNA replication (Ziegelin et al. 1993).

A common characteristic of initiator proteins is that they often oligomerize upon binding to the DNA. The DnaA and λ O initiator proteins form complexes of 8 or about 20 proteins bound to *OriC* or *ori*λ, respectively, even though there are only four DNA recognition sequences in the replicator (Echols 1986; Kelman and O'Donnell 1994). Similarly, the SV40 T antigen has four recognition sites in the replicator, yet two hexamers of T antigen bind to the SV40 *ori* in an ATP-dependent manner (Dean et al. 1987). The oligomerization probably acts to stabilize the initiator-replicator interactions and to allow efficient interaction with other proteins. In the case of SV40 T antigen, the functional helicase unit is a hexamer. In many respects, the T-antigen hexamer resembles the RuvB ATPase that is involved in branch migration of DNA at a recombination Holliday junction (Parsons et al. 1995).

Oligomerization of an initiator protein can prevent DNA replication. The bacteriophage P1 genome can exist in a lysogenic state in its host bacteria as a plasmid, and initiation of DNA replication requires the *OriP1* replicator and the plasmid-encoded RepA initiator (Baker and Wickner 1992). The RepA protein exists as a stable dimer when not bound to the DNA. The *E. coli* chaperone proteins DnaJ, DnaK, and GrpE combine to break apart the RepA dimer into RepA monomers, thereby allowing RepA to bind to *OriP1* in cooperation with the DnaA protein to form a productive replicator-initiator complex (Baker and Wickner 1992).

In many cases, but not all, the binding of the initiator protein to the replicator causes local unwinding of a region of DNA at the origin of DNA replication. The DnaA protein induces unwinding of the 13-mer elements at *OriC*, and SV40 T antigen causes unwinding of 8 bp in the partial "early" palindrome within the SV40 *ori* (Bramhill and Kornberg 1988; Borowiec et al. 1990; Hassell and Brinton; Brush and Kelly; both this volume). Unwinding of the DNA at the origin of DNA replication is a key step in the pathway leading to initiation of DNA synthesis at the origin. This is facilitated by regions within the origin of DNA replication that have a low helical stability and thus have a tendency to unwind when placed under torsional strain (Kowalski and Eddy 1989; Natale et al. 1993; Lin and Kowalski 1994).

The assembly and disassembly of the multiprotein complex on the replicator and the ability of the complex to function as a landing pad are highly regulated processes. The function of the *E. coli* DnaA complex is regulated by membrane-bound acidic phospholipids (Crooke et al. 1992). The ATP form, but not the ADP form, of DnaA is able to unwind the 13-mers in *OriC*, and membrane-bound acidic phospholipids greatly stimulate the exchange of ADP to ATP bound to DnaA. Exchange to the active ATP·DnaA complex requires the DnaA to be bound to *OriC* because the acidic phospholipids inactivate DNA-free DnaA protein so that it cannot bind *OriC*. These results suggest a link between DnaA activity and the membrane localization of *OriC*.

Phosphorylation of proteins is not known to be a direct mechanism for control of DNA replication in bacteria, but it is a major mechanism for regulation of some eukaryotic virus initiator proteins. For example, the ability of SV40 T antigen to unwind the origin of DNA replication is positively regulated by phosphorylation of a single threonine by a cyclin-dependent protein kinase (Ismail et al. 1993; McVey et al. 1993). On the other hand, the DNA-binding activity is negatively regulated by phosphorylation of two serine residues by casein kinase I, which can be reversed by the action of phosphatase 2A (Cegielska et al. 1994a,b). In eukaryotes, phosphorylation can also control the cellular localization of proteins. For example, phosphorylation by casein kinase II of serine residues that lie adjacent to the nuclear localization signal in SV40 T antigen greatly facilitates the transport of the protein from the cytoplasm to the nucleus. It is easy to imagine that this may play a regulatory role for the initiation of DNA replication. Thus, in eukaryotes, unlike bacteria, control of the intracellular locale of proteins may affect their function. A discussion of the role of the nuclear envelope in the initiation of cellular DNA replication can be found in Laskey and Madine (this volume).

UNWINDING AT THE ORIGIN

Following DNA sequence recognition and establishment of the initiator-protein complex at the origin of DNA replication, the next important events are unwinding of the template DNA and priming DNA replication. Initiator proteins such as DnaA protein (Bramhill and Kornberg 1988) and SV40 T antigen (Borowiec et al. 1990) can locally unwind a region of the origin of DNA replication, and this region of single-stranded DNA is presumed to be the entry site for a DNA helicase and single-stranded DNA-binding proteins such as the *E. coli* DnaB helicase and SSB, respectively. A DNA helicase has to be loaded onto the initiator-DNA complex for more extensive unwinding to occur. SV40 T antigen is an unusual case because it is a DNA helicase as well as an initiator protein; thus, the helicase is automatically located in a prime site for extensive unwinding. Indeed, T antigen unwinds almost the entire SV40 genome in cooperation with the eukaryotic cell single-stranded DNA-binding protein RP-A and a topoisomerase (Borowiec et al. 1990; see Borowiec; Hassell and Brinton; both this volume). In most cases, however, the helicase is a separate polypeptide from the initiator protein and needs to be loaded onto the DNA.

Loading of the DNA helicase is a complicated affair. The DnaB helicase from *E. coli* is used to initiate DNA replication from *OriC* and is also required for replication of the bacteriophage λ genome (Baker and Wickner 1992; Stillman 1994a). In neither case can the DnaB helicase recognize the initiator protein (DnaA or λ O protein) by itself, but needs to be chaperoned by a specialized protein. The *E. coli* DnaC and the phage λ P proteins function as helicase-loading proteins because they both bind to the DnaB helicase and also bind either to the DnaA or λ O proteins, respectively (Fig. 2). These helicase-loading proteins can inhibit the DNA helicase activity, and it is necessary to remove them before the helicase can be activated to unwind the template DNA (Stillman 1994a). For initiation at the phage λ origin (*ori*λ), the *E. coli* heat shock chaperone proteins DnaJ, DnaK, and GrpE cooperate to remove λ P protein and thus activate the DnaB helicase (Baker and Wickner 1992; Stillman 1994a). Here is another example from the prokaryotic world where chaperone proteins pull apart protein complexes that are involved in the initiation of DNA replication. To date, chaperones have not been shown to play a role in disassembling protein complexes in eukaryotes, but little is known about the eukaryotic pre-replicative complex. It is likely that similar assembly and disassembly of multiprotein complexes occur at eukaryotic replicators.

The helicase-loading proteins can also facilitate the assembly of a DNA helicase onto the template DNA in the absence of an origin of DNA replication. For example, the bacteriophage T4 gene 59 protein (gp59) binds to both single-stranded DNA and the phage single-strand-binding protein, gp32 (Barry and Alberts 1994; Morrical et al. 1994).

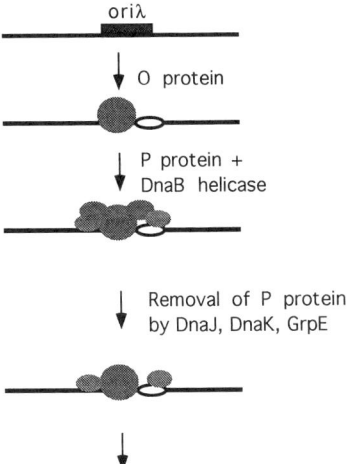

PRIMING AND INITIATION OF DNA REPLICATION
Figure 2 Loading of a DNA helicase. The bacteriophage λ P protein loads the DNA helicase onto the DNA. The λ P protein must be removed before the helicase (DnaB) can function to unwind the DNA prior to DNA replication. This is facilitated by the chaperone proteins DnaJ, DnaK, and GrpE. (Reprinted, with permission, from Stillman 1994a.)

When gp32 coats a single-stranded DNA template, the phage helicase gp41 cannot bind to the DNA. The gp59 protein loads the gp41 helicase onto the gp32-coated DNA, and the helicase can now move rapidly along the template in search of duplex DNA to unwind. This mechanism of helicase loading is almost certainly used during the elongation stages of DNA replication. In a similar manner, SV40 T antigen can bind the RP-A protein, the eukaryotic SSB, facilitating loading of a helicase onto the RP-A-coated single-stranded DNA (Collins and Kelly 1991; Melendy and Stillman 1993).

Although many DNA helicases have been identified in eukaryotes, it is not clear which are required for the initiation of DNA replication (Borowiec, this volume). A good candidate is the helicase B from a murine cell line that harbors a temperature-sensitive helicase which causes a defect in DNA replication at the nonpermissive temperature (Seki et al. 1995). Moreover, although proteins such as Cdc6p that interact with the eukaryotic initiator protein ORC are beginning to be identified (Liang et al. 1995), it is not yet clear whether the same mechanism of helicase loading occurs in eukaryotic cells as occurs in bacteria.

PRIMING DNA SYNTHESIS

Following unwinding of the template DNA surrounding the replicator, priming of DNA synthesis occurs, usually by a DNA primase. As discussed by others in this volume (Brush and Kelly; Salas et al.), priming can occur by a number of different mechanisms, including priming by a nucleotide linked to a protein such as occurs during Φ29 phage and adenovirus DNA replication. For initiation of chromosomal DNA replication in bacteria and almost certainly in eukaryotes, priming involves the assembly of a primase on the DNA. In *E. coli*, the primase exists as a free enzyme that is not always bound to the DNA polymerase (Baker and Wickner 1992; Kornberg and Baker 1992), whereas in eukaryotes, the primase enzyme is always in a complex with DNA polymerase-α (Wang, this volume).

Herein lies a significant difference between prokaryotes and eukaryotes: Bacteria contain only one principal replicative DNA polymerase (DNA polymerase III in *E. coli*) that must function on both the leading and lagging strands at a replication fork, but in eukaryotes, the DNA polymerizing activity is shared by three DNA polymerases, pol-α, pol-δ, and pol-ε (Kornberg and Baker 1992; Wang, this volume). Thus, the pol III enzyme in *E. coli* needs to be free of the primase when replicating the continuously synthesized leading strand, but must cyclically interact with the primase during lagging-strand replication (Hacker and Alberts 1994a,b; Stillman 1994b; Stukenberg et al. 1994). Eukaryotes have devised a DNA polymerase switching mechanism that involves the pol-α:primase complex in priming DNA replication at the origin of DNA replication and for every Okazaki fragment, and then a second DNA polymerase that is not associated with the primase (pol-δ or ε) completing the synthesis (Figs. 3 and 4) (Nethanel and Kaufmann 1990; Tsurimoto et al. 1990; Waga and Stillman 1994; Stillman 1994b).

A key question is how the primase finds either the origin of DNA replication or the site at which Okazaki fragments must begin. Bacterial and phage primases can exist in two different classes: those that are part of the helicase polypeptide and others that exist as proteins separate from the helicase polypeptide (Ilyina et al. 1992). If the primase activity coexists in the same polypeptide as the helicase, then the primase is associated with the replication fork by default. For bacterial chromosome replication and for the replication of some eukaryote virus genomes such as SV40 and herpesvirus, the primase-loading problem has been solved by the primase having an affinity either for the helicase that has already been loaded onto the template DNA or, alternatively, for a structure in the DNA created by the helicase (Baker and Wickner 1992; Marians

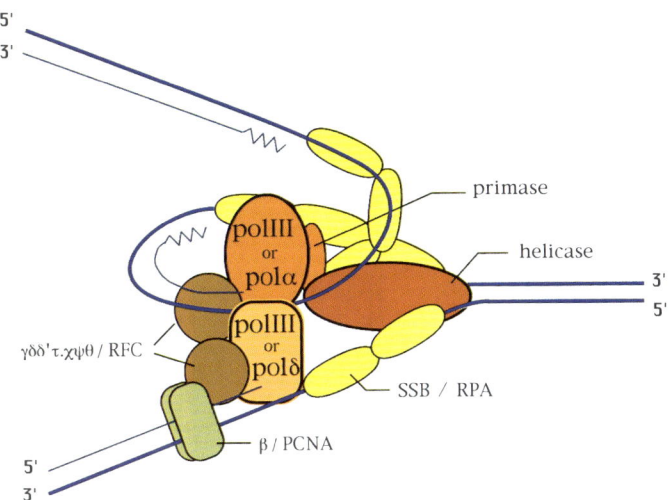

Figure 3 Replication fork proteins. Model for a dimeric DNA polymerase at the replication fork. (Reprinted, with permission, from Stillman 1994b [copyright Cell Press].)

1992; Kelman and O'Donnell 1994; Stillman 1994a; Hassell and Brinton; Challberg; both this volume). For example, the SV40 T antigen binds directly to the pol-α:primase complex and can load the primase onto an RP-A-coated template DNA by direct interaction with RP-A (Collins and Kelly 1991; Melendy and Stillman 1993; Dornreiter et al. 1993). In *E. coli,* the DnaB helicase activates the DnaG primase (Baker and Wickner 1992; Marians 1992). Once located at the priming site, the primase in both prokaryotes and eukaryotes synthesizes a short (8–12 nucleosides) RNA primer that is utilized by the DNA polymerase for synthesis of DNA.

ASSEMBLY OF THE DNA POLYMERASE HOLOENZYME

A common feature of the DNA replication fork in bacteria and eukaryotes is that DNA is synthesized in a semidiscontinuous manner, with one strand, the leading strand, polymerized in a continuous manner and the other lagging strand formed in a discontinuous manner via the synthesis of Okazaki fragments (Brush and Kelly, this volume). The DNA polymerase is loaded onto the template DNA at the site of the primer by a mechanism that is remarkably conserved between prokaryotes and eukaryotes (Table 1) (Kelman and O'Donnell 1994; Stillman 1994b). The process is accomplished by polymerase accessory

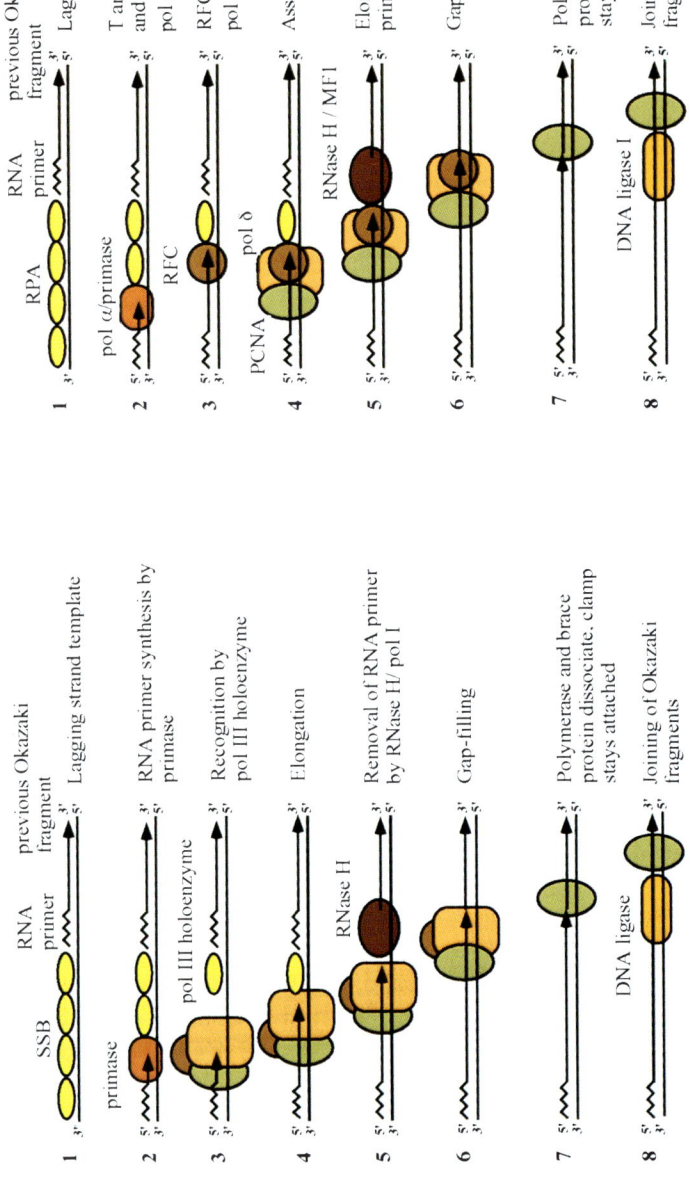

Figure 4 Replication fork proteins. Comparison between the synthesis of Okazaki fragments in bacteria and eukaryotes. (Reprinted, with permission, from Stillman 1994b [copyright Cell Press].)

Table 1 Functions of DNA replication fork proteins

E. coli	Phage T4	SV40/human	Functions
DnaB	41	T antigen	DNA helicase; stimulates priming on ssDNA
DnaC	59	T antigen	allows loading of helicase and primase onto SSB-coated DNA (primosome assembly)
SSB	32	RP-A	single-stranded DNA-binding; stimulates DNA polymerase; facilitates helicase loading
γ-Complex (γδ.δ'χψ)	44/62	RF-C	DNA-dependent ATPase; primer-template binding; stimulates DNA polymerase
τ	43	?	dimerization of holoenzyme
β	45	PCNA	stimulates DNA polymerase; stimulates DNA-dependent ATPase
pol III core (αθε)	43	pol-δ[a]	DNA polymerase 3'→5' exonuclease
DnaG	61	primase pol-α[b]	primase
Ligase	T4 ligase	ligase I	ligation of DNA
DNA pol I	43	FEN-1/MF1	nuclease for removal of RNA primers
RNase H	RNase H	RNase H1	nuclease for removal of RNA primers

(Reprinted, with permission, from Stillman 1994b [copyright Cell Press].)
[a]DNA polymerase-δ has been shown to function in SV40 DNA replication; an essential DNA polymerase, DNA polymerase-ε, has not yet been assigned a specific function in DNA replication.
[b]The human DNA polymerase-α and primase activities function as a multiprotein complex to synthesize RNA/DNA primers. (RP-A) Replication protein A; (RF-C) replication factor C; (PCNA) proliferating cell nuclear antigen; (SSB) single-strand DNA-binding protein.

proteins. The eukaryotic proteins are discussed by Hübscher et al. (this volume). Similar mechanisms and enzymatic activities in these eukaryote proteins and the bacteriophage T4 counterparts were recognized first (Tsurimoto and Stillman 1990; Tsurimoto et al. 1990), and later, some sequence similarities were observed (O'Donnell et al. 1993).

The polymerase accessory proteins from *E. coli*, bacteriophage T4, and eukaryotes include the single-strand-binding protein (SSB, gp32, or RP-A), a polymerase clamp protein (β subunit of pol III, gp45, or PCNA), and a DNA-dependent ATPase that loads the polymerase clamp onto the DNA (γδδ'τχψ, gp44/62, and RF-C). Only in the case of the clamp-loading proteins have significant DNA sequence similarities been observed (O'Donnell et al. 1993), but there is remarkable structural similarity between the polymerase clamp proteins even though there is no primary amino acid sequence similarity (Fig. 5) (Krishna et al. 1994). The β subunit of the pol III enzyme from *E. coli* and the *S. cerevisiae* PCNA have three-dimensional structures that can be almost superimposed one on top of the other. Both form a donut or torus-like shape, and the DNA passes through the hole in the middle of the structure. The β subunit contains 366 amino acids, whereas the PCNA contains only 258 amino acids, approximately two-thirds the size of the bacterial counterpart. Two monomers of the β subunit form the torus structure, whereas three monomers of the smaller PCNA combine to form an identical structure. Each torus ring has an internal sixfold symmetry. The 228-amino acid phage T4 gp45 protein is presumed to also form a trimer.

Interestingly, the carrot plant has two highly related PCNA proteins that are similar in sequence to PCNA from human and yeast; one is 264 amino acids, similar to the "eukaryotic short" form and the other is 365 amino acids, similar to the "bacterial long" form (Kelman and O'Donnell 1995). It has been suggested that these plants have one dimer PCNA and one trimer PCNA that may function at different times during development when DNA replication rates vary. Alternatively, since PCNA also functions in DNA repair (Nichols and Sancar 1992; Shivji et al. 1992), it is possible that one PCNA is involved in DNA repair and the other in DNA replication.

Polymerase III β subunit

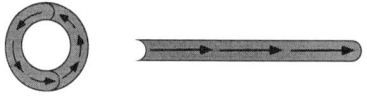

Proliferating Cell Nuclear Antigen (PCNA)

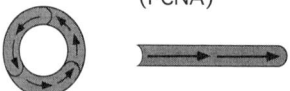

Figure 5 Comparison of the structures of the clamp proteins from bacteria and yeast. Both proteins form a torus or donut-shaped molecule with sixfold symmetry. The *E. coli* pol III β subunit forms a dimer and PCNA forms a trimer. The arrows represent a repeated structural motif in the proteins. The DNA passes through the center of the structure.

The mechanism of polymerase loading is very similar in bacteria and eukaryotes and is discussed in detail elsewhere in this volume (Hübscher et al.) and in other reviews (Kelman and O'Donnell 1994; Stillman 1994b). In both systems, an ATP-dependent clamp-loading protein recognizes a primer-template junction and assembles the polymerase clamp onto the DNA. The single-stranded DNA-binding protein that coats the single-stranded template DNA facilitates the recognition of the primer-template junction by preventing the clamp-loading protein from binding to single-stranded DNA. Once the clamp complex is assembled, a DNA polymerase is assembled onto the primer ready for DNA synthesis.

TOPOLOGY AND DNA REPLICATION

Topoisomerases in both prokaryotes and eukaryotes deal with the topological problems associated with unwinding the double helix during DNA replication and with separating the daughter chromosomes after DNA replication is completed. The different types of topoisomerase have been reviewed recently (Roca 1995), and a discussion of the eukaryotic topoisomerases appears in Hangaard Andersen et al. (this volume). The role of topoisomerases from bacteria and eukaryotes in DNA replication has been well discussed in a recent review (Ullsperger et al. 1995), and only some general points are noted here. Bacterial chromosome topology is controlled by a balance between topoisomerases that relieve torsional strain on the one hand and gyrases that induce negative supercoiling on the other. There is net negative supercoiling in the bacterial chromosome, and these negative supercoils are available to facilitate unwinding of the double helix for processes such as transcription and the initiation of DNA replication. In contrast, eukaryotic chromosomes have more complex topology. Eukaryotic chromosomal DNA has the same supercoil density as found in bacteria, but all the negative supercoils are believed to be constrained in nucleosomes (Ullsperger et al. 1995). Furthermore, the eukaryotic chromosome is folded into complex higher-order structures that the DNA replication apparatus must deal with. Thus, the DNA replication machineries in the two cell types must deal with the very different templates, although as indicated above, the basic replication proteins are quite similar.

Unwinding of the DNA helix during replication fork progression generates positive supercoils ahead of the replication fork in the unreplicated DNA. There are different mechanisms to remove these positive supercoils (Ullsperger et al. 1995). The first is to actively generate negative

Table 2 Topoisomerase activities

	Prokaryotes				Eukaryotes		
	topo I	topo II	topo III	topo IV	topo I	topo II	topo III
Name	omega	gyrase	–	–	–	–	–
Type	1	2	1	2	1	2	1
Genes	*topA*	*gyrA, gyrB*	*topB*	*parC, parE*	*top1*	*top2*	*top3*
(+) Supercoil removal in vivo	no	yes	no	yes?	yes	yes	no
Decatenation	no	yes	no	yes	no	yes	no
Relaxes only (–) super-coils	yes	no	yes	no	no	no	yes

(Adapted from Ullsperger et al. 1995 [copyright Springer-Verlag GmbH].)

supercoils in the unreplicated DNA before the positive supercoils accumulate, a role for the bacterial type II topoisomerase commonly referred to as DNA gyrase (Table 2). The bacterial gyrase can keep up with the generation of positive supercoils ahead of the replication fork by introducing negative supercoils. A second mechanism is to remove directly the positive supercoils in the unreplicated DNA; both the eukaryotic type I and type II topoisomerases can perform this function (Table 2). This mechanism is thought to occur in the early stages of replication from an origin when there is a relatively large amount of unreplicated DNA. In the later stages of DNA replication, it is more difficult to introduce positive supercoils into the unreplicated DNA, and the replicated DNA tends to become interwound or catenated (Ullsperger et al. 1995). Thus, a third mechanism for relieving the positive supercoils generated by replication fork progression is to remove catenated DNA, a job for different types of topoisomerases in bacteria and eukaryotes. The bacterial type I topoisomerase, topo III, binds to single-stranded DNA and thus can operate behind the replication fork to decatenate the replicated DNA. In contrast, the type II topoisomerases, topo II in eukaryotes and type IV in bacteria, can decatenate the replicated duplex DNA and thus segregate completely replicated daughter chromosomes (Ullsperger et al. 1995).

Finally, as pointed out by Ullsperger et al. (1995), if the replication fork contains a dimeric DNA polymerase with physically linked polymerase molecules of the type shown in Figure 3, this creates even more problems with DNA topology because the DNA polymerases are not free to rotate about the template DNA. If the DNA polymerases are connected to each other, as seems likely in bacteria and eukaryotes, then enzymatic mechanisms must exist that deal with the passage of the double helix through the polymerase complex.

FIDELITY AND DNA REPAIR

Both bacteria and eukaryotic cells have mechanisms for repairing DNA (Friedberg et al. 1995), and it is beyond the scope of this discussion to review these varied mechanisms here (for eukaryotes, see Roberts and Kunkel; Friedberg and Wood; both this volume). As discussed by Nasmyth (this volume) and others elsewhere (Li and Deshaies 1993; Murray 1994; Nurse 1994), eukaryotic cells have a monitoring system that links DNA replication with cell-cycle progression, cell growth, and DNA repair. The existence of DNA damage-checkpoint mechanisms in bacteria that temporally delay DNA replication and cell-cycle progression in response to DNA damage is not as clear as has become apparent in eukaryotes, but inhibition of DNA synthesis and recovery are known to occur (Bridges 1995). Although the mechanisms of DNA repair are similar between bacteria and eukaryotes, the mechanisms that control the cell division cycle are very different. Thus, simple parallels between prokaryotes and eukaryotes have not emerged.

CONTROL OF INITIATION

Because the pathways that control the cell division cycle are very different between bacteria and eukaryotes, the mechanisms for controlling the frequency of initiation are also different. Methylation of the bacterial origin of DNA replication by the dam methylase causes the bacterial DNA to become hemi-methylated immediately after it has replicated (Baker 1994). In the *E. coli OriC* replicator, the Dam methylation sequences (5'-GATC-3') are clustered in the 13-mer repeats that unwind in the presence of DnaA and ATP. *OriC* is negatively regulated by a process called sequestration. For some time after the initiation of DNA replication, the hemi-methylated DNA is unable to reinitiate, thereby limiting multiple rounds of DNA replication in a cell. The product of the SeqA gene is required for sequestration, and SeqA interacts genetically with DnaA (Campbell and Kleckner 1990; Baker 1994). Recently, it has been shown that the SeqA protein binds to hemi-methylated DNA from either the *E. coli OriC* 13-mer region or from the bacteriophage P1 origin, *P1oriR* (Brendler et al. 1995). Binding to these replication origins is preferred because these sequences are somehow slow to be fully methylated and therefore remain in the hemi-methylated state for some time. These data suggest that the SeqA protein might prevent initiation of DNA replication by interacting with the newly hemi-methylated replicator, thereby preventing productive binding of the replicator by the initiator DnaA.

Methylation probably does not play a role in the regulation of initiation of DNA replication in eukaryotes. More likely, a multiprotein complex is formed on the DNA at the site of the replicator to produce a "competent" pre-replicative complex. This complex can then be acted on by other regulators of DNA replication that coordinate the initiation of DNA replication with progression through the cell division cycle (see Nasmyth, this volume). In many ways, this type of regulation may be similar to the assembly and disassembly of proteins at the phage λ replicator (Stillman 1994a). This type of regulation may also facilitate the coordination of replication from multiple origins of DNA replication that does not occur in bacteria.

TERMINATION OF DNA REPLICATION

The termination of DNA replication has been well studied in bacteria, including *E. coli* and *Bacillus subtilis* (Baker 1995). Replication of the bacterial chromosome terminates when the two replication forks that diverged from *OriC* converge in a region of genome that is 180° from the origin. There are multiple *cis*-acting sites in the DNA called Ter sites in *E. coli* and imperfect repeats (IR-I and IR-II) in *B. subtilis*. These DNA elements have directionality with respect to the advancing replication fork and interact with a protein called Tus in *E. coli* and RTP in *B. subtilis*. The Ter-Tus protein complex functions as an anti-helicase complex because it blocks the progression of DNA unwinding by a DNA helicase in the absence of DNA replication. The Tus-Ter complex has been shown to block the DnaB helicase (Khatri et al. 1989; Lee et al. 1989) and, interestingly, it can also inhibit the SV40 T-antigen helicase (Bedrosian and Bastia 1991). A more thorough description of termination is given by Bastia and Mohanty (this volume).

Replication termination in eukaryotes is not well studied, but it is clear that specific DNA sequences are not generally required. For example, the SV40 origin of DNA replication has been cloned into many plasmids, and these plasmids replicate very well when they are introduced into primate cells, yet the DNA sequences located in the region opposite the origin are very different (Lai and Nathans 1975). Similarly, replication forks that meet each other following initiation at adjacent origins in eukaryotic chromosomes are believed to terminate when the two forks meet each other, but in a sequence-independent manner.

There are, however, DNA sequences in eukaryotic genomes that cause DNA replication to arrest and occasionally terminate. For example, repeated DNA recognition sites for the Epstein-Barr virus (EBV) nuclear antigen 1 (EBNA1) cause replication forks to pause or arrest completely (Little and Schildkraut 1995). Furthermore, DNA replication fork barriers have been found in the ribosomal DNA repeats in *S. cerevisiae* (Brewer et al. 1992). Thus, it is possible that sequence-specific DNA-binding proteins may affect the passage of the eukaryotic DNA replication fork in much the same way as the Tus-Ter system in *E. coli* and that these sites may play some regulatory role that has yet to be appreciated.

REFERENCES

Adachi, Y. and U.K. Laemmli. 1994. Study of the cell cycle-dependent assembly of the DNA prereplication centres in *Xenopus* egg extracts. *EMBO* 13: 4153–4164.

Baker, T.A. 1994. A new controller in *Escherichia coli*. *Curr. Opin. Genet. Dev.* 4: 945–946.

Aladjem, M.I., M. Groudine, L.L. Brody, E.S. Dieken, R.E.K., Fournier, G.M. Wahl, and E.M. Epner. 1995. Participation of the human β-globin locus control region in initiation of DNA replication. *Science* 270: 815–819.

———. 1995. Replication arrest. *Cell* 80: 521–524.

Baker, T.A. and A. Kornberg. 1988. Transcriptional activation of initiation of replication from the *E. coli* chromosomal origin: An RNA-DNA hybrid near oriC. *Cell* 55: 113–123.

Baker, T.A. and S.H. Wickner. 1992. Genetics and enzymology of DNA replication in *Escherichia coli*. *Annu. Rev. Genet.* 26: 447–477.

Barry, J. and B. Alberts. 1994. Purification and characterization of bacteriophage T4 gene 59 protein. *J. Biol. Chem.* 269: 33049–33062.

Bedrosian, C.L. and D. Bastia. 1991. *Escherichia coli* replication terminator protein impedes simian virus 40 (SV40) DNA replication fork movement and SV40 large tumor antigen helicase activity *in vitro* at a prokaryotic terminus sequence. *Proc. Natl. Acad. Sci.* 88: 2618–2622.

Bell, S.P. and B. Stillman. 1992. ATP-dependent recognition of eukaryotic origins of DNA replication by a multiprotein complex. *Nature* 357: 128–134.

Blumenthal, A.B., H.J. Kriegstein, and D.S. Hogness. 1974. The units of DNA replication in *Drosophila melanogaster* chromosomes. *Cold Spring Harbor Symp. Quant. Biol.* 38: 205–223.

Borowiec, J.A., F.B. Dean, P.A. Bullock, and J. Hurwitz. 1990. Binding and unwinding—How T-antigen engages the SV40 origin of DNA replication. *Cell* 60: 181–184.

Bramhill, D. and A. Kornberg. 1988. A model for initation at origins of DNA replication. *Cell* 54: 915–918.

Brendler, T., A. Abeles, and S. Austin. 1995. A protein that binds to the P1 origin core and the *oriC* 13-mer region in a methylation-specific fashion is the product of the host *seqA* gene. *EMBO J.* 14: 4083–4089.

Brewer, B., D. Lockshon, and W. Fangman. 1992. The arrest of replication forks in the rDNA of yeast occurs independently of transcription. *Cell* **71:** 267-276.

Bridges, B.A. 1995. Are there DNA damage checkpoints in *E. coli*? *BioEssays* **17:** 63-70.

Bruckner, R.C., J.J. Crute, M.S. Dodson, and I.R. Lehman, 1991. The herpes simplex virus origin binding protein: A DNA helicase. *J. Biol. Chem.* **266:** 2669-2674.

Callan, H.G. 1974. DNA replication in the chromosomes of eukaryotes. *Cold Spring Harbor Symp. Quant. Biol.* **38:** 195-203.

Campbell, J.L. and N. Kleckner. 1990. *E coli oriC* and the *dnaA* gene promoter are sequestered from *dam* methyltransferase following the passage of the chromosomal replication fork. *Cell* **62:** 967-979.

Campbell, J.L. and C.S. Newlon. 1991. Chromosomal DNA replication. In *The molecular and cellular biology of the yeast* Saccharomyces: *Genome dynamics, protein synthesis, and energetics* (ed. J.R. Broach), pp. 41-146. Cold Spring Harbor Laboratory Press. Cold Spring Harbor, New York.

Cegielska, A., I. Moarefi, E. Fanning, and D.M. Virshup. 1994a. T-antigen kinase inhibits simian virus 40 DNA replication by phosphorylation of intact T antigen on serines 120 and 123. *J. Virol.* **68:** 269-275.

Cegielska, A., S. Shaffer, R. Derua, J. Goris, and D.M. Virshup. 1994b. Different oligomeric forms of protein phosphatase 2A activate and inhibit simian virus 40 DNA replication. *Mol. Cell. Biol.* **14:** 4616-4623.

Collins, K.L. and T.J. Kelly. 1991. Effects of T antigen and replication protein A on the initiation of DNA synthesis by DNA polymerase α-primase. *Mol. Cell. Biol.* **11:** 2108-2115.

Crooke, E., C. Castuma, and A. Kornberg. 1992. The chromosome origin of *Escherichia coli* stabilizes DnaA protein during rejuvenation by phospholipids. *J. Biol. Chem.* **267:** 16779-16782.

Dean, F.B., M. Dodson, H. Echols, and J. Hurwitz. 1987. ATP-dependent formation of a specialized nucleoprotein structure by simian virus 40 (SV40) large tumor antigen at the SV40 replication origin. *Proc. Natl. Acad. Sci.* **84:** 8981-8985.

DePamphilis, M.L. 1993. How transcription factors regulate origins of DNA replication in eukaryotic cells. *Trends Cell Biol.* **3:** 1161-163.

Dershowitz, A. and C.S. Newlon. 1993. The effect on chromosome stability of deleting replication origins. *Mol. Cell. Biol.* **13:** 391-398.

Diffley, J.F.X., J.H. Cocker, S.J. Dowell, and A. Rowley. 1994. Two steps in the assembly of complexes at yeast replication origins in vivo. *Cell* **78:** 303-316.

Dodson, M., J. Roberts, R. McMacken, and H. Echols. 1985. Specialized nucleoprotein structures at the origin of replication of bacteriophage λ: Complexes with λO protein and with λO, λP, and *Escherichia coli* DnaB proteins. *Proc. Natl. Acad. Sci.* **82:** 4678-4682.

Dornreiter, I., W.C. Copeland, and T.S.-F. Wang. 1993. Initiation of simian virus 40 DNA replication requires the interaction of a specific domain of human DNA polymerase α with large T antigen. *Mol. Cell. Biol.* **13:** 809-820.

Echols, H. 1986. Bacteriophage W development: Temporal switches and the choice of lysis or lysogeny. *Trends Genet.* **2:** 26-30.

Fangman, W. and B. Brewer. 1992. A question of time: Replication origins of eukaryotic chromosomes. *Cell* **71:** 363-366.

Friedberg, E.C., G.C. Walker, and W. Siede. 1995. *DNA repair and mutagenesis.* ASM

Press, Washington, D.C.

Gesteland, R.F. and J.F. Atkins, eds. 1993. *The RNA world.* Cold Spring Harbor Laboratory Press, Cold Spring Harbor, New York.

Hacker, K.J. and B.M. Alberts. 1994a. The rapid dissociation of the T4 DNA polymerase holoenzyme when stopped by a DNA hairpin helix. *J. Biol. Chem.* **269:** 24221–24228.

———. 1994b. The slow dissociation of the T4 DNA polymerase holoenzyme when stalled by nucleotide omission. *J. Biol. Chem.* **269:** 24209–24220.

Hand, R. 1978. Eucaryotic DNA: Organization of the genome for replication. *Cell* **15:** 317–325.

Hirota, Y., S. Yasuda, M. Yamada, A. Nishimura, K. Sugimoto, H. Sugisaki, A. Oka, and M. Takanami. 1979. Structural and functional properties of the *Escherichia coli* origin of DNA replication. *Cold Spring Harbor Symp. Quant. Biol.* **43:** 129–138.

Hwang, D. and A. Kornberg. 1992. Opening of the replication origin of *Escherichia coli* by DnaA protein with protein HU or IHF. *J. Biol. Chem.* **267:** 23083–23086.

Ilyina, T.V., A.E. Gorbalenya, and E.V. Koonin 1992. Organization and evolution of eubacterial and bacteriophage primase-helicase systems. *J. Mol. Evol.* **34:** 351–357.

Ismail, F., D. Small, I. Gilbert, M. Hopfner, S. Randall, C. Schneider, A. Russo, U. Ramsperger, A. Arthur, H. Stahl, T. Kelly, and E. Fanning. 1993. Mutation of the cyclin-dependent kinase phosphorylation site in simian virus 40 (SV40) large T antigen specifically blocks SV40 origin DNA unwinding. *J. Virol.* **67:** 4992–5002.

Jacob, F., S. Brenner, and F. Cuzin. 1964. On the regulation of DNA replication in bacteria. *Cold Spring Harbor Symp. Quant. Biol.* **28:** 329–348.

Kelly, T.J., G.S. Martin, S.L. Forsburg, R.J. Stephen, A. Russo, and P. Nurse. 1993. The fission yeast *cdc18+* gene product couples S. phase to START and mitosis. *Cell* **74:** 371–382.

Kelman, Z. and M. O'Donnell. 1994. DNA replication: Enzymology and mechanisms. *Curr. Opin. Genet. Dev.* **4:** 185–195.

———. 1995. Embryonic PCNA: A missing link? *Curr. Biol.* **5:** 814.

Khatri, G.S., T. MacAllister, P.R. Sista, and D. Bastia. 1989. The replication terminator protein of *E. coli* is a DNA sequence-specific contra-helicase. *Cell* **59:** 667–674.

Kornberg, A. and T.A. Baker. 1992. *DNA replication*, 2nd edition. W.H. Freeman, New York.

Kowalski, D. and M.J. Eddy. 1989. The DNA unwinding element: A novel, *cis*-acting component that facilitates opening of the *Escherichia coli* replication origin. *EMBO J.* **8:** 4335–4344.

Krishna, T.S.R., X.-P. Kong, S. Gary, P.M. Burgers, and J. Kuriyan. 1994. Crystal structure of the eukaryotic DNA polymerase processivity factor PCNA. *Cell* **79:** 1233–1243.

Lai, C.J. and D. Nathans. 1975. Non-specific termination of simian virus 40 DNA replication. *J. Mol. Biol.* **97:** 113–118.

Learn, B., A.W. Karzai, and R. McMacken. 1993. Transcription stimulates the establishment of bidirectional λ DNA replication in vitro. *Cold Spring Harbor Symp. Quant. Biol.* **58:** 389–402.

Lee, E.H., A. Kornberg, M. Hidaka, T. Kobayashi, and T. Horiuchi. 1989. *Escherichia coli* replication termination protein impedes the action of helicases. *Proc. Natl. Acad. Sci.* **86:** 9104–9108.

Li, J.J. and R.J. Deshaies. 1993. Exercising self-restraint: Discouraging illicit acts of S and M in eukaryotes. *Cell* **74:** 223–226.

Liang, C., M. Weinreich, and B. Stillman. 1995. ORC and Cdc6p interact and determine the frequency of initiation of DNA replication in the genome. *Cell* **81:** 667-676.

Lin, S. and D. Kowalski. 1994. DNA helical instability facilitates initiation at the SV40 replication origin. *J. Mol. Biol.* **235:** 496-507.

Little, R.D. and C.L. Schildkraut. 1995. Initiation of latent DNA replication in the Epstein-Barr virus genome can occur at sites other than the genetically defined origin. *Mol. Cell. Biol.* **15:** 2893-2903.

Marahrens, Y. and B. Stillman. 1992. A yeast chromosomal origin of DNA replication defined by multiple functional elements. *Science* **255:** 817-823.

Marczynski, G.T., K. Lentine, and L. Shapiro. 1995. A developmentally regulated chromosomal origin of replication uses essential transcription elements. *Genes Dev,* **9:** 1543-1557.

Marians, K.J. 1992. Prokaryotic DNA replication. *Annu. Rev. Biochem.* **61:** 673-719.

McVey, D., S. Ray, Y. Gluzman, L. Berger, A.G. Wildeman, D.R. Marshak, and P. Tegtmeyer. 1993. cdc2 phosphorylation of threonine 124 activates the origin-unwinding functions of simian virus 40 T antigen. *J. Virol.* **67:** 5206-5215.

Melendy, T. and B. Stillman. 1993. An interaction between replication protein A and SV40 T antigen appears essential for primosome assembly during SV40 DNA replication. *J. Biol. Chem.* **268:** 3389-3395.

Mensa-Wilmot, K., K. Carroll, and R. McMacken. 1989. Transcriptional activation of bacteriophage λ DNA replication *in vitro*: Regulatory role of histone-like protein HU of *Escherichia coli*. *EMBO J.* **8:** 2393-2402.

Messer, W., M. Meijer, H.E.N. Bergmans, F.G. Hanse, K. von Meyenburg, E. Beck, and H. Schaller. 1979. Origin of replication, *oriC*, of the *Escherichia coli* K12 chromosome: Nucleotide sequence. *Cold Spring Harbor Symp. Quant. Biol.* **43:** 139-145.

Morrical, S.W., K. Hempstead, and M.D. Morrical. 1994. The gene 59 protein of bacteriophage T4 modulates the intrinsic and single-stranded DNA-stimulated ATPase activities of gene 41 protein, the T4 replicative DNA helicase. *J. Biol. Chem.* **269:** 33069-33081.

Murray, A. 1994. Cell cycle checkpoints. *Curr. Opin. in Cell Biol.* **6:** 872-876.

Natale, D.A., R.M. Umek, and D. Kowalski. 1993. Ease of DNA unwinding is a conserved property of yeast replication origins. *Nucleic Acids Res.* **21:** 555-560.

Nethanel, T. and G. Kaufmann. 1990. Two DNA polymerases may be required for synthesis of the lagging DNA strand of simian virus 40. *J. Virol.* **64:** 5912-5918.

Nichols, A.F. and A. Sancar. 1992. Purification of PCNA as a nucleotide excision repair protein. *Nucleic Acids Res.* **20:** 2441-2446.

Nurse, P. 1994. Ordering S phase and M phase in the cell cycle. *Cell* **79:** 547-550.

O'Donnell, M., R. Onrust, F.B. Dean, M. Chen, and J. Hurwitz. 1993. Homology in accessory proteins of replicative polymerase—*E. coli* to humans. *Nucleic Acids Res.* **21:** 1-3.

Parsons, C.A., A. Stasiak, R.J. Bennett, and S.C. West. 1995. Structure of a multisubunit complex that promotes DNA branch migration. *Nature* **374:** 375-378.

Parsons, R., M.E. Anderson, and P. Tegtmeyer. 1990. Three domains in the simian virus 40 core origin orchestrate the binding, melting, and DNA helicase activities of T antigen. *J. Virol.* **64:** 509-518.

Rao, H. and B. Stillman. 1995. The origin recognition complex interacts with a bipartitie DNA binding site within yeast replicators. *Proc. Natl. Acad. Sci.* **92:** 2224-2228.

Roca, J. 1995. The mechanism of DNA topoisomerases. *Trends Biochem. Sci.* **20:** 156–160.

Rowley, A., J.H. Cocker, J. Harwood, and J.F.X. Diffley. 1995. Initiation complex assembly at budding yeast replication origins begins with the recognition of a bipartitie sequence by limiting amounts of the initiator, ORC. *EMBO J.* **14:** 2631–2641.

Seki, M., T. Kohda, T. Yano, S. Tada, J. Yanagisawa, T. Eki, M. Ui, and T. Enomoto. 1995. Characterization of DNA synthesis and DNA-dependent ATPase activity at a restrictive temperature in temperature-sensitive tsFT848 cells with thermolabile DNA helicase B. *Mol. Cell. Biol.* **15:** 165–172.

Sekimizu, K. and A. Kornberg. 1988. Cardiolipin activation of dna A protein, the initiation protein of replication in *Escherichia coli*. *J. Biol. Chem.* **263:** 7131–7135.

Sekimizu, K., D. Bramhill, and A. Kornberg. 1988a. Sequential early stages in the in vitro initiation of replication at the origin of the *Escherichia coli* chromosome. *J. Biol. Chem.* **263:** 7124–7130.

Sekimizu, K., B.Y. Yung, and A. Kornberg. 1988b. The dna A protein of *Escherichia coli*: Abundance, improved purification and membrane binding. *J. Biol. Chem.* **263:** 7136–7140.

Shivji, M.K.K., M.K. Kenny, and R.D. Wood. 1992. Proliferating cell nuclear antigen is required for DNA excision repair. *Cell* **69:** 367–374.

Stahl, H., P. Droge, and R. Knippers. 1986. DNA helicase activity of SV40 large tumor antigen. *EMBO J.* **5:** 1939–1944.

Stillman, B. 1989. Initiation of eukaryotic DNA replication in vitro. *Annu. Rev. Cell. Biol.* **5:** 197–245.

———. 1994a. Initiation of chromosomal DNA replication in eukaryotes. Lessons from lambda. *J. Biol. Chem.* **269:** 7047–7050.

———. 1994b. Smart machines at the DNA replication fork. *Cell* **78:** 725–728.

Stukenberg, P.T., J. Turner, and M. O'Donell. 1994. An explanation for lagging strand replication: Polymerase hopping among DNA sliding clamps. *Cell* **78:** 877–887.

Taylor, J.H. 1977. Increase in DNA replication sites in cells held at the beginning of S phase. *Chromosoma* **62:** 291–300.

Tsurimoto, T. and B. Stillman. 1990. Functions of replication factor C and proliferating cell nuclear antigen: Functional similarity of DNA polymerase accessory proteins from human cells and bacteriophage T4. *Proc. Natl. Acad. Sci.* **87:** 1023–1027.

Tsurimoto, T., T. Melendy, and B. Stillman. 1990. Sequential initiation of lagging and leading strand synthesis by two DNA polymerase complexes at the SV40 DNA replication origin. *Nature* **346:** 534–539.

Ullsperger, C.J., A.V. Vologodskii, and N.R. Cozzarelli. 1995. Unlinking of DNA by topoisomerases during DNA replication. *Nucleic Acids Mol. Biol.* **9:** 115–142.

von Meyenburg, K., F.G. Hansen, E. Riise, H.E.N. Bergmans, M. Meijer, and W. Messer. 1979. Origin of replication, *oriC*, of the *Escherichia coli* K12 chromosome: Genetic mapping and minichromosome replication. *Cold Spring Harbor Symp. Quant. Biol.* **43:** 121–128.

Waga, S. and B. Stillman. 1994. Anatomy of a DNA replication fork revealed by reconstitution of SV40 DNA replication in vitro. *Nature* **369:** 207–212.

Wellinger, R.J. and V.A. Zakian. 1989. Lack of positional requirements for autonomously replicating sequence elements on artificial yeast chromosomes. *Proc. Natl. Acad. Sci.* **86:** 973–977.

Woese, C.R. and N.R. Pace. 1993. Probing RNA structure, function, and history by comparative analysis. In *The RNA world* (ed. R.F. Gesteland and J.F. Atkins), pp. 91–117. Cold Spring Harbor Laboratory Press, Cold Spring Harbor, New York.

Ziegelin, G., E. Scherzinger, R. Lurz, and E. Lanka. 1993. Phage P4 α protein is multifunctional with origin recognition, helicase and primase activities. *EMBO J.* **12:** 3703.

15
DNA Replication in Eukaryotic Cells: 1996 to 1998

Melvin L. DePamphilis
National Institute of Child Health and Human Development
National Institutes of Health
Bethesda, MD 20892-2753

The preceding chapters summarize a wealth of information and ideas on DNA replication, DNA repair, and chromatin assembly in eukaryotic cells that was available in 1995–1996. Although this information remains as relevant today as it was then, a number of important advances have occurred, particularly in the area of initiation of DNA replication. Therefore, I have attempted to summarize some of the notable developments in the study of DNA replication that have occurred since *DNA Replication in Eukaryotic Cells* went to press. This is not intended as a critical review, but simply as a guide to the literature that appeared from 1996 to 1998. Wherever possible, the bibliography has been restricted to recent reviews and publications in which the reader will find a more detailed discussion of the subject along with more extensive literature citations. I have tried to avoid repeating what has already been discussed in the preceding chapters, while at the same time organizing the information in a way that will be understandable to the average reader.

- **INITIATION OF DNA REPLICATION IN EUKARYOTIC GENOMES**

Initiation of DNA replication in most, if not all, DNA genomes occurs in three steps. First, one or more "origin recognition proteins" bind to specific *cis*-acting DNA sequences referred to as the "origin of replication." Replication origins are defined either as genetically required sequences that bind initiation proteins and exhibit autonomously replicating sequence (ARS) activity when transferred to a foreign DNA molecule (the "replicator") or as specific sites where DNA replication begins ("initiation sites"). So far, both definitions map to the same locus ("replication origin") in all bacteria, bacterial plasmids, bacteriophage, viral, yeast, and protozoan genomes that have been extensively characterized.

Concepts in Eukaryotic DNA Replication
© 1999 Cold Spring Harbor Laboratory Press 0-87969-557-9/99

Second, DNA unwinding begins at an easily unwound site within the replicator. In some cases, this step is carried out by a helicase activity inherent in the origin recognition protein (e.g., papovavirus T antigen, papillomavirus E1 protein, and adenovirus preterminal protein/DNA polymerase). In other cases, the origin recognition protein may contribute to the unwinding process by distorting the DNA, but it must recruit other proteins to complete the job (e.g., Epstein-Barr virus EBNA-1 protein, yeast origin recognition complex, bacteriophage λ O protein, and *Escherichia coli* DnaA protein). Finally, DNA synthesis is initiated on one or both strands of the origin using either an RNA primer (papovaviruses, cellular genomes), a protein-dNTP primer (adenoviruses, φ29 bacteriophage), a DNA primer (parvoviruses, geminiviruses), or a cellular RNA molecule (mitochondria, retroviruses). Replication is generally bidirectional, but in some cases, it can be unidirectional (e.g., φ29 bacteriophages, adenoviruses, parvoviruses, and geminiviruses).

A unique feature of eukaryotic cell genomes is that they replicate once and only once per cell division cycle (the S phase), although specific regions may undergo endoreduplication (i.e., gene amplification) in specialized cells during animal development. In contrast, viral genomes (with the apparent exception of latent Epstein-Barr virus) can replicate multiple times during a single S phase or, in the case of mitochondrial genomes, throughout the cell cycle. During the past 3 years, significant advances have been made in understanding the mechanism by which eukaryotic cell genomes initiate DNA replication.

Initiation Proteins

It now seems clear that most, if not all, of the proteins used to initiate DNA replication in yeast also are used to initiate DNA replication in the chromosomes of many, perhaps all, eukaryotes. Homologs for most of the proteins required to initiate DNA replication in the budding yeast, *Saccharomyces cerevisiae*, have been identified in other yeast, fungi, frogs, flies, or mammals. These include three of the six origin recognition proteins (ORC1, 2, and 4), proteins encoded by cell division cycle (Cdc) genes *6* and *45*, six minichromosome maintenance (Mcm) proteins, and the Cdc7 protein kinase and its cofactor Dbf4 (Dutta and Bell 1997; Rowles and Blow 1997; Sato et al. 1997; Hua and Newport 1998; Saha et al. 1998). In addition, several of these homologs have been shown to be required for DNA replication. In *Drosophila*, the *ORC2* gene is required for chorion gene amplification and animal development (Landis et al. 1997). In *Xenopus*, immunodepletion of ORC, Cdc6, or Mcm proteins from an egg extract prevents initiation of DNA replication in sperm chromatin, whereas replacement of the depleted protein restores replication activity (Rowles and Blow 1997). In human cells, injection of antibodies against Mcm (Todorov et al. 1994), Cdc6 (Yan et al. 1998), or Dbf4 (H. Masai, pers. comm.) proteins prevents cells from entering S phase. Thus, the mechanism for initiation of eukaryotic DNA replication appears to be highly conserved.

In fact, biochemical fractionation and immunodepletion studies have revealed that assembly and activation of prereplication complexes (pre-RCs) in *Xenopus* egg extract, a system that can initiate DNA replication in vitro, appear to follow the same sequential pathway observed in *S. cerevisiae*, a system that has been defined by genetic analysis (Fig. 1) (for review, see Dutta and Bell 1997; Rowles and Blow 1997; Newlon 1997; Hua and Newport 1998). ORC binds to chromatin first, followed by Cdc6 and then the six Mcm proteins to produce a pre-RC. In yeast, ORC requires ATP in order to bind specifically to yeast replication origins. The ATP-binding motif in ORC1 is conserved in metazoan homologs, but the sequence specificity of ORC in metazoa has not yet been determined. Once ORC is bound to DNA in yeast, it appears to remain bound throughout the cell cycle, but in *Xenopus* and mammals, ORC appears to be absent from chromatin during mitosis and reassociates with chromatin during the G_1 phase (Coleman et al. 1996; Romanowski et al. 1996; Abdurashidova et al. 1998; Hua and Newport 1998). In *Xenopus*, binding of Mcm proteins to chromatin requires replication licensing factor B (RLF-B), a step that is sensitive to protein kinase inhibitors such as 6-dimethylaminopurine (6-DMAP) (Rowles and Blow 1997). In yeast, Cdc6 is then replaced by Cdc45 through the action of the cyclin-dependent protein kinase Cdc28/Clb5, 6 (also called Cdk1/Clb5, 6) to form a preinitiation complex in which Cdc45 physically interacts with chromatin-bound Mcm proteins (Zou and Stillman 1998). Presumably, metazoan Cdc45 has the same role. DNA replication is then initiated by the action of the protein kinase Cdc7/Dbf4. Cdc7 activates early and late origins separately, with late origins requiring Cdc7 later in S phase to permit replication initiation (Bousset and Diffley 1998; Donaldson et al. 1998a). Activation of pre-RCs does not require the presence of either ORC or Cdc6 proteins, confining their roles to pre-RC assembly factors (Donovan et al. 1997; Hua and Newport 1998). The requirement for at least two protein kinase activities to initiate DNA replication accounts for its sensitivity to p21, a specific inhibitor of cyclin-dependent protein kinases during G_1 and S phases, and to 6-DMAP and related inhibitors. In yeast, Cdc45 and some Mcm proteins that initially associate with chromatin at

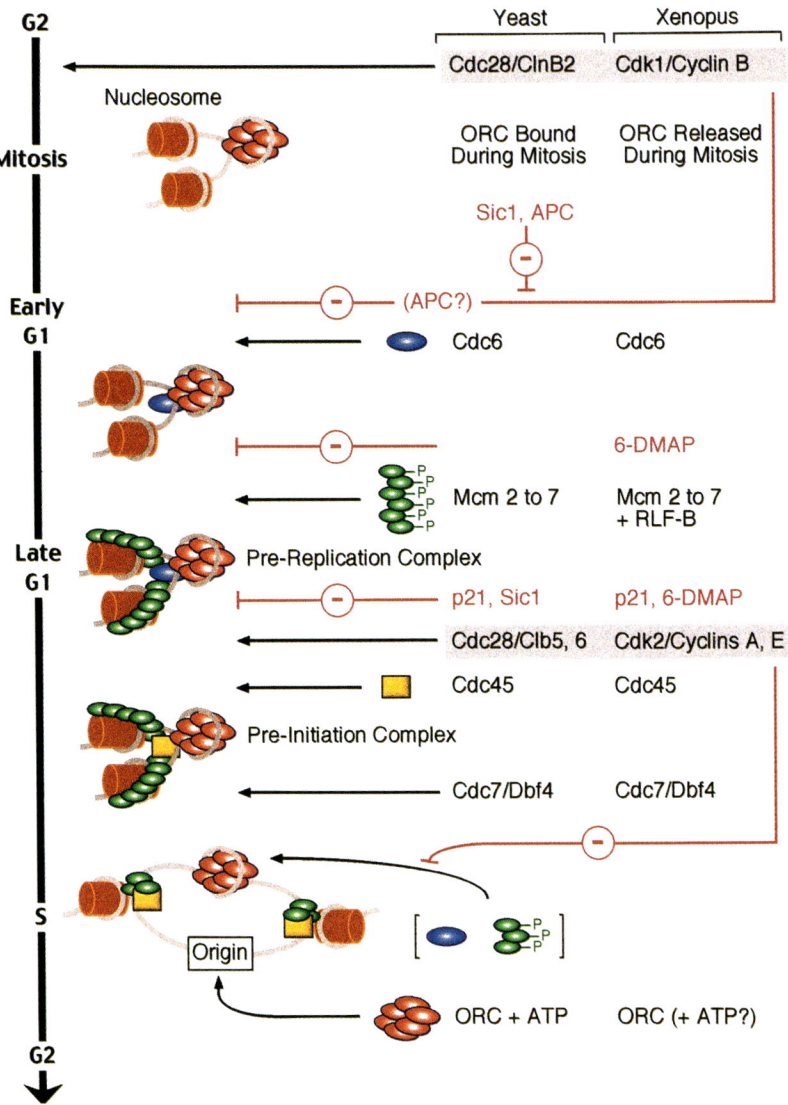

Figure 1 Assembly and activation of prereplication complexes in the budding yeast *S. cerevisiae* and in eggs from the frog *Xenopus*. See text for description. Nomenclature is that of *S. cerevisiae*. ORC is the origin recognition complex. APC is the yeast "anaphase promoting complex." Cdc are cell division cycle genes. Cdc28 is also called Cdk1. 6-DMAP is the general protein kinase inhibitor 6-dimethylaminopurine.

ORC/Cdc6 sites migrate into the surrounding chromatin (Aparicio et al. 1997). Since a mammalian Mcm4/6/7 complex exhibits DNA helicase activity (Ishimi 1997), Mcm proteins may be involved in DNA unwinding at replication forks.

Regulatory Mechanisms

Several features of the initiation mechanism contribute to the general observation that initiation of DNA replication in eukaryotes is limited to once per origin per S phase. First, Cdk1/Cyclin B, the protein kinase required for mitosis in yeast and in metazoa, also prevents assembly of pre-RCs, thus ensuring that DNA replication cannot begin until mitosis is complete (see Chapter 11 and Dutta and Bell 1997). Therefore, proteins that specifically inhibit Cdk1, such as *S. cerevisiae* Sic1 and *Schizosaccharomyces pombe* Rum1, can promote assembly of pre-RCs as well as prevent their activation. They are present at the beginning of G_1 phase and are targeted for phosphorylation by G_1 cyclin-dependent kinase and then degraded via the ubiquitination pathway to allow entry into S phase (Verma et al. 1997). The anaphase-promoting complex (APC), which is essential to separate sister chromatids and to destroy many different proteins, including M-phase cyclins, could also relieve inhibition of pre-RC assembly by cyclin-dependent protein kinases by degrading ClnB (Irniger and Nasmyth 1997). Surprisingly, a functional APC appears to be required to prevent reinitiation during G_2 and M phases (Heichman and Roberts 1998), although this conclusion is disputed (Pichler et al. 1997). Perhaps the APC also degrades a factor required for pre-RC assembly or activation.

The need for additional regulatory mechanisms is demonstrated by the fact that a mutation in yeast Cdc6 that prevents release of chromatin-bound Mcm proteins during S phase also results in reinitiation of DNA replication in the absence of cell division (Liang and Stillman 1997). Therefore, reassociation of Mcm proteins with replication origins must be prevented after initiation of DNA replication. This is accomplished in four ways. First, at the onset of S phase, and perhaps prior to DNA replication, Cdc6 in both yeast and *Xenopus* is lost from chromosomes and does not reappear until early G_1 phase (Drury et al. 1997; Jallepalli and Kelly 1997; Hua and Newport 1998; Zou and Stillman 1998). However, although Cdc6 protein is rapidly degraded in yeast, it appears to be stable in metazoa, suggesting that additional mechanisms are necessary to prevent reassembly of pre-RCs.

Second, in yeast (Piatti et al. 1996), frogs (Hua et al. 1997; Mahbubani et al. 1997; Coverley et al. 1998), and flies (Su and O'Farrell 1998), the same cyclin-dependent protein kinase that activates pre-RCs at the beginning of S phase (Cdk2/cyclin A, E in metazoa and Cdk1/Clb5, 6 in yeast) also inhibits association of Mcm proteins with ORC chromatin complexes to prevent reformation of pre-RCs during S phase. They do not prevent binding of ORC to chromatin. When these protein kinases are inhibited, reinitiation of DNA replication occurs without passing through mitosis. This mechanism would explain why stable expression of Cdc6 protein in yeast does not induce rereplication or assembly of pre-RCs after S phase (Drury et al. 1997).

Third, Mcm proteins are generally associated with chromatin during G_1 phase and released from chromatin during S phase (Dutta and Bell 1997). Since they are absent from active replication sites, only a small fraction of Mcm proteins can travel with replication forks, or Mcm proteins at replication forks do not remain for extensive periods of time. The Mcm proteins that are released form aggregates and are phosphorylated, and genetic studies suggest that phosphorylation by Cdc7/Dbf4 may facilitate their release. Perhaps additional phosphorylation by Cdks prevents them from reassociating with chromatin during S phase.

Although this mechanism may be adequate in yeast where Mcm proteins are released into the cytoplasm during S phase, a fourth mechanism appears to be required in metazoa where Mcm proteins remain within nuclei during S phase (Rowles and Blow 1997). This is provided by the disappearance of RLF-B after S phase begins and the observation that new RLF-B cannot reenter nuclei unless they are permeabilized. Thus, assembly of Mcm proteins onto pre-RCs in metazoa would be delayed until after mitosis.

Replication Origins in Simple Genomes

It is evident from comparison of bacteriophage, bacterial plasmids, and bacterial cells that the essential features of replication origins that function in prokaryotic cells are conserved (see Chapter 14 and Kornberg and Baker 1992). Comparison of viruses that replicate in mammalian nuclei (e.g., simian virus 40 [SV40]) with yeast (e.g., *S. cerevisiae*) leads to the same conclusion in eukaryotes (see Chapters 2, 3, and 14). Replication origins in animal virus genomes generally consist of a core component that contains an origin recognition element that binds a specific origin recognition protein (e.g., SV40 T antigen), an easily unwound DNA sequence (the DNA-unwinding element [DUE]) that is the initial target for a DNA helicase (e.g., SV40 T-antigen hexamer), and an A:T-rich element with thymines on one strand and adenines on the other (Fig. 2). A:T elements are part of the origin recognition element or they facilitate DNA unwinding. The origin of bidirectional DNA replication (OBR) is the site where the two replication forks are born and is identified by the transition from continuous to discontinuous DNA synthesis that occurs on each template. Most origins contain auxiliary components that are not required, but that facilitate replication when the ratio of initiation proteins to DNA is low. Auxiliary components bind specific transcription factors that can facilitate either binding of ORC or DNA unwinding. Viral origins impart ARS activity to plasmid DNA when it is provided with the cognate origin recognition protein and appropriate replication factors. However, viral origins differ from yeast origins in three principal ways: Viral origins initiate replication many times per cell cycle, viral origins exhibit a rigid modular anatomy in which the sequence elements require a specific spacing and orientation with respect to one another, and viral origins function independently of their DNA context.

The most well-characterized eukaryotic replication origins are found in the budding yeast, *S. cerevisiae* (diagrammed in Fig. 2; reviewed in Chapter 2 and Marahrens and Stillman 1996). Replication begins at specific DNA sequences consisting of 100–150 bp that exhibit ARS activity. The core component consists of an ORC-binding site composed of elements A (an A:T-rich element corresponding to the 11-bp ARS consensus sequence) and B1, as well as a DUE (~50–100 bp) that usually encompasses a genetically identifiable element, B2. The B2 element exhibits some of the characteristics of a DUE, and a DUE that contains a B2 element can be replaced by a DUE from an origin that does not (Lin and Kowalski 1997). Coordinate action of five of the six ORC subunits is required for origin-specific DNA binding in which ORC interacts preferentially with one strand (Lee and Bell 1997). ORC protects approximately 50 bp but makes critical contacts with a region of approximately 30 bp. Some origins also contain a binding site for transcription factor Abf-1 (~22 bp). The OBR (~18 bp) overlaps the B1 element and therefore lies adjacent to the ORC-binding site and the DUE (Bielinsky and Gerbi 1998). Both the size of this OBR and its location relative to origin recognition proteins are remarkably similar to the situation in SV40 and polyomavirus (Fig. 2) (DePamphilis et al. 1988).

Despite the fact that only approximately 15% of the sequences in *S. cerevisiae* origins are shared in common (parts of elements A and B1), they exhibit a flexible modular anatomy in which homologous elements from different origins are interchangeable. Furthermore, the same elements identified in plasmid ARS assays also function as elements of replication origins in the natural context of a yeast chromosome (Huang and Kowalski 1996). Yeast origins vary considerably in the frequency at which they are activated during cell proliferation and in the temporal order they are activated during S phase (Friedman et al. 1996; Yamashita et al. 1997). Sensitivity to their DNA (chromatin?) context most likely accounts for the facts that not all ARS elements function as replication origins in yeast chromosomes and that some origins can bind ORC and Cdc6, but still not initiate replication (Santocanale and Diffley 1996). Sequence context is critical in determining the temporal order of origin activation (Friedman et al. 1996), and this order is determined between mitosis and Start in the subsequent G_1 phase (Raghuraman et al. 1997) through the action of Cdk1 (Donaldson et al. 1998b). Either Cdk1/Clb5 or Cdk1/Clb6 can activate early origins, but only Cdk1/Clb5 can activate late origins.

Figure 2 Replication origins that function in the nuclei of eukaryotic cells. (*Red*) Sequence elements required under all conditions (core components); (*yellow*) sequence elements (transcription-factor-binding sites) that facilitate replication under some conditions (auxiliary components). The SV40 core origin consists of an origin recognition element (ORE) that is required for T-antigen binding, an easily unwound sequence (DUE) where DNA unwinding begins, and an A:T-rich element containing adenines on one strand and thymines on the other. The site from which replication forks move out in opposite directions is called the origin of bidirectional replication. Auxiliary elements bind a T-antigen dimer (aux-1) and transcription factor Sp1 (aux-2). The spacing and orientation of these elements are critical. The budding yeast *S. cerevisiae* origins consist of two core elements (A and B1) that constitute the binding site for the six protein ORC, and a DUE that generally contains a genetically defined B2 element. Some origins contain an auxiliary element (B3) that binds transcription factor Abf-1. Each element is interchangeable with homologous elements from other *S. cerevisiae* origins. The fission yeast *S. pombe* origins consist of at least one autonomously replicating sequence (ARS) element that is much larger than those in *S. cerevisiae*. In some cases, multiple ARS elements in close proximity form an initiation zone in which replication bubbles appear to occur randomly, but in fact, originate from specific ARS elements. In mammalian cells, some origins consist of a large intergenic initiation zone (*shaded region*) defined by two-dimensional gel origin mapping methods, and one or more high-frequency initiation sites (OBRs) defined by nascent strand origin mapping methods (e.g., the *DHFR* and rRNA gene region). Other origins such as the lamin B2 gene region appear to consist of a single OBR that exhibits a cell cycle-dependent DNA footprint (ORC?) reminiscent of those produced by ORC at yeast origins. The lamin B2 origin also overlaps three transcription-factor-binding sites belonging to the promoter of a downstream gene. Initiation events in the intergenic region (0.6 kb) downstream from the lamin B gene have not been analyzed by two-dimensional gel analysis.

In contrast to the small replication origins in *S. cerevisiae*, replication origins in *S. pombe* and *Yarrowia lipolytica* are 0.5 to approximately 1 kb long and lack a consensus sequence that is essential for replication. However, they do exhibit ARS activity, and they do contain large A:T-rich elements similar to those in centromeres that can bind proteins with a strong affinity for A:T sequences (Vernis et al. 1997; Sanchez et al. 1998b). In *Tetrahymena*, the minimal *cis*-acting sequences required for replication of the rRNA gene region are located within a 1-kb segment just upstream of the rRNA gene promoter (Blomberg et al. 1997). Thus, although origin functions are conserved, origin characteristics, even among unicellular organisms, can differ significantly.

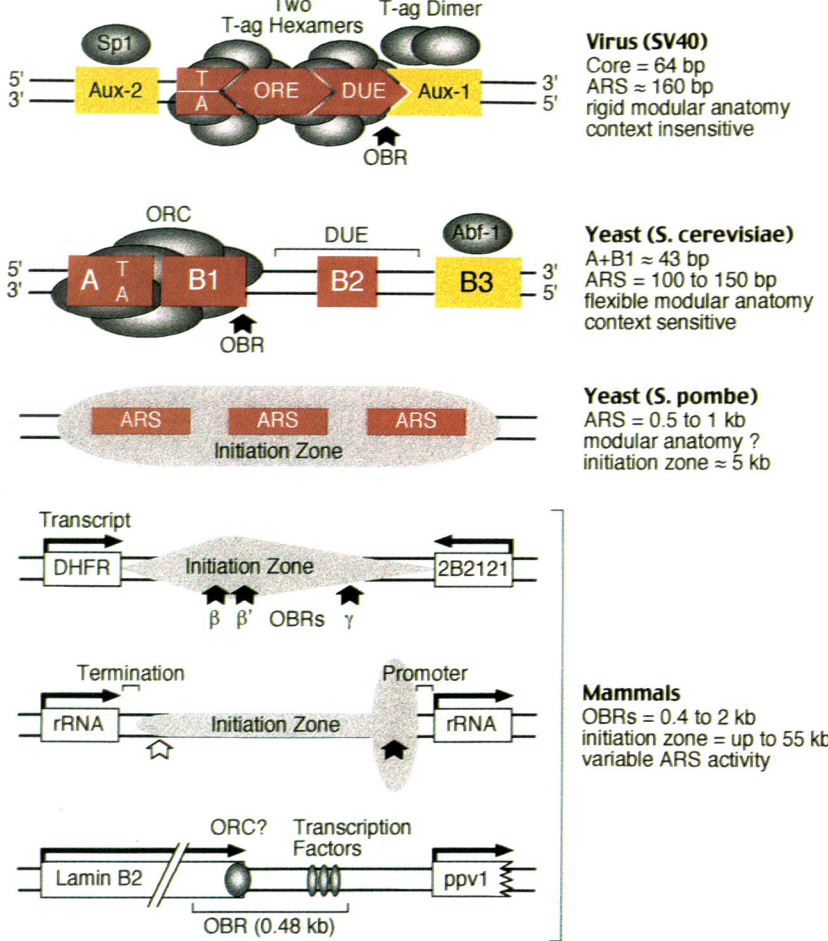

Figure 2 (See facing page for legend.)

Replication Origins in Metazoan Genomes

Given the remarkable conservation in eukaryotes of the proteins used to initiate replication (see Fig. 1), and the fundamental need for replication origins to ensure efficient genome replication, determine the length and replication pattern of S phase, and prevent replication from interfering with transcription, one would expect initiation of DNA replication in metazoa, like initiation in unicellular organisms, to occur at specific DNA sequences. Although this appears to be the case, their characteristics have proven elusive. Yeast origins exhibit ARS activity, but ARS elements that function in mammalian cells have been difficult to identify (see Chapter 2). Furthermore, in early embryos undergoing rapid cell cleavage (e.g., frogs, flies, sea urchin, and fish), initiation of DNA replication appears neither to require specific DNA sequences nor to occur at specific DNA sites (Coverley and Laskey 1994). However, in differentiated cells of flies, frogs, and mammals, it is clear that DNA replication begins at specific genomic loci (see Chapter 2). For example, a 200-kb region at the human β-globin gene (Kitsberg et al. 1993; Aladjem et al. 1995) and a 500-kb region at the mouse *Igh* gene (Michaelson et al. 1997) are both replicated from a single initiation locus. What has been difficult to nail down is the size of initiation loci and the nature of specific replication origins.

At least eight different strategies have been used to map the sites where DNA replication begins in eukaryotic chromosomes (DePamphilis 1997), and at least 23 specific initiation sites have been reported in the chromosomes of flies, frogs, and mammals (see Table 4 in Chapter 2 and Zhao et al. 1994; Hyrien et al. 1995; Gencheva et al. 1996; Gogel et al. 1996; Lu and Tower 1997; Kobayashi et al. 1998; Toledo et al. 1998). Although all of the data are consistent with bidirectional replication involving classical replication bubbles and forks, a complex and sometimes contradictory view of replication origins has emerged. When cellular DNA is first fractionated by two-dimensional gel electrophoresis in order to identify replication bubble and fork structures, and then hybridized with sequence-specific radioactively labeled probes in order to identify their genomic location ("two-dimensional gel" mapping protocols), initiation events often appear to be distributed throughout large (from 4 to 55 kb) DNA regions ("initiation zones") with no clear preference for one site over another. In contrast, when newly synthesized mammalian DNA is labeled with nucleotide precursors during its biosynthesis and then its origin mapped by various strategies ("nascent strand analysis" mapping protocols), most initiation events (85% to >95%) appear to occur at specific sites of 0.5–2 kb (OBRs) analogous to those found in single-cell eukaryotes such as yeast, *Tetrahymena*, and *Physarum*. This paradox is particularly evident when both two-dimensional gel analyses and nascent strand analyses are applied to the same loci (e.g., *DHFR* and rRNA gene regions in mammalian cells, Fig. 2) (for discussion, see Kobayashi et al. 1998).

A relevant paradigm is provided by yeast. Yeast replication origins generally consist of a single ARS element, but two-dimensional gel analyses of the *ura4* region in *S. pombe* initially suggested that initiation events were distributed throughout an approximately 5-kb initiation zone (Fig. 2). However, when combined with genetic analyses, it became clear that the replication bubbles were coincident with three separate ARS elements (Dubey et al. 1994). Thus, what at first appeared to be a continuous initiation zone is actually a cluster of three replication origins. This phenomenon can be understood from observations that origins in close proximity (~6 kb) function at reduced efficiency, and only one of the origins is activated in each cell during each S phase (Brewer and Fangman 1993; Marahrens and Stillman 1994). Dominance of one origin over another most likely results from the fact that origin activity is strongly affected by its surrounding sequences ("sequence context") (Brewer and Fangman 1994; Marahrens and Stillman 1994).

A similar phenomenon appears to occur in mammals. The relative number of replication bubbles has recently been quantified throughout the hamster *DHFR* gene region using the "nascent strand abundance assay." This assay and its close relative the "nascent strand length assay" are based on the concepts that sequences closest to an OBR are represented more frequently in nascent DNA strands than sequences further away and that nascent DNA strands approximately 1 kb long originate from newly formed replication bubbles. These assays have been used to map at least ten different replication origins in mammalian cells to sites as small as 0.5 kb (for review, see Yoon et al. 1995; Pelizon et al. 1996; Staib and Grummt 1997; Kobayashi et al. 1998). Results from the nascent strand abundance assay confirmed that initiation occurred at ori-β 10–12 times more frequently than at distal sites such as the *DHFR* gene where two-dimensional gels do not detect replication bubbles. These results, together with 11 other nascent strand analyses (summarized in Chapter 2 and in Kobayashi et al. 1998), demonstrate that ori-β is a primary initiation site and that most (>90%) initiation events at ori-β occur within a 2-kb locus centered at the ori-β OBR. The fact that the same OBR has been identified in unsynchronized cells (Vassilev et al. 1990; Pelizon et al. 1996) as well as synchronized cells (Kobayashi et al. 1998) confirms that it is a primary initiation site, one that is used in most of the cells during each cell division. In addition, a previously unrecognized initiation site (ori-β´) was identified about 5 kb further downstream from ori-β (Kobayashi et al. 1998). Thus, the *DHFR* initiation zone contains at least three primary initiation sites (ori-β, β´, and γ) within a 28-kb locus (Fig. 2). These sites could account for most of the initiation events in this region, because the intensity of replication bubble structures in the *DHFR* gene initiation zone appears to be greatest within a 12-kb region containing ori-β and β´. Similarly, the fraction of replication bubbles in the rRNA gene initiation zone was reported to vary approximately tenfold (Little et al. 1993) and was greatest in the regions containing primary initiation sites (Yoon et al. 1995). This site is greater than tenfold more active than distal sites, and it appears in rRNA genes of all eukaryotes examined so far (Sanchez et al. 1998a). A second, weaker site has sometimes been observed downstream from the 3´ end of the gene.

Most of the reasons offered for why two-dimensional gel analyses do not detect the same primary initiation sites identified by nascent DNA strand analyses are based on differences in data quantification and resolution of closely spaced origins (Kobayashi et al. 1998). Nevertheless, at least some mammalian replication origins consist of a single high-frequency initiation site (1.7 kb) that can be detected by two-dimensional gel analysis of replication fork migration and confirmed by measuring the relative abundance of nascent DNA strands (Toledo et al. 1998)

- **PARAMETERS THAT DETERMINE INITIATION SITES**

Initiation of DNA replication in metazoan chromosomes follows the Jesuit dictum that "many are called, but few are chosen" (see Fig. 3 in Chapter 2). Analyses of DNA replication in embryos undergoing rapid cell cleavage (see Chapters 2 and 4) and reversion of ARS mutations in yeast (Kipling and Kearsey 1990) clearly demonstrate that DNA contains many potential initiation sites. These sites may include potential DUEs or weak ORC-binding sites that can be activated either by the high concentration of initiation proteins in rapidly cleaving embryos or by a less restrictive chromatin structure on plasmid DNA. However, when cells undergo differentiation, most initiation sites may be repressed while a few are activated. For example, gene transcription, chromatin structure, and nuclear organization could prevent access of initiation proteins to some DNA sites while making other sites either more accessible or easier to unwind. This could create initiation zones consisting of one or more high-frequency initiation sites (OBRs) and perhaps several low-frequency initiation sites that escape detection by biolabeling methods but not by two-dimensional gel methods. The low-frequency sites may represent accessible DUEs, whereas the high-frequency sites may represent strong ORC-binding sites adjacent to a DUE. As development progresses from rapid cell cleavage stages to differentiated cells, the ratio of initiation factors to DNA is reduced, and those sites with the greatest affinity for ORC and the greatest accessibility to initiation factors will be selected. This process provides animals with the flexibility to change the rate of cell division by changing the number of active origins, while at the same time selecting initiation sites that do not interfere with gene expression as cells undergo differentiation.

Nuclear Structure

An intact nucleus is generally required to observe initiation DNA replication in extracts from *Xenopus* eggs (Wu et al. 1997), human cells (Krude et al. 1997), and yeast cells (Pasero et al. 1997), demonstrating that nuclear structure has one or more critical roles in regulating eukaryotic DNA replication. One role is regulating the access of replication factors to the DNA substrate. For example, replication licensing factor activity is absent from nuclei during the G_2 phase of the cell cycle, but it can be introduced by permeabilizing the nuclei with a detergent and then incubating them in *Xenopus* egg extract (see Chapter 4). Recently, a single round of ORC-dependent DNA replication has been achieved in a *Xenopus* egg extract in the absence of nuclear structure by substituting a concentrated nuclear extract (Walter et al. 1998). This result suggests that the primary role of the nucleus in DNA replication is to concentrate replication factors and implies that any role for nuclear structure in establishing replication forks or in selecting initiation sites will be facilitative rather than obligatory, because these functions can be achieved in the apparent absence of nuclear structure. In fact, the ability of SV40 chromosomes to complete replication outside the nucleus (Su and DePamphilis 1976) and the ability to assemble active DNA replication forks from soluble factors (Stillman 1989) demonstrate that nuclear structure is not required for replication fork activity. Nevertheless, the evidence is compelling that nuclear structure is involved either in the assembly or in the maintenance of cellular replication forks.

Several studies have shown that newly synthesized DNA is preferentially associated with nuclear structure in the form of "nuclear matrix," "nuclear scaffold," and "nucleoskeleton" (experimental definitions of a network of filaments within the nucleus) and that replication forks are colocalized in "replication foci" or "replication factories" distributed throughout the nucleus (see Chapter 4) (Hozák et al. 1996). Although formation of replication foci does not appear to require specific DNA sequences, it does require a specific protein activity (Yan and Newport 1995). A functional requirement for replication foci is suggested by the fact that DNA must first be assembled into chromatin, and then into nuclei, before DNA replication can begin in *Xenopus* eggs or egg extracts. If nuclear assembly is prevented by omitting the vesicular fraction from the extract, or if preformed, non-S-phase nuclei are permeabilized before adding them to a vesicle-free extract, then the ability of the extract to replicate DNA is lost (see Chapter 4). More specifically, formation of a nuclear lamina is required for DNA replication. Nuclei assembled in lamin-B3-depleted *Xenopus* egg extracts do not assemble a nuclear lamina (a network of filaments underneath the nuclear membrane) and do not replicate DNA, but the ability to replicate DNA can be rescued by restoring lamin B3 to the depleted extract (Goldberg et al. 1995). Moreover, perturbation of nuclear lamina organization by introduction of truncated lamin proteins also inhibits DNA replication (Ellis et al. 1997; Spann et al. 1997). Nuclei assembled in the absence of lamin B3 still contain nuclear pores and continue to accumulate a variety of karyophylic proteins, but they do not form replication factories (Jenkins et al. 1993). Thus, nuclear lamina is required for DNA replication because it may be required for correct assembly of a nuclear matrix (Zhang et al. 1996).

Another role suggested for nuclear structure is in establishing initiation sites. This hypothesis is based on the observation that site-specific initiation of DNA replication can be achieved in a frog egg extract if intact nuclei are used as the substrate rather than DNA (Gilbert et al. 1995). When either sperm chromatin or DNA is added to *Xenopus* egg extracts, replication is initiated at many sites along the DNA molecule, regardless of whether or not it contains specific prokaryotic or eukaryotic replication origins. However, when intact nuclei are isolated from differentiated mammalian cells in the G_1 phase of their cell cycle and then incubated in a *Xenopus* egg extract, DNA replication is initiated at or close to the same replication origins normally utilized by this cell in vivo. Initiation under these conditions requires neither the ability to assemble nuclei nor *Xenopus* ORC, but it does require late G_1 phase nuclei that have not been permeabilized and a protein kinase activity (final 6-DMAP sensitive step; Fig. 1). Nuclei from early G_1 phase can also initiate DNA replication under these conditions, but initiation occurs "randomly" throughout the genome (Wu and Gilbert 1996). Therefore, establishment of specific initiation sites requires both nuclear structure and a cell-cycle-dependent event, the "origin decision point" (Wu and Gilbert 1996). Since the origin decision point is sensitive to protein kinase inhibitors and independent of *Xenopus* ORC2 (Wu et al. 1997; Wu and Gilbert 1997), establishment of site specificity in mammalian chromosomes may occur during assembly of the RLF-B-dependent pre-RCs in late G_1 phase, a step that is also dependent on a protein kinase activity (Fig. 1; inhibition by 6-DMAP).

Chromatin Structure

Chromatin structure can inhibit both promoter and origin activity by blocking access of transcription factors to specific DNA sites (see Chapter 2). Moreover, transcriptional insulator elements that organize transcriptionally active domains in *Drosophila* chromosomes can also protect a DNA replication origin in *Drosophila* chromosomes from position effects (Lu and Tower 1997). Thus, it appears that the variable activity of replication origins observed at different chromosomal positions is due largely to differences in chromatin organization throughout the genome. For example, histone H1 can reduce the frequency of initiation in *Xenopus* egg extract by limiting the assembly of prereplication complexes on sperm chromatin (Lu et al. 1998), and changes in chromosome structure can determine which initiation sites will be used (Lawlis et al. 1996). Incubation of condensed chromosomes from metaphase-arrested hamster cells in *Xenopus* egg extract elicits assembly of nuclei and initiation at a novel site (ori-δ) that is not used when these cells enter S phase, whereas initiation at ori-β, a site normally used to initiate DNA replication at the beginning of S phase, is repressed (Fig. 3). Thus, changes in chromosome structure that occur during a cell cycle can create new initiation sites while eliminating others, presumably by affecting their accessibility to initiation factors.

DNA Sequence

Although DNA sequence is clearly a determinant of origin activity in yeast cells, its role in metazoan cells has been more elusive. The simple fact that mammalian origins map to specific sites that replicate at specific times during S phase demonstrates that origins of replication are inherited from one cell division to the next. This conclusion is reinforced by reports that the same OBR identified in cells containing two copies per diploid genome are also identified in cells containing many tandem

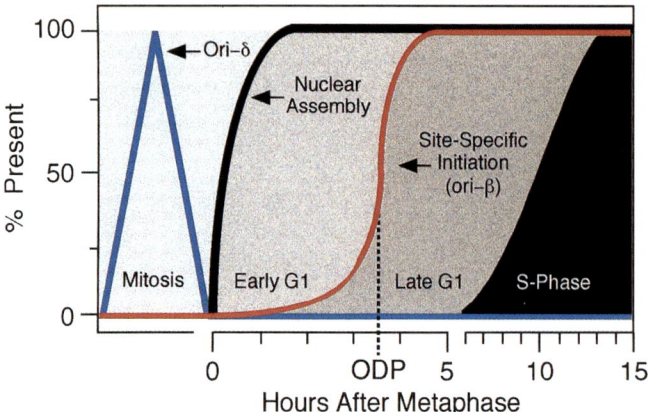

Figure 3 Specific initiation sites in mammalian chromosomes are reestablished during G_1 phase of each cell division cycle. When nuclei from G_1 phase hamster cells are exposed to S phase factors in a soluble *Xenopus* egg extract, DNA replication begins at "randomly" chosen sites in nuclei from early G_1 phase cells, but at specific sites (e.g., ori-β) in nuclei from late G_1 phase cells (Wu and Gilbert 1996, 1997). Specific initiation sites are also present in metaphase chromosomes (e.g. ori-δ), but these sites disappear during G_1 phase (Lawlis et al. 1996).

copies of either chromosomal or extrachromosomal (episomal) sequences (see Table 3 in Chapter 2). Moreover, the same replication origin is used by other cells of the same species but of different derivation (Kumar et al. 1996). Therefore, OBRs are specified by DNA sequences.

Specific protein:DNA interactions have been detected at three mammalian OBRs. Only two micrococcal-nuclease-hypersensitive sites appear in the hamster *DHFR* gene initiation zone as cells enter S phase, one at ori-β and one at ori-γ (J. Hamlin, pers. comm.). The ori-β-hypersensitive site maps to the A:T element within the high-density cluster of ^{m}CpG dinucleotides flanking the OBR. A 78-nucleotide protein footprint has been observed in vivo at the human lamin B origin (Fig. 2). It is highly asymmetric on the two strands (Abdurashidova et al. 1998), similar to that of other origin recognition proteins such as yeast ORC, *E. coli* DnaA, and SV40 T antigen (Lee and Bell 1997), and the A:T element upstream of the footprint becomes DNase-I-hypersensitive, reminiscent of yeast origin:ORC complexes (Abdurashidova et al. 1998). Cell-cycle-dependent changes in this footprint are consistent with assembly of a pre-RC during G_1 phase that is modified to a postreplicative state during S phase and then lost completely from DNA during mitosis.

Direct evidence that metazoan replication origins are genetically determined comes from genetic analyses of origin activity. To date, six reports of ARS elements that function in mammalian cells and cell extracts have been documented in detail and shown to correspond to sites where replication occurs in mammalian chromosomes (see Table 3 in Chapter 2) (Zhao et al. 1994). ARS activity in mammalian cells may depend on a number of factors (for discussion, see DePamphilis 1996), such as how easily their DNA can be unwound or whether they require specific associations with nuclear structure or particular chromosome structures that are not easily reproduced in small episomal DNAs. Conversely, as with yeast, many sequences that do not exhibit origin activity in the context of a "real" chromosome do exhibit ARS activity in the context of a plasmid, making it difficult to distinguish between a real replication origin and nonspecific initiation events. This concern has been addressed by measuring the ability of origins to function at ectopic chromosomal sites.

A 1.2-kb fragment containing the major initiation site (ori-β) in the *Drosophila* chorion gene amplification locus and a *cis*-acting DNA element required for chorion gene amplification (*ACE3*) are sufficient for high-level gene amplification in *Drosophila* (Lu and Tower 1997; J. Tower, pers. comm.). Moreover, a transcriptional insulator element that can protect this fly amplification locus from sensitivity to its chromosomal position can also prevent replication when inserted between *ACE3* and ori-β, suggesting that proteins must interact with both *ACE3* and ori-β sequences in order to initiate DNA replication. It seems likely that parameters which determine amplification origins will also determine S phase origins, because the same replication origins that direct rRNA gene amplification in *Tetrahymena* also function during S phase (Zhang et al. 1997). In mammals, DNA fragments of 2.4 kb containing the human c-Myc origin (M. Leffak, pers. comm.), 6–18 kb containing the *DHFR* ori-δ OBR (Handeli et al. 1989; T. Kobayashi and M. DePamphilis, unpubl.), and 4–8 kb containing the human β-globin OBR (Aladjem et al. 1998) exhibit origin activity when transferred to ectopic chromosomal locations, whereas fragments that lack origin activity in situ exhibit little if any ectopic activity. In addition, deletions from 2.7 kb to 8 kb at the human β-globin OBR eliminate the only initiation site detected within a 200-kb region (Kitsberg et al. 1993; Aladjem et al. 1998). Taken together, these data reveal that metazoan origins of replication are determined by specific sequences encompassed within approximately 2 kb, perhaps similar to those in *S. pombe* (Dubey et al. 1996) and *Tetrahymena* (Blomberg et al. 1997). In addition, origin activity can also be regulated from locus control regions that are many kilobases distal to the OBR but affect the accessibility of initiation sites to replication proteins (Aladjem et al. 1995).

Transcription or transcription factors are often considered to have a role in the initiation of cellular DNA replication, because active genes generally are replicated early in S phase, whereas inactive genes are replicated late (see Chapter 12), and transcription factor DNA-binding sites are components of many viral replication origins (see Chapters 2 and 3). However, transcription is not required for DNA replication, because the first S phase in mammalian embryos and the first few S phases in embryos of frogs, flies, and fish occur in the absence of transcription. Moreover, only a small number of yeast origins contain transcription-factor-binding sites, and only 45% of metazoan origins map to a transcription promoter or enhancer. In fact, the activity of several replication origins is independent of local transcription units (Aladjem et al. 1995; Blomberg et al. 1997; Lu and Tower 1997). Thus, although some replication origins such as lamin B2 (see Fig. 2) contain transcription-factor-binding sites, a role for transcription in determining metazoan replication origins remains speculative.

DNA methylation regulates replication origins in *E. coli* by increasing their efficiency and reducing their rate of reinitiation. The presence of large densely methylated islands in which all cytosines were methylated, regardless of their dinucleotide composition, suggested a role for DNA methylation in metazoan replication origins as well (see Chapter 2). Although these results were subsequently shown to be artifacts of the methods used to detect them, at least two mammalian OBRs (hamster *DHFR* ori-β and RPS14) lie adjacent to a 350-bp sequence in which the frequency of ^{m}CpG is approximately ten times above average, and all the CpG dinucleotides within 2-kb regions containing these two OBRs remain methylated throughout the cell cycle (Rein et al. 1997a,b). Analysis of newly replicated DNA confirmed that these highly methylated origins are active.

Mammalian replication origins are rapidly remethylated following initiation of DNA replication (Araujo et al. 1998; T. Rein and M. DePamphilis, unpubl.), making it unlikely that hemimethylated DNA has a role in regulating the rate of reinitiation in mammalian cells as it does in *E. coli*. However, analysis of methyl-deficient cell lines revealed that the ^{m}CpGs at *DHFR* ori-β were conserved, and in one cell line that was 50% deficient in methylation at ori-β, site-specific initiation was lost (T. Rein and M. DePamphilis, unpubl.). These results suggest either that a cluster of ^{m}CpGs can contribute to assembly of pre-RCs at specific sites or that DNA methylation affects expression of a gene whose product is required at the origin decision point (Fig. 3). It is unlikely, however, that methylation has a role at all mammalian replication origins, because some origins either lack a significant number of CpGs or lie within the sparsely methylated CpG islands that surround many transcription promoters (Delgado et al. 1998).

Animal Development

During development (see Chapter 13), frogs, flies, and mammals are able to change the number of replication origins used in a single S phase (thus regulating the length of S phase), as well as the number of times each replication origin is used in a single S phase (thus producing polyploid cells or amplified DNA loci). Thus, the metazoa, in contrast to simpler organisms, exhibit a great deal of flexibility in regulating initiation of DNA replication.

Many studies have concluded that initiation of DNA replication can occur in virtually any DNA sequence that is injected into rapidly cleaving frog eggs or incubated in frog egg extracts (Wu et al. 1997). Consistent with these observations, replication bubbles in rapidly cleaving *Drosophila* and *Xenopus* embryos are distributed uniformly with a periodic spacing of approximately 10 kb, suggesting that the periodic structure of chromatin imposes a periodic pattern of initiation sites for replication (Hyrien and Mechali 1993; Blumenthal et al. 1974). Specific initiation sites first appear in the rRNA gene repeats in *Xenopus* (Hyrien et al. 1995) and in the DNA polymerase α gene region in *Drosophila* (Sasaki et al. 1998) only after zygotic gene expression begins. Thus, site specificity can be developmentally acquired.

What regulates the frequency of initiation sites? Titration of *Xenopus* egg extract with sperm chromatin reduces the frequency of initiation in the sperm chromatin without affecting the rate of fork migration (Walter and Newport 1997). As the ratio of initiation factors to DNA substrate decreases, those sites where pre-RCs are assembled most easily will be selected. In addition, changes occur in nuclear lamin proteins and histone H1 appears once zygotic gene expression begins. Thus, weaker origins could be more accessible to higher concentrations of initiation proteins in rapidly cleaving embryos and less accessible to lower concentrations of initiation proteins in adult tissues. Analogous changes also occur in mammals. DNA injected into the nuclei of preimplantation mouse embryos, where the rate of cell division is similar to adult tissues, does not initiate DNA replication unless it contains a strong replication origin and its cognate origin recognition protein (DePamphilis et al. 1988). Even so, auxiliary origin sequences such as enhancers are not required to activate replication origins until the 2-cell to 4-cell stage (Martinez-Salas et al. 1988), at which time a coactivator appears that allows enhancers to relieve chromatin-mediated repression (Majumder et al. 1997).

Reinitiation of DNA replication without first passing through mitosis ("endoreduplication") can occur throughout the genome in *Drosophila* salivary glands and at specific loci such as the *Drosophila* chorion genes in follicle cells. These phenomena appear to result from altering the normal use of S phase replication origins. Cyclins A and B are absent in cells as they enter endocycles, and fluctuations in cyclin E levels are required for multiple rounds of endocycle S phases (Follette et al. 1998). Mcm proteins cycle between chromatin-bound and -unbound states, despite the absence of mitosis, apparently in response to oscillations in cyclin E levels (Su and O'Farrell 1998). Amplification of the chorion gene begins within the S phases of synchronously cycling follicle cells and then undergoes a developmentally regulated transition into a late-stage continuous process that requires cyclin E as well as ORC2 (Calvi et al. 1998; Landis et al. 1997). It appears that high levels of Cdk2/cyclin E normally inhibit activation of S phase origins, but additional factors associated with the chorion gene amplification origin permit it to continually reinitiate even in the presence of high levels of cyclin E.

- **DNA REPLICATION FORKS**

Events at DNA replication forks in both prokaryotic and eukaryotic genomes remain essentially as described in Chapters 1, 5, and 14. The proteins, their organization at replication forks, and the sequence of events at replication forks are illustrated in Chapter 1, Figure 5, and Chapter 14, Figures 3 and 4. Mechanisms for priming DNA synthesis are described in Chapter 5, with two views of Okazaki fragment synthesis outlined in Figure 8.

Most, if not all, of the proteins required for replication fork function in vitro and in vivo have been identified and characterized biochemically and genetically. It is clear that both the sequence of events at replication forks and the functions provided by replication proteins are highly conserved. Current research focuses on the interactions between these proteins, their organization of replication forks and their three-dimensional structure (Bambara et al. 1997; Baker and Bell 1998). The DNA loop frequently postulated by models in which leading- and lagging-strand DNA synthesis are coupled has recently been observed by electron microscopy at replication forks generated by bacteriophage T7 proteins (Park et al. 1998).

Replication Proteins

Replication proteins in prokaryotes and eukaryotes exhibit nine principal activities:

DNA Helicase

DNA helicase is required to unwind the two DNA templates. Analyses of the *E. coli* replicative helicase DnaB, the SV40 T-antigen helicase, and the helicases encoded by bacteriophages T7 and T4 reveal a common hexameric architecture with similar biochemical properties, including ATP dependence, high processivity, and association with their cognate replicative DNA polymerases (Baker and Bell 1998). ATP-dependent DNA unwinding by SV40 T antigen strongly correlates with formation of a double hexamer, and the double hexamer binds preferentially to a 50-nucleotide replication bubble (Smelkova and Borowiec 1997). These data support previous electron microscopic evidence that a double hexamer helicase bridges two replication forks within a single replication bubble, simultaneously activating DNA unwinding and drawing the DNA substrate through the DNA-unwinding machine. The DNA helicase used at eukaryotic cell replication forks is still not identified, but Mcm proteins are leading candidates because some of them exhibit a limited helicase activity and some of them appear to travel with replication forks. The existence of hexameric Mcm complexes is consistent with their proposed role as a mammalian helicase.

DNA Primase

DNA primase is required to initiate RNA primer synthesis on each template of cellular replication origins. In eukaryotes, the two polypeptides that comprise this enzyme are tightly bound to DNA polymerase α. Selection of primase initiation sites is strongly influenced by nucleoside triphosphate (NTP) concentrations (Kirk et al. 1997). At low NTP concentrations primers are synthesized only at pyrimidine-rich sites, whereas at "physiological" NTP concentrations, primers are synthesized at all regions of the template, characteristic of Okazaki fragment synthesis in vivo.

DNA Polymerases

The separate tasks of initiation of DNA synthesis, elongation of nascent DNA chains, and filling in the gaps left by RNA primer excision are carried out by three different DNA polymerases. DNA polymerase α is required to initiate DNA synthesis on RNA primers at replication origins as well on the retrograde (lagging-strand) template throughout the genome. Once DNA synthesis has progressed from 30 to 300 nucleotides, DNA polymerase δ/proliferating cell nuclear antigen (PCNA) takes over. This highly processive enzyme can synthesize thousands of nucleotides without falling off the template and is therefore responsible primarily for DNA synthesis on the forward (leading-strand) template. In addition to the two catalytic subunits previously identified for this enzyme in mammals and yeast, two additional subunits have been identified in S. pombe pol-δ (Zuo et al. 1997). pol-δ also is required in DNA repair (Longley et al. 1997). DNA polymerase ε is required for DNA replication in yeast (D'Urso and Nurse 1997) where it probably is used to fill in the gaps that result from excision of RNA primers (Mozzherin and Fisher 1996). However, although pol-ε can be cross-linked to replication forks in yeast and mammalian chromosomes, it has not been trapped at replication forks in SV40 chromosomes (Zlotkin et al. 1996; Aparicio et al. 1997).

Proofreading Exonuclease

A proofreading 3′→5′ exonuclease is required to excise mismatched nucleotides. In eukaryotes, this activity is associated with DNA polymerases δ and ε.

Sliding Clamp

PCNA acts as a special "sliding clamp" to attach the leading-strand DNA polymerase to its template. PCNA is a doughnut-shaped homotrimeric protein that encircles the DNA and thereby slides along its surface (Schurtenberger et al. 1998). Proteins such as DNA polymerases δ and ε that interact with distinct sites in PCNA (Eissenberg et al. 1997) are thus fastened to the DNA template and thereby become highly processive. However, although increased processivity facilitates extensive DNA replication, it can also promote misincorporation of the wrong nucleotide and reduce the ability of polymerases to recognize DNA damage (Mozzherin et al. 1997). The carboxyl terminus of PCNA also binds to DNA polymerase δ, replication factor C (RF-C), Fen1, p21, and DNA ligase I (Jonsson et al. 1998). Thus, the dynamic movement of PCNA on and off the DNA appears to render this protein an ideal communicator for a variety of proteins essential for DNA metabolic events in eukaryotic cells.

Sliding Clamp Assembly Factor

RF-C is required to load the circular sliding clamp (PCNA) onto replication forks or internal DNA repair sites. All five subunits of yeast and human RF-Cs have now been cloned and can be assembled into an active complex (Gerik et al. 1997; Podust and Fanning 1997; Ellison and Stillman 1998). RF-C interacts with the carboxyl terminus of PCNA (Mossi et al. 1997), and phosphorylation of the PCNA-binding domain in the large subunit of RF-C reduces its binding to PCNA and consequently inhibits DNA synthesis by polymerases δ and ε (Maga et al. 1997).

Single-stranded DNA-binding Protein

The heterotrimeric single-strand-specific replication factor A (RP-A) binds DNA through two single-stranded DNA-binding domains in the 70-kD subunit, as well as through another single-stranded DNA binding domain in the 32-kD + 14-kD complex (Bochkareva et al. 1998). The structure of the single-stranded DNA-binding domains in human RP-A is similar to those encoded by bacteriophage. RP-A facilitates DNA unwinding (Iftode and Borowiec 1997), stabilizes the exposed template, and attracts to it other replication proteins such as DNA polymerase α and SV40 T antigen (Wold 1997). RP-A may itself be recruited to replication origins by association with the activation domains of transcription factors such as VP16. Phosphorylation of RP-A modulates its ability to participate in protein-protein interactions; interactions with T antigen are reduced and interactions with DNA polymerase α are disrupted (M. Wold, pers. comm.).

RNA Primer Excision

RNase H1, the enzyme that specifically cleaves off initiator RNA from Okazaki fragments at the RNA-DNA junction, recognizes the transition from RNA to DNA on a single-stranded substrate (Murante et al. 1998). A single ribonucleotide remains attached to the 5′ end of the Okazaki fragment that is then removed by the 5′ to 3′ exonuclease/ endonuclease, FEN-1. Since both enzymes strongly prefer single-stranded structures, it appears that RNA primers are displaced from the template prior to their excision.

DNA Ligase

Five DNA ligase activities have been purified from mammalian cells and three DNA ligase genes (I, III, and IV) have been cloned (Lindahl et al. 1997; Tomkinson and Mackey 1998). DNA ligase I joins Okazaki fragments during discontinuous DNA synthesis. The amino-terminal region of DNA ligase I interacts specifically with PCNA and therefore cannot be replaced by other DNA ligases in the joining of Okazaki fragments. DNA ligase II is a processed form of DNA ligase III. DNA ligase III occurs in two alternatively spliced forms (α and β); III-α interacts specifically with the DNA repair protein XRCC1. DNA ligase IV is bound tightly to XRCC4, a protein involved in the repair of DNA double-strand breaks. Thus, both ligases III and IV are heterodimers. In addition to its ability to stimulate DNA-dependent protein kinase, Ku protein promotes joining at double-strand breaks by mammalian DNA ligases. DNA ligases I, III-α, and IV have distinct roles, guided by their unique noncatalytic domains which specifically interact with other replication, repair, or recombination proteins. *S. cerevisiae* has counterparts to mammalian DNA ligases I and IV, but not III. Ligase IV in yeast is not essential and therefore not directly involved in DNA replication, but it is required for joining of double-strand breaks.

Replication Fork Barriers

When replication forks approach rRNA and tRNA gene transcription units, forks traveling opposite to the direction of transcription are arrested at the normal transcription termination site, whereas forks traveling in the same direction as transcription are not arrested (see Chapter 6 and MacAlpine et al. 1997). Thus, specific DNA sequence signals at the ends of genes can prevent DNA replication from interfering with DNA transcription. In pol-I-transcribed rRNA genes, this phenomenon does not require transcription, whereas in pol-III-transcribed tRNA genes it does (Deshpande and Newlon 1996; Lopez-estrano et al. 1998). The replication fork barrier in mouse rRNA genes simply requires binding of the normal transcription termination protein (TTF-1) to its specific transcription termination sequence in the 3´-terminal spacer region (Gerber et al. 1997). This is analogous to replication fork barriers that occur in the termination site for bacterial DNA replication (see Chapter 6). Conversely, bacterial replication terminator proteins can arrest RNA polymerases arriving from one direction but not the other (Mohanty et al. 1998).

It may be that many proteins that bind DNA tightly can arrest replication forks. For example, EBNA-1, a transcription factor and origin recognition protein encoded by Epstein-Barr virus, causes replication forks to accumulate at the EBNA-1 family of repeated binding sites, apparently by blocking the ability of helicase to unwind DNA in an orientation-independent manner (Ermakova et al. 1996).

Chromatin Assembly

Nucleosome assembly at replication forks requires at least two "chromatin assembly complexes." The first to arrive consists of the three polypeptides of CAF-1 and a special population of histones H3 and H4, with H4 acetylated at Lys-5 and Lys-12. The second to arrive consists of another assembly factor, NAP-1, associated with histones H2A and H2B (Chang et al. 1997; Verreault et al. 1998). The major histone acetyltransferase in eukaryotes (yeast HAT-1 ≈ human HAT-B) can diacetylate histone H4 in the cytoplasm, but not after it is assembled into nucleosomes. Both HAT-1 and CAF-1 each contains a polypeptide that binds to histone H4, presumably binding these two activities to their histone substrates. The role of H4 acetylation in chromatin assembly is not clear, but it may serve to target NAP-1/H2A, H2B to the replication fork. Histones are then rapidly deacetylated following nucleosome assembly onto newly replicated DNA.

- **DNA REPLICATION AND CHECKPOINT CONTROLS**

Checkpoints are regulatory mechanisms that interrupt progression through the cell cycle when certain cell-cycle-specific events have not occurred, or in response to DNA damage or damage to the mitotic spindle apparatus (Elledge 1996). Both types of checkpoints can require the function of the DNA replication machinery. For example, regulation of the onset of S phase (i.e., activation of pre-RCs) is accomplished through specific inhibitors of cyclin-dependent protein kinases such as p21 whose synthesis is regulated by a variety of stimuli, including p53, a transcription factor that is mobilized by DNA damage and stimulates production of p21, which in turn blocks the onset of S phase (see Fig. 1) until the damage is repaired (see Chapter 11). Regulation of S phase by p21 requires a functional retinoblastoma protein (Niculescu et al. 1998), a protein that regulates expression of E2F-dependent genes during the G_1 to S phase transition (see Chapter 11). The Cdk inhibitor p21 also has been suggested as an intra-S-phase checkpoint control by virtue of its ability to inhibit SV40 DNA replication in vitro through its interaction with PCNA. However, a similar inhibitory effect of p21 on S phase progression in vivo has not been detected (Niculescu et al. 1998). Intra-S-phase checkpoints (sometimes referred to as S/M checkpoints; Paulovich et al. 1997) avoid the catastrophic consequences of progression through mitosis in the absence of completely replicated DNA by sensing the presence of replication structures containing pol-ε, Dbf11p, and Rfc5p in *S. cerevisiae*

(Paulovich et al. 1997) or pol-α, Orp1p, and Cdc18p in *S. pombe* (Huberman 1998). In *S. cerevisiae*, this (or a similar) checkpoint may also inhibit activation of late-firing origins in early S phase cells that either are treated with hydroxyurea or are actively replicating DNA (Santocanale and Diffley 1998; W. Burhans, pers. comm.). Damage checkpoints inhibit DNA replication presumably to protect cells from the deleterious consequences of replicating DNA with damage lesions. The G_1 checkpoint blocks entry into S phase if DNA damage is sensed prior to S phase, whereas the intra-S-phase checkpoint blocks progression through S phase if the damage occurs after replication begins. In mammals, the intra-S-phase checkpoint inhibits both initiation and elongation of nascent DNA chains (Painter and Young 1980), and inhibition of initiation in response to DNA damage has recently been detected in *S. cerevisiae* (W. Burhans, pers. comm.; Shirahige et al. 1998).

Elements of checkpoint controls that regulate DNA replication in response to DNA damage appear to be extraordinarily conserved among eukaryotic organisms as diverse as yeast and humans. Both the G_1 and the intra-S-phase checkpoints are defective in mammalian cells containing mutations in the *ATM* gene (Paulovich et al. 1997). Mutations in *MEC1*, an *S. cerevisiae* homolog of *ATM*, produce similar phenotypes, whereas mutations in *rad3+*, the *S. pombe* homolog, result in an intra-S-phase checkpoint phenotype (Elledge 1996; Paulovich et al. 1997); a G_1 checkpoint has not yet been reported in *S. pombe*. All three genes are part of a large group of highly conserved phosphatidylinositol kinases that have roles in DNA damage responses in yeast, *Drosophila*, and mammals, most likely by transducing signals from damage to downstream target molecules (Elledge 1996).

The specific roles played by the products of these genes are not clear. The *ATM* gene, which does not appear to be absolutely required for the G_1 checkpoint in mammals, functions upstream of p53 (Elledge 1996). In addition, mutations in the *ATM* gene (Paulovich et al. 1997) are responsible for the "radio-resistant DNA synthesis" (RDS) phenotype of cells from patients with the cancer-predisposing hereditary disorder ataxia telangectasia (AT). RDS corresponds to a reduced inhibitory effect of DNA damage on both the initiation and elongation phases of DNA replication as a result of a defective intra-S-phase DNA damage checkpoint. However, the *ATM* gene is not required for some aspects of the intra-S-phase checkpoint. Specifically, the initiation-inhibitory effects of higher doses of radiation occur independently of *ATM*.

In *S. cerevisiae*, Mec1p phosphorylates Rad53p, in response to DNA damage (Sanchez et al. 1996). Although the *S. cerevisiae* intra-S-phase checkpoint requires *MEC1*, two recent studies show that, similar to *ATM* in mammals, it is not necessarily required for the initiation-inhibitory component of this checkpoint (W. Burhans, pers. comm.; Shirahige et al. 1998). One possibility is that Atmp/Mec1p is required for inhibition of initiation in response to only some types of damage lesions or when the amount of DNA damage is low. Alternatively, the role of Mec1p in the intra-S-phase checkpoint may be limited to the elongation phase of DNA replication.

In both yeast and mammals, the highly conserved single-stranded DNA-binding protein RP-A may be a downstream effector of the intra-S-phase checkpoint or, given its roles in both initiation and elongation, perhaps the checkpoint's ultimate target. In mammals, RP-A is phosphorylated in response to DNA damage, and this phosphorylation is modulated by *ATM* (Liu and Weaver 1993). In some, but not all, cases, DNA-damage-induced phosphorylation of RP-A results in reduced replication activity (Carty et al. 1994). A remarkably similar scenario occurs in *S. cerevisiae*, where DNA-damage-induced phosphorylation of RP-A is dependent on *MEC1* (Brush et al. 1996). It is not yet known whether or not DNA-damage-induced phosphorylation of *S. cerevisiae* modulates its function in DNA repair or transcription instead of, or perhaps in addition to, a checkpoint response that inhibits DNA replication. However, mutations in a gene encoding a subunit of RP-A, RFA-M2, cause defects in the intra-S-phase checkpoint (Longhese et al. 1996), which directly shows that RP-A has a role in this checkpoint.

Although the indirect evidence described above suggests that RP-A is involved in the intra-S-phase checkpoint in mammals as well, *ATM* remains the only mammalian gene for which a role in this checkpoint has been firmly established. In contrast, no less than 12 genes have been shown to be required for this checkpoint in *S. cerevisiae*, including several that, like *MEC1* and RP-A, have homologs in mammals. These latter genes include *PRI1*, a subunit of DNA primase (Marini et al. 1997), *RFC1*, a subunit of RF-C (Sugimoto et al. 1997), and several genes encoding subunits of ORC. For instance, mutations in *ORC1, ORC2*, and *ORC5* can also produce defects in the intra-S-phase checkpoint in *S. cerevisiae* (W. Burhans, pers. comm.; Shirahige et al. 1998). Strains harboring these mutations also have defects in the replication checkpoint that inhibits firing of origins late in S phase in cells blocked in early S phase with hydroxyurea (W. Burhans, pers. comm.). Similar to the G_1 and intra-S-phase DNA damage checkpoints, replication checkpoints that block late-firing origins (Santocanale and Diffley 1998) or mitosis (Paulovich et al. 1997) also require *MEC1* and *RAD53*, and *RAD53* also clearly functions upstream of ORC in the intra-S-phase DNA damage checkpoint (Shirahige et al. 1998).

- **VIRAL DNA REPLICATION**

Papovaviruses and Papillomaviruses

The mechanisms for replicating the genomes of small DNA tumor viruses (pavovaviruses and papillomaviruses) are adequately described in the CSH monograph, *Replication in Eukaryotic Cells* (see Chapters 22 and 23). It has become even clearer that their mechanisms for initiation of DNA replication are strikingly similar (Chen and Stenlund 1998). Each virus encodes a single origin recognition protein (ORP) that binds specifically to a palindromic sequence within the core origin. Despite differences in the sequences among these origins, both the ORP-binding patterns and the arrangement of DNA-binding motifs in the origin are strikingly similar. Each ORP exhibits DNA helicase activity and initiates DNA-unwinding specifically at its cognate replication origin. In addition to its essential ORP-binding site, each origin contains one or more transcription factor DNA-binding sites (auxiliary sequences, Fig. 2). Host-range specificity for papovavirus DNA replication is mediated through specific associations between T antigen and the host cell's DNA polymerase α:DNA primase (Stadlbauer et al. 1996). Perhaps the same will be true for papillomaviruses.

Papovavirus origins bind cellular transcription factors that facilitate DNA unwinding, apparently by recruiting RP-A. The BPV origin, however, binds its own transcription factor, E2, that facilitates binding of the BPV ORP, E1, to the BPV replication origin. First, cooperative DNA binding of E1 and E2 generates a sequence-specific ori-recognition complex, and then, in an ATP-dependent step, E2 is displaced and additional E1 molecules are incorporated (Sanders and Stenlund 1998). Thus, a highly sequence-specific transcription factor (E2) facilitates binding of an ORP with low sequence specificity to a specific site on the BPV chromosome.

Adenoviruses

Adenovirus (Ad) DNA replication also is adequately described in the CSH monograph (Chapter 24), although several new insights are noteworthy. Adenovirus encodes three proteins that are required for adenovirus DNA replication: DNA polymerase (Ad pol), preterminal protein (pTP) that primes DNA synthesis, and DNA-binding protein (DBP). Both adenovirus and φ29 DNA polymerases initiate DNA replication in the form of a tight heterodimer with its cognate pTP-dNTP. Shortly after initiation, the viral DNA polymerase dissociates from its primer protein, making it a much better enzyme for chain elongation and proofreading (King et al. 1997; Mendez et al. 1997). This transition from an optimal initiation to an optimal elongation stage is analogous to the transition from DNA pol α to pol δ at replication forks in cellular and papovavirus DNA replication.

In contrast to other replication systems, adenovirus DNA replication does not require a DNA helicase to unwind the double-stranded template. Instead, elongation is dependent on the adenovirus DNA-binding protein which has helix-destabilizing properties. DNA-binding protein binds cooperatively to single-stranded DNA in a nonsequence-specific manner and stimulates both initiation and elongation of DNA replication. Multimerization of DNA-binding protein is the driving force for ATP-independent DNA unwinding during strand-displacement synthesis (Dekker et al. 1998).

Initiation of adenovirus DNA replication is strongly enhanced by two cellular transcription factors, NFI and Oct-1, that bind to the origin auxiliary component and tether the pTP-dNTP/Ad pol complex to the origin core (van Leeuwen et al. 1997). NFI and Oct-1 enhance initiation synergistically by touching different targets in the preinitiation complex and dissociate independently after initiation. NFI acts through a direct contact with the DNA polymerase. Oct-1 acts through contact with pTP. NFI dissociates very early in initiation, whereas Oct-1 dissociates only when the binding site is rendered single-stranded upon translocation of the replication fork.

Herpes Simplex Virus

The mechanism of herpes simplex virus (HSV) DNA replication has been reviewed recently (Boehmer and Lehman 1997) and remains essentially as described in the CSH monograph, although progress has been made in understanding how DNA unwinding begins at an HSV replication origin. HSV encodes seven proteins that are required to replicate HSV DNA. However, despite its intrinsic helicase activity and ability to bind to oriS, the HSV-encoded origin recognition protein, UL9, is unable to unwind oriS, either in the presence or in the absence of the HSV-encoded single-stranded DNA-binding protein, ICP8. Recently, a complex of the UL9 protein and ICP8 was found to unwind oriS if it contained a 3′ single-stranded tail at least 18 nucleotides in length positioned downstream from box I (Lee and Lehman 1997). This suggests that UL9 bound to box I, and ICP8 bound to single-stranded DNA generated at the A:T-rich origin region, perhaps as a consequence of transcription, can unwind an HSV-1 origin of replication and permit initiation of DNA synthesis by HSV DNA primase and DNA polymerase. The active role in DNA unwinding proposed for ICP8 is similar to that for RP-A-mediated denaturation of the SV40 replication origin (Iftode and Borowiec 1997) and for DBP-mediated denaturation of the adenovirus replication origin (Dekker et al. 1998).

Epstein-Barr Virus

Latent replication of the 170-kb linear Epstein-Barr virus (EBV) genome in the nuclei of infected cells remains a plausible model for regulation of DNA copy number control in mammalian cells. EBNA-1, the only EBV-encoded protein required for viral DNA replication during its latent period, binds to a unique dyad symmetry region (DS) consisting of two EBNA-1 DNA-binding sites as well as a family of 20 repeated EBNA-1-binding sites (FR) approximately 1 kb distal to the DS. When the genome replicates as an episomal form, both the DS and FR regions are required (oriP), and the DS sequence is coincident with the origin of bidirectional replication. Since no enzymatic activities have been discovered in EBNA-1, all subsequent steps in viral DNA replication must be carried out by the cell, including presumably recruitment of a cellular helicase to unwind the origin. However, EBNA-1 binding to the DS component produces a strong DNA bend, and both the bend and oriP function require that the two EBNA-1 DNA-binding sites exist in a precise spatial relationship (J. Yates and C. Schildkraut, pers comm.). Furthermore, the crystal structure of the EBNA-1 DNA-binding domain bound to DNA suggests that it may facilitate DNA unwinding and DNA bending (Edwards et al. 1998). The ability of EBNA-1 to recruit RP-A to oriP may facilitate this process (Zhang et al. 1998). Thus, EBNA-1, like other origin recognition proteins (papovavirus T antigen, HSV UL9), may facilitate initiation of replication by providing torsional stress to the replication origin, making it easier to unwind.

Although initiation was detected at oriP in all latent EBV genomes examined by two-dimensional gel electrophoresis, some replication forks appear to originate from alternative initiation sites (Little and Schildkraut 1995). In the Raji EBV genome, the majority of initiation events appeared to be distributed over a broad region to the left of oriP, resembling initiation zones detected by two-dimensional gel analysis of mammalian chromosomal loci. In fact, the DS component of oriP can be deleted from viral chromosomes and viral episomes can be recovered with initiation events mapped to the initiation zone outside of oriP (J. Yates and C. Schildkraut, pers. comm.). Under these conditions, replication is limited to a few cell division cycles and other sequences appear to substitute for DS (Kirchmaier and Sugden 1998). Therefore, initiation at oriP may not be required for viral DNA replication, but it probably is required for orderly episomal replication synchronized with the cell cycle. Both EBNA-1 and the FR component remain essential under all conditions, perhaps as a nuclear retention function. oriP is sufficient for stable episomal maintenance of yeast artificial chromosomes (YACs) up to 660 kb in human cells expressing EBNA-1, and these oriPYACs are closely associated with host-cell chromosomes (Simpson et al. 1996). Thus, it is possible that cellular initiation proteins are able to identify initiation sites in the EBV genome and induce once per S phase replication, whereas oriP/EBNA-1 ensure that the resulting episomes remain within the nucleus throughout cell division. oriP/EBNA-1 may represent a strongly preferred initiation site that uses the same pathway as a cellular DNA/ORC (see Fig. 1).

Parvoviruses

Replication of linear, single-stranded parvovirus DNA proceeds by a rolling-hairpin mechanism that generates long, palindromic, duplex concatamers. Processing to monomer length requires initiation from origins of DNA replication located at the 3´ and 5´ ends of each embedded monomer. These reactions can be recapitulated in vitro for minute virus of mice (MVM). The 3´ origin is a 50-bp sequence containing three distinct recognition elements that are critical for replication (Brunstein and Astell 1997), an NS1-binding site, a site at which NS1 nicks the DNA to generate the priming 3´ OH, and a region containing a consensus-activated transcription factor (ATF)-binding site. All protein-binding patterns were directly adjacent to or overlapping with these sequence elements. A novel 110-kD parvovirus initiation factor (PIF) has been purified from human cells that binds to a unique 20-bp sequence in the MVM replication origin. PIF functions as an essential cofactor in the replication initiation process and is required for the viral replicator protein NS1 to nick and become covalently attached to a specific origin sequence (Christensen et al. 1997a,b). DNA polymerase δ appears to be primarily responsible for DNA synthesis (Cossons et al. 1996).

Geminiviruses

Several advances have been made in geminivirus DNA replication, primarily in the areas of (1) characterization of the plus-strand replication origin, (2) functional domains of the initiator Rep protein, and (3) dependence of viral DNA replication on cellular events. The geminivirus DNA replication cycle contains at least two distinct phases: conversion of viral genomic single-stranded DNA into double-stranded DNA intermediates, a process carried out entirely by cellular proteins, and conversion of double-stranded DNA into new single-stranded DNA and double-stranded DNA molecules by a rolling-circle type of mechanism. In both cases, replicator sequences (minus- and plus-strand origins, respectively) direct initiation of DNA replication. The nature of the minus-strand origin is still poorly understood, in particular, the differences among the three geminivirus subgroups. In the case of the plus-strand origin, special DNA structures affect viral DNA replication. First, the stem-loop structure contained within the previously mapped origin of tomato golden mosaic virus (TGMV) is essential for DNA replication with stem sequences specifically contributing to replication efficiency (Orozco and Hanley-Bowdoin 1996). Since all three geminivirus subgroups have the potential to form this stem-loop structure, its importance likely extends to the entire family. Second, a static bent DNA locus, directed by the presence of several phased A/T tracts, exists within the large intergenic region (LIR) of wheat dwarf virus (WDV) and other subgroup I viruses (Suárez-López et al. 1995). Two groups of phased A/T tracts conform the bent locus, with one at approximately 40 bp downstream from the DNA replication initiation site and the other centered at approximately 40 bp upstream of the TATA box for virion-sense transcription. Only the A/T tracts close to the initiation site have a small stimulatory effect on viral DNA replication (Sanz-Burgos and Gutierrez 1998).

The minimal virus replicator element encompasses 200 bp in WDV (subgroup I; Sanz-Burgos and Gutierrez 1998), 120 bp in tomato leaf curl virus (TLCV) (subgroup II; Akbar Behjatnia et al. 1998), and 89 bp in TGMV (subgroup III; Orozco et al. 1998). In each case, the replicator includes the stem-loop structure and a binding site for the initiator Rep protein. The organization of the geminivirus plus-strand origin is reminiscent of the organization of replication origins in animal viruses. It appears to consist of an essential core element flanked by stimulatory (auxilliary) DNA sequences (Sanz-Burgos and Gutierrez 1998).

All geminiviruses encode an approximately 40-kD initiator Rep protein that is required for viral DNA replication. Rep protein contains blocks of amino acids conserved among the three subgroups, suggesting a similar domain organization. The amino-terminal half of Rep is responsible for both DNA binding and the initiation reaction involving DNA cleavage at the initiation site (Orozco et al. 1997). The 116 amino-terminal amino acids of Rep confer DNA replication specificity, whereas two domains within the amino-terminal half of Rep mediate origin recognition (Gladfelter et al. 1997). In addition to its initiation (nicking) activity, for which a tyrosine residue within the highly conserved motif III is required (Hoogstraten et al. 1996), Rep possesses a DNA joining activity (Laufs et al. 1995), Rep exists as a multimeric complex in solution, and the oligomerization domain has been mapped to the middle of the protein (Orozco et al. 1997). Hetero-oligomerization between Rep and AL3 proteins also has been observed, although its significance is unclear (Settlage et al. 1996).

Rep is the only viral protein required for completion of the viral replicative cycle. Therefore, cellular proteins must have important roles in viral DNA replication. For example, plant cells contain DNA polymerases α and δ (Benedetto et al. 1996; Garcia et al. 1997), and PCNA, a DNA polymerase δ accessory protein that is absent in differentiated noninfected plant cells but present in virus-infected cells, suggesting that geminiviruses, like animal DNA tumor viruses, can induce expression of cellular proteins required for their replication. The identification of a retinoblastoma (Rb)-binding motif (LXCXE) in the primary sequence of WDV (subgroup I) RepA protein that can mediate binding to human and plant Rb proteins supports this hypothesis (Xie et al. 1995, 1996; Grafi et al. 1996).

RepA is a virus-encoded protein that can be translated from the unspliced mRNA and shares the amino-terminal approximately 200 amino acids (including the LXCXE motif) with Rep. Interaction between WDV Rep with human Rb also has been reported (Collin et al. 1996), but these observations have not been reproduced (Horvath et al. 1998; Q. Xie and C. Gutiérrez, unpubl.). Geminiviruses from the other two subgroups do not encode RepA, and their Rep proteins do not contain a LXCXE motif. However, TGMV Rep (subgroup III) does interact with plant Rb (Ach et al. 1997), suggesting that different geminiviruses use different proteins to interfere with plant Rb activity. Thus, the geminiviruses appear to have developed a strategy similar to that of SV40, adenoviruses, and papillomaviruses in that they encode a protein that can bind Rb and block its normal function. Presumably, this results in release of Rb-bound E2F-like transcription factors which, in turn, induce S phase functions that are required for viral DNA replication. Alternatively, such interactions may relieve a negative effect of Rb on viral gene expression.

Geminivirus DNA also can replicate in an Agrobacterium, and this replication depends on the presence of the gemini origins of replication and virus-encoded Rep protein, indicating that an active viral replication process is occurring in the bacterial cell (Rigden et al. 1996). The ability of bacterial cellular machinery to support geminivirus DNA replication suggests that these circular single-stranded DNA viruses have evolved from prokaryotic episomal replicons.

Baculoviruses

The sequences of several baculovirus genomes have recently been completed, leading to identification of several genes that may influence viral DNA replication (G.F. Rohrmann, unpubl.). Genes encoding both the large and small subunits of ribonucleotide reductase were identified in the genomes of the *Orgyia pseudotsugata* multicapsid nucleopolyhedrovirus (OpMNPV) and the *Lymantria dispar* MNPV (LdMNPV). Both of these viruses also encode a protein related to dUTPase. In addition, LdMNPV encodes a type I DNA ligase that can form a covalent adenylate complex and can join nicked or linear DNA. LdMNPV also encodes a gene with helicase motifs, but neither the helicase gene nor the ligase gene is essential for DNA replication in transient assays.

Retroviruses

Retroviruses, as well as retrotransposons, recruit preexisting, host-encoded tRNAs to prime DNA synthesis by reverse transcriptase (see Chapter 5). Three new developments have occurred in our understanding of this process. First, both reverse transcriptase and viral nucleocapsid protein appear to be involved in loading the tRNA primer onto the viral RNA during formation of the initiation complex (H. Li et al. 1996; Chan and Musier-Forsyth 1997) (see Fig. 5 in Chapter 5). Second, in several studies on human immunodeficiency virus type 1 (HIV-1) in which the RNA primer-binding site was changed so that it is complementary to a different tRNA primer, the mutant viruses are functional but slower to replicate. With passage, the altered primer-binding site reverts to wild type (Lund et al. 1997). In contrast, if compensatory mutations in the U5-inverted repeat sequence that restores interactions with the anticodon loop of the tRNA primer are also present, then the new primer tRNAs are stable and do not revert to wild-type (X. Li et al. 1997; Y. Li et al. 1997). Conversely, deletion of the $(A)_4$ sequence in the HIV-1 U5-IR loop, which is proposed to pair with the $(U)_3$ sequence in the tRNA anticodon loop, decreased replication in vivo (Liang et al. 1997). Therefore, interactions between the U%-IR loop and tRNA as well as between tRNA and its template-binding site are required to form a stable initiation complex. The T-ψC interaction region in avian sarcoma virus (ASV) appears to serve the same function as the U5-inverted repeat in HIV-1 RNA (J. Leis, pers. comm.). There is now good evidence that secondary structures in the Tf1 transposable element that are equivalent to the T-ψC interaction region in the ASV RNA are needed for efficient initiation of reverse transcription in Tf1 as well (Lin and Levin 1997).

- **MITOCHONDRIAL AND KINETOPLAST DNA REPLICATION**

Although the material described in the corresponding CSH monograph chapters remain accurate, the proposed models are now supported by additional experimental data (for review, see Shadel and Clayton 1997). The precursor primer RNA for mammalian mitochondrial DNA leading-strand replication remains as a persistent R loop formed during transcription through the mitochondrial DNA control region. Mouse mitochondrial RNA processing ribonuclease (MRP) accurately cleaves an R loop containing the mouse mitochondrial DNA origin, suggesting that RNase MRP alone is capable of generating virtually all of the leading-strand replication primers (Lee and Clayton 1997).

Kinetoplast DNA, the mitochondrial DNA of trypanosomatids, consists of a network of several thousand intact, topologically interlocked DNA minicircles. Replication involves the release of minicircles from the interior of the network, the synthesis of nicked or gapped progeny minicircles, and the reattachment of the progeny to the network periphery. A DNA primase has been purified from *Crithidia fasciculata* that synthesizes RNA primers approximately ten nucleotides in length (Li and Englund 1997). This enzyme localizes to specific regions of the cell's single mitochondrion, above and below the condensed kinetoplast DNA. It does not colocalize with the mitochondrial topoisomerase II and DNA polymerase β, both of which are associated with two protein complexes positioned on opposite sides of the kinetoplast disc. These results suggest a compartmentalization of function in which minicircle DNA replication occurs in the region above and below the kinetoplast disc while replication is completed (i.e., gap-filling) at the two antipodal protein complexes.

ACKNOWLEGMENTS

I am indebted to the following people for their advice, and in particular, to William Burhans for the section on "DNA Replication and Checkpoint

Controls" and to Crisanto Gutierrez for the section on "Geminivirus DNA Replication."

David Bisaro, Ohio State University, Columbus, OH
Julian Blow, University of Dundee, Dundee, Scotland
William Burhans, Roswell Park Cancer Institute, Buffalo, NY
Walter Fangman, University of Washington, Seattle, WA
Lori Frappier, University of Toronto, Toronto, Ontario, Canada
Susan Gerbi, Brown University, Providence, RI
Crisanto Gutierrez, Centro de Biologia Molecular "Severo Ochoa," Madrid, Spain
Joel Huberman, Roswell Park Cancer Institute, Buffalo, NY
Ulrich Hübscher, University of Zürich, Zürich, Switzerland
David Kowalski, Roswell Park Cancer Institute, Buffalo, NY
Jonathan Leis, Case Western Reserve University Medical School, Cleveland, OH
Cong-Jun Li, NICHD, Bethesda, MD
Tomas Lindahl, ICRF, Herts, UK
Carole Newlon, New Jersey Medical School, Newark, NJ
Joseph Pagano, University of North Carolina, Chapel Hill, NC
George Rohrmann, Oregon State University, Corvallis, OR
Bruce Stillman, Cold Spring Harbor Laboratory, Cold Spring Harbor, NY
Fydor Urnov, Brown University, Providence, RI
Peter van der Vliet, Utrecht University, Utrecht, The Netherlands
Teresa Wang, Stanford University Medical School, Stanford, CA
Marc Wold, University of Iowa Medical School, Iowa City, IA
John Yates, Roswell Park Cancer Institute, Buffalo, NY

REFERENCES

Abdurashidova, G., S. Riva, G. Biamonti, M. Giacca, and A. Falaschi. 1998. Cell cycle modulation of protein-DNA interactions at a human replication origin. *EMBO J.* **7:** 2961–2969.

Ach, R.A., T. Durfee, A.B. Miller, P. Taranto, L. Hanley-Bowdoin, P.C. Zambryski, and W. Gruissem. 1997. RRB1 and RRB2 encode maize retinoblastoma-related proteins that interact with a plant D-type cyclin and geminivirus replication protein. *Mol. Cell Biol.* **17:** 5077–5086.

Akbar Behjatnia, S.A., I.B. Dry, and M. Ali Rezaian. 1998. Identification of the replication-associated protein binding domain within the intergenic region of tomato leaf curl geminivirus. *Nucleic Acids Res.* **26:** 925–931.

Aladjem, M.I., L.W. Rodewald, J.L. Kolman, and G.M. Wahl. 1998. Genetic dissection of a mammalian replicator in the human beta-globin locus. *Science* **281:** 1005–1009.

Aladjem, M.I., M. Groudine, L.L. Brody, E.S. Dieken, R.E. Fournier, G.M. Wahl, and E.M. Epner. 1995. Participation of the human β-globin locus control region in initia-

tion of DNA replication. *Science* **270**: 815–819.
Aparicio, O.M., D.M. Weinstein, and S.P. Bell. 1997. Components and dynamics of DNA replication complexes in *S. cerevisiae:* Redistribution of MCM proteins and Cdc45p during S phase. *Cell* **91**: 59–69.
Araujo, F.D., J.D. Knox, M. Szyf, G.B. Price, and M. Zannis-Hadjopoulos. 1998. Concurrent replication and methylation at mammalian origins of replication. *Mol. Cell Biol.* **18**: 3475–3482.
Baker, T.A. and S.P. Bell. 1998. Polymerases and the replisome: Machines within machines. *Cell* **92**: 295–305.
Bambara, R.A., R.S. Murante, and L.A. Henricksen. 1997. Enzymes and reactions at the eukaryotic DNA replication fork. *J. Biol. Chem.* **272**: 4647–4650.
Benedetto, J.P., R. Ech-Chaoui, J. Plissonneau, P. Laquel, S. Litvak, and M. Castroviejo. 1996. Changes of enzymes and factors involved in DNA synthesis during wheat embryo germination. *Plant Mol. Biol.* **31**: 1217–1225.
Bielinsky, A.K. and S.A. Gerbi. 1998. Discrete start sites for DNA synthesis in the yeast ARS1 origin. *Science* **279**: 95–98.
Blomberg, P., C. Randolph, C.H. Yao, and M.C. Yao. 1997. Regulatory sequences for the amplification and replication of the ribosomal DNA minichromosome in *Tetrahymena thermophila*. *Mol. Cell Biol.* **17**: 7237–7247.
Blumental, A.B., H.J. Kriegstein, and D.S. Hogness. 1974. The units of DNA replication in *Drosophila melanogaster* chromosomes. *Cold Spring Harbor Symp. Quant. Biol.* **38**: 205–223.
Bochkareva, E., L. Frappier, A.M. Edwards, and A. Bochkarev. 1998. The RPA32 subunit of human replication protein A contains a single-stranded DNA-binding domain. *J. Biol. Chem.* **273**: 3932–3936.
Boehmer, P.E. and I.R. Lehman. 1997. Herpes simplex virus DNA replication. *Annu. Rev. Biochem.* **66**: 347–384.
Bousset, K. and J.F. Diffley. 1998. The Cdc7 protein kinase is required for origin firing during S phase. *Genes Dev.* **12**: 480–490.
Brewer, B.J. and W.L. Fangman. 1993. Initiation at closely spaced replication origins in a yeast chromosome. *Science* **262**: 1728–1731.
———. 1994. Initiation preference at a yeast origin of replication. *Proc. Natl. Acad. Sci.* **91**: 3418–3422.
Brunstein, J. and C.R. Astell. 1997. Analysis of the internal replication sequence indicates that there are three elements required for efficient replication of minute virus of mice minigenomes. *J. Virol.* **71**: 9087–9095.
Brush, G.S., D.M. Morrow, P. Hieter, and T.J. Kelly. 1996. The ATM homologue MEC1 is required for phosphorylation of replication protein A in yeast. *Proc. Natl. Acad. Sci.* **93**: 15075–15080.
Calvi, B.R., M.A. Lilly, and A.C. Spradling. 1998. Cell cycle control of chorion gene amplification. *Genes Dev.* **12**: 734–744.
Carty, M.P., M. Zernik-Kobak, S. McGrath, and K. Dixon. 1994. UV light-induced DNA synthesis arrest in HeLa cells is associated with changes in phosphorylation of human single-stranded DNA-binding protein. *EMBO J.* **13**: 2114–2123.
Chan, B. and K. Musier-Forsyth. 1997. The nucleocapsid protein specifically anneals tRNALys-3 onto a noncomplementary primer binding site within the HIV-1 RNA genome in vitro. *Proc. Natl. Acad. Sci.* **94**: 13530–13535.
Chang, L., S.S. Loranger, C. Mizzen, S.G. Ernst, C.D. Allis, and A.T. Annunziato. 1997.

Histones in transit: Cytosolic histone complexes and diacetylation of H4 during nucleosome assembly in human cells. *Biochemistry* **36:** 469–480.

Chen, G. and A. Stenlund. 1998. Characterization of the DNA-binding domain of the bovine papillomavirus replication initiator E1. *J. Virol.* **72:** 2567–2576.

Christensen, J., S.F. Cotmore, and P. Tattersall. 1997a. A novel cellular site-specific DNA-binding protein cooperates with the viral NS1 polypeptide to initiate parvovirus DNA replication. *J. Virol.* **71:** 1405–1416.

———. 1997b. Parvovirus initiation factor PIF: A novel human DNA-binding factor which coordinately recognizes two ACGT motifs. *J. Virol.* **71:** 5733–4741.

Coleman, T.R., P.B. Carpenter, and W.G. Dunphy. 1996. The *Xenopus* Cdc6 protein is essential for the initiation of a single round of DNA replication in cell-free extracts. *Cell* **87:** 53–63.

Collin, S., M. Fernandez-Lobato, P.S. Gooding, P.M. Mullineaux, and C. Fenoll. 1996. The two nonstructural proteins from wheat dwarf virus involved in viral gene expression and replication are retinoblastoma-binding proteins. *Virology* **219:** 324–329.

Cossons, N., E.A. Faust, and M. Zannis-Hadjopoulos. 1996. DNA polymerase δ-dependent formation of a hairpin structure at the 5′ terminal palindrome of the minute virus of mice genome. *Virology* **216:** 258–264.

Coverley, D. and R.A. Laskey. 1994. Regulation of eukaryotic DNA replication. *Annu. Rev. Biochem.* **63:** 745–776.

Coverley, D., H.R. Wilkinson, M.A. Madine, A.D. Mills, and R.A. Laskey. 1998. Protein kinase inhibition in G2 causes mammalian Mcm proteins to reassociate with chromatin and restores ability to replicate. *Exp. Cell Res.* **238:** 63–69.

D'Urso, G. and P. Nurse. 1997. *Schizosaccharomyces pombe cdc20⁺* encodes DNA polymerase epsilon and is required for chromosomal replication but not for the S phase checkpoint. *Proc. Natl. Acad. Sci.* **94:** 12491–12496.

Dekker, J., P.N. Kanellopoulos, J. van Oosterhout, G. Stier, P.A. Tucker, and P.C. van der Vliet. 1998. ATP-independent DNA unwinding by the adenovirus single-stranded DNA binding protein requires a flexible DNA binding loop. *J. Mol. Biol.* **277:** 825–838.

Delgado, S., M. Gomez, A. Bird, and F. Antequera. 1998. Initiation of DNA replication at CpG islands in mammalian chromosomes. *EMBO J.* **17:** 2426–2435.

DePamphilis, M.L. 1996. Origins of DNA replication. In *DNA replication in eukaryotic cells* (ed. M.L. DePamphilis), pp. 45–86. Cold Spring Harbor Laboratory Press, New York.

——— 1997. Identification and analysis of replication origins in eukaryotic cells. *Methods* 209–324.

DePamphilis, M.L., E. Martínez-Salas, D.Y. Cupo, E.A. Hendrickson, C.E. Fritze, W.R. Folk, and U. Heine. 1988. Initiation of polyomavirus and SV40 DNA replication, and the requirements for DNA replication during mammalian development. In *Eukaryotic DNA replication* (ed. B. Stillman and T. Kelly), pp. 165–175. Cold Spring Harbor Laboratory Press, Cold Spring Harbor, New York.

Deshpande, A.M. and C.S. Newlon. 1996. DNA replication fork pause sites dependent on transcription. *Science* **272:** 1030–1033.

Donaldson, A.D., W.L. Fangman, and B.J. Brewer. 1998a. Cdc7 is required throughout the yeast S phase to activate replication origins. *Genes Dev.* **12:** 491–501.

Donaldson, A.D., M.K. Raghuraman, K.L. Friedman, F.R. Cross, B.J. Brewer, and W.L. Fangman. 1998b. CLB5-dependent activation of late replication origins in *S. cerevisiae*. *Mol. Cell* (in press).

Donovan, S., J. Harwood, L.S. Drury, and J.F. Diffley. 1997. Cdc6p-dependent loading of Mcm proteins onto pre-replicative chromatin in budding yeast. *Proc. Natl. Acad. Sci.* **94:** 5611–5616.

Drury, L.S., G. Perkins, and J.F. Diffley. 1997. The Cdc4/34/53 pathway targets Cdc6p for proteolysis in budding yeast. *EMBO J.* **16:** 5966–5976.

Dubey, D.D., S.M. Kim, I.T. Todorov, and J.A. Huberman. 1996. Large, complex modular structure of a fission yeast DNA replication origin. *Curr. Biol.* **6:** 467–473.

Dubey, D.D., J. Zhu, D.L. Carlson, K. Sharma, and J.A. Huberman. 1994. Three ARS elements contribute to the ura4 replication origin region in the fission yeast, *Schizosaccharomyces pombe. EMBO J.* **13:** 3638–3647.

Dutta, A. and S.P. Bell. 1997. Initiation of DNA replication in eukaryotic cells. *Annu. Rev. Cell. Dev. Biol.* **13:** 293–332.

Edwards, A.M., A. Bochkarev, and L. Frappier. 1998. Origin DNA-binding proteins. *Curr. Opin. Struct. Biol.* **8:** 49–53.

Eissenberg, J.C., R. Ayyagari, X.V. Gomes, and P.M. Burgers. 1997. Mutations in yeast proliferating cell nuclear antigen define distinct sites for interaction with DNA polymerase delta and DNA polymerase epsilon. *Mol. Cell. Biol.* **17:** 6367–6378.

Elledge, S.J. 1996. Cell cycle checkpoints: Preventing an identity crisis. *Science* **274:** 1664–1672.

Ellis, D.J., H. Jenkins, W.G. Whitfield, and C.J. Hutchison. 1997. GST-lamin fusion proteins act as dominant negative mutants in *Xenopus* egg extract and reveal the function of the lamina in DNA replication. *J. Cell. Sci.* **110:** 2507–2518.

Ellison, V. and B. Stillman. 1998. Reconstitution of recombinant human replication factor C (RFC) and identification of an RFC subcomplex possessing DNA-dependent ATPase activity. *J. Biol. Chem.* **273:** 5979–5987.

Ermakova, O.V., L. Frappier, and C.L. Schildkraut. 1996. Role of the EBNA-1 protein in pausing of replication forks in the Epstein-Barr virus genome. *J. Biol. Chem.* **271:** 33009–33017.

Follette, P.J., R.J. Duronio, and P.H. O'Farrell. 1998. Fluctuations in cyclin E levels are required for multiple rounds of endocycle S phase in *Drosophila. Curr. Biol.* **8:** 235–238.

Friedman, K.L., J.D. Diller, B.M. Ferguson, S.V. Nyland, B.J. Brewer, and W.L. Fangman. 1996. Multiple determinants controlling activation of yeast replication origins late in S phase. *Genes Dev.* **10:** 1595–1607.

Garcia, E., D. Orjuela, Y. Camacho, J.J. Zuniga, J. Plasencia, and J.M. Vazquez-Ramos. 1997. Comparison among DNA polymerases 1, 2 and 3 from maize embryo axes. A DNA primase activity copurifies with DNA polymerase 2. *Plant Mol. Biol.* **33:** 445–455.

Gencheva, M., B. Anachkova, and G. Russev. 1996. Mapping the sites of initiation of DNA replication in rat and human rRNA genes. *J. Biol. Chem.* **271:** 2608–2614.

Gerber, J.K., E. Gogel, C. Berger, M. Wallisch, F. Muller, I. Grummt, and F. Grummt. 1997. Termination of mammalian rDNA replication: Polar arrest of replication fork movement by transcription termination factor TTF-I. *Cell* **90:** 559–567.

Gerik, K.J., S.L. Gary, and P.M. Burgers. 1997. Overproduction and affinity purification of *Saccharomyces cerevisiae* replication factor C. *J. Biol. Chem.* **272:** 1256–1262.

Gilbert, D.M., H. Miyazawa, and M.L. DePamphilis. 1995. Site-specific initiation of DNA replication in *Xenopus* egg extract requires nuclear structure. *Mol. Cell. Biol.* **1555:** 2942–2954.

Gladfelter, H.J., P.A. Eagle, E.P. Fontes, L. Batts, and L. Hanley-Bowdoin. 1997. Two domains of the AL1 protein mediate geminivirus origin recognition. *Virology* **239:** 186–197.

Gogel, E., G. Langst, I. Grummt, E. Kunkel, and F. Grummt. 1996. Mapping of replication initiation sites in the mouse ribosomal gene cluster. *Chromosoma* **104:** 511–518.

Goldberg, M., H. Jenkins, T. Allen, W.G. Whitfield, and C.J. Hutchison. 1995. *Xenopus* lamin B3 has a direct role in the assembly of a replication competent nucleus: Evidence from cell-free egg extracts. *J. Cell. Sci.* **108:** 3451–3461.

Grafi, G., R.J. Burnett, T. Helentjaris, B.A. Larkins, J.A. DeCaprio, W.R. Sellers, and W.G. Kaelin, Jr. 1996. A maize cDNA encoding a member of the retinoblastoma protein family: Involvement in endoreduplication. *Proc. Natl. Acad. Sci.* **93:** 8962–8967.

Handeli, S., A. Klar, M. Meuth, and H. Cedar. 1989. Mapping replication units in animal cells. *Cell* **57:** 909–920.

Heichman, K.A. and J.M. Roberts. 1998. CDC16 controls initiation at chromosome replication origins. *Mol. Cell.* **1:** 457–463.

Hoogstraten, R.A., S.F. Hanson, and D.P. Maxwell. 1996. Mutational analysis of the putative nicking motif in the replication-associated protein (AC1) of bean golden mosaic geminivirus. *Mol. Plant-Microbe Interact.* **9:** 594–599.

Horvath, G., V. Pettko-Szandtner, K. Kelemen, M. Bilgin, M. Boulton, J. Davies, C. Gutiérrez, and D. Dudits. 1998. Prediction of functionally significant regions of the maize streak virus Rep protein by protein-protein interaction analysis. *Plant Mol. Biol.* (in press).

Hozák, P., D.A. Jackson, and P.R. Cook. 1996. Role of nuclear structure in DNA replication. In *Eukaryotic DNA replication* (ed. J.J. Blow), pp. 124–142. IRL Press, Oxford.

Hua, X.H. and J. Newport. 1998. Identification of a preinitiation step in DNA replication that is independent of origin recognition complex and cdc6, but dependent on cdk2. *J. Cell. Biol.* **140:** 271–281.

Hua, X.H., H. Yan, and J. Newport. 1997. A role for Cdk2 kinase in negatively regulating DNA replication during S phase of the cell cycle. *J. Cell. Biol.* **137:** 183–192.

Huang, R.Y. and D. Kowalski. 1996. Multiple DNA elements in ARS305 determine replication origin activity in a yeast chromosome. *Nucleic Acids Res.* **24:** 816–823.

Huberman, J. 1998. DNA damage and replication checkpoints in the fission yeast, *Schizosaccharomyces pombe*. *Prog. Nucleic Acids Res. Mol. Biol.* (in press).

Hyrien, O. and M. Mechali. 1993. Chromosomal replication initiates and terminates at random sequences but at regular intervals in the ribosomal DNA of *Xenopus* early embryos. *EMBO J.* **12:** 4511–4520.

Hyrien, O., C. Maric, and M. Mechali. 1995. Transition in specification of embryonic metazoan DNA replication origins. *Science* **270:** 994–997.

Iftode, C. and J.A. Borowiec. 1997. Denaturation of the simian virus 40 origin of replication mediated by human replication protein A. *Mol. Cell. Biol.* **17:** 3876–3883.

Irniger, S. and K. Nasmyth. 1997. The anaphase-promoting complex is required in G1 arrested yeast cells to inhibit B-type cyclin accumulation and to prevent uncontrolled entry into S-phase. *J. Cell. Sci.* **110:** 1523–1531.

Ishimi, Y. 1997. A DNA helicase activity is associated with an MCM4, -6, and -7 protein complex. *J. Biol. Chem.* **272:** 24508–24513.

Jallepalli, P.V. and T.J. Kelly. 1997. Cyclin-dependent kinase and initiation at eukaryotic origins: A replication switch? *Curr. Opin. Cell. Biol.* **9:** 358–363.

Jenkins, H., T. Holman, C. Lyon, B. Lane, R. Stick, and C. Hutchison. 1993. Nuclei that

lack a lamina accumulate karyophilic proteins and assemble a nuclear matrix. *J. Cell. Sci.* **106:** 275–85.
Jonsson, Z.O., R. Hindges, and U. Hubscher. 1998. Regulation of DNA replication and repair proteins through interaction with the front side of proliferating cell nuclear antigen. *EMBO J.* **17:** 2412–2425.
Katsuhiko, S. and H. Yoshikawa. 1997. Replicon dynamics of chromosome VI of *S. cerevisiae*. *Genes Cells* **2:** 655–665.
King, A.J., W.R. Teertstra, and P.C. van der Vliet. 1997. Dissociation of the protein primer and DNA polymerase after initiation of adenovirus DNA replication. *J. Biol. Chem.* **272:** 24617–24623.
Kipling, D. and S.E. Kearsey. 1990. Reversion of autonomously replicating sequence mutations in *Saccharomyces cerevisiae:* Creation of a eucaryotic replication origin within procaryotic vector DNA. *Mol. Cell. Biol.* **10:** 265–272.
Kirchmaier, A.L. and B. Sugden. 1998. Rep*: A viral element that can partially replace the origin of plasmid DNA synthesis of Epstein-Barr virus. *J. Virol.* **72:** 4657–4666.
Kirk, B.W., C. Harrington, F.W. Perrino, and R.D. Kuchta. 1997. Eucaryotic DNA primase does not prefer to synthesize primers at pyrimidine rich DNA sequences when nucleoside triphosphates are present at concentrations found in whole cells. *Biochemistry* **36:** 6725–6731.
Kitsberg, D., S. Selig, I. Keshet, and H. Cedar. 1993. Replication structure of the human β-globin gene domain. *Nature* **366:** 588–590.
Kobayashi, T., T. Rein, and M.L. DePamphilis. 1998. Identification of primary initiation sites for DNA replication in the hamster dihydrofolate reductase gene initiation zone. *Mol. Cell. Biol.* **18:** 3266–3277.
Kornberg, A. and T.A. Baker. 1992. *DNA replication*. W.H. Freeman, New York.
Krude, T., M. Jackman, J. Pines, and R.A. Laskey. 1997. Cyclin/Cdk-dependent initiation of DNA replication in a human cell-free system. *Cell* **88:** 109–119.
Kumar, S., M. Giacca, P. Norio, G. Biamonti, S. Riva, and A. Falaschi. 1996. Utilization of the same DNA replication origin by human cells of different derivation. *Nucleic Acids Res.* **24:** 3289–3294.
Landis, G., R. Kelley, A.C. Spradling, and J. Tower. 1997. The k43 gene, required for chorion gene amplification and diploid cell chromosome replication, encodes the *Drosophila* homolog of yeast origin recognition complex subunit 2. *Proc. Natl. Acad. Sci.* **94:** 3888–3892.
Laufs, J., S. Schumacher, N. Geisler, I. Jupin, and B. Gronenborn. 1995. Identification of the nicking tyrosine of geminivirus Rep protein. *FEBS Lett.* **377:** 258–262.
Lawlis, S.J., S.M. Keezer, J.-R. Wu, and D.M. Gilbert. 1996. Chromosome architecture can dictate site-specific initiation of DNA replication in *Xenopus* egg extracts. *J. Cell Biol.* **135:** 1207–1218.
Lee, D.G. and S.P. Bell. 1997. Architecture of the yeast origin recognition complex bound to origins of DNA replication. *Mol. Cell. Biol.* **17:** 7159–7168.
Lee, D.Y. and D.A. Clayton. 1997. RNase mitochondrial RNA processing correctly cleaves a novel R loop at the mitochondrial DNA leading-strand origin of replication. *Genes Dev.* **11:** 582–592.
Lee, S.S. and I.R. Lehman. 1997. Unwinding of the box I element of a herpes simplex virus type 1 origin by a complex of the viral origin binding protein, single-strand DNA binding protein, and single-stranded DNA. *Proc. Natl. Acad. Sci.* **94:** 2838–2842.
Li, C. and P.T. Englund. 1997. A mitochondrial DNA primase from the trypanosomatid

Crithidia fasciculata. J. Biol. Chem. **272:** 20787–20792.

Li, X., C. Liang, Y. Quan, R. Chandok, M. Laughrea, M.A. Parniak, L. Kleiman, and M.A. Wainberg. 1997. Identification of sequences downstream of the primer binding site that are important for efficient replication of human immunodeficiency virus type 1. *J. Virol.* **71:** 6003–6010.

Li, X., Y. Quan, E.J. Arts, Z. Li, B.D. Preston, H. de Rocquigny, B.P. Roques, J.L. Darlix, L. Kleiman, M.A. Parniak, and M.A. Wainberg. 1996. Human immunodeficiency virus type 1 nucleocapsid protein (NCp7) directs specific initiation of minus-strand DNA synthesis primed by human tRNA(Lys3) in vitro: Studies of viral RNA molecules mutated in regions that flank the primer binding site. *J. Virol.* **70:** 4996–5004.

Li, Y., Z. Zhang, J.K. Wakefield, S.M. Kang, and C.D. Morrow. 1997. Nucleotide substitutions within U5 are critical for efficient reverse transcription of human immunodeficiency virus type 1 with a primer binding site complementary to tRNA(His). *J. Virol.* **71:** 6315–6322.

Liang, C. and B. Stillman. 1997. Persistent initiation of DNA replication and chromatin-bound MCM proteins during the cell cycle in cdc6 mutants. *Genes Dev.* **11:** 3375–3386.

Lin, J.H., and H.L. Levin. 1997. A complex structure in the mRNA of Tf1 is recognized and cleaved to generate the primer of reverse transcription. *Genes Dev.* **11:** 270-285.

Lin, S. and D. Kowalski 1997. Functional equivalency and diversity of *cis*-acting elements among yeast replication origins. *Mol. Cell. Biol.* **17:** 5473–5484.

Lindahl, T., D.E. Barnes, A. Klungland, V.J. Mackenney, and P. Schar. 1997. Repair and processing events at DNA ends. *Ciba Found. Symp.* **211:** 198–205 (discussion 205–208).

Little, R.D. and C.L. Schildkraut. 1995. Initiation of latent DNA replication in the Epstein-Barr virus genome can occur at sites other than the genetically defined origin. *Mol. Cell. Biol.* **15:** 2893–2903.

Little, R.D., T.H. Platt, and C.L. Schildkraut. 1993. Initiation and termination of DNA replication in human rRNA genes. *Mol. Cell. Biol.* **13:** 6600–6613.

Liu, V.F. and D.T. Weaver. 1993. The ionizing radiation-induced replication protein A phosphorylation response differs between ataxia telangiectasia and normal human cells. *Mol. Cell. Biol.* **13:** 7222–7231.

Longhese, M.P., H. Neecke, V. Paciotti, G. Lucchini, and P. Plevani. 1996. The 70 kDa subunit of replication protein A is required for the G1/S and intra-S DNA damage checkpoints in budding yeast. *Nucleic Acids Res.* **24:** 3533–3537.

Longley, M.J., A.J. Pierce, and P. Modrich. 1997. DNA polymerase δ is required for human mismatch repair in vitro. *J. Biol. Chem.* **72:** 10917–10921.

Lopez-estrano, C., J.B. Schvartzman, D.B. Krimer, and P. Hernandez. 1998. Co-localization of polar replication fork barriers and rRNA transcription terminators in mouse rDNA. *J. Mol. Biol.* **277:** 249–256.

Lu, L. and J. Tower. 1997. A transcriptional insulator element, the su(Hw) binding site, protects a chromosomal DNA replication origin from position effects. *Mol. Cell. Biol.* **17:** 2202–2206.

Lu, Z.H., D.B. Sittman, P. Romanowski, and G.H. Leno. 1998. Histone H1 reduces the frequency of initiation in *Xenopus* egg extract by limiting the assembly of prereplication complexes on sperm chromatin. *Mol. Biol. Cell.* **9:** 1163–1176.

Lund, A.H., M. Duch, J. Lovmand, P. Jorgensen, and F.S. Pedersen. 1997. Complementation of a primer binding site-impaired murine leukemia virus-derived retroviral vector by a genetically engineered tRNA-like primer. *J. Virol.* **71:** 1191-1195.

MacAlpine, D.M., A. Zhang, and G.M. Kapler. 1997. Type I elements mediate replication fork pausing at conserved upstream sites in the *Tetrahymena thermophila* ribosomal DNA minichromosome. *Mol. Cell. Biol.* **17:** 4517-4525.

Maga, G., R. Mossi, R. Fischer, M.W. Berchtold, and U. Hubscher. 1997. Phosphorylation of the PCNA binding domain of the large subunit of replication factor C by Ca^{2+}/calmodulin-dependent protein kinase II inhibits DNA synthesis. *Biochemistry* **36:** 5300-5310.

Mahbubani, H.M., J.P. Chong, S. Chevalier, P. Thommes, and J.J. Blow. 1997. Cell cycle regulation of the replication licensing system: Involvement of a Cdk-dependent inhibitor. *J. Cell. Biol.* **136:** 125-135.

Majumder, S., Z. Zhao, K. Kaneko, and M.L. DePamphilis. 1997. Developmental acquisition of enhancer function requires a unique co-activator activity. *EMBO J.* **16:** 1721-1731.

Marahrens, Y. and B. Stillman. 1996. Initiation of DNA replication in the yeast *Saccharomyces cerevisiae*. In *Eukaryotic DNA replication* (ed. J.J. Blow), pp. 66-95. IRL Press.

Marini, F., A. Pellicioli, V. Paciotti, G. Lucchini, P. Plevani, D.F. Stern, and M. Foiani. 1997. A role for DNA primase in coupling DNA replication to DNA damage response. *EMBO J.* **16:** 639-650.

Martinez-Salas, E., D.Y. Cupo, and M.L. DePamphilis. 1988. The need for enhancers is acquired upon formation of a diploid nucleus during early mouse development. *Genes Dev.* **2:** 1115-1126.

Mendez, J., L. Blanco, and M. Salas. 1997. Protein-primed DNA replication: A transition between two modes of priming by a unique DNA polymerase. *EMBO J.* **16:** 2519-2527.

Michaelson, J.S., O. Ermakova, B.K. Birshtein, N. Ashouian, C. Chevillard, R. Riblet, and C.L. Schildkraut. 1997. Regulation of the replication of the murine immunoglobulin heavy chain gene locus: Evaluation of the role of the 3′ regulatory region. *Mol. Cell. Biol.* **17:** 6167-6174.

Mohanty, B.K., T. Sahoo, and D. Bastia. 1998. Mechanistic studies on the impact of transcription on sequence-specific termination of DNA replication and vice versa. *J. Biol. Chem.* **273:** 3051-3159.

Mossi, R., Z.O. Jonsson, B.L. Allen, S.H. Hardin, and U. Hubscher. 1997. Replication factor C interacts with the C-terminal side of proliferating cell nuclear antigen. *J. Biol. Chem.* **272:** 1769-1776.

Mozzherin, D.J. and P.A. Fisher. 1996. Human DNA polymerase ε: Enzymologic mechanism and gap-filling synthesis. *Biochemistry* **35:** 3572-3577.

Mozzherin, D.J., S. Shibutani, C.K. Tan, K.M. Downey, and P.A. Fisher. 1997. Proliferating cell nuclear antigen promotes DNA synthesis past template lesions by mammalian DNA polymerase δ. *Proc. Natl. Acad. Sci.* **94:** 6126-6131.

Murante, R.S., L.A. Henricksen, and R.A. Bambara. 1998. Junction ribonuclease: An activity in Okazaki fragment processing. *Proc. Natl. Acad. Sci.* **95:** 2244-2249.

Newlon, C.S. 1997. Putting it all together: Building a prereplicative complex. *Cell* **91:** 717-720.

Niculescu, A.B.R., X. Chen, M. Smeets, L. Hengst, C. Prives, and S.I. Reed. 1998. Effects of p21(Cip1/Waf1) at both the G1/S and the G2/M cell cycle transitions: pRb is a critical determinant in blocking DNA replication and in preventing endoreduplication (erratum *Mol. Cell. Biol.* [1998] **18:** 1763). *Mol. Cell. Biol.* **18:** 629–643.

Orozco, B.M. and L. Hanley-Bowdoin. 1996. A DNA structure is required for geminivirus replication origin function. *J. Virol.* **70:** 148–158.

Orozco, B.M., A.B. Miller, S.B. Settlage, and L. Hanley-Bowdoin. 1997. Functional domains of a geminivirus replication protein. *J. Biol. Chem.* **272:** 9840–9846.

Orozco, B.M., H.J. Gladfelter, S.B. Settlage, P.A. Eagle, R.N. Gentry, and L. Hanley-Bowdoin. 1998. Multiple *cis* elements contribute to geminivirus origin function. *Virology* **242:** 346–356.

Painter, R.B. and B.R. Young. 1980. Radiosensitivity in ataxia-telangiectasia: A new explanation. *Proc. Natl. Acad. Sci.* **77:** 7315–7317.

Park, K., Z. Debyser, S. Tabor, C.C. Richardson, and J.D. Griffith. 1998. Formation of a DNA loop at the replication fork generated by bacteriophage T7 replication proteins. *J. Biol. Chem.* **273:** 5260–5270.

Pasero, P., D. Braguglia, and S.M. Gasser. 1997. ORC-dependent and origin-specific initiation of DNA replication at defined foci in isolated yeast nuclei. *Genes Dev.* **11:** 1504–1518.

Paulovich, A.G., D.P. Toczyski, and L.H. Hartwell. 1997. When checkpoints fail. *Cell* **88:** 315–321.

Pelizon, C., S. Diviacco, A. Falaschi, and M. Giacca. 1996. High-resolution mapping of the origin of DNA repication in the hamster dihdrofolate reductase gene domain by competitive PCR. *Mol. Cell. Biol.* **16:** 5358–5364.

Piatti, S., T. Bohm, J.H. Cocker, J.F. Diffley, and K. Nasmyth. 1996. Activation of S-phase-promoting CDKs in late G1 defines a "point of no return" after which Cdc6 synthesis cannot promote DNA replication in yeast. *Genes Dev.* **10:** 1516–1531.

Pichler, S., S. Piatti, and K. Nasmyth. 1997. Is the yeast anaphase promoting complex needed to prevent re-replication during G2 and M phases? *EMBO J.* **16:** 5988–5997.

Podust, V.N. and E. Fanning. 1997. Assembly of functional replication factor C expressed using recombinant baculoviruses. *J. Biol. Chem.* **272:** 6303–6310.

Raghuraman, M.K., B.J. Brewer, and W.L. Fangman. 1997) Cell cycle-dependent establishment of a late replication program. *Science* **276:** 806–809.

Rein, T., H. Zorbas, and M.L. DePamphilis. 1997a. Active mammalian replication origins are associated with a high-density cluster of mCpG dinucleotides. *Mol. Cell. Biol.* **17:** 416–426.

Rein, T., D.A. Natale, U. Gartner, M. Niggemann, M.L. DePamphilis, and H. Zorbas. 1997b. Absence of an unusual "densely methylated island" at the hamster dhfr ori-beta. *J. Biol. Chem.* **272:** 10021–10029.

Rigden, J.E., I.B. Dry, L.R. Krake, and M.A. Rezaian. 1996. Plant virus DNA replication processes in *Agrobacterium:* Insight into the origins of geminiviruses? *Proc. Natl. Acad. Sci.* **93:** 10280–10284.

Romanowski, P., M.A. Madine, A. Rowles, J.J. Blow, and R.A. Laskey. 1996. The *Xenopus* origin recognition complex is essential for DNA replication and MCM binding to chromatin. *Curr. Biol.* **6:** 1416–1425.

Rowles, A. and J.J. Blow. 1997. Chromatin proteins involved in the initiation of DNA replication. *Curr. Opin. Genet. Dev.* **7:** 152–157.

Saha, P., K.C. Thome, R. Yamaguchi, Z. Hou, S. Weremowicz, and A. Dutta. 1998. The human homolog of Saccharomyces cerevisiae CDC45. *J. Biol. Chem.* **273:** 18205–18209.

Sanchez, J.A., S.M. Kim, and J.A. Huberman. 1998a. Ribosomal DNA replication in the fission yeast, *Schizosaccharomyces pombe*. *Exp. Cell Res.* **238:** 220–230.

Sanchez, J.P., Y. Murakami, J.A. Huberman, and J. Hurwitz. 1998b. Isolation, characterization, and molecular cloning of a protein (Abp2) that binds to a *Schizosaccharomyces pombe* origin of replication (ars3002). *Mol. Cell. Biol.* **18:** 1670–1681.

Sanchez, Y., B.A. Desany, W.J. Jones, Q. Liu, B. Wang, and S.J. Elledge. 1996. Regulation of RAD53 by the ATM-like kinases MEC1 and TEL1 in yeast cell cycle checkpoint pathways (see comments). *Science* **271:** 357–360.

Sanders, C.M. and A. Stenlund. 1998. Recruitment and loading of the E1 initiator protein. An ATP-dependent process catalysed by a transcription factor. *EMBO J.* (in press).

Santocanale, C. and J.F.X. Diffley. 1998. A Mec1- and Rad53-dependent checkpoint controls late firing origins of DNA replication. *Nature* (in press).

Santocanale, C. and J.F. Diffley. 1996. ORC- and Cdc6-dependent complexes at active and inactive chromosomal replication origins in *Saccharomyces cerevisiae*. *EMBO J.* **15:** 6671–6679.

Sanz-Burgos, A.P. and C. Gutierrez. 1998. Organization of the *cis*-acting element required for wheat dwarf geminivirus DNA replication and visualization of a rep protein-DNA complex. *Virology* **243:** 119–129.

Sasaki, T., T. Sawado, M. Yamaguchi, and T. Shinomiya. 1998. Specification of initiation regons of DNA replication during embryogenesis in the 65 kb DNApola-dE2F locus of *Drosophila melanogaster*. *Mol. Cell. Biol.* (in press).

Sato, N., K. Arai, and H. Masai. 1997. Human and *Xenopus* cDNAs encoding budding yeast Cdc7-related kinases: In vitro phosphorylation of MCM subunits by a putative human homologue of Cdc7. *EMBO J.* **16:** 4340–4351.

Schurtenberger, P., S.U. Egelhaaf, R. Hindges, G. Maga, Z.O. Jonsson, R.P. May, O. Glatter, and U. Hubscher. 1998. The solution structure of functionally active human proliferating cell nuclear antigen determined by small-angle neutron scattering. *J. Mol. Biol.* **275:** 123–132.

Settlage, S.B., A.B. Miller, and L. Hanley-Bowdoin. 1996. Interactions between geminivirus replication proteins. *J. Virol.* **70:** 6790–6795.

Shadel, G.S. and D.A. Clayton. 1997. Mitochondrial DNA maintenance in vertebrates. *Annu. Rev. Biochem.* **66:** 409–435.

Shirahige, K., Y. Hori, K. Shiraishi, M. Yamashita, K. Takahashi, C. Obuse, T. Tsurimoto, and H. Yoshikawa. 1998. Links between cell cycle progression and the activation of DNA replication origins. *Nature* (in press).

Simpson, K., A. McGuigan, and C. Huxley. 1996. Stable episomal maintenance of yeast artificial chromosomes in human cells. *Mol. Cell. Biol.* **16:** 5117–5126.

Smelkova, N.V. and J.A. Borowiec. 1997. Dimerization of simian virus 40 T-antigen hexamers activates T-antigen DNA helicase activity. *J. Virol.* **71:** 8766–8773.

Spann, T.P., R.D. Moir, A.E. Goldman, R. Stick, and R.D. Goldman. 1997. Disruption of nuclear lamin organization alters the distribution of replication factors and inhibits DNA synthesis. *J. Cell. Biol.* **136:** 1201–1212.

Stadlbauer, F., C. Voitenleitner, A. Bruckner, E. Fanning, and H.P. Nasheuer. 1996. Species-specific replication of simian virus 40 DNA in vitro requires the p180 subunit of human DNA polymerase α-primase. *Mol. Cell. Biol.* **16:** 94–104.

Staib, C. and F. Grummt. 1997. Mapping replication origins by nascent DNA strand length. *Methods* **13:** 293–300.

Stillman, B. 1989. Initiation of eukaryotic DNA replication in vitro. *Annu. Rev. Cell. Biol.*

5: 197–245.
Su, R.T. and M.L. DePamphilis. 1976. In vitro replication of simian virus 40 DNA in a nucleoprotein complex. *Proc. Natl. Acad. Sci.* **73:** 3466–3470.
Su, T.T. and P.H. O'Farrell. 1998. Chromosome association of minichromosome maintenance proteins in *Drosophila* endoreplication cycles. *J. Cell. Biol.* **140:** 451–460.
Suárez-López, P., E. Martínez-Salas, P. Hernandez, and C. Gutiérrez. 1995. Bent DNA in the large intergenic region of wheat dwarf geminivirus. *Virology* **208:** 303–311.
Sugimoto, K., S. Ando, T. Shimomura, and K. Matsumoto. 1997. Rfc5, a replication factor C component, is required for regulation of Rad53 protein kinase in the yeast checkpoint pathway. *Mol. Cell. Biol.* **17:** 5905–5914.
Todorov, I.T., R. Pepperkok, R.N. Philipova, S.E. Kearsey, W. Ansorge, and D. Werner. 1994. A human nuclear protein with sequence homology to a family of early S phase proteins is required for entry into S phase and for cell division. *J. Cell. Sci.* **107:** 253–365.
Toledo, F., B. Baron, M.A. Fernandez, A.M. Lachages, V. Mayau, G. Buttin, and M. Debatisse. 1998. oriGNAI3: A narrow zone of preferential replication initiation in mammalian cells identified by 2D gel and competitive PCR replicon mapping techniques. *Nucleic Acids Res.* **26:** 2313–2321.
Tomkinson, A.E. and Z.B. Mackey. 1998. Structure and function of mammalian DNA ligases. *Mutat. Res.* **407:** 1–9.
van Leeuwen, H.C., M. Rensen, and P.C. van der Vliet. 1997. The Oct-1 POU homeodomain stabilizes the adenovirus preinitiation complex via a direct interaction with the priming protein and is displaced when the replication fork passes. *J. Biol. Chem.* **272:** 3398–3405.
Vassilev, L.T., W.C. Burhans, and M.L. DePamphilis. 1990. Mapping an origin of DNA replication at a single-copy locus in exponentially proliferating mammalian cells. *Mol. Cell. Biol.* **10:** 4685–4689.
Verma, R., R.S. Annan, M.J. Huddleston, S.A. Carr, G. Reynard, and R.J. Deshaies. 1997. Phosphorylation of Sic1p by G1 Cdk required for its degradation and entry into S phase. *Science* **278:** 455–460.
Vernis, L., A. Abbas, M. Chasles, C.M. Gaillardin, C. Brun, J.A. Huberman, and P. Fournier. 1997. An origin of replication and a centromere are both needed to establish a replicative plasmid in the yeast *Yarrowia lipolytica*. *Mol. Cell. Biol.* **17:** 1995–2004.
Verreault, A., P.D. Kaufman, R. Kobayashi, and B. Stillman. 1998. Nucleosomal DNA regulates the core-histone-binding subunit of the human Hat1 acetyltransferase. *Curr. Biol.* **8:** 96–108.
Walter, J. and J.W. Newport. 1997. Regulation of replicon size in *Xenopus* egg extracts. *Science* **275:** 993–995.
Walter, J., L. Sun, and J. Newport. 1998. Regulated chromosomal DNA replication in the absence of a nucleus. *Mol. Cell* (in press).
Wold, M.S. 1997. Replication protein A: A heterotrimeric, single-stranded DNA-binding protein required for eukaryotic DNA metabolism. *Annu. Rev. Biochem.* **66:** 61–92.
Wu, J.R. and D.M. Gilbert. 1996. A distinct G1 step required to specify the Chinese hamster DHFR replication origin. *Science* **271:** 1270–1272.
———. 1997. The replication origin decision point is a mitogen-independent, 2-aminopurine-sensitive, G1-phase event that precedes restriction point control. *Mol. Cell. Biol.* **17:** 4312–4321.
Wu, J.-R., G. Yu, and D.M. Gilbert. 1997. Origin-specific initiation of mammalian nuclear

DNA replication in a *Xenopus* cell-free system. *Methods* **13:** 313–324.

Xie, Q., P. Suarez-Lopez, and C. Gutierrez. 1995. Identification and analysis of a retinoblastoma binding motif in the replication protein of a plant DNA virus: Requirement for efficient viral DNA replication. *EMBO J.* **14:** 4073–4082.

Xie, Q., A.P. Sanz-Burgos, G.J. Hannon, and C. Gutierrez. 1996. Plant cells contain a novel member of the retinoblastoma family of growth regulatory proteins. *EMBO J.* **15:** 4900–4908.

Yan, H. and J. Newport. 1995. FFA-1, a protein that promotes the formation of replication centers within nuclei. *Science* **269:** 1883–1885.

Yan, Z., J. DeGregori, R. Shohet, G. Leone, B. Stillman, J.R. Nevins, and R.S. Williams. 1998. Cdc6 is regulated by E2F and is essential for DNA replication in mammalian cells. *Proc. Natl. Acad. Sci.* **95:** 3603–3608.

Yamashita, M., Y. Hori, T. Shinomiya, C. Obuse, T. Tsurimoto, H. Yoshikawa, and K. Shirahige. 1997. The efficiency and timing of initiation of replication of multiple replicons of *Saccharomyces cerevisiae* chromosome VI. *Genes Cells* **2:** 655–665.

Yoon, Y., J.A. Sanchez, C. Brun, and J.A. Huberman. 1995. Mapping of replication initiation sites in human ribosomal DNA by nascent-strand abundance analysis. *Mol. Cell. Biol.* **15:** 2482–2489.

Zhang, D., L. Frappier, E. Gibbs, J. Hurwitz, and M. O'Donnell. 1998. Human RPA (hSSB) interacts with EBNA1, the latent origin binding protein of Epstein-Barr virus. *Nucleic Acids Res.* **26:** 631–6317.

Zhang, Z., D.M. Macalpine, and G.M. Kapler. 1997. Developmental regulation of DNA replication: Replication fork barriers and programmed gene amplification in *Tetrahymena thermophila*. *Mol. Cell. Biol.* **17:** 6147–6156.

Zhao, Y., R. Tsutsumi, M. Yamaki, Y. Nagatsuka, S. Ejiri, and K. Tsutsumi. 1994. Initiation zone of DNA replication at the aldolase B locus encompasses transcription promoter region. *Nucleic Acids Res.* **22:** 5385–5390.

Zlotkin, T., G. Kaufmann, Y. Jiang, M.Y., Lee, L. Uitto, J. Syvaoja, I. Dornreiter, E. Fanning, and T. Nethanel. 1996. DNA polymerase epsilon may be dispensable for SV40—but not cellular—DNA replication. *EMBO J.* **15:** 2298–2305.

Zuo, L. and B. Stillman. 1998. Formation of a preinitiation complex by S-phase cyclin CDK-dependent loading of Cdc45p onto chromatin. *Science* **280:** 593–596.

Zuo, S., E. Gibbs, Z. Kelman, T.S. Wang, M. O'Donnell, S.A. MacNeill, and J. Hurwitz. 1997. DNA polymerase delta isolated from *Schizosaccharomyces pombe* contains five subunits. *Proc. Natl. Acad. Sci.* **94:** 11244–11249.

Index

AAF adducts. *See* N-2-Acetylaminofluorine adducts
AAV. *See* Adeno-associated viruses
ABF1, 50, 88, 105–107, 441
ACE3, 428, 474–475
N-2-Acetylaminofluorine (AAF) adducts, mutagenenic translesion replication, 238
Adeno-associated viruses (AAV)
  DNA replication
    replication proteins, 9–10
    Rep role, 7, 9, 190, 489–490
    rolling hairpin model, 3, 7, 9
  genome, 7
Adenovirus
  DNA replication. *See also* Terminal protein
    elongation, 11–13
    initiation, 10–11, 91, 485–486
    protein factors, 10, 142, 485–486
    protein primer for DNA replication, 3
    sliding-back mechanism for transition from initiation to elongation, 145–146
    transcription factor stimulation
      nuclear factor 1, 91–92, 142
      octamer-binding transcription factor 1, 93–94, 142
  genome, 10, 141
AL3, 490
Amplification, developmental
  Drosophila chorion genes, 427–428, 474
  Sciara DNA puffs, 428
Anaphase-promoting complex (APC), 372, 465
APC. *See* Anaphase-promoting complex
AP endonuclease, base excision repair role, 252–253

APEX, 252
ARS. *See* Autonomously replicating sequence
Ataxia telangectasia, gene mutations, 483
ATM, 483–484
A/T-rich element, 54
Autonomously replicating sequence (ARS). *See also* Origin
  complex origins, 57, 65–66, 468–469, 474
  simple origins, 50, 412, 440–441, 465–467
Aux-1, 95, 98
aux-1, 49
Aux-2, 95, 97–99
aux-2, 49, 55–56
5-Azacytidine, effect on replication timing, 396

B1 protein kinase. *See* Vaccinia virus
Baculovirus, DNA replication, 491
BAP1, 252
Base excision repair, enzymes
  AP endonuclease, 252–253
  DNA deoxyribophosphodiesterase, 253
  DNA glycosylase, 250–252
  DNA ligase, 253
  DNA polymerases, 253
  FEN-1, 253
  poly(ADP-ribose) polymerase, 254
BHLF1, 103
BHRF1, 103–104
BPV-E1, 88, 97
BPV-E2, 87, 99, 104, 109
BZLF1, 103–104

CAF-1. *See* Chromatin assembly factor 1
CaMKII, 416

509

C_1C_2, 158
Ccl1, 340
Cdc2, 33, 332, 338–339, 350, 355
cdc2, 367
Cdc4, 339, 347, 412
Cdc6, 305–306, 338, 351, 365, 437, 445, 462–463, 465–466, 472–473
Cdc7, 347–349, 412, 462–463
Cdc10, 350–351
cdc10, 372
Cdc13, 361
cdc13, 350, 367
Cdc18, 305, 351, 437
cdc18, 350
Cdc25, 355
Cdc28, 339–340, 410–412, 463
cdc28, 412
CDC33, 335
Cdc34, 347
CDC34, 339
Cdc45, 463
CDC46, 305–306
CDC46, 338–339, 363
CDC47, 338
Cdc53, 347
CDC53, 339
CDC54, 338
Cdk1, 332, 334, 339–340, 348–352, 361–362, 367–368, 372, 374–375, 464, 466
Cdk2, 76, 312–313, 334, 353–354, 368, 417, 465, 473, 477
Cdk4, 334
cdt1, 350
Cell cycle. *See individual phases*
Cell growth
  coordinating DNA replication and cell growth, 331–332, 482–485
  independence from chromosome cycle, 332–333, 335
  S phase dependence, 333–334
CEN3 sequence, 201, 206
Chromatin
  activated transcriptional states in yeast, 283–286
  assembly
    assay, 281
    chaperones, 287, 482
    replication-dependent pathways, 279–281
    replication-independent pathways, 279
  DNA replication effect on structure
    histone oligomers, 274–276
    nucleosome, 273–276
    regulatory complexes, 277–279
    effect on replication speed, 19–20, 473
  nucleoprotein complex and active/repressed states of genes, 272–273
  replication templates, 21
  replication timing role, 286–287
  repressed transcriptional states in yeast, 282–283
  structure during cell cycle, 364, 473, 475
  transcription factors, effect on structure, 88–89, 473
Chromatin assembly factor 1 (CAF-1), 280–281, 287, 482
Chromosome. *See also* X chromosome
  structure and replication
    eukaryotes, 436–438
    prokaryotes, 436–439
    supercoil density, 451
cig1, 350, 369
cig2, 350, 369
Clb cyclins
  activation by Cln cyclins, 342–343
  Clb5, 340–342, 346, 358–359, 368, 463
  Clb6, 340–342, 346, 358–359, 463
  genes, 340–342, 346, 368
  M phase promotion, 341–342
  regulation during cell cycle, 341, 343–344
  similarity with cyclin E, 358–359
Cln cyclins
  CLN1, 340–346, 348–349, 361, 375
  CLN2, 340–346, 348–349, 361, 375
  CLN3, 340–341, 343, 345, 348–350, 357–358, 375
  regulation during cell cycle, 341, 343–344, 358
  S phase promotion, 341–344
  synthesis, 356–357
CMV. *See* Cytomegalovirus
Cyclin A, 370
Cyclin B, 362–363, 370, 372, 425
Cyclin D1
  cell cycle regulation, 352–353, 358
  regulation by Rb, 358
  synthesis, 356–357
Cyclin-dependent kinases (Cdks). *See also specific kinases*

Index 511

G_1
  arrest with inhibitors, 333, 354–355
  regulation, 346, 352–354
G_2-specific kinases, 360–361
p40Sic1 inhibitor and S phase onset, 346
phosphorylation, 355
primordial cell cycle regulation, 370–371
rereplication prevention in yeast, 367–369
S phase promotion, 339–342, 360–361, 374–375
Cyclin E
  cell cycle regulation, 353–354, 359, 362, 370, 411, 422–423, 425
  similarity with Clb cyclins, 358–359
Cystic fibrosis gene, replication, 393

D5 NTPase. *See* Vaccinia virus
DBF4, 305, 338–339, 348, 412, 462
DBP. *See* Adenovirus
Development, control of replication initiation, 476–477
DHFR. *See* Dihydrofolate reductase
dif, 202
Dihydrofolate reductase (DHFR)
  gene replication initiation zone, 469–470, 474
  *ori*β region, 63–65, 67–68, 72
Dislocation mutagenesis, DNA polymerase fidelity, 229
DnaA, initiation of replication, 442–443, 453
DnaB, 161, 188, 192, 195–197, 444, 478
DnaC, 444
DNA deoxyribophosphodiesterase, base excision repair role, 253
dnaE, 224
DNA glycosylase, base excision repair role, 250–252
DNA helicase. *See* Helicase
DNA ligase
  base excision repair role, 253
  types and functions in replication, 26, 481
DNA methylation, control of replication initiation, 68, 453–454, 475–476
DNA-PK, 313
DNA polymerase I. *See* Klenow polymerase

DNA polymerase-α:primase
  accessory proteins, 158
  error rate, 222–225, 230, 239
  exonucleolytic proofreading, 224
  expression in cell cycle, 307
  inhibitors, 157
  nucleoside triphosphate concentration dependence, 479
  phosphorylation, 307–310
  primer synthesis, 3, 24, 45, 157–158, 446, 479, 481
  recognition sites, 158–159
  site-directed mutagenesis, 224–225
  subunits, 157, 307
DNA polymerase-β
  accessory proteins, 225
  error rate, 222, 225–226, 230, 239
  metal binding, 175
  structure, 226
DNA polymerase-δ
  auxiliary proteins, 24–25, 29, 226
  error rate, 222, 226, 230, 239
  phosphorylation, 314
  strand elongation, 5, 17
  subunits, 314, 479
DNA polymerase-ε
  auxiliary proteins, 24–25
  error rate, 222, 226, 230, 239
  exonucleolytic proofreading, 226, 479
  phosphorylation, 314
  S phase checkpoint role, 25
  strand elongation, 5
  subunits, 314
  trapping at replication forks, 479
DNA polymerase-γ
  error rate, 222, 226–227, 230, 239
  exonucleolytic proofreading, 226
dnaQ, 224
DNA repair. *See* Base excision repair; Mismatch repair; Nucleotide excision repair
DNA replication
  checkpoint controls, 485–488
  common steps in mechanisms
  opening at origins, 1–2
  priming of synthesis, 2–4
  strand elongation, 4–5
  strand maturation, 5–6
  unwinding at replication forks, 2
  continuous replication, 3, 6
  discontinuous replication. *See* Okazaki fragment
  factories in nuclei, 121–123

DNA replication (*continued*)
fidelity. *See* DNA polymerases;
Mismatch repair
forks, 480
linear genomes, 133–134, 140–141
nuclear structure importance, 69–72, 120–121
semidiscontinuous replication
chromosomal DNA, 18–20
proofreading, 31, 217
protein-protein interactions, 27–30
replication centers, 31–32
replication fork enzymes, 21–27
viruses, 20–21
termination
eukaryotes, 177–179, 203–207
prokaryotes, 177–181, 183–188, 190–202
viral
adenoviruses, 488–489
baculoviruses, 494
Epstein-Barr virus, 490
geminiviruses, 492–494
herpes simplex, 489–490
papillomaviruses, 488
papovaviruses, 488
parvoviruses, 491
retroviruses, 494–495
DNase I, sensitivity studies of genome accessibility, 392
DNase IV. *See* FEN-1
DNase V, 225
DNA telomerase. *See* Telomerase
DNA topoisomerase. *See* Topoisomerase
DNA unwinding element (DUE), 53–54, 67–68, 465–466, 470
Drosophila
chorion gene amplification, 427–428
developmental alteration of replication, 410–411, 414–420, 422–428
endo cell cycle, 423–425
replication
cell-free replication system, 120
inhibition in egg, 414–415
rapid embryonic cycles, 416–420
replication fork clustering, 123
restart at egg fertilization, 415–416
DUE. *See* DNA unwinding element

E1
initiation of replication, 442, 462, 485
phosphorylation role in papillomavirus replication, 304

E2
E1 binding, 100–101
stimulation of papillomavirus replication, 100–101, 485
E2F, 356–358, 404, 409, 423
EBNA1, 87, 102–104, 203, 462, 482, 487–488
EBV. *See* Epstein-Barr virus
eIF-4e, 348
Embryogenesis, replication changes
addition of G_1 in *Drosophila*, 422–423
gene amplification, 427–428
inhibition of replication in oocyte, 414–415
MBT in *Xenopus*, 421–422
mice, 420–421
polyploidy regulation, 423–427
posttranscriptional control, 417–418
regulation of entry into S phase, 411–412
replication origin control, 412–413, 418–419
restart of S phase at fertilization, 415–417
slowdown of S phase, 419–420
Epstein-Barr virus (EBV). *See also*
EBNA1; Herpesvirus
DNA replication
arrest sites, 203
origins, 56–57, 102–104, 203, 487–488
transcription factor stimulation, 102–104
Erythrocyte, growth and cell cycle, 336
escargot, 425
ESP1, 366–367
Exonuclease
effect on fidelity of replication, 233–236, 239
proofreading, 220–222, 482

FEN-1, 253, 481
base excision repair role, 253
exonuclease processing of Okazaki fragments, 25
proofreading role, 31
Fertilization, restart of S phase, 415–416
Frameshift
error rates of DNA polymerases, 230
initiation mechanisms
misinsertion, 229
template-primer slippage, 227–228
processivity effects, 230–231

G_1
 addition in *Drosophila* embryogenesis, 422–423
 coordinating DNA replication and cell growth, 331–332
 cyclin-dependent kinases
  arrest with inhibitors, 333, 354–355
  regulation, 346, 352–354, 360–362, 374, 409
 late transcription
  E2F activation, 356–357
  factors, 344, 355–356
  genes, 355
 rationale for arrest, 374, 483
G_2, DNA replication
 cyclin-dependent kinase role, 360–362
 licensing factor, 362–363
 prevention of rereplication, 367–369
GAL4, 106, 311
Geminivirus, DNA replication
 host proteins, 490–491
 origins, 489
 Rep protein, 489–490
giant nuclei, 416–417
β-Globin, gene replication, 395–396, 438
Gp32, 444–445
Gp41, 445
Gp45, 449
Gp59, 444

HAP1, 252–253
HAT. *See* Histone acetyltransferase
HCS26, 344
HeLa cell, effects on replication fidelity in extracts
 damaged substrates, 236–238
 exonucleolytic proofreading, 233–236
 mismatch repair, 232–233
Helicase
 mitochondria, 17
 polarity, 2, 22
 replication initiation, overview, 478
 viral helicases. *See individual viruses*
Helicase II, 190
Hepadnavirus, DNA replication
 protein priming of minus-strand synthesis, 143
 reverse transcription, 143
Hepatitis B virus. *See* Hepadnavirus
Hereditary nonpolyposis colon cancer (HNPCC), gene mutations, 261–263
Herpesvirus, DNA replication
 DNA primase, 160–161
 herpes simplex virus, 486–487
 transcription factor stimulation, 102–104
 UL9, 442, 486
H19, imprinting, 399
Histone. *See* Chromatin
Histone acetyltransferase (HAT), 482
HIV. *See* Human immunodeficiency virus
HML, 388
HMR, 388
HNPCC. *See* Hereditary nonpolyposis colon cancer
HOT 1 sequence, 204
HP1, 272
HSSB. *See* Human single-stranded DNA-binding protein
HU, 441
Human immunodeficiency virus (HIV)
 minus-strand priming, 147
 reverse transcriptase
  closed conformation, 219
  discrimination against base mispairs, 218–219
  processivity and frameshift fidelity, 230–231
 RNA secondary structure and minus-strand DNA synthesis, 149
Human single-stranded DNA-binding protein (HSSB). *See* Replication protein A

ICP8, 486–487
Igf2, replication timing, 397, 399–400
Imprinted genes, replication timing, 397–400
Initiation proteins, 462–465
Initiator RNA (iRNA)
 length, 156, 160
 Okazaki fragment priming, 153–154
 radiolabeling, 154–156
 site specificity for priming, 156–157
iRNA. *See* Initiator RNA

Kinetoplast, DNA replication, 491–492
Klenow polymerase
 conformational change, 220
 discrimination against base mispairs, 218–219
 exonucleolytic proofreading, 220–222

Lamin B, 475
LCR. *See* Locus control region
Licensing. *See* Replication licensing
Linear DNA, protein priming of DNA replication
  chromosomes, 141
  plasmids, 140–141
Locus control region (LCR), 395–396, 438

M2, 135
Maturation factor I (MFN-I). *See* FEN-1
Mbp1, 344
MBT. *See Xenopus*
MCM proteins. *See* Minichromosome maintenance proteins
MEC1, 483–484
O^6-Methylguanine-DNA methyltransferase, mismatch repair, 263
MFN-I. *See* Maturation factor I
Minichromosome maintenance (MCM) proteins
  helicase activity, 478
  nuclear membrane translocation, 127–128
  phosphorylation, 306–307
  replication initiation, 462–464, 473, 478
  replication licensing, 126–128, 306
Minute virus of mice (MVM), DNA replication by rolling hairpin model, 3, 7, 9, 488
MIP1, 17
Mismatch repair
  contribution to replication fidelity, 232–233, 239
  defective cell tolerance to methylating agents, 263
  inactivation and hereditary nonpolyposis colon cancer, 261–263
  mechanisms of mismatch formation, 260
  prokaryotes, 261
  yeast, 261
Mitochondria
  DNA replication
    arrest sites, 204–205
    chain elongation, 17–18
    continuous mechanism, 13–14
    priming, 14, 16, 152, 491
    strand origins, 14, 16
  genome, 13

Mos, 414–415
M phase promoting factors
  cell fusion studies, 337–338
  Clb cyclins in yeast, 341–343
  cyclin-dependent kinases, 360–361
  M-phase promoting factor, 337–338, 344–345, 372
MPF. *See* M phase promoting factors
MutH, 261
MutL, 261
mutL, 238
MutS, 261
MVM. *See* Minute virus of mice

NAP-1, 482
NFI. *See* Nuclear factor 1
NS1, 488
Nuclear factor 1 (NFI), 10, 486
  DNA-binding specificity, 91–92
  domains, 91
  stimulation of adenovirus replication initiation, 91–92, 142
Nuclear structure, role in initiation of DNA replication, 69–72, 471–473
Nucleosome. *See* Chromatin
Nucleotide excision repair
  comparison with RNA polymerase II transcription initiation, 255, 257
  defects in xeroderma pigmentosum, 255
  human repair genes, 255–256
  proteins
    DNA-binding proteins, 259
    DNA helicases, 259
    endonucleases, 257–259
    replication protein A, 260
    TFIIH role, 255, 257–258

O protein, 441–442
Oct-1. *See* Octamer-binding transcription factor 1
Octamer-binding protein-1, 10
Octamer-binding transcription factor 1 (Oct-1), 486
  DNA-binding specificity, 93
  stimulation of adenovirus replication initiation, 93–94, 142
Okazaki fragment
  length, 153

processing, 18, 25
synthesis
  cycling of pol-α:primase, 30–31
  initiation, 152–153
  initiation zone model, 161–162
  nested discontinuity model, 162–163
  primer. See Initiator RNA
ORC. See Origin recognition complex
OriC, Escherichia coli replicator, 141, 439
Ori-DBP, 133
Origin. See also organisms
  classification, 47
  complex (metazoan) origins
    amplification promoting element, 67
    autonomously replicating sequences, 57, 65–66, 469
    chromosome mapping, 57, 59–61, 63, 468
    densely methylated island, 68
    DNA unwinding element, 67–68
    heredity, 64–65
    initiation zone, 61, 63, 66, 73–75
    models for replication
      Jesuit model, 73–76
      migrating replication complex model, 73
      strand separation model, 72–73
    nuclear scaffold attachment sites, 68–69
    nuclear structure role in initiation of replication, 69–72
    origin of bidirectional replication, 59, 61–64, 73, 474
    palindromic sequences, 68–69
    quiet zone, 75
    replication complex, 75–76
    transcription factor-binding sites, 67
  distribution in genome, 46–48, 437
  meiotic origins
    spatial control, 412–413
    temporal control, 413
  mitochondria, 14, 16
  opening of DNA, 1–2, 45, 441
  simple genome origins
    A/T-rich element, 54
    autonomously replicating sequences, 50, 412, 465–467
    auxiliary components, 54–56
    DNA unwinding element, 53–54
    genetic versus functional origin, 50

origin core, 49
origin recognition element, 51
origin recognition proteins, 51–52, 485
transcription factors
  binding sites, 49–50
  stimulation of origin core, 55–56
viruses, 56–57, 104
yeast. See Origin recognition complex
Origin recognition complex (ORC)
  components, 305, 462
  footprinting, 105, 305
  phosphorylation, 305–306
  replication initiation, 32, 286, 462–465
  repression of transcription, 282–283
  sequence recognition, 105, 305
  transcription factor stimulation, 105–107
oriP. See Epstein-Barr virus
8-Oxo-deoxyguanosine, effect on replication fidelity, 237

p5, φ29, 138–139, 142
p6, φ29, 137–138, 142
p16INK4, 354, 418
p21, 463
p21WAF1, 354–355
p27KIP1, 354–355
p40Sic1, 346–347, 366, 375
p53, 311
pan gu, 416–417
Papillomavirus. See also E1; E2
  DNA replication
    origin, 100
    required proteins, 99–100, 488
    transcription factor stimulation, 100–101, 488
Papovavirus. See Papillomavirus; Polyomavirus; SV40
parC, 200
parE, 200
PARP. See Poly(ADP-ribose) polymerase
Parvovirus. See also Adeno-associated viruses; Minute virus of mice
  classes, 7
  DNA replication
    origin, 56–57, 488
    rep gene requirement, 6
  rolling hairpin model, 3, 7, 9, 488
Parvovirus initiation factor (PIF), 488
Pause site
  biological roles, 179–180
  sequences, 177–179

PBS. *See* Primer-binding site
PCNA. *See* Proliferating cell nuclear antigen
PCR. *See* Polymerase chain reaction
Phage Cp-1, protein priming of DNA replication, 140
Phage φ29, DNA replication
  fidelity, 144
  ion stimulation, 135
  origin, 137
  pause sites, 179
  polymerase-terminal protein interactions, 136–137
  requirements for initiation, 134–136
  sliding-back mechanism for transition from initiation to elongation, 143–146
  stimulating proteins
    p5, 138–139
    p6, 137–138
  terminal protein priming, 132–133
Phage PRD1
  protein priming of DNA replication, 139
  sliding-back mechanism for transition from initiation to elongation, 144–145
PIF. *See* Parvovirus initiation factor
Pif1, 17–18
plutonium, 416–418
Poly(ADP-ribose) polymerase (PARP), base excision repair role, 254
Polycomb, 272
Polymerase chain reaction (PCR), chromosome mapping, origins of replication, 59
Polyomavirus. *See also* SV40
  T antigen, phosphorylation role in replication, 303–304
  transcription factor stimulation of replication, 98–99
Polyteny
  endo cell cycle regulation, 424–426
  heterochromatin underrepresentation, 426–427
  S phase regulation in embryogenesis, 423–424
POU domain transcription factors, 93
PP2A. *See* Protein phosphatase 2A
PPR1, 403
Precursor terminal protein (pTP), 91–92, 108, 142, 145–146, 485
PriA, 188, 192, 197

Primase. *See* DNA polymerase-α:primase
Primer
  mitochondria, 147–150, 152
  Okazaki fragment. *See* Initiator RNA
  prokaryotes, 446–447
  protein primer for virus DNA replication
    adenovirus, 3, 141–142
    hepadnavirus, 143
    linear chromosomes, 141
    linear plasmids, 140–141
    phage Cp-1, 140
    phage φ29, 132–139
    phage PRD1, 139
  removal from DNA strands, 5
  RNA-DNA primer synthesis, 159–160
  synthesis. *See* DNA polymerase-α:primase
  transfer RNA primer encapsidation by retroviruses, 146–147
  types, 131–132
Primer-binding site (PBS), 147, 151–152
Processivity, effect on frameshift fidelity, 230–231
Proliferating cell nuclear antigen (PCNA)
  binding to DNA polymerase, 5, 24–25, 29, 449–450, 479
  detection at replication centers, 31–32
  DNA repair role, 450
  homology with DNA polymerase III, 19
  phosphorylation, 33
  sliding clamp mechanism, 479–480
Protein phosphatase 2A (PP2A), 33, 299, 303, 443
pTP. *See* Precursor terminal protein
Pyrimidine dimer, replication of DNA, 237–238

Rad1, 258–259
Rad2, 253, 257–259
Rad3, 259
Rad4, 257
RAD7, 260
Rad10, 257–259
Rad14, 257
RAD16, 260
RAD23, 260
Rad27, 253
RAP-1, 106
Rb. *See* Retinoblastoma protein
Reb1, 203
Rep. *See* Adeno-associated viruses

RepA, 442, 490
Replication factor C (RF-C)
  PCNA loading, 24
  phosphorylation, 314–315
  primer binding, 24
  sliding clamp mechanism, 480
Replication fork
  arrest at replication termini in prokaryotes. *See also* Replication terminator protein; Ter
    activation under stringent conditions, 197–198
    chromosomal location of arrest sequences, 184
    detection of termini, 180–181, 183–184
    physiological roles, 198–199
    structure of arrest sequences, 184–186
  barriers in eukaryotes, 481–482
  clustering in nucleus, 121–123
  organization, 27–32
  pause sites
    biological roles, 179–180
    sequences, 177–179
  polarity, 60
  rate of movement, 19–20, 387–388, 436
  transcription factor displacement, 278
Replication licensing
  factor B (RFL-B), 463
  factor, S phase control, 362–363
  nuclear structure role, 125–128, 286–287
  one-time initiation during cell cycle, 77, 119, 123, 125
  rereplication
    cyclin-dependent kinase prevention, 367–369
    yeast mutants, 366–367
Replication origin. *See* Origin
Replication protein A (RP-A)
  checkpoint role, 484
  detection at replication centers, 31–32
  DNA binding in replication, 23–24, 45, 480
  DNA recombination role, 311
  phosphorylation, 33, 312–314
  protein interactions in SV40 replication, 28–30, 96–97
  role in nucleotide excision repair, 260
  subunits, 23
Replication terminator protein (RTP)
  helicase inhibition, 190–192, 207
  interaction with replication arrest sites, 191
  purification, 186
  structure
    crystal structure, 192–193
    dimer-dimer interaction domain, 194–196
    dimerization domain, 194
    DNA-binding domain, 193–194
    helicase-blocking surface, 196–197
    subunits, 191
Replication terminus
  eukaryotes
    arrest sites, 203–205
    sibling molecule separation at termination sites, 205–207
  prokaryotes
    activation under stringent conditions, 197–198
    chromosomal location, 184
    detection of termini, 180–181, 183–184
    physiological roles, 198–199
    recombination activity, 202
    structure, 184–186
Replication timing
  assay, 388, 392–393
  effect on gene expression
    chromatin structure role, 403
    developmental changes, 390
    housekeeping genes, 388–390
    imprinted genes, 397–400
    mechanism in *Xenopus*, 401–403
    prokaryotes, 387
    X chromosome, 395–397, 399–400
    yeast, 388, 403
  genome reproduction time, 19–20, 387–388
  prokaryotes, 387
  regulation
    allelic exclusion, 400–401
    cis-acting elements, 394–395
    locus control region, 395–396
    methylation of DNA, 397–398
    regional regulation, 398–400
    telomere, 395
  replicon structure, 393–394
  time zones in mammalian cells, 390, 392–393, 401, 404
Replicon
  Drosophila, 123
  replication timing, 393–394

Res1, 350–351
Res2, 350–351
Retinoblastoma protein (Rb)
  cell cycle regulation, 356–357
  cyclin regulation, 358
  E2F association, 356
  phosphorylation, 356, 409
Retrotransposon
  primer-binding site, 150, 152
  reverse transcription initiation, 150–152
Retrovirus. *See also* Human immunodeficiency virus
  minus-strand priming, 147
  plus-strand priming, 149–150
  RNA secondary structure and minus-strand DNA synthesis, 147–149
  transfer RNA primer encapsidation by retroviruses, 146–147
Reverse transcriptase. *See* Human immunodeficiency virus
RF-C. *See* Replication factor C
RFA1, 260
Ribonuclease, mitochondrial RNA processing, 152
Ribonuclease H, primer removal from DNA strands, 5, 297, 480–481
Ribonuclease MRP, 16
Ribosomal RNA (rRNA), replication arrest sites in yeast genes, 203–204
RIM1, 17–18
5S RNA, replication timing, 401–403
RP-A. *See* Replication protein A
rRNA. *See* Ribosomal RNA
RTP. *See* Replication terminator protein
rum1, 351
RuvB, 442

Saccharomyces cerevisiae. *See* Yeast
Schizosaccharomyces pombe. *See* Yeast
Sciara puff, developmental amplification, 428
SeqA, 453–454
SIC1, 347
Sic1, 465
Single-stranded DNA-binding protein (SSB)
  effect on polymerase fidelity, 231–232
  role in SV40 replication, 23–24
Sliding clamp, 482–483

Snrpn, replication timing, 397, 399
SPF. *See* S phase, promoting factors
S phase
  checkpoint role of DNA polymerase-ε, 25, 482
  dependence on anaphase, 371–372
  dependence on cell growth, 333–334
  developmental meiosis
    inhibition of replication in oocyte, 414–415
    MBT in *Xenopus*, 421–422
    mice, 420–421
    polyploidy regulation, 423–427
    posttranscriptional control, 417–418
    regulation of entry into S phase, 411–412
    replication origin control, 412–413, 418–419
    restart of S phase at fertilization, 415–417
    slowdown of S phase, 419–420
  genetic analysis in animals, 359–360
  initiation of replication, dependence on phase, 47–48, 77, 119
  premeiotic S phase, 411–413
  promoting factors
    animal cells, 351–352
    cell fusion studies, 337–338
    mechanism of promotion, 348–349
    Saccharomyces cerevisiae factors
      Cdc6, 365
      Cdc7, 347–348
      Cln cyclins, 340–344
      cyclin-dependent kinases, 339–342, 345, 374–375, 409–410
      genes, 338–339
      G_1-specific transcription factors, 344
      meiosis control, 410–412
      p40Sic1, 346–347
      S-phase promoting factor, 337, 344–345, 348, 372
    S. pombe, 350–351
    rereplicating yeast mutants, 366–367
    triggering by growth-related protein synthesis, 334–336
    two-step mechanism for onset
      chromatin structure during cell cycle, 364
      feasibility, 372
      licensing factor, 362–363
      pre-RC formation, 365–366, 369–370, 373–374

yeast, 363–367
SSB. *See* Single-stranded DNA-binding protein
Ssl2, 259
string, 424
SV40
  DNA replication
    cell-free system, 20
    DNA ligase, 26
    DNA polymerases, 24–25
    nucleases, 25
    Okazaki fragment synthesis, 30–31
    origin, 94–95
    ori replicator, 439–440
    pause sites, 179
    primase, 24
    protein phosphorylation role, 32–33, 297–300, 302–303
    protein-protein interactions, 27–30, 96
    replication proteins, 21–22, 296–297
    S phase dependence, 296
    single-stranded DNA-binding protein, 23–24
    T antigen role, 297–300, 302–303
    topoisomerases, 26–27
    transcription factor stimulation, 95–98
  genome, 296
Swi4, 344, 356
Swi6, 344, 356

T4 DNA polymerase, discrimination against base mispairs, 218–219
T7 DNA polymerase
  closed conformation, 219
  discrimination against base mispairs, 218–219
  processivity and frameshift fidelity, 231
  thioredoxin role, 231
T antigen
  initiation of replication, 442
  phosphorylation state in DNA replication
    analysis, 316–317
    kinases, 302–303
    polyomavirus, 303–304
    SV40, 298–300, 302–303, 443
TATA-binding protein (TBP), 106, 109, 272

TBP. *See* TATA-binding protein
Ter, 198–199, 454–455
  helicase inhibition, 188, 190–191, 207
  interaction with replication arrest sites, 187–188
  regulation of synthesis, 198
  structure, 186
Terminal protein (TP)
  deoxynucleotide complexes
    dAMP, 135, 144
    dCMP, 142, 145
    dGMP, 139
  DNA polymerase interactions, 136–137
  priming of DNA synthesis in viruses, 132–133
Terminal region recognition factor 1 (TRF1), 140
Termination. *See* Replication terminator protein; Replication terminus
TFIIA, 108, 272
TFIIB, 109, 272
TFIID, 272
TFIIH, 109, 255, 257–259
TFIIK, 255
Thioredoxin, DNA polymerase accessory protein, 231
Thymine mismatch-DNA glycosylase, 250–252
Timing. *See* Replication timing
Topoisomerase
  decatenation of daughter chromosomes in prokaryotes
    multimer resolution, 202
    topo II, 201
    topo IV, 200
  phosphorylation, 315–316
  prokaryotes, 451–452
  role in SV40 replication, 26–27
  topoisomerase I, 26, 206, 298, 315, 452
  topoisomerase II, 26, 201, 206, 298, 315–316, 452, 492
  topoisomerase III, 452
  topoisomerase IV, 177, 200
TP. *See* Terminal protein
Transcription factors. *See also individual transcription factors*
  binding sites
    complex origin, 67
    simple genome origin, 49–50
    viruses, 87
  enhancement of replication

Transcription factors (*continued*)
adenovirus, 91–94
comparison with transcriptional activation, 109–110
herpesvirus, 102–104
mechanisms, 88–89
papillomavirus, 99–101
polyomavirus, 98–99
rationale, 107–109
stimulation origin core, 55–56
SV40, 94–98
yeast, 104–107
Transfer RNA (tRNA), primer encapsidation by retroviruses, 146–147
Transposable elements, yeast, 150–151
TRF1. *See* Terminal region recognition factor 1
tRNA. *See* Transfer RNA
TTF-1, 481–482
Tus, 198–199, 454–455
Two-dimensional gel electrophoresis
chromosome mapping, origins of replication, 60–61, 63–64, 468
replication termini in prokaryotes, detection, 183–184

U5-IR stem, 148–149
U5-leader stem, 148–149
UL proteins. *See* Herpesvirus
Ultraviolet damage, effect on replication fidelity, 237–238
ung, 250
URA3, 403
ura4, 469
Uracil-DNA glycosylase, 250

VP16, 99, 311

WGA. *See* Wheat germ agglutinin
Wheat germ agglutinin (WGA), protein import and DNA replication inhibition, 121

X chromosome, replication timing, 395–397, 399–400

Xenopus
cell cycle control, 362–363
chromatin assembly, 279–280
DNA replication
cell-free replication system, 120
developmental alteration of replication, 410–411, 414, 416–422, 476–477
licensing factor control, 362–363
nuclear structure effects on replication initiation, 471–473
replication fork clustering, 123
MBT, 421–422
Xeroderma pigmentosum (XP), gene defects, 255–256
Xic, 400
Xist, 400
XP. *See* Xeroderma pigmentosum

Yeast
chromatin transcriptional states, 282–286
cyclin-dependent kinases, rereplication prevention, 367–369
DNA replication. *See also* Origin recognition complex
enhancement by transcription factors, 104–107
timing, gene expression effects, 388, 403
mismatch repair, 261
S phase onset
mechanism, 363–367
Saccharomyces cerevisiae factors
Cdc6, 365
Cdc7, 347–348
Cln cyclins, 340–344
cyclin-dependent kinases, 339–342, 345, 374–375, 409–410
G_1-specific transcription factors, 344
genes, 338–339
meiosis control, 410–412
p40Sic1, 346–347
Schizosaccharomyces pombe factors, 350–351

QP 624 .D125 1999
DePamphilis, Melvin L.
Concepts in eukaryotic DNA
replication